the energy resources center illustrated guide to

home retrofitting for energy savings

the energy resources center illustrated guide to home retrofitting for energy savings

paul a. knight energy resources center,
university of illinois at chicago circle

john m. porterfield • technical consultant
paul s. galen • technical and editorial assistance

HEMISPHERE PUBLISHING CORPORATION
Washington New York London

McGRAW-HILL BOOK COMPANY
New York St. Louis San Francisco Auckland Bogotá
Hamburg Johannesburg London Madrid Mexico
Montreal New Delhi Panama Paris São Paulo
Singapore Sydney Tokyo Toronto

Energy Resources Center, University of Illinois at Chicago Circle

Dr. James P. Hartnett • Director
Gary L. Fowler • Associate Director
Paul S. Galen • Area Manager-Residential & Commercial Energy Conservation

The Energy Resources Center is a University of Illinois interdisciplinary research and public service organization. It was established in September 1973 by the University's Board of Trustees to conduct studies in the field of energy and to provide local, state and federal governments and the public with current information and advice on energy technology and policy.

The Center's professional research staff is drawn from a variety of university disciplines. Full-time staff members come from the fields of engineering, architecture, computer-sciences, economics, political science, geography and urban planning. When other specialized skills are required, the ERC draws upon the University of Illinois faculty for support.

An academic advisory committee, drawn from all University of Illinois campuses, and an advisory council, with representatives from energy industries; local, state and federal agencies; and public interest groups, annually review the work of the Center and advise on future directions.

Energy Resources Center
University of Illinois at Chicago Circle
Box 4348
Chicago, Illinois 60680

THE ENERGY RESOURCES CENTER ILLUSTRATED GUIDE TO HOME RETROFITTING FOR ENERGY SAVINGS

1 2 3 4 5 6 7 8 9 0 H D H D 8 9 8 7 6 5 4 3 2 1

Library of Congress Cataloging in Publication Data

Knight, Paul A date
The Energy Resources Center Illustrated Guide to Home Retrofitting for Energy Savings
Bibliography: p.
Includes index.
1. Dwellings—Energy conservation—Amateurs' manuals. 2. Dwellings—Insulation—Amateurs' manuals.
I. Title.
TJ163.5.D86K59 693.8'32 80-23568

ISBN 0-07-019490-4

to MOM and DAD

contents

Preface xi

Section I **A Word About . . ." 1**

1 Heat Loss 3
2 Insulation 9
3 Vapor Barriers 19
4 Condensation 23
5 Caulks 25
6 Safety 33

Section II **Window Retrofits 43**

7 Replace Broken Glass 45
8 Re-set Glass 51
9 Weatherstrip Windows 57
10 Pack and/or Caulk Windows and Doors 73

11 Change Window Operation 81
12 Install Plastic Storm Windows 87
13 Install Glass Storm Windows 99
14 Install Window Insulating Shutters 107
15 Install Window Insulating Panels 123
16 Replace Existing Window 127

Section III Door Retrofits 139

17 Install Door Threshold and/or Bottom Seal 141
18 Weatherstrip Doors 151
19 Install Storm Door 161
20 Replace Existing Door 165

Section IV Insulation Retrofits 175

21 Insulate Basement Walls on the Interior 177
22 Insulate Basement Walls on the Exterior 187
23 Insulate Crawl Space Walls on the Interior 193
24 Insulate Crawl Space Walls on the Exterior 203
25 Insulate Floor Above Crawl Space 209
26 But What About Accessing the Crawl Space 215
27 Insulate Slab on Grade 221
28 Insulate Masonry Wall on the Interior 227
29 Insulate Masonry Wall on the Exterior 239
30 Insulate Open Frame Wall 247
31 Insulate Closed Frame Wall 253
32 Insulate Unfinished Attic 261
33 Insulate Finished Attic 271
34 Insulate Finished Attic Outer Sections 281
35 But What About Accessing the Attic 287
36 But What About Attic Ventilation 295
37 Insulate Finished Ceiling Against Roof 305

Section V Hole/Crack Retrofits 313

38 Pack, Caulk, and Seal Structural Cracks 315
39 Seal Air Bypasses 321
40 Repair Structural Hole 327
41 Seal Foundation Crawl Space

Vents 333
42 Install Skirting Around Building 339
43 Insulate Exposed Ducts and/or Pipes 345

Glossary 351

Index 359

preface

Houses can be "sick". Not in the sense that they look sick requiring only superficial treatment such as painting, cleaning, or repair, but in a less visible but much more serious manner. This illness pertains to a home's energy consumption. Even homes that appear cosmetically fine may be suffering from an energy disorder. If your home is uncomfortably cool, drafty, and expensive to heat, you already know that there is something wrong with it. However, it may not be so apparent to you why your home has such an acute energy demand or what can be done to reduce your home's energy fever.

A home is made up of various component parts. Any one or any combination of parts may be responsible for your energy woes. You must keep in mind even if you live in a recently built home that design and construction methods have presumed that energy would forever be cheap and abundant. With this presumption, it made little sense to spend the extra time and money to build energy efficient homes. We have the opportunity, with new homes, to build in prevention of energy problems. With existing homes, we must seek a cure. It is now up to you to go back and install the "remedies" to bring down your home's energy fever.

These "remedies" are also called "retrofits" (pronounced "re- trō- fits"). The word "retrofit" means to "go back to an item and apply something that was not available or needed when the item was first manufactured or built". In this case, you are going back to the various parts of your home ("items") and applying energy conservation measures or devices. All of the "remedies" in this guide are referred to as "retrofits".

There are 34 energy retrofits discussed in this guide and they are divided into five sections. In addition, various weatherization basics, materials, and concepts are discussed in the explanatory section entitled "A Word about ...". The retrofits fall into the following major categories:

- Window Retrofits
- Door Retrofits
- Insulation Retrofits
- Hole/Crack Retrofits
- Mechanical Retrofit

It is not the intent of this guide to estimate the energy savings you can expect from installing these retrofits. That estimation is dependent upon the physical characteristics and mechanical efficiency of your home and would require an evaluation or energy audit of your home. The purpose of this guide is to provide you with the necessary information for doing the retrofit work yourself or, at least, to familiarize you with the installation techniques should you hire someone to do the work.

You will find detailed descriptions, primarily in graphic form, of retrofits in this guide. In addition, you will find information on needed materials, preparation procedures, and installation techniques.

Each retrofit is formatted in the same fashion:

description

This is a brief introduction of what the retrofit entails in terms of materials and installation techniques. If any special labor skills are required, they are noted here.

materials

A list of common material types that can be used for the retrofit.

preparation

Steps needing to be done prior to installing the retrofit are outlined.

installation procedures

A step-by-step methodology for installing the retrofit. In most cases, two or more procedures are outlined for different material types and/or installation techniques.

This guide was produced by staff members at the Energy Resources Center of the University of Illinois at Chicago Circle.

The guide was originally developed as one component of a weatherization methodology demonstration program conducted under the direction of Paul S. Galen and John M. Porterfield. This program consisted of a training and technical assistance package for the Community Services Administration (CSA) and the National Bureau of Standards (NBS) for use in their low-income weatherization programs. The guide has also been used by NBS in its Optimum Weatherization Program and I am grateful to Richard Crenshaw at NBS for his contributions to the guide as a result of its use in this program. In addition, I would like to thank the Energy Resources Center for being given the opportunity to produce this guide.

The guide was produced by

Paul A. Knight • Author

and John M. Porterfield
• Technical Consultant

Paul S. Galen
• Technical and Editorial Assistant

with the assistance of
Ronald Kirsininkas Graphic/Research Aide

and Dave Balderas, Paul Vecchiet
• Graphic Aides

the energy resources center illustrated guide to

home retrofitting for energy savings

a word about . . .

Heat loss

Insulation

Vapor barriers

Condensation

Caulks

Safety

1

heat loss

description

Heat is thermal energy that is transferred between two objects at different temperatures. The direction of this heat transfer is always from the object with the higher temperature to the object with the lower temperature. When both objects reach, or are at the same temperature there is no further heat transfer between them. In the winter, when heat flows between the inside of a house and the outdoors, it is called "heat loss". In this situation, the indoor air temperature is at a higher temperature than the outdoor air, consequently, heat is lost to the outside. During the summer, the process is reversed and is called "heat gain".

The driving force behind heat loss is the temperature difference between the inside and outside (or any two objects). The greater the temperature difference, the greater the rate of heat transfer. Thus, our basic heat loss problem occurs whenever the outdoor air temperature is lower than the inside temperature of the home. Thus, heat loss occurs at a greater rate as the difference between these two temperatures increase. That's why heat loss is a greater problem in northern climates than southern climates.

HEAT LOSS

There are three mechanisms of heat transfer:

- conduction
- convection
- radiation

One form of convective heat transfer is infiltration. This occurs when outdoor air replaces air in the home and must then be heated to room air temperature.

conduction

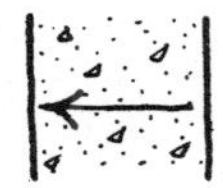

Conduction is the transfer of energy through a body by direct molecular action. In other words, conduction is the transfer of heat through a solid object, or from one solid object to another by direct contact. In general, heat moves more quickly through denser materials.

convection

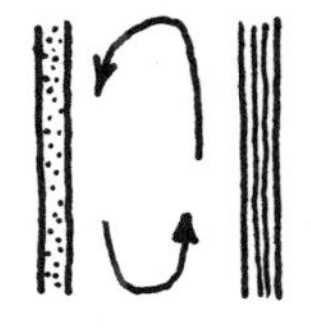

Convection is the transfer of heat by liquids and gases in contact with solids, fluids, or gases. Hot air rises and, as it gives off its heat to surrounding surfaces, cools and drops back down again. This is known as a convective current and is the cause of significant heat loss in floor, wall, and ceiling sections. Air moving about in a house causes convective heat transfer. Such convection "currents" can be tracked down by watching smoke or dust floating in a house.

In small spaces or enclosures where convection currents can't develop, air is a good insulator. In fact, insulation materials are those having a low density and many tiny air compartments where convection can't develop.

radiation

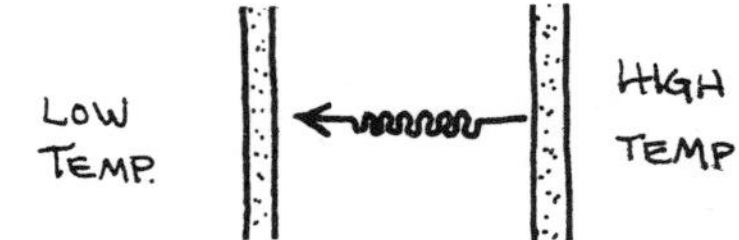

Radiation is the transfer of heat by electromagnetic waves. This means that heat moves from a warm object to a cooler object without appreciably warming the air in between (and without the objects touching one another). This is one reason why snow will melt on a clear, sunny day, even though the temperature is below 32° F.

Just like conductive heat transfer, radiation of energy goes up as the

difference in temperature of two objects increases. The cooler object always receives heat from the hotter object.

infiltration

Infiltration occurs when heated air leaves the home and is replaced by cold air, which, in turn, must be heated to maintain a comfortable temperature. Some amount of infiltration takes place around the edges and joints of windows and doors, but nearly all parts of a house leak air a little. The majority of air leakage occurs at locations other than windows and doors, such as vents, fireplaces, electrical outlets, and foundation cracks (see Section V, "Hole/Crack Retrofits"). One effective way to locate these leaks is a special test using a high powered fan to exhaust air from the house.* It is very unlikely that you could cause a problem of lack of fresh air if all windows and doors were sealed shut.

* The Center for Energy and Environmental Studies at Princeton University has done work in testing homes with this method.

putting it all together

Let's look at a typical wall section and how heat moves through it.

Heat in a room moves to the interior wall surfaces by convection and radiation. It flows through the interior wall surface by conduction.

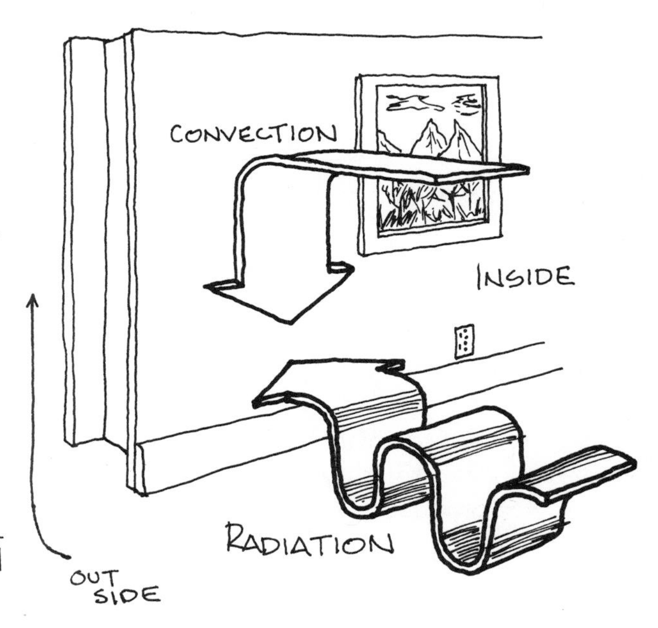

Heat travels from the interior wall surface to the exterior wall surface by convection, radiation, and through adjoining framing members, by conduction.

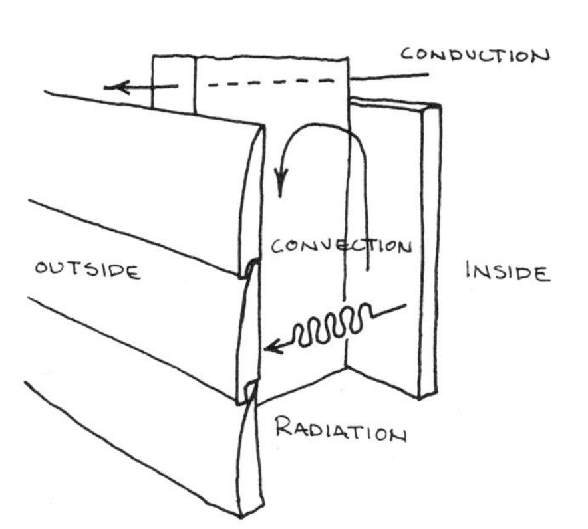

Heat is lost from the exterior wall surface by convection and radiation.

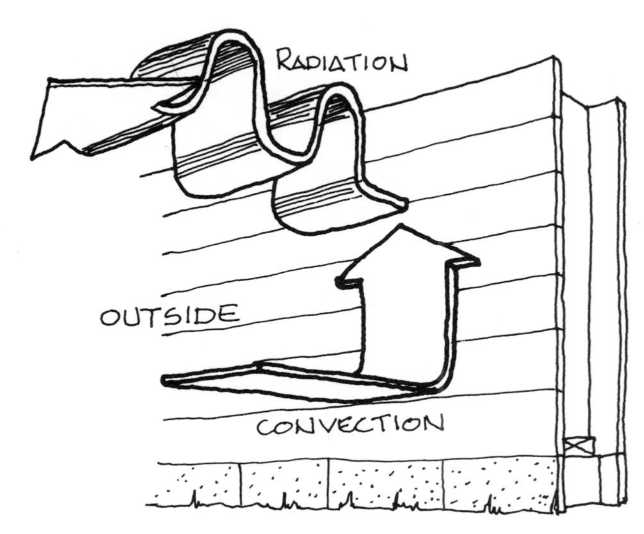

MEASURING HEAT LOSS

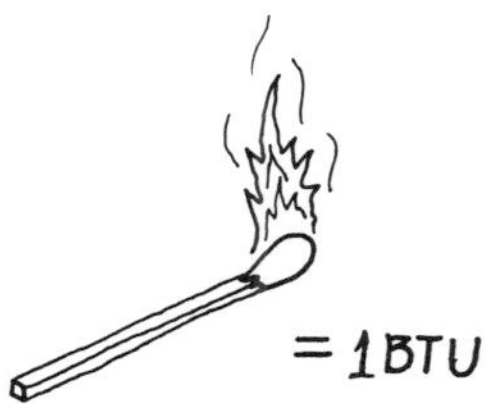

Heat is expressed in Btu's (British Thermal Units). 1 Btu is the amount of heat needed to raise 1 pound of water 1° F. It's roughly equivalent to the heat produced by one blue tip match.

The rate of heat loss is then expressed in Btu's per hour per degree of temperature difference (Btu/hr./F). In dealing with heat loss in buildings, it is useful to think of heat loss rate on a square foot basis of surface area (Btu/hr-ft^2-F).*

---

*This means that for every degree of temperature difference, one square foot of material will let so many Btu's pass through it every hour. If 5 Btu's pass through one square foot of material every hour for a 1 degree temperature difference, and the temperature difference is 70°F, 350 Btu's will pass through that one square foot of material every hour.

But all materials do not conduct heat at the same rate. For example, one square foot of concrete, 4" thick, allows 3 Btu's/hr. to pass through it.

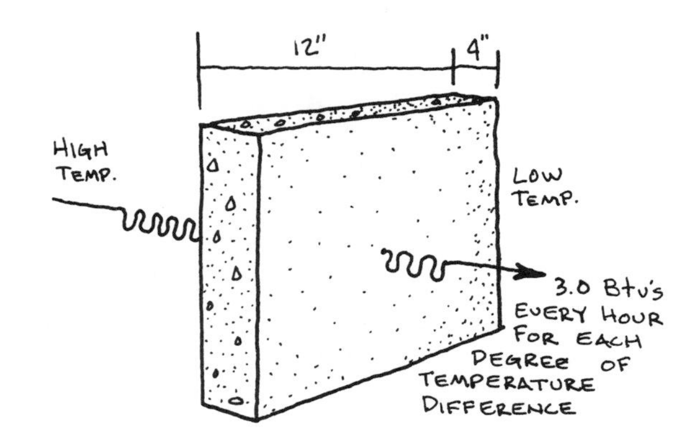

On the other hand, one square foot of fiberglass insulation, 3-1/2" thick, allows about .10 Btu's/hr. to pass through it.

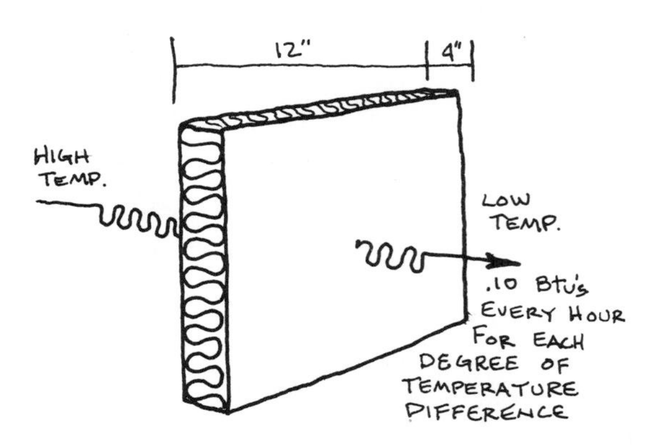

It's easy to see why people use fiberglass (and other types of insulation) to insulate rather than concrete.

When weatherizing homes, we like to think of how well a material resists heat flow, rather than conducts it. The more resistance the material offers, the slower the rate of heat loss through it. The higher the resistance, the better the insulating quality of the material. This resistance, or "R" value, is simply the reciprocal of how fast a material conducts heat.

$$\text{RESISTANCE} = \frac{1}{\text{HEAT CONDUCTION}}$$

Let's go back to our 1 square foot, 4" thick, piece of concrete. Its conductance is 3.0. Its resistance, or "R" value, is .33 (1/3 = .33). On the other hand, the "R" value of the insulation is 11.11 (1/.09 = 11.11). The important thing to remember here is that the higher the "R" value, the better the insulating quality of the material. Thus, our 3-1/2" piece of insulation is about 34 times better in resisting the flow of heat than the concrete. Or, in other words, 34 times as much heat moves through the concrete as through the insulation.

"R" values for materials can be added together to obtain a total "R" value of a construction section. Let's look at the "R" values of some typical construction sections.*

walls

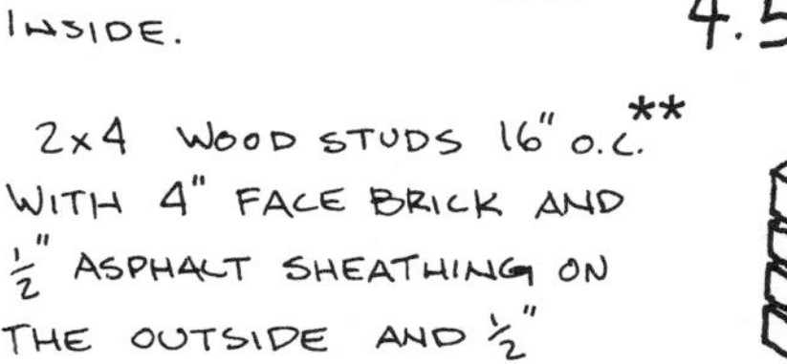

2x4 WD. STUDS 16" O.C.* WITH LAPPED WOOD SIDING AND 1/2" ASPHALT SHEATHING ON THE OUTSIDE AND 1/2" GYPSUM BOARD ON THE INSIDE. 4.5

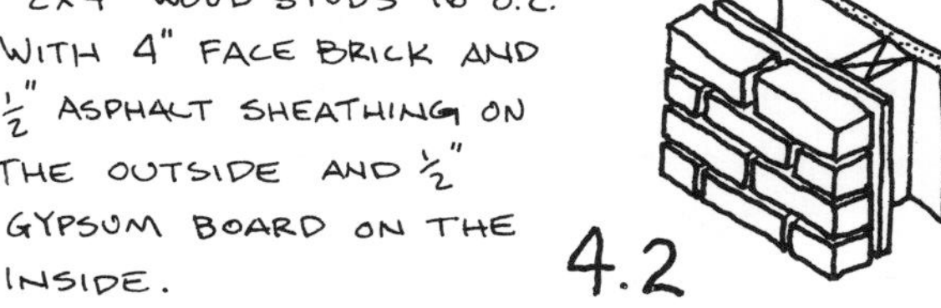

2x4 WOOD STUDS 16" O.C.** WITH 4" FACE BRICK AND 1/2" ASPHALT SHEATHING ON THE OUTSIDE AND 1/2" GYPSUM BOARD ON THE INSIDE. 4.2

---

*Source: ASHRAE Handbook of Fundamentals (1972), American Society of Heating, Refrigerating and Air Conditioning Engineers.

** o.c., "on center" - the measurement of spacing for studs, rafters, joists, and the like in a building from the center of one member to the center of the next.

4" FACE BRICK, 4" COMMON BRICK, 3/4" AIR SPACE, AND 1/2" GYPSUM BOARD ON THE INSIDE. 3.5

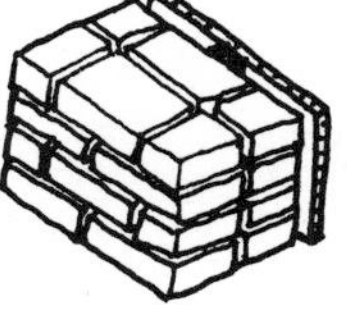

8" CONCRETE BLOCK, 1" AIRSPACE, 1/2" FOIL BACKED GYPSUM BOARD ON WOOD FURRING STRIPS. 5.5

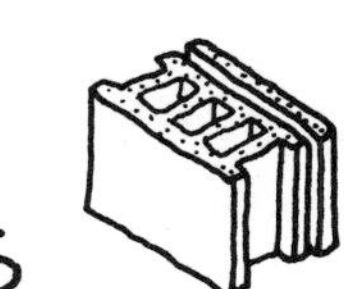

4" COMMON BRICK, 4" AIRSPACE, 4" CONCRETE BLOCK, 1" AIRSPACE, AND 1/2" GYPSUM BOARD. 4.8

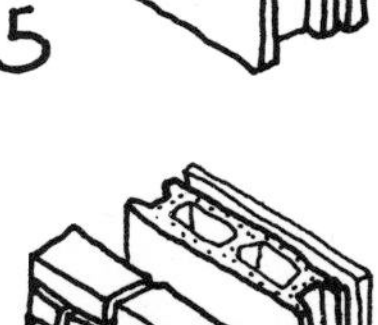

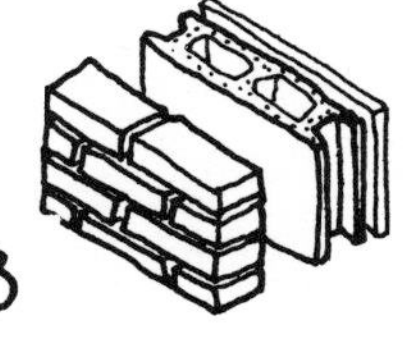

4" FACE BRICK, 2" AIRSPACE, AND 4" FACE BRICK. 2.7

roofs

ASPHALT SHINGLE PITCHED ROOF, 5/8" PLYWOOD DECK SUPPORTED BY WOOD FRAMING, 3-1/2" AIRSPACE, AND 1/2" GYPSUM BOARD. 4.5

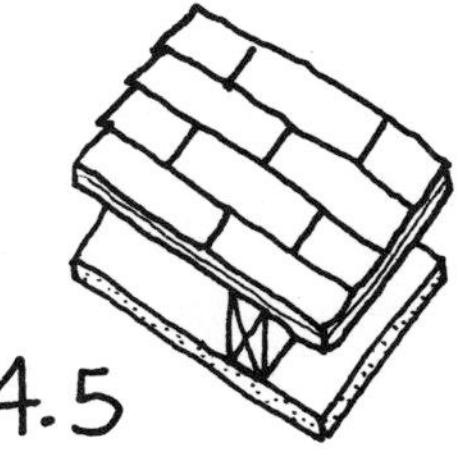

3/8" BUILT-UP ROOFING ON 1" STYROFOAM INSULATION, 4" CONCRETE SLAB, AND 3/4" PLASTER ON METAL LATH 7.7

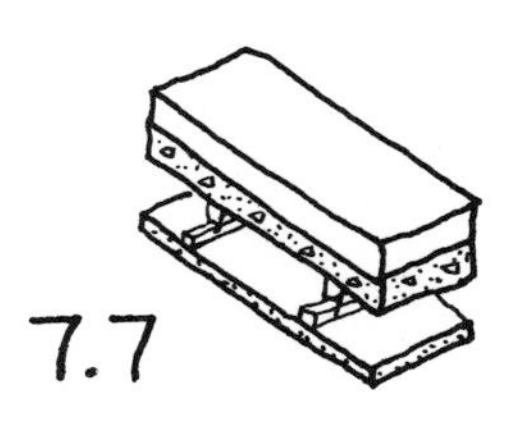

floor

2x10 WOOD JOISTS 16" O.C. WITH HARDWOOD AND WOOD SUBFLOOR AND 3/8" GYPSUM BOARD. 3.7

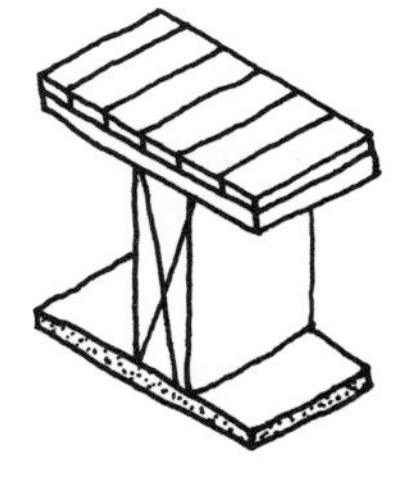

doors

1-3/4" SOLID WOOD DOOR WITH WEATHERSTRIPPING 2.8

SAME AS ABOVE, BUT WITH WOOD STORM DOOR

3.8

windows

1/8" TO 1/2" THICK GLASS

.89

TWO PANES OF GLASS WITH 1/4" AIR SPACE

1.54

THREE PANES OF GLASS WITH 1/4" AIR SPACES

2.13

1/8" TO 1/4" THICK ACRYLIC PLASTIC SHEET

.99

The previous examples represent heat loss by conduction. They were given to signify the importance of "resistance". That is, heat loss by either conduction, convection, and radiation, can be reduced by adding resistance to the path of heat flow. This resistance may take the form of insulation (for conduction, convection, and radiation heat loss), weatherstripping (for convection or infiltration heat loss), metal foil (for radiation heat loss) and other "resistance"materials (storm windows, doors, caulk, etc.). The important thing to remember here is that regardless of how heat is lost (conduction, convection, and/or radiation) and how much heat is lost, it must be replaced if the home is to remain comfortable, and this means using energy.

When selecting weatherization materials, you will undoubtedly hear claims of very high thermal resistances from some insulations and very low leakage rates for window and door retrofit materials. Remember that field measurements have demonstrated that window and door air leakage is a minor percentage of that of the whole house. Refer to Chapter 2, "A Word about Insulation", for tested resistances of insulation materials.

YOU AND HEAT LOSS

When weatherizing, or retrofitting your home, your work will involve adding these "resistance" materials to existing home components (windows, doors, walls, attics, floors, etc.) to reduce heat loss. Keep that in mind as you read through the different retrofit chapters. Some retrofits are specifically for reducing heat loss by infiltration, others are for reducing conductive heat loss, and still others, the emphasis is on a combination of the various heat loss mechanisms. But it all boils down to adding "resistance units" - the more resistance units, the smaller the amount of heat loss. Fewer Btu's of heat will be needed for heating as a result, so in effect, weatherizing your home is another way of "buying" needed heat.

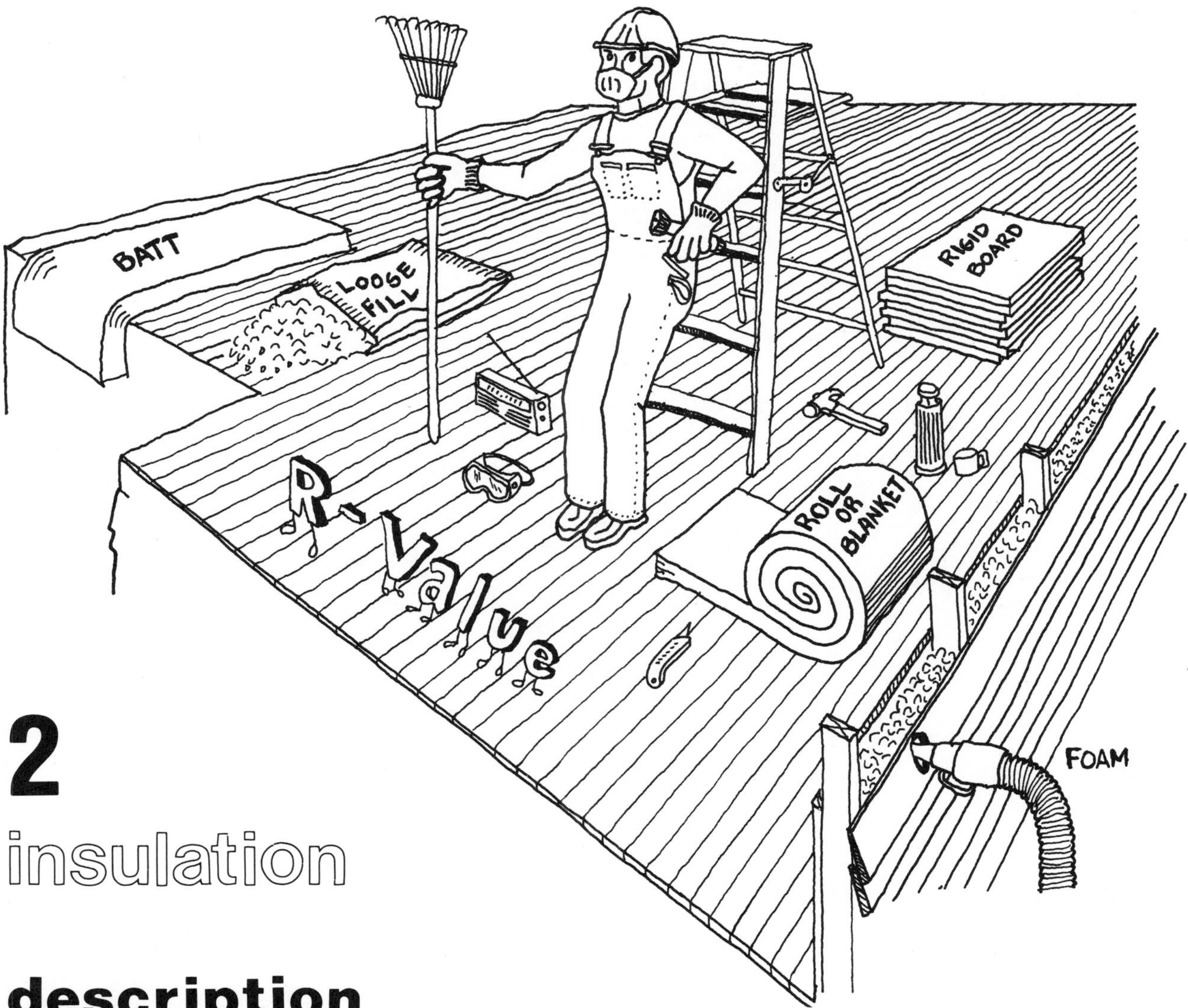

2 insulation

description

What makes insulation "insulation"? Oddly enough, it's air. Air trapped in and between the fibers and particles of the material, and not the fibers and particles themselves, gives a material insulating qualities. These small air spaces do not allow convective currents, and consequently convective heat loss, to occur. It's these "dead air" spaces that give insulation its high resistance value. For this reason, if you compress insulation to "make it fit", you reduce the amount of air in the insulation and the resistance is reduced somewhat. As a matter of fact, it's good practice to fluff insulation before installing it to insure plenty of dead air spaces in the insulation.

Think of insulation as resistance--it resists heat flow. Because of its resistance value, insulation is an important tool in reducing the energy used in your home.

Where in your home will insulation reduce this heat loss? Wherever there is a building section (wall, floor, ceiling) with different temperatures on either side. That is, insulation can reduce heat loss wherever heat could move to a cooler area - either to the

WHERE INSULATION CAN BE INSTALLED

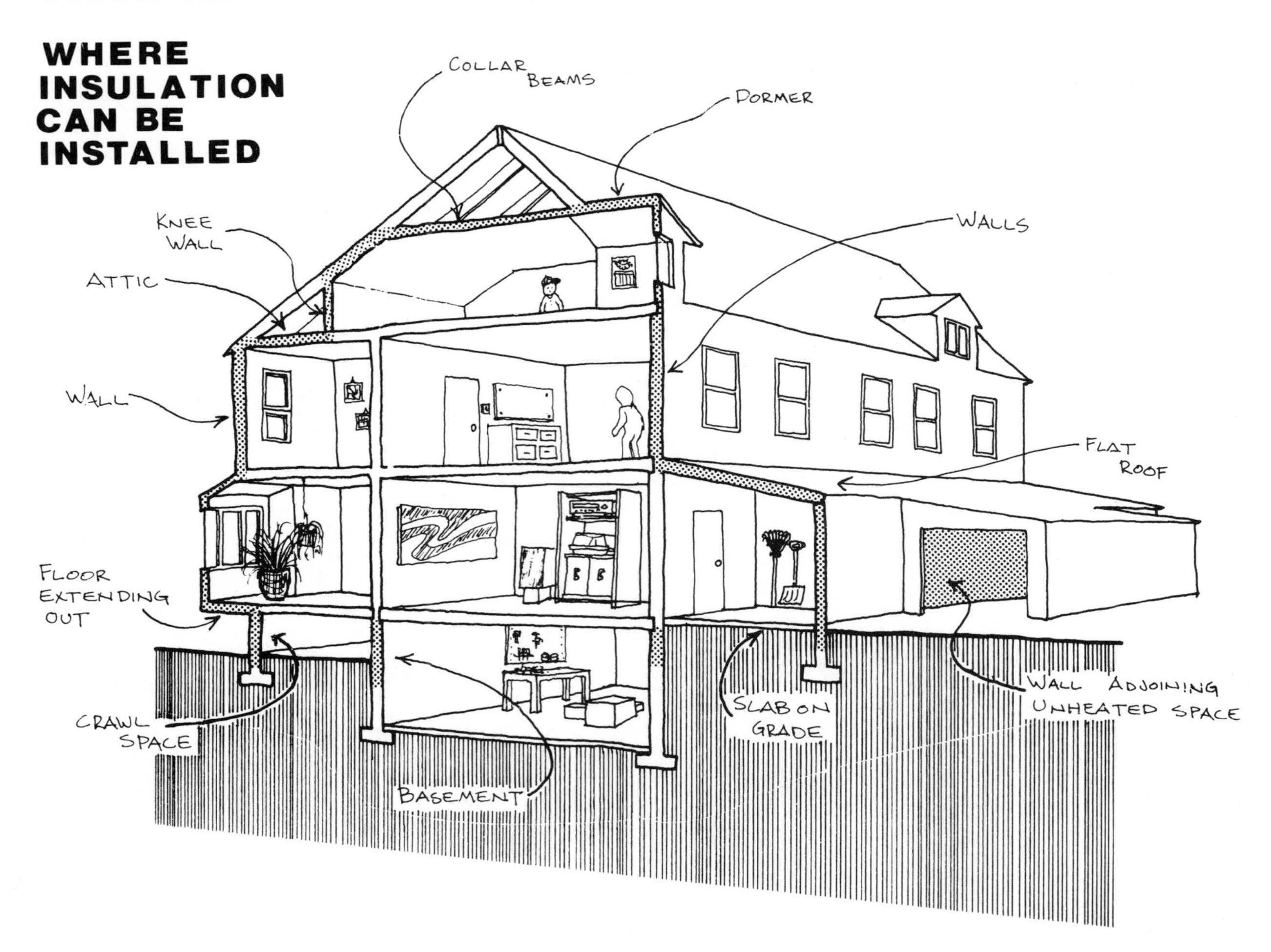

outside or to an unheated space such as a garage.

TYPES OF INSULATION

glass fiber

Also referred to as "fiberglass", it is available in blankets, batts, rigid board, and loose fill. It is relatively inexpensive and easy to install. It is available with a kraft paper or foil facing (which serves as a vapor retardant) or unfaced (for installation over existing insulation). Glass fiber can be highly irritating when handling (wear gloves and long sleeve shirts).

rock wool

Rock wool, which is spun from molten slag rock, is very similar to glass fiber and has a slightly higher "R" value. Rock wool is also available in blankets, batts, rigid board, and loose fill. Though irritating to the skin, rock wool is less bothersome than glass fiber.

cellulose

Cellulose is generally available as a loose fill. Look for certification that it has been treated for fire (fire treated cellulose should have a "flame spread rating" of 25 or less) and rodent resistance. Because of its fine consistency, cellulose can be blown into wall cavities and attics. A blower that fluffs the insulation with air and sprays the insulation like a garden hose can be rented from many lumber yards or machine rental companies. If a blowing machine is not available, cellulose can be installed in the attic by shaking the insulation in a container to "fluff" it, pouring, and raking to an even depth.

Although non-irritating to the skin, it can be quite dusty. Cellulose insulation is usually made from recycled newspaper.

vermiculite

Vermiculite, made from expanded mica, is available as loose fill granules. It can be blown or raked into place. If it gets wet, vermiculite will absorb moisture, lowering its already relatively low "R" value. This insulation is non-combustible.

perlite

Perlite, made from volcanic rock, is similar to vermiculite but has a slightly higher "R" value. This insulation is also non-combustible.

urea-formaldehyde

A "foam in" type insulation, it is used for insulating frame building sections finished on both sides. The equipment and procedures used for installing this insulation are specialized and should be done by highly skilled personnel. The insulation has a high "R" value; however, it has been observed to shrink slowly over time and may afterwards have a resistance no higher than other insulation types. This insulation may produce an odor that lingers for a week or two after installation and if installed when weather conditions will not promote drying, will emit moisture for a long period of time.

polystyrene

This rigid insulation board is used to insulate foundation walls (either on the interior or exterior) and slabs on grade. It dents and is highly combustible. Therefore, it must be covered with a fire rated material.

polyurethane

Polyurethane is similar to polystyrene, but has a higher "R" value. It too must be covered with a fire rated material.

INSULATION 'R' VALUES

Insulation is often discussed in terms of its resistance, or "R" value. (If you recall from the previous chapter, the higher the "R" value, the better the insulating quality of the material). Since there are many types of insulation, it's helpful to compare them by their respective "R" values and thickness required to obtain a certain "R" value. To compare the cost of two different types of insulation, compare costs for the same "R" value, not for thickness.

RIGID INSULATION BOARDS

POLYSTYRENE

THICKNESS (in.)	3/4	1	1-1/2	2
R-VALUE	4.0	5.4	8.1	10.8

URETHANE

THICKNESS (in.)	1/2	3/4	1	1-1/2	1-3/4	2	2-1/4
R-VALUE	4	6	8	12	14	16	18

	BATTS/BLANKETS		LOOSE & BLOWN FILL*				
R-VALUE	GLASS FIBER	ROCK WOOL	GLASS FIBER	ROCK WOOL	CELLULOSE FIBER	VERMICULITE	PERLITE
R-11	3 1/2 in.	3 in.	5 in.	4 in.	3 in.	5 in.	4 in.
R-13	4	3 1/2	6	4 1/2	3 1/2	6	5
R-19	6	5	8 1/2	6 1/2	5	9	7
R-22	7	6	10	7 1/2	6	10 1/2	8
R-26	8	7	12	9	7	12 1/2	9 1/2
R-30	9 1/2	8	13 1/2	10	8	14	11
R-33	10 1/2	9	15	11	9	15 1/2	12
R-38	12	10 1/2	17	13	10	18	14

* The "R" value for urea-formaldehyde is 4.2 per inch of thickness. However, Materials Bulletin No. 74, Sept. 15, 1977, from the Dept. of Housing and Urban Development (HUD) indicates that the effective "R" value of this type of insulation may only be 3.3 per inch when installed, due to a 6% average linear shrinkage. Therefore, urea-formaldehyde foam in a 3-1/2" cavity wall may have an "R" value of 10.5.

Source: Dept. of Energy Fact Sheet
DOE/CS-0017

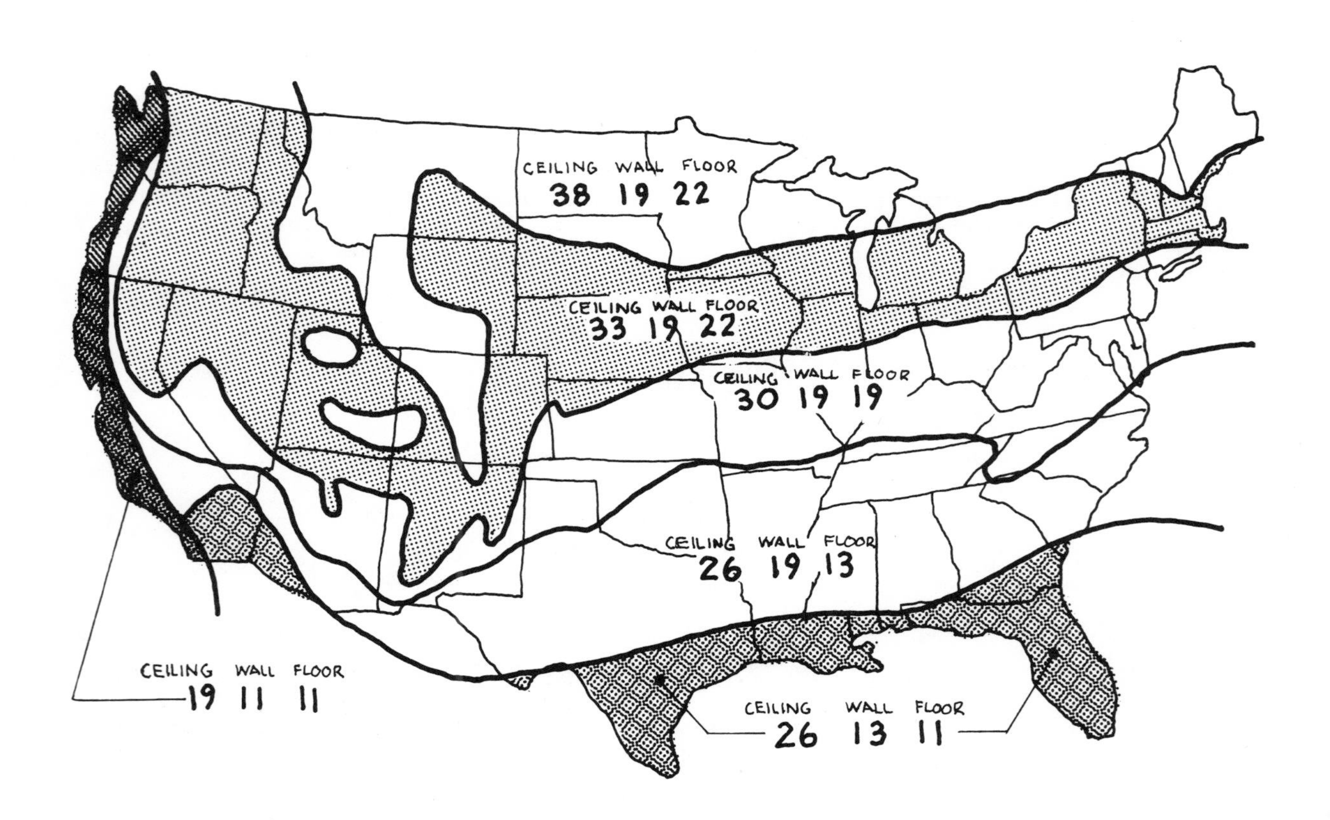

HOW MUCH

The map indicates recommended "R" values for ceilings (attics), walls, and floors. These are good general rules, but do not necessarily represent what would be best for a particular home.

The amount of insulation for new homes is now mandated by standards in some building codes. How much insulation a home should have is not perfectly clear. We hear about millions of American homes which are not "up to standard". Certainly, most homes have places where installing insulation would prove a good investment.

It is usually better to install some insulation where none exists rather than to add more insulation where even a "sub-standard" amount exists.

SQUARE FOOT COVERAGE

Information is included here to assist you indetermining the amount of insulation that you'll need to achieve a certain "R" value for a given insulation type. For example, if your attic is 800 square feet and you have decided to use cellulose and want to add R-30 (8"), you will require approximately 44 bags of insulation.

$$\frac{54.6 \text{ BAGS}}{1000 \text{ FT}^2} \times \frac{800 \text{FT}^2}{1} = 44 \text{ BAGS}$$

Or, if you decided to use glass fiber (R-30) and the ceiling joists are 16" o.c., you will need 14 rolls of 15" x 48" insulation.

$$\frac{\text{ROLL}}{60 \text{FT}^2} \times \frac{800 \text{FT}^2}{1} = 14 \text{ ROLLS}$$

Wastage factors are not included and that could amount to an additional 10%. Consequently, you may need about 16 rolls.

LOOSE FILL

GLASS FIBER

R VALUES	SQ. FT. / BAG	BAG / 1000 Ø	LBS. / SQ. FT.	THICK.
R-11	90	11	.28	5"
R-19	50	20	.50	8"-9"
R-22	45	22	.56	10"
R-30	33	30	.76	13"-14"

NOTE: 25 lb. BAG

CELLULOSE FIBER

R VALUES	SQ. FT. / BAG	BAG / 1000 Ø	LBS. / SQ. FT.	THICK.
R-11	53.8	18.6		3"
R-19	29.9	33.4	1.1	5"
R-22	24.5	40.8		6"
R-30	18.3	54.6		8"
R-40	13.4	74.6	2.24	11¼"

NOTE: 30 lb. BAG

ROCK WOOL

R VALUES	SQ. FT. / BAG	BAG / 1000 Ø	LBS. / SQ. FT.	THICK.
R-11	36.4	22	0.75	4"
R-19	21.1	37	1.3	6"-7"
R-22	18.2	43	1.5	7"-8"
R-30	13.3	58	2.0	10"-11"

NOTE: 30 lb. BAG

BATTS

GLASS FIBER - FOIL		
R VALUES	WIDTH / LENGTH	SQ. FT. /ROLL
R-11	15" x 40"	200
	23" x 48"	306.6
R-19	15" x 48"	120
	23" x 48"	184
R-22	15" x 48"	80
	23" x 48"	122.6

GLASS FIBER - KRAFT		
R VALUES	WIDTH / LENGTH	SQ. FT. /ROLL
R-11	15" x 96"	200
	23" x 48"	306.6
R-19	15" x 48"	120
	23" x 48"	184
R-22	15" x 48"	80
	23" x 48"	122.6
R-30	15" x 48"	60
	23" x 48"	92

GLASSFIBER - UNFACED		
R VALUES	WIDTH / LENGTH	SQ. FT. /ROLL
R-11	15" x 48"	200
	23" x 48"	306.6
R-19	15" x 48"	120
	23" x 48"	184
R-22	15" x 48"	80
	23" x 48"	122.6
R-30	15" x 48"	60
	23" x 48"	92

ROCK WOOL		
R VALUES	WIDTH / LENGTH	SQ. FT. /ROLL
R-11	15" x 48"	INFORMATION NOT AVAILABLE
	23" x 48"	
R-13	15" x 48"	
	23" x 48"	
R-19	15" x 48"	
	23" x 48"	
R-22	15" x 48"	
	23" x 48"	

NOTE: SQUARE FOOT/ROLL QUANTITIES MAY VARY WITH DIFFERENT MANUFACTURERS. CHECK WITH YOUR SUPPLIER FOR ACTUAL SQUARE FOOT COVERAGE.

	batts/ blankets	loose fill	blown fill	rigid board
method of installation	• FITTED BETWEEN STUDS, JOISTS AND BEAMS	• POURED BETWEEN ATTIC JOISTS	• BLOWN IN PLACE USING SPECIAL EQUIPMENT	• MUST BE COVERED WITH A FIRE RATED MATERIAL; ½" GYP. BRD. • ADHESIVE, NAILS, PLACED BETWEEN FURRING STRIPS
where applicable	• UNFINISHED WALLS, FLOORS, AND CEILINGS INCLUDING KNEE WALLS AND BETWEEN RAFTERS WHERE APPROPRIATE	• UNFINISHED ATTIC FLOORS AND HARD-TO-REACH PLACES • IRREGULARLY SHAPED AREAS AND AROUND OBSTRUCTIONS	• ANYWHERE THAT FRAME IS COVERED ON BOTH SIDES (SIDE WALLS) • UNFINISHED ATTIC FLOORS AND HARD-TO-REACH PLACES	• FOUNDATION WALLS, INTERIOR AND EXTERIOR • CEILING IS FINISHED AGAINST ROOF • "FURRING IN", A FINISHED ATTIC WHERE CEILING IS FINISHED AGAINST RAFTERS
advantages	• DO-IT-YOURSELF • SUITED FOR STANDARD STUD/JOIST SPACINGS WHICH ARE FREE FROM OBSTRUCTIONS • BLANKETS: LITTLE WASTE • BATTS: MORE WASTE, BUT EASIER TO HANDLE	• DO-IT-YOURSELF • CONFORMS TO SHAPE OF AREA TO BE INSULATED	• INSULATION CAN BE USED IN FINISHED AREAS • EASY TO USE FOR IRREGULARLY SHAPED AREAS AND AROUND OBSTRUCTIONS	• HIGH "R" VALUE FOR RELATIVELY LITTLE THICKNESS
materials and cost per resistance unit per square foot	• GLASS FIBER (1.09¢) • ROCK WOOL (1.09¢)	• GLASS FIBER (2.8¢) • ROCK WOOL (2.8¢) • CELLULOSE (0.68¢) • VERMICULITE (9.67¢) • PERLITE (9.12¢)	• GLASS FIBER (2.8¢) • ROCK WOOL (3.08¢) • CELLULOSE (0.68¢) • VERMICULITE (7.67¢) • PERLITE (9.12¢) • UREA-FORMALDEHYDE (3.36¢) [NOT FOR UNFINISHED AREAS]	• POLYSTYRENE (7.67¢) • URETHANE (2.02¢) • GLASS FIBER (7.20¢)

SOME TERMS

blankets/batts

This insulation comes with various facings, and widths corresponding to standard stud and joist spacing for easy installation. Blankets are continous rolls which can be hand cut to desired lengths. Batts are precut to 4 and 8 foot lengths.

loose fill

This is loose material that can be poured into place.

blown in place

These are loose fibers or plastic foams that are blown into building sections with special equipment.

rigid board

These are plastics, fibers, or particles that are pressed into rigid boards. Some are composites of the above.

REMEMBER, SAFETY FIRST

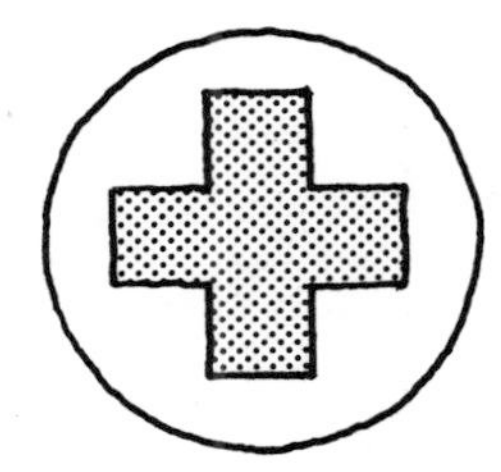

glass fiber

For comprehensive information, request "Criteria for a Recommended Standard ...Occupational Exposure to Fibrous Glass", NIOSH (National Institute for Occupational Safety and Health), 1977. Available from:

Robert A. Taft Laboratories
4676 Columbia Parkway
Cincinnati, Ohio 45226

Or, for more general information, refer to "Good Practice Manual for Insulation Installers". (Available, free of charge, from above address). Generally speaking, both glass fiber and rock wool require heavy clothing and gloves to avoid skin irritation and a "dust respirator" and goggles to protect against eye and throat irritation. Dust respirators which are NIOSH approved follow numbers TC-21C-132 through TC-21C-189 (see "Cumulative Supplement June 1977 NIOSH Certified Equipment"--address as above). Other desirable precautions include ventilation of work space.

rock wool (mineral fiber)

Basically, the same precautions as for glass fiber.

cellulose

For installation of blown cellulose where workers are exposed to suspended dust, dust respirators and goggles should be worn. Check area to be insulated for sources of ignition and use only grounded electrical equipment to prevent air/dust mixture from possible explosion. Ventilate space if possible. For additional information regarding cellulose information, contact:

Cellulose Manufacturers' Association
PO Box 4045
Falls Church, Virginia 22044

703 - 931 - 5200

The Department of Energy is currently considering the use of UF foam insulation for inclusion in its Residential Conservation Service Program (RCS).* There is some question that long-term, low-level exposure to formaldehyde may have acute adverse health effects (formalde-

---

* The RCS program is intended to encourage the installation of energy conservation measures, including home insulation, in existing homes.

hyde is an ingredient found in many different products, including textiles, paper, cosmetics, and wood products). The Consumer Product Safety Commission has received more than 500 complaints about the health effects of formaldehyde exposure. More than 500,000 homes, however, have been insulated with UF foam since 1976. For more information regarding the health effects of formaldehyde, refer to the February, 1980, "CPSC Memo", from:

The Consumer Product Safety
Commission
Washington, D.C. 20207

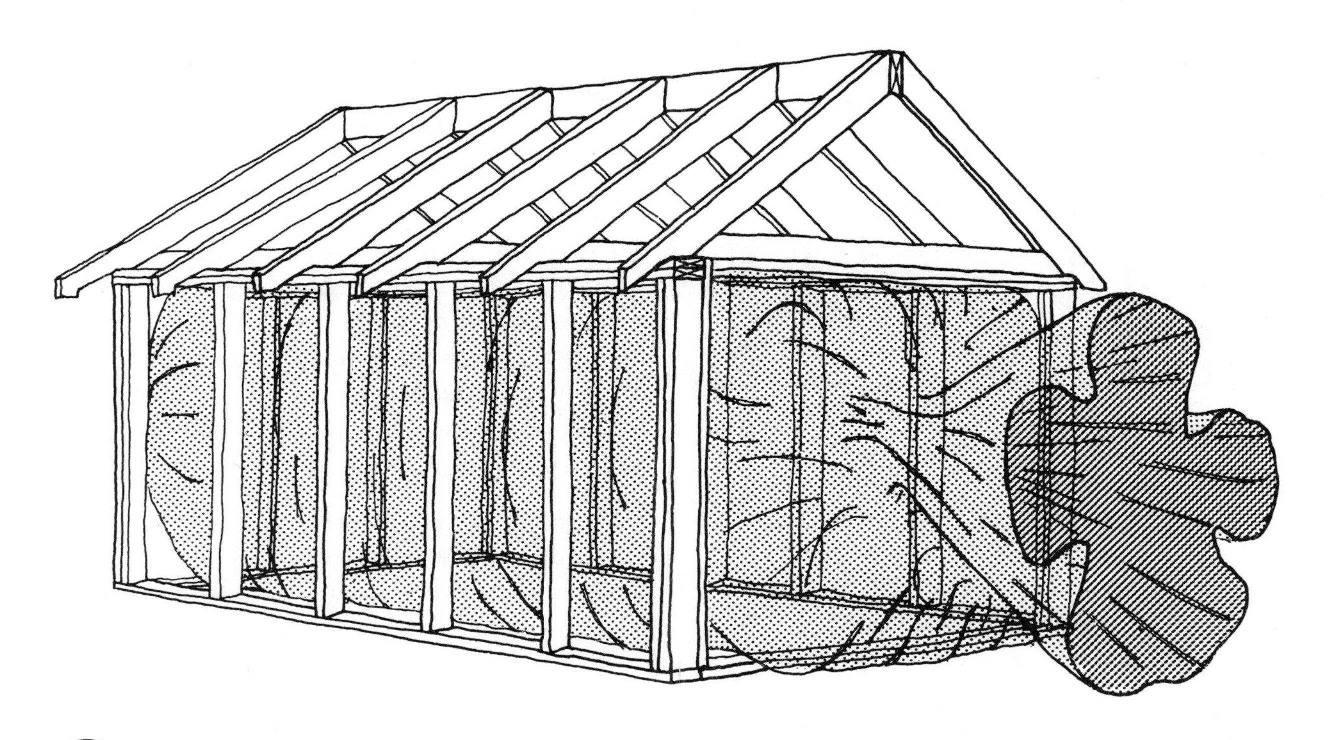

3

vapor barriers

description

Vapor barriers prevent moisture from passing through walls, ceilings, and floors, and condensing within these sections (see Chapter 4, "A Word about Condensation"). The effectiveness of vapor barriers is determined by the amount of water vapor permitted to pass through it and is measured in perms*. The higher the perm rating, the more porous the material and conversely, the lower the perm rating, the more effective the material will be as a vapor barrier.

* A perm (for permeability) is defined as the number of grains of water vapor that pass through a square foot of material per inch of mercury difference in vapor pressure.

Technically, a vapor barrier is a material that has a perm rating of "1.0" or less. Some materials, although referred to as being vapor barriers are actually "vapor retarders"; i.e., the perm ratings are low, but greater than 1. For the sake of this discussion (and throughout this guide), vapor barriers and vapor retarders will be called vapor barriers. Perm ratings for various materials are given later in this chapter.

Vapor barriers are usually installed in conjunction with insulation. If moisture condenses within insulation, the effectiveness (the "R" value) of the insulation is greatly reduced. Therefore, VAPOR BARRIERS ARE ALWAYS INSTALLED ON

THE WARM SIDE OF INSULATION, since the moisture is moving from the warm interior to the exterior.

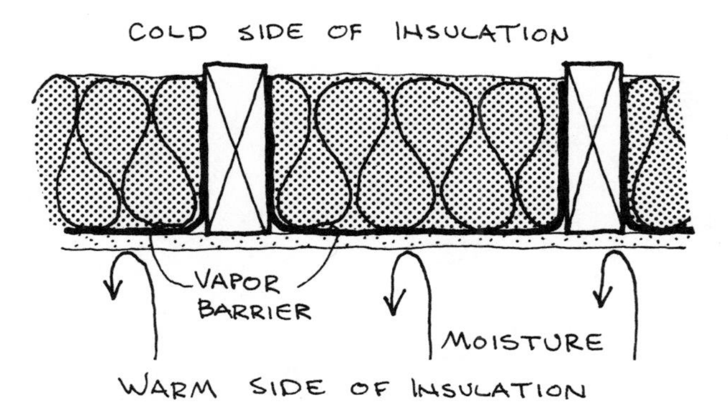

Ventilation plays an important role in deciding whether or not a vapor barrier is needed. For example, if an attic is properly vented, a vapor barrier is not needed since the moisture will be vented out as fast as it condenses.

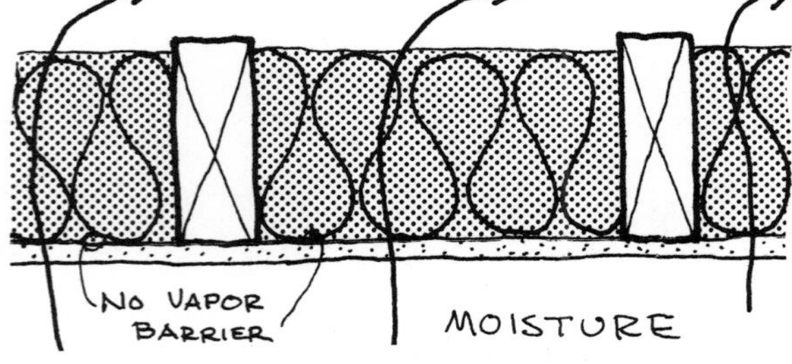

Some batt/blanket insulations have their own attached vapor barriers in the form of foil or asphalt impregnated kraft paper backing. This is known as "faced" insulation.

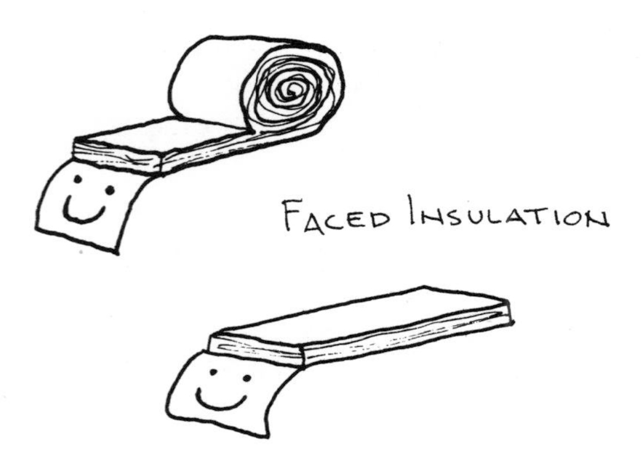

NOTE: 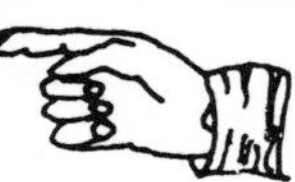

In some cases (when there is exterior plywood sheathing or bituminous roofing, for example), the house has a vapor barrier on the COLD SIDE. Seek expert advice before placing insulation where this is the case. Where there is a vapor barrier already present on the warm side, refer to appropriate chapter of this guide for instructions.

Remember, always protect the integrity of vapor barriers. If you should accidentally rip the vapor barrier, use duct or electricians tape to mend it.

A separate vapor barrier is needed for "unfaced" (no attached vapor barrier) batt/blanket insulation and loose fill insulation. This is usually a polyethylene sheet. The polyethylene should be at least 2 mil (0.002") thick for walls, floors, and ceilings and at least 6 mil (0.006") thick for a ground cover in crawl spaces and beneath slabs on grade.

Polyethylene sheeting is more effective as a vapor barrier than a vapor barrier that is integral to "faced" insulation since the sheeting covers the entire surface with no spaces at framing members. Foil backed drywall may be used in place of polyethylene sheeting with the foil placed against wall studs or ceiling joists. Plastic faced panels and vinyl wall coverings may also be used as vapor barriers (they're also decorative).

Vapor protection can also be achieved by painting the interior finish with latex base paint especially formulated to provide vapor protection. The paint

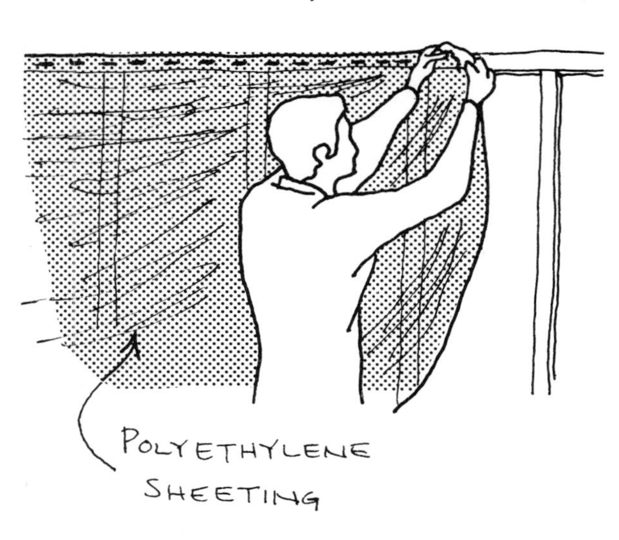

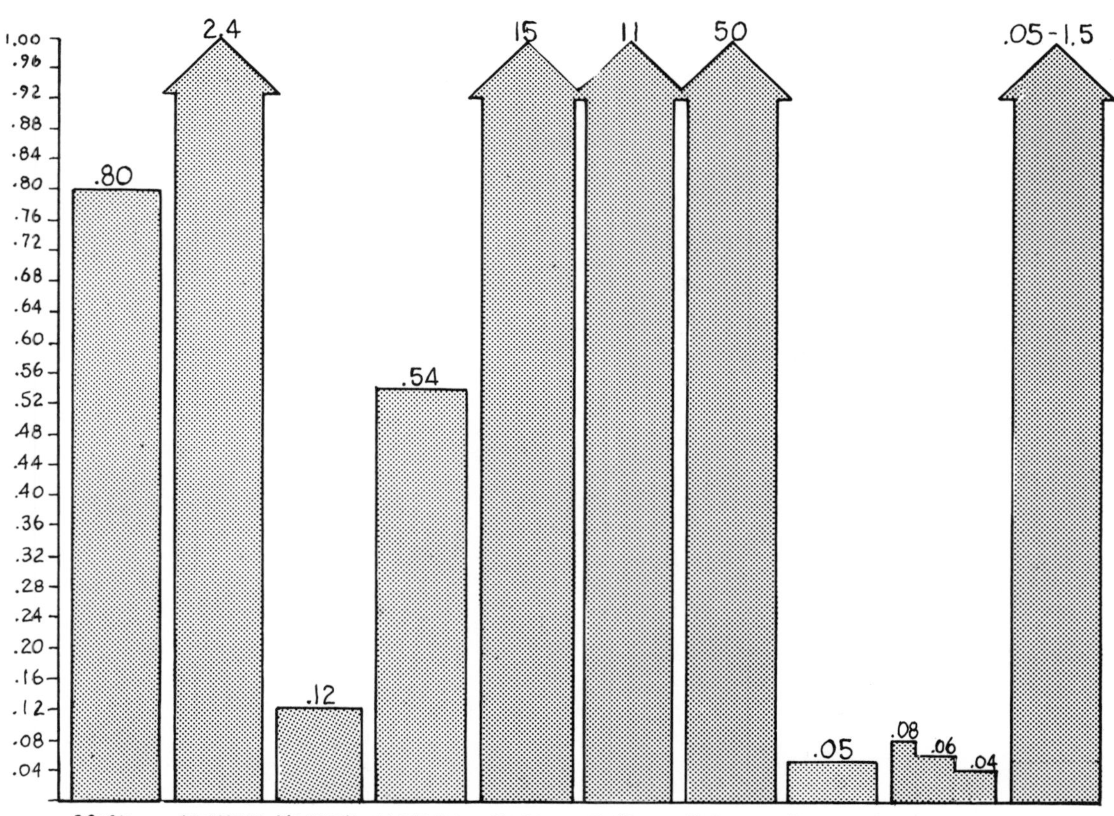

should have a low permeability as listed on the can. Two coats of an aluminized or oil base enamel can also be used but may not present the appearance of a standard interior paint.

Vapor barrier installation techniques for the different building sections are presented along with the appropriate insulation retrofits.

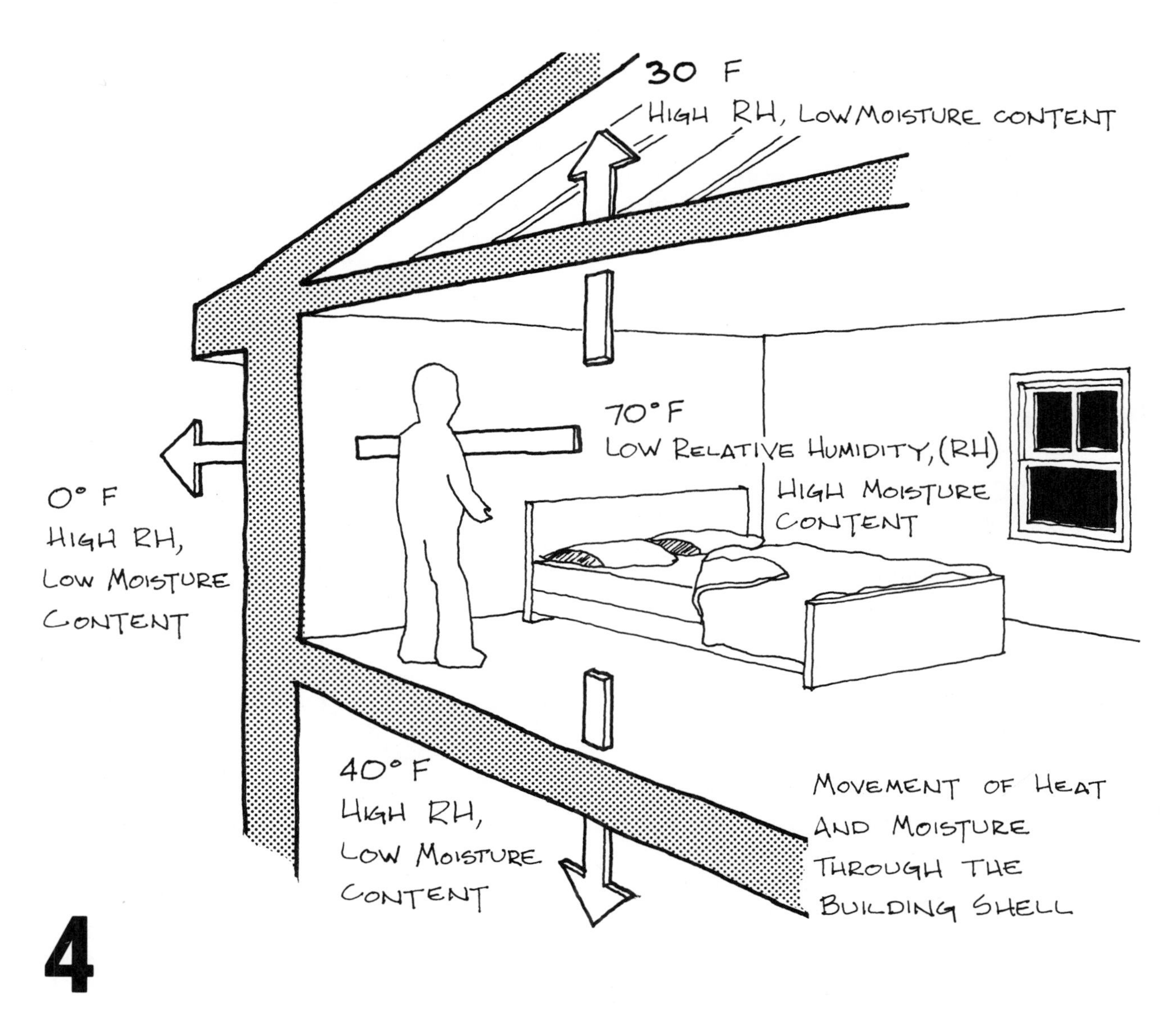

4

condensation

description

Just as heat flows from a high temperature to a low temperatue, water vapor moves from a high vapor pressure to a low pressure. Both of these points are important during the winter when the indoor temperature and moisture content are low.

Consequently, as heat is lost from a home, so is moisture (sources of moisture on the interior are cooking, dishwashing, bathing, clothes washing, drying, and the occupants). The moisture level of the air is expressed by "relative humidity" (RH) and is defined as the percentage of the maximum amount of water vapor that the air can hold at a given temperature.

Warm air holds more moisture than cold air. Thus, as the temperature of the air decreases, the relative humidity increases although the actual amount of moisture in the air remains constant. When the temperature decreases to the point that the relative humidity becomes 100% (the air is "saturated"), the "dew point" is said to have been reached. If the temperature drops below the dew point temperature, moisture will condense out of it.

A volume of air at a temperature of T_1 has a relative humidity of 50% (A). If the temperature is lowered to T_2 and the actual amount of moisture remains the same, the RH is now 75% (B). If the temperature is lowered to T_3, the air is saturated (RH is 100%). T_3 represents the dew point temperature (C). If the temperature is lowered to T_4, the amount of moisture represented by ▒ is forced out of the air in the form of condensation (D).

Let's apply this to a home. During the winter, for example, the temperature in an uninsulated attic is relatively warm due to heat loss from the living space. With insulation, however, this heat loss is reduced and the air temperature in the attic goes down, as does its ability to hold moisture (or the relative humidity goes up). Consequently, there is a greater chance of moisture that could eventually damage the structure and/or the interior finishes condensing in the attic. Vapor barriers (see Chapter 3, "A Word about Vapor Barriers") and proper ventilation (see Chapter 36, "But What about Attic Ventilation?") are techniques to prevent moisture build-up and possible condensation. Methods for installing vapor barriers and providing adequate ventilation are discussed with the appropriate retrofits.

If the situation pictured in (D) happens as heat and moisture move through the exterior of a home, the condensed moisture, left on wall, floor, and ceiling surfaces, can cause peeling paint, mildew, rotting, and damage to interior finish materials.

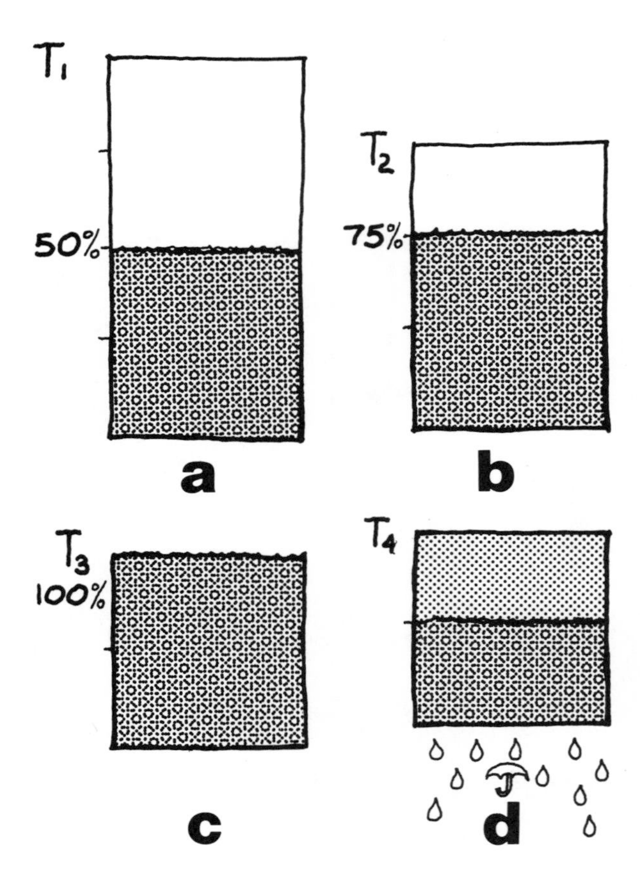

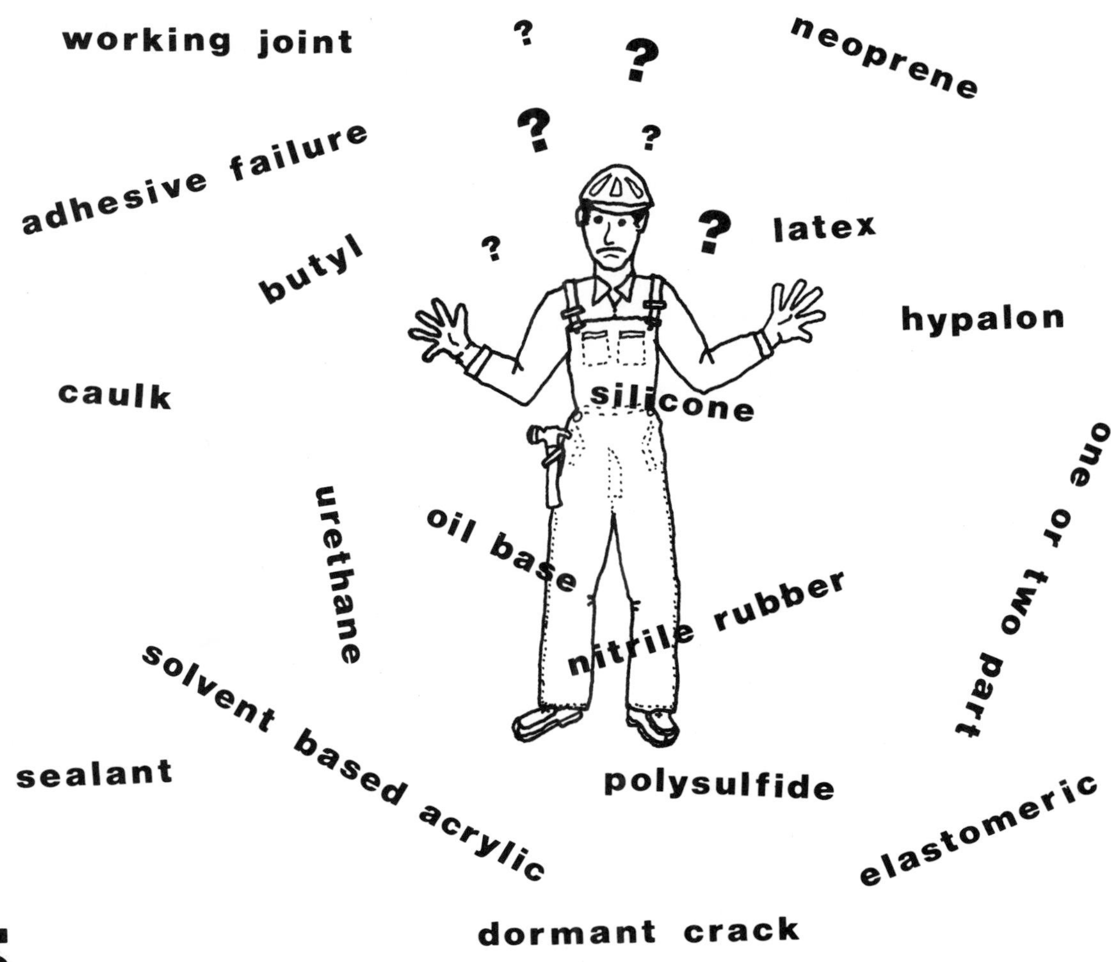

5 caulks

description

what...

You'll see the word "caulk" used for most of the retrofits described in this guide, even though only two chapters, 10 and 38, deal specifically with caulking. Caulk is a thick, mastic material that is designed to remain flexible and never harden completely. By its nature, it's a sealant (the words caulk and sealant are interchangeable) and some varieties have excellent adhesive qualities, so they can be used like glue when necessary for other weatherization retrofits.

why...

The building exterior needs to be weatherproofed to keep out wind and rain. Different parts or components of buildings such as walls and roofs are built to fit tightly together and they may have been caulked as they were put together. But keeping a weatherproof seal is especially hard to do where various building compo-

nents come together. Wind can cause cold, uncomfortable drafts, and water leakage can cause structural and unsightly damage. One way to get an air and watertight seal is to seal these joints with caulk.

It is good to think of caulks and sealants as a way to provide a first line of defense and a second line of defense. You'll see in the illustrations for various retrofits that there are times when it is better not to "stop air leakage cold" on the very outside of the building, and that air leakage inside walls and other places inside buildings need your attention too.

Cracks can open in the building shell due to the movement of the building or any of its various parts. This movement can be very slow as the building settles and as materials shrink and expand. Open cracks can be classified as "working" or "dormant". Both types may be designed into the building as an "expansion joint" to deliberately allow for movement. Older buildings, however, often have cracks that are the result of settling, contraction, aging materials, etc.

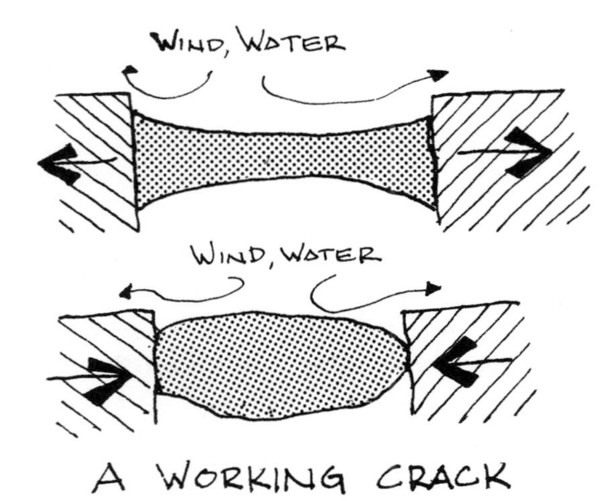

where...

Listed below are areas, by no means limited, where caulk can be used to seal cracks and joints.

- wall joints and cracks
- porch-to-house joints
- joints between two different materials (such as, where a chimney meets siding)
- joint between a wood sill and foundation
- loose siding cracks
- corners formed by siding
- window frames
  - between drip caps and siding
  - between frame and siding
  - between sill and siding
- door frames
  - between drip caps and siding
  - between frame and siding
  - between sill and siding
- cracks around pipes and wires in floors, walls, and ceilings
- flashing
- leaky gutters and downspouts
- skylights
- between permanently mounted storm windows and prime window frames
- an adhesive for weatherstrips

when...

Caulk should be applied when the home is built, and it generally is. However, anywhere from 1 to 20 years (see table) after application, caulk will dry, crack, and need to be replaced because it is no longer providing the proper weatherseal (most caulks used on homes will wear out after 5 years). Of course, the durability of the caulk is dependent

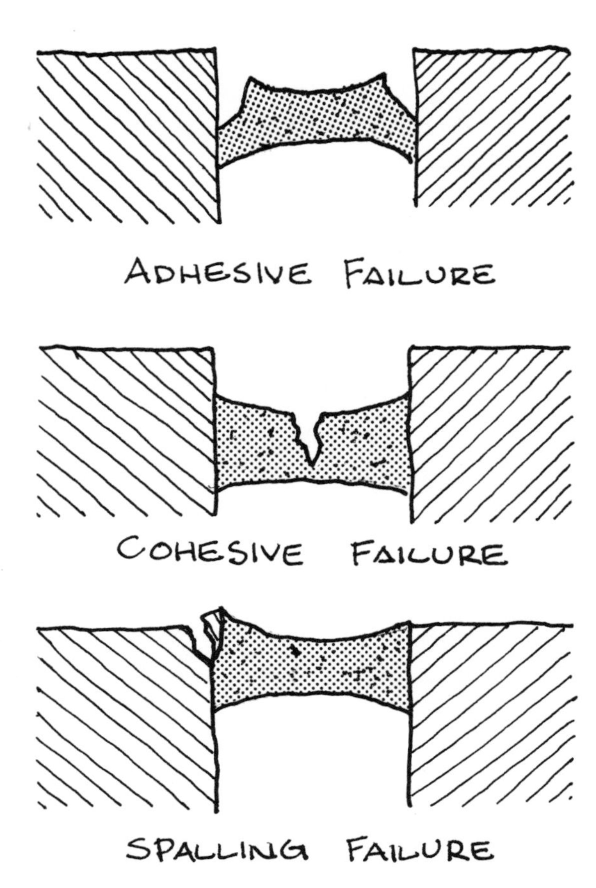

upon the type of caulk used and how it was applied.

You'll know if the caulk needs to be replaced by looking for cracks in it.

how...

The success or failure of sealing cracks and joints is dependent upon how the surfaces are prepared, how the caulk is applied, and on the physical properties of the caulk.

Preparing the surfaces is probably the most important part of doing an effective caulking job. Some caulks do not stick well (do not adhere) without proper surface preparation. (Detailed Preparation Procedures are outlined in Chapters 10 and 38).

"Drawing a good bead"* of caulk (applying the caulk) is also of paramount importance. The bead should be "concave" rather than "convex".

There are a multitude of caulks on the market today. Where continued crack movement seems likely, such as where cracks have opened due to settling, be sure to select a caulk that adheres well to the materials between which the crack has opened and that offers good elasticity or stretchability (known as an elastomeric caulk). Where the crack does not appear to have moved or where movement could only be slight, a caulk with little elasticity will usually do the job.

Some caulks are known for their durability, others are not. Some caulks are for general use, others are for a very specific use. Before selecting a caulk, carefully study the properties to determine the most suitable caulk for a particular application.

---

*A "bead", sometimes referred to as "bed", is a continuous strip of caulk.

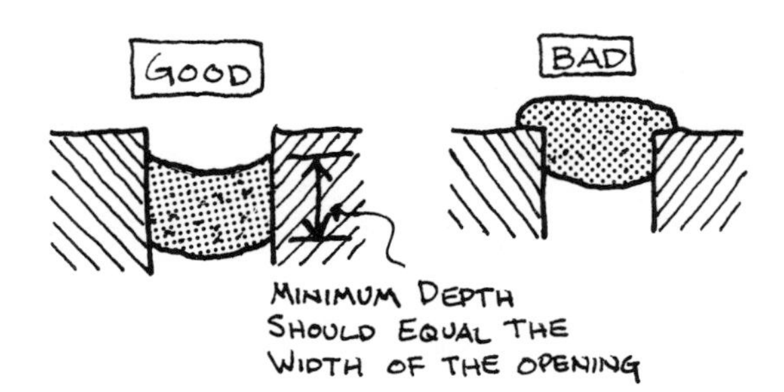

	How Long Will It Last?	Elongation	Min. Temp. for Use	Particular Good Features	Particular Problems	Relative Cost ($/11 oz. tube)	Staining
Oil Base Caulk	∘ High quality last 5-10 yrs ∘ Low quality last 3-5 yrs	• Little elongation and no recovery		∘ Good adhesion to most surfaces while there is little or no movement	∘ Frequently used for glazing; with disappointing results	∘ 0.75	∘ Staining & bleeding of most substrotes
Latex Caulks	∘ 8-10 yrs. perhaps longer	∘ Approximate 60% when formed into ½" × ½" specimens	40°F	∘ May be applied to absortive surfaces - brick, wood, good adhesion	∘ Adhesive property good w/little or no joint movement	∘ 1.50	∘ Non-staining to any type of substrate
Solvent Based Acrylic Sealants	∘ 10-20 years	∘ Below average (25-60%)		∘ Excellent adhesion, can be used for corner beads, & other odd shaped joints	∘ Strong odors - must ventilate interiors, vapors are contominent to foods	- 3.25	∘ Non-staining to most materials
Butyl Caulks	∘ 10 years	• Average (75-125%)	40°F	∘ Particularily good adhesion metal to masonry & for sidings - alum, vinyl, etc.	∘ Only for joints with moderote movement	∘ 1.25	∘ Non-staining
Poly-Sulfide Caulks	• One component has good durability ∘ Two component 20-30 yrs.	∘ Superior (150-200%) ∘ Superior (150-350%)	• Cannot be applied below 40°F • Will not cure below 40°F	∘ Good ∘ Good performance on wide joints & working joints		∘ 7.30	∘ Non-staining to masonry
Urethane Sealants	∘ 20-30 years	∘ Excellent (300-450%)	32°F	∘ Excellent adhesion to wide variety of materials ∘ Abrasion resistant	∘ May adhere so strongly to weak masonry; spalling occurs	∘ 7.50	∘ Non-staining
Silicone Sealants	∘ 30 years	∘ (100-200%)	∘ Can be applied down to -35°F	∘ Good adhesion when applied to metal & glass	∘ Poor adhesion when applied to concrete; cannot be painted	∘ 4.50	∘ Non-staining on most materiols
Nitrile Rubber	∘ 15-20 years	∘ (75-125%)		∘ Adheres well to metal, masonry & high moisture areas	∘ Poor performance on moving joints or wide cracks	∘ 3.25	
Neoprene	∘ 15-20 years	∘ (20-40%)	∘ Difficult to apply at low temps.	∘ Good with asphalt or concrete (foundations)	∘ Difficult to apply		
Chlorosulfonated Poly-Ethylene	∘ 15-20 years	∘ (15-20%)	∘ Difficult to apply at low temps.	∘ Working joints	∘ Very slow cure (3-4 mths.)	∘ 3.50	

Actual performance of caulks and adhesives depends upon many factors, including application technique, condition of crack or joint, and climate, and may vary significantly from those figures listed in this table. Refer to manufacturers' literature for more specific information concerning the product that you use.

	VERTICAL SAGGING	TEMPERATURE RESISTANCE LOW-HIGH	SURFACE PREPARATION	EASE OF APPLICATION	SHELF LIFE	EASE OF PREPARATION	SHRINKAGE
OIL BASE CAULK		o POOR LOW TEMPERATURE PERFORMANCE -10° → 180°F	o QUICK DUSTING OF JOINT	o APPLY AND TOOL EASILY	o GOOD	o EASILY MIXED	o LOW INITIAL SHRINKAGE 5-25%
LATEX CAULKS	o SUITABLE FOR SMALLER JOINTS (3/8" MAX.)	-20°F → 150°F	o CLEAN THE JOINT, BUT NOT AS THOROUGHLY AS HIGH GRADE SEALANTS	o EASY APPLICATION CLEAN UP W/ DAMP CLOTH	o GOOD	o GOOD, AVAILABLE IN 1/10 GALLON SQUEEZE TUBES.	o HIGH DEGREE OF SHRINKAGE CAUSES UNSIGHTLY WRINKLED APPEARANCE
SOLVENT BASED ACRYLIC SEALANTS	o CAN BE OBTAINED IN NON-SAGGING FORM	o POOR LOW TEMPERATURE ELASTICITY -20° → 150°F	o MINIMUM CLEANING REQUIRED	o VENTILATE WHEN APPLYING IN INTERIORS - VAPORS CAN CONTAMINATE FOOD	o GOOD	o RECOMMENDED TO BE HEATED TO ABOUT 120°F BEFORE APPLICATION	o NEGLIGIBLE - HOWEVER LOW RECOVERY CAUSES WRINKLED APPEARANCE
BUTYL CAULKS	o AVAILABLE IN LOW-SAG FORM	-25° → 210°F	o CLEAN CONCRETE WIPE: METAL W/ OIL FREE CLOTH & SOLVENT, WOOD W/ CLOTH OR SOFT BRUSH, GLASS W/ CLOTH	o CLEAN UP W/ SOLVENT - BELOW 40°F, CAULK STIFFENS & IS DIFFICULT TO APPLY	o VERY LONG	o NO MIXING OR HEATING	o HIGH
POLY-SULFIDE CAULKS		-60° → 250°F			o LOW, POOR	o MESSY MIXING	o LOW, NEGLIGIBLE
URETHANE SEALANTS		-60° → 275°F	o MUCH: REQUIRES VERY CLEAN SURFACE JOINTS TO OBTAIN ADHESION	o ABOVE 70%RH CAN REDUCE EFFECTIVENESS OF SEAL	o VERY SHORT	o MIXING IS EXTREMELY CRITICAL	o NEGLIGIBLE
SILICONE SEALANTS	o NON-SAGGING ON VERTICAL WALLS	o HIGH - -90° → 400°F	o SURFACE REQUIRES CAREFUL PREPARATION		o VERY SHORT		o NONE
NITRILE RUBBER							o HIGH
NEOPRENE							o HIGH
CHLOROSULFONATED POLY-ETHYLENE			o POROUS SURFACES REQUIRE PRIMING		o EXCELLENT		o HIGH

Information from Sealants, edited by Adolfas Damusis, Reinhold '67, and personal communication with manufacturers.

Physical properties for the following caulks are summarized here and in tabular form:

- oil based caulks
- latex caulks
- solvent based acrylic sealants
- butyl caulks
- polysulfide sealants
- urethane sealants
- silicone sealants
- nitrile rubber
- neoprene
- chlorosulfonated polyethylene

(Detailed installation procedures are outlined in Chapters 10 and 38.)

OIL-BASED CAULKS

- general purpose sealant designed for exterior or interior joints; responds to small joint movements; can be painted almost immediately after application (30 min.); available in cartridges and in cans; can be applied on damp surfaces

LATEX CAULK

- general purpose sealant designed for exterior or interior joints; responds to small joint movements; can be painted almost immediately after application (30 min.); available in cartridges and in cans; can be applied on damp surfaces

SOLVENT BASED ACRYLIC SEALANTS

- for general purpose caulking in joints with moderate movement; excellent adhesion to most building materials without the use of primer; estimated service life up to 20 years; strong odor restricts use to exterior joints; available in cartridges and cans in numerous colors

BUTYL CAULKS

- an all purpose caulk with greater service life and performance than the oil base caulks and about equal in overall quality to the latex/acrylic types; forms a skin which can be painted (one week later); responds to joint movements up to about 10% of the original width; adheres to all building surfaces (especially masonry to metal) and retains a high degree of flexibility; available in cartridges and cans in numerous colors

POLYSULFIDE SEALANTS

- 1 and 2 part; cures to a rubbery compound with exceptional flexibility, recovery and adhesion to most building surfaces; responds to joint movements up to 25%; service life is 20 years or more; cured sealant has practically no shrinkage; primer may be required; available in cartridges and cans in various colors

URETHANE SEALANTS

- performance characteristics similar to polysulfide; high abrasion and tear strength; available in single and multi-compenent types

SILICONE SEALANTS

- class of single component sealants known for its versatility as a joint filler with exceptional performance under very high and low temperatures; exceptionally high adhesion, ease of application from the caulking gun is unaffected by cold temperature; forms a skin in less than an hour and cures in a few days to an elastic compound, available in cartridges, cans, and tubes in several colors including clear; cannot be painted, except with a silicone paint

NITRILE RUBBER

- extremely good for use in high moisture areas; not for use on moving joints or wide cracks; life expectancy of 15-20 years

NEOPRENE RUBBER

- especially good for use on concrete walls and foundations; can be used on narrow (less than 1/4" wide) moving joints; forms a skin in one hour; life expectancy of 15-20 years

HYPALON

- can be used for moving joints; requires priming before use on porus surfaces; lasts up to 20 years

6

safety

description

Just as it is important to do a thorough and complete job when installing energy retrofits, it is equally (if not more) important to maintain a safe working environment at the same time. Some principles to keep in mind are;

- be aware of people who are around you
- keep the work area clean and free of potential hazards
- use tools and equipment properly
- understand the nature and limitations of the materials that you're working with

Obviously, these principles are not being adhered to in the adjacent sketch. Always remember: It's good to work efficiently. But if the fast or easy way to do the job is hazardous, then it is not the right way.

Safety pointers, guidelines, and ideas that are important to keep in mind while working are discussed in this chapter. It is by no means comprehensive. However, you may find some points that you haven't thought of or didn't think were important.

ATTIC

1 safety gear

A "dust filter mask" and safety goggles should be worn at all times while blowing insulation, loading the insulation blower, or laying cellulosic or fibrous insulation. Additionally, gloves and long sleeve shirts should be worn when handling fibrous insulation to prevent skin irritation. It's also a good idea to wear a hard hat, especially if there are protruding nails in the roof. Be sure to provide adequate lighting in the work area.

Refer to Chapter 2, "A Word about Insulation", for additional information on the safe handling and use of various insulation types.

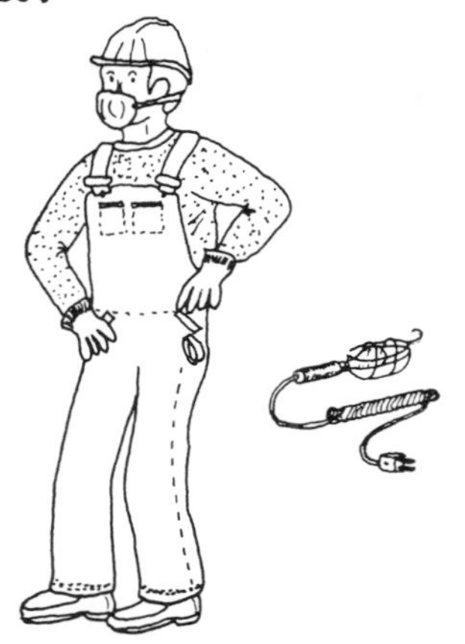

2 stairs

If there are stairs leading up to the attic, make sure they are secure. Replace missing, broken, rotted, or loose treads with minimum 1-1/4" thick by 8" wide treads. Stairs not in a suitable condition should not be used. Access the attic by a different route. (See Chapter 35, "But What About Accessing the Attic?" for details.)

PROTECTIVE GEAR SHOULD MEET THE FOLLOWING STANDARDS;

Hardhats	-	Safety Requirements for Industrial Head Protection, American National Standards Institute, 289.1 - 1969
Safety Goggles	-	Practice for Occupational and Educational Eye and Face Protection, American National Standards Institute, 287.1 - 1968

Respirators should conform to NIOSH (National Institute for Occupational Safety and Health), Publication #77 - 195, Cumulative Supplement June 1977 NIOSH Certified Equipment, for "Dust, Fume, and Mist Respirators".

It's suggested that rugged ankle high boots, preferably with steel toe and sole, be worn to prevent puncture wounds from nails.

3 structural strength

Check the condition of the roof structural members and those structural members that you'll use for support. Areas which are structurally unsafe should not be insulated until the situation is corrected. In unfloored attics, lay 1/2" x 24" x 98" (minimum dimensions) plywood or plank "catwalk" running perpendicular to the joists. Floored areas should be unobstructed - pay particular attention to loose wire, nails, and other valuables ("junk") stored in the attic.

4 electrical

If you find exposed wiring or wiring with loose insulation, turn off the circuit and either replace or wrap the wire with electricians' tape. This work should be done by someone experienced in electrical work. Likewise, check for loose connections, ungrounded receptacles or outlets, short circuits, or motors. Circuits to these devices should also be turned off, especially if you will be blowing insulation and suspended house dust or cellulose dust concentrations will be present. High temperatures or sparks can ignite suspended dusts.

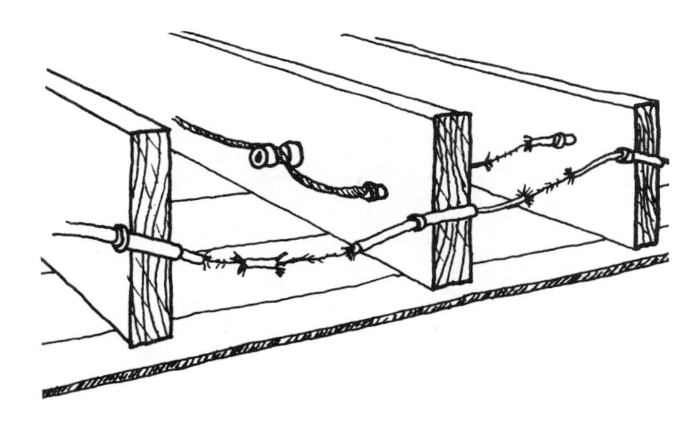

5 flue/chimney

Check the condition of the flue or chimney which runs through the attic. If it appears that flue gases could be introduced into the attic, sparks have escaped the flue or chimney into the attic, or dust from insulation could enter the flue or chimney, insulation should not be installed until these cracks/holes in the flue or chimney have been sealed.

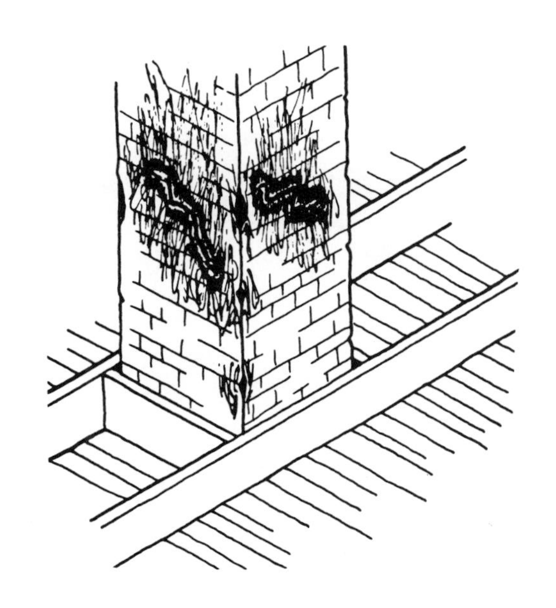

6 animals

Check for the presence of rats, stray animals, insects (wasps, hornets, spiders, etc.) or other hazards from animals. Insulation should not be installed until these pests are removed.

7 high temperatures

Take special precautions if the attic air temperature is greater than 100° F. With this situation, it is advisable to spend no more than 25 minutes of each 30 minute period in the attic. Insulation should not be installed in the attic if the attic air temperature cannot be reduced to 115° F.

Under hot conditions, watch yourself and your co-workers for muscle cramps, "flush" face, weak rapid pulse, giddiness, and dry skin following heavy sweating. These are signs of heat fatigue and heat stroke that require medical attention. Take the person with these symptoms to a shaded cool area, loosen the clothing, expose the chest, arms, and legs, and bathe with cool water until medical attention has been obtained.

8 signalling

It is preferable to have three people cooperate when blowing insulation - one person with the hose, the second person feeding insulation into the blower, and the third person to insure communication between the other two people.

If only two people are blowing insulation, use an alternate signalling system. For example, tie a rope off on a rafter (when insulating an attic, for example) and extend it out to the blower. A suggested signalling system would be as follows:

1 tug - OK
2 tugs - Stop
3 tugs - Need one more bag

LADDERS

9 Before using, check ladders for broken or damaged rungs, steps, and rails. Damaged ladders should not be used.

Ladders should meet the following standards:

- Safety Code for Portable Wood Ladders
  American National Standards
  Institute, A 14.1 - 1968

- Safety Code for Portable Metal Ladders
  American National Standards
  Institute, A 14.2 - 1956

10 Carry ladders over your shoulder with the front end elevated. Ladders should not be dropped as that can weaken them.

11 Before setting up a ladder, make sure it will clear trees and wires (wires are of primary concern, especially when using aluminum ladders). The National Safety Council recommends that all ladders be kept a minimum of ten feet from wires. Where possible, ladders should not be set up near corners or in front of doors (if you must set the ladder up in front of a door, lock the door first).

12 Set ladders on a firm, level base. The horizontal distance between the foot of a straight ladder and the house should be about one fourth of the length of the ladder (from the foot of the ladder to the point at which it rests against the house).

13 A step-ladder is not to be used as a straight ladder.

14 If both hands are needed while working on a straight ladder, hook a leg over

one of the rungs. It is advisable to keep both hips between the side rails. Tools should not be kept on top of, or strung from, ladders. Face the ladder both when going up and coming down.

15 ROOFS

When working atop roofs with a pitch greater than 30° (6 in 12), it is advisable to use a safety line tied around the waist and anchored to the roof.

NOTE:

Exercise extreme caution when placing materials and tools on pitched roofs so they don't slide off.

16 Usually, rafters of a pitched roof will be of smaller dimensions than attic joists. They will probably not be as strong. An excellent "tool" to assure level footing and distribution of body weight is a platform with planking on angled supports which is long enough to rest on several rafters.

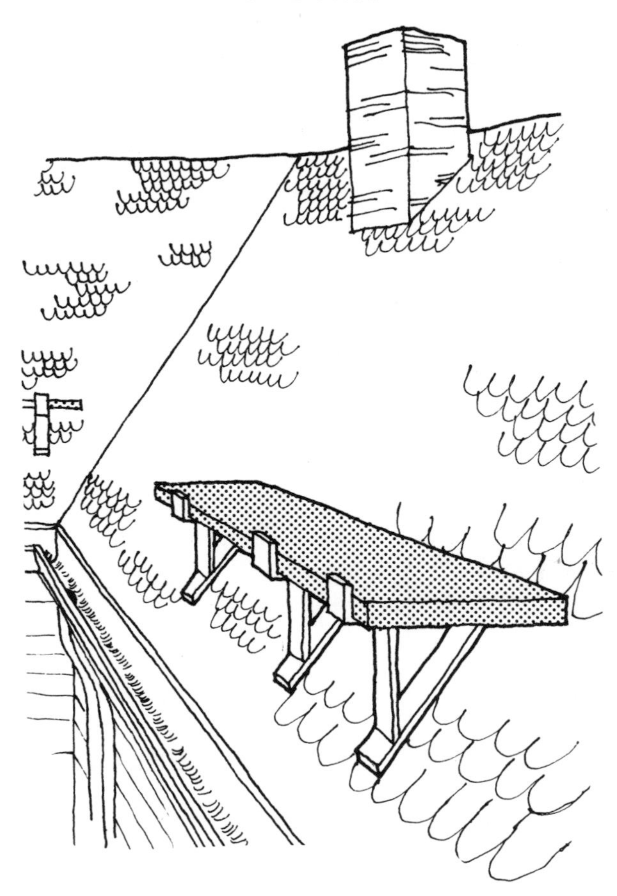

NOTE:

Precautions 1, 3, and 7 also apply for roof work.

17 SIDEWALLS

Safety goggles and a dust mask should be worn while drilling access holes and/or removing siding. Impact resistant eye gear is strongly recommended when breaking through masonry.

NOTE: 

Precautions 1, 4, 6, 7, 8, and 9 through 14 apply to side wall work.

18 BASEMENT interior

When blowing insulation near the heating plant or hot water heater, turn off the pilot light to eliminate hazard of dust combustion.

20 BASEMENT
exterior

Trenches around the perimeter of the basement should be a minimum 18" wide to install exterior insulation. Shore with plywood "batter board" if soil will not hold a clean edge or if the trench will be deeper than 3'.

21 If you must trench deeper than 5' to install exterior basement insulation, slope, shore, sheet, brace, or otherwise support the edge of the trench. When trenches are more than 4' deep, it's recommended that steps or ladders be located no more than 50' apart.

NOTE:

Refer to the Occupational Safety and Health Administration (OSHA) publication, #2226, "Excavating and Trenching Operations", for additional details.

---

19 If there is standing water on the basement floor, or if the walls are moist, electrically powered equipment should not be used until the area is dry (mop and/or let air dry).

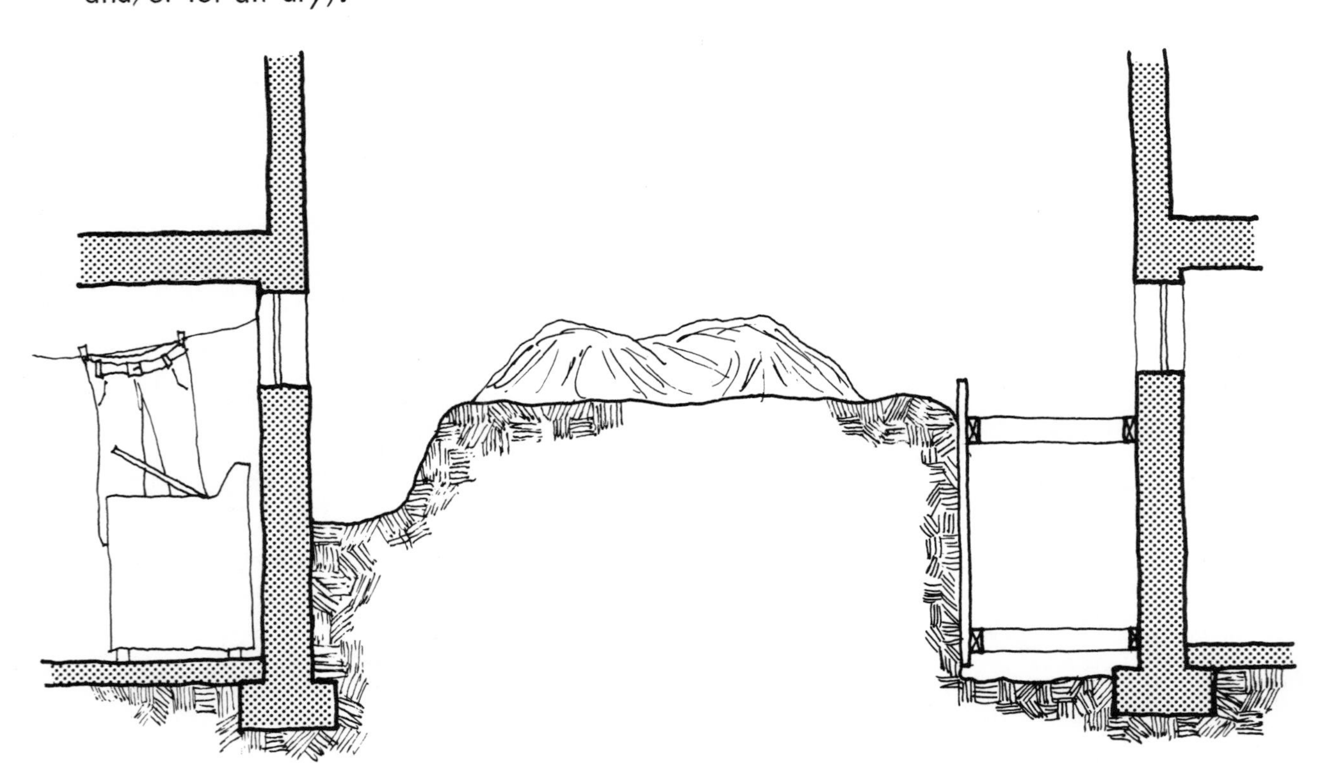

22 Mark the perimeter of the trench with 2'-6" stakes with two lines of twine strung between the stakes. Tie white "flags" to the twine approximately 6'-0" apart.

CRAWL SPACE

Precautions 1, 4, 6, 17, 20, 21, and 22 apply.

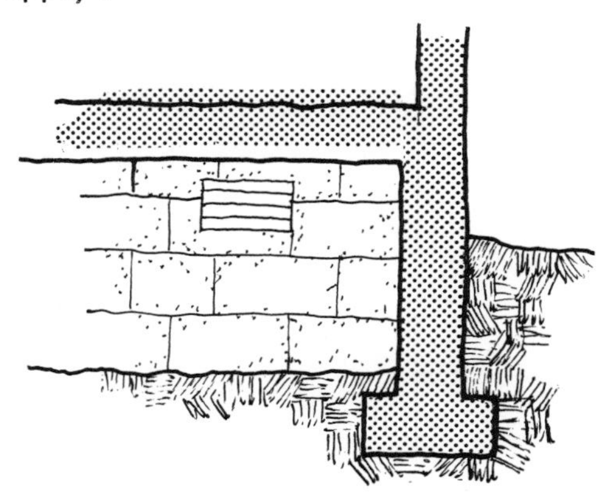

SLAB-ON-GRADE

Precautions 1, 6, 20, 21, and 22 apply.

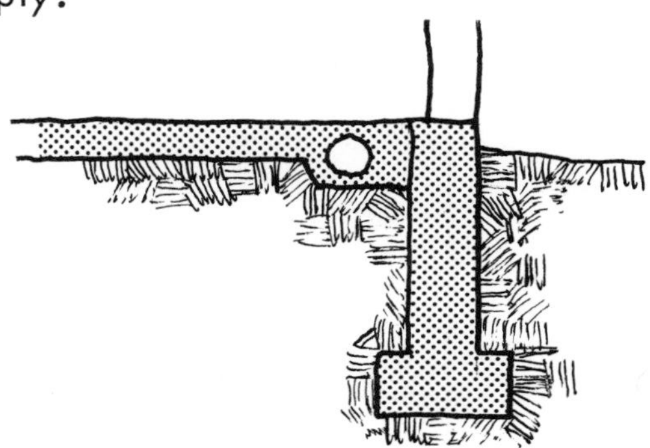

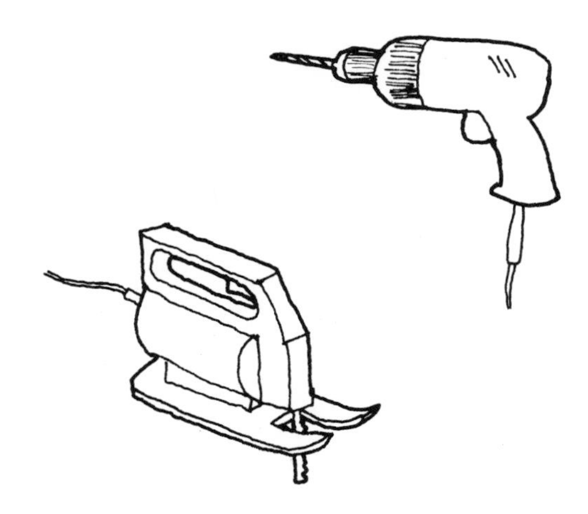

23 POWER TOOLS

Only power tools that are grounded or have double insulated cords should be used. Use an adaptor plug for connecting a grounded tool to a two prong receptacle and attach the ground wire (the free wire) of the adaptor to the screw in the middle of the outlet cover. Better yet, connect the tool to a grounded circuit. Do not cut off the ground from a grounded plug to accommodate a two prong receptacle. For additional information, refer to the following publications;

- "On the Job Safety Rules for Power Tools" (NIOSH)
- "Ground Fault Protection on Construction Sites" (OSHA), #3007

NOTE: 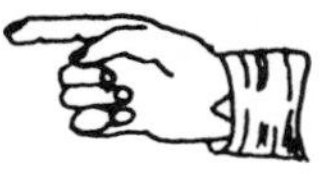

Precaution 1 applies.

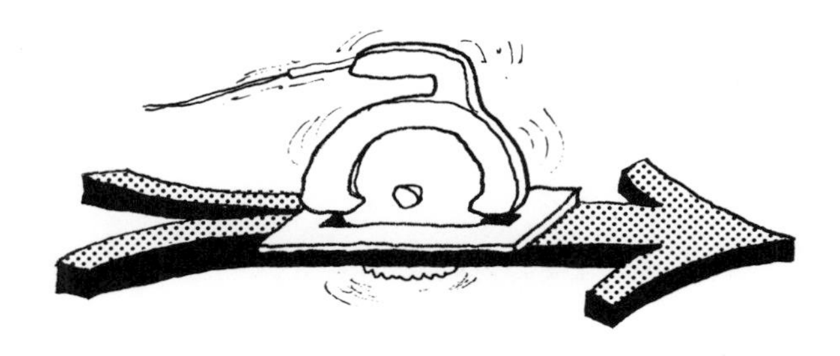

24 Power tools should only be operated by persons who are familiar in their use and safe operation. This applies also to persons who are "helping" in the power tool operation. It is suggested that:

1) Specific instructions included with tools be studied by operator and helper and kept for reference.

2) Power tools be checked for safety on a regular basis.

3) Power tools with inadequate safety guard protection, overheating, or problems with vibration or poor operation due to lack of maintenance should not be used.

25 To the extent possible, check building sections for wiring, ducts, and/or pipes before cutting into the sections.

for your information...

U. S. Department of Labor
Occupational Safety and Health Administration
Washington, D.C. 20210

National Institute for Occupational Safety and Health
Robert A. Taft Laboratories
4676 Columbia Parkway
Cincinnati, Ohio 45226

American National Standards Institute
1430 Broadway
New York, New York 10018

II

window retrofits

Replace broken glass

Re-set glass

Weatherstrip windows

Pack and/or caulk windows and doors

Change window operation

Install plastic storm windows

Install glass storm windows

Install window insulating shutters

Install window insulating panels

Replace existing window

WINDOW DETAIL

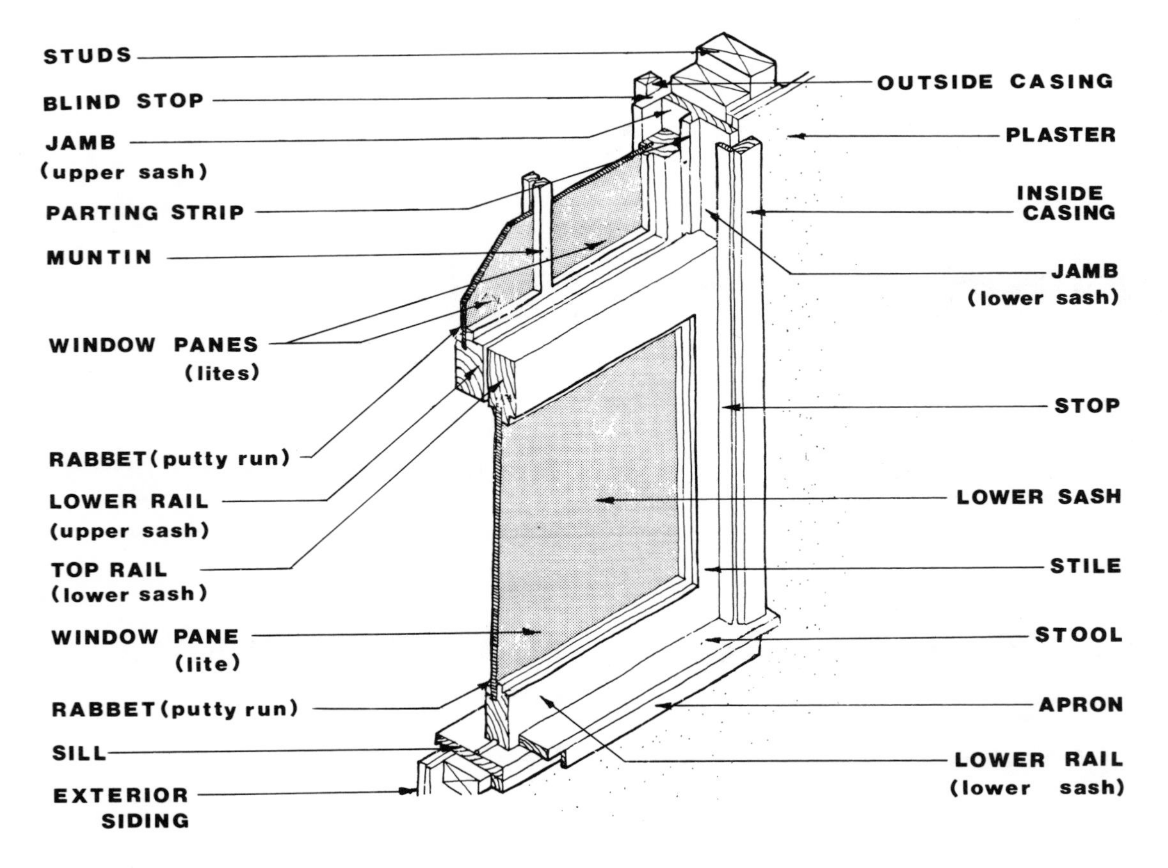

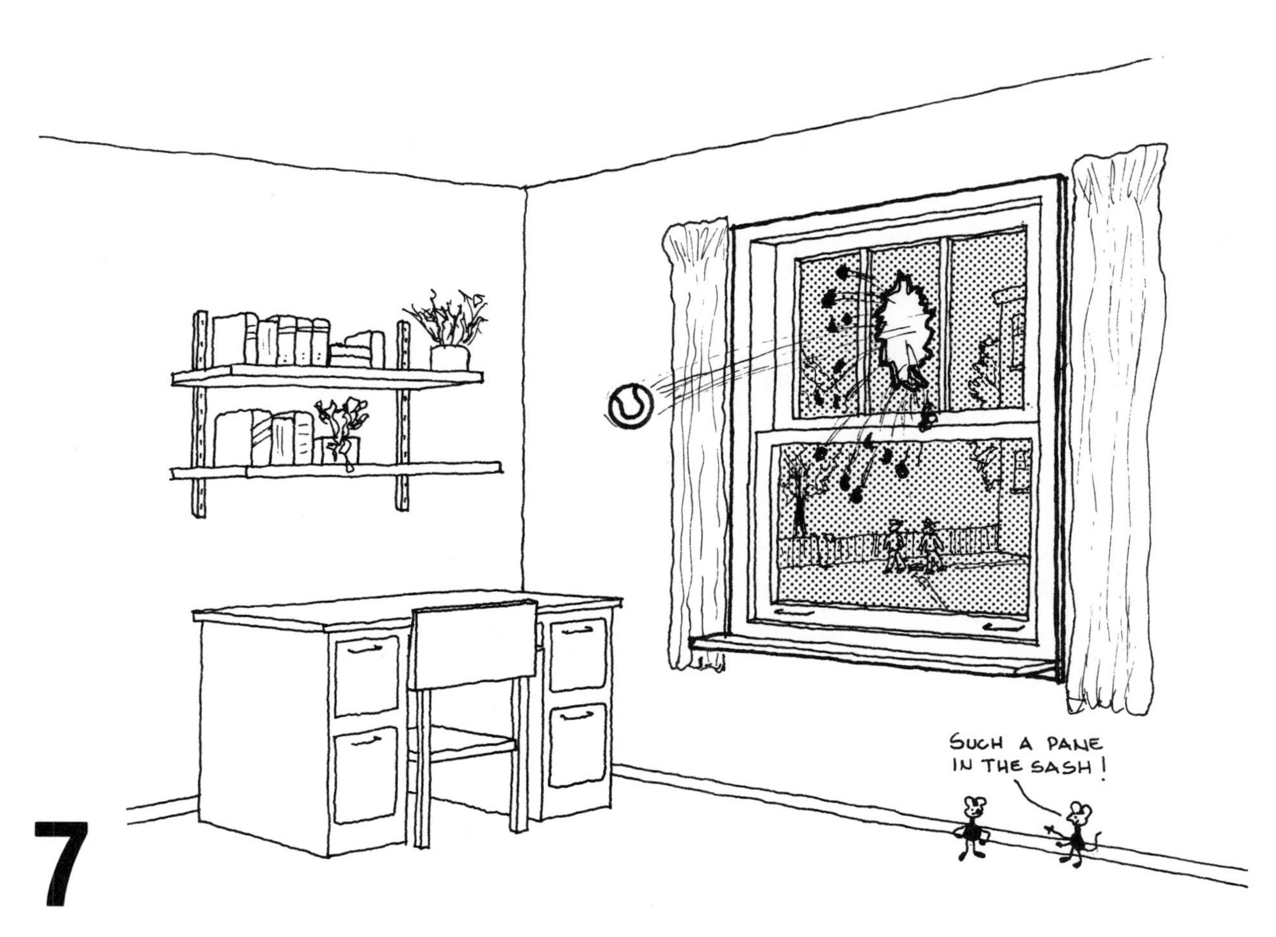

7

replace broken glass

description

Replacing broken glass is not as difficult as you may think. With only a few tools and materials, you can replace glass in most wood and metal sashes.

If the pane of glass is just cracked, it may be better to tape over the crack rather than replace the pane (use polyethylene or "weatherstrip" tape). If, however, there are extensive cracks, or there's a hole in the pane, the glass will have to be replaced.

This retrofit includes all cracked or broken glass in windows, doors, sidelites, storm windows, and storm doors.

materials

OF COURSE, THERE'S GLASS,
Window glass is available in three grades based on strength and clarity:

AA - Superior
A - Very good
B - General or utility

Generally, glass should not be installed in sizes exceeding 7'0" width and 10'0" height. Manufacturer's recommendations for glass thicknesses based on size, location of lites, and wind loading supercede the following:

- single strength (.1" thick) - do not use where lite exceeds 25 square feet

- double strength (.133") - use in large lites up to 40 square feet and/or where glass will be exposed to high winds
- plate (1/8 - 1 1/4") - use for large window areas where clarity and strength are required
- safety (tempered) - use for public doors, in locations adjacent to or near public doors, and for storm doors
- sheet (3/16 - 1/4") - slightly wavy and may cause some distortion of view; for general glazing use
- wireglass - use for security, for fire regulation applications, and for skylights
- acrylic plastic sheet - use in cellar windows, storm doors, and other locations subject to severe impact, where safety is required, or to prevent glass breakage by vandalism
- translucent fiberglass - may be suitable where a direct view is not required
- insulating (3/8 - 1" with welded or sealed edges) - available with double or triple layers of glass from manufacturer in stock sizes and from manufacturer or glass specialty company in any size up to 6' 0" in width and 8' 0" in height

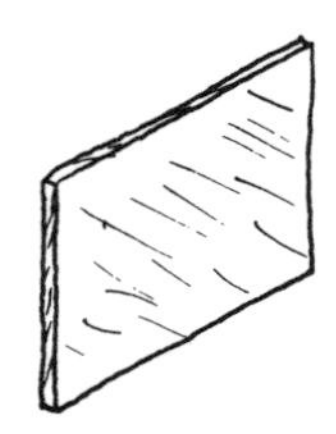

SOMETHING TO SEAL THE GLASS,

- wood sash putty (oil base)
- glazing compound - generally superior application and service performance than oil base putty

AND SOMETHING TO HOLD THE GLASS IN PLACE

- glazier points - galvanized metal (triangle or diamond shaped), spring steel clips (for metal sash)

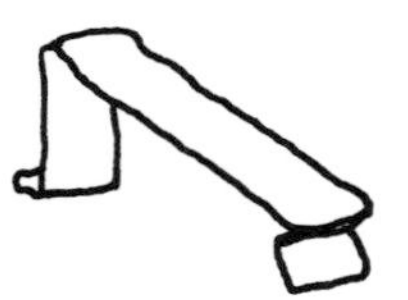

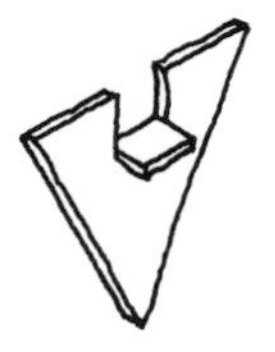

- setting blocks - rubber, neoprene, vinyl, or teflon
- wood or metal stops
- sash bead - rubber bead is used to prevent pressure cracks when re-glazing metal sash

preparation

1 The first thing you'll have to do is remove all the broken glass from the frame. Protect your hands by wearing gloves or by using rags to pull out the broken glass. If you can't pull out a piece, slowly move it back and forth to loosen the old putty and then pull the glass out. Place the broken glass in a box, not in

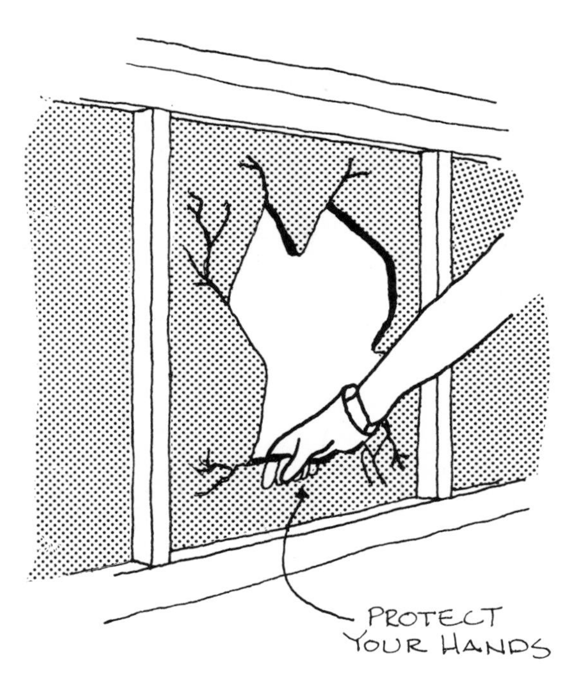

a paper bag, and discard the glass (or recycle it!) when all is removed.

2 Scrape out the old putty with a chisel or screwdriver. If the putty is extra

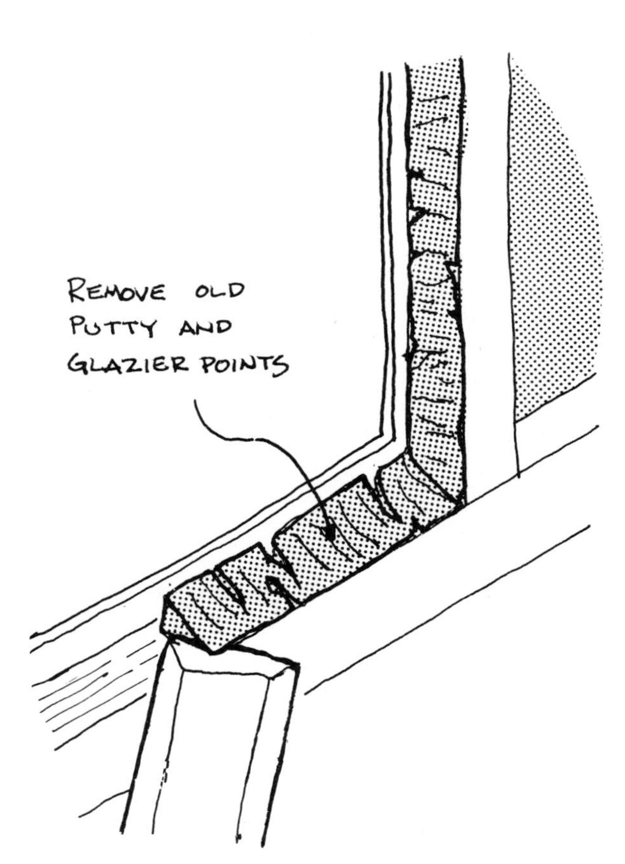

hard, tap your chisel or screwdriver with a hammer to loosen the putty. Be careful not to gouge the molding strip in which the glass fits. Also, remove the old glazier points. Be extra cautious if you use a torch to loosen the putty.

3 Once you've removed all the putty, clean the wood with a wire brush or with steel wool and mineral spirits.

4 Then seal the wood by painting the recess (rabbet) with primer. When using oil base putty, wipe on linseed oil so that

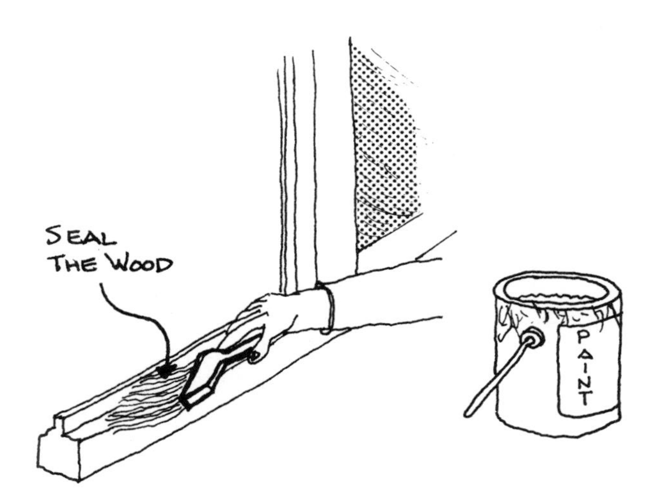

the pores in the wood will not draw the oils out of the new putty, leaving the putty dry and brittle. Do not use shellac or varnish for priming.

installation procedures

1 Now measure the opening in which the new glass will fit. When measuring, the piece of glass should be about 1/8" less in height and width than the actual opening to allow for clearance. If you're using plastic, leave slightly more than 1/8" to allow for its expansion.

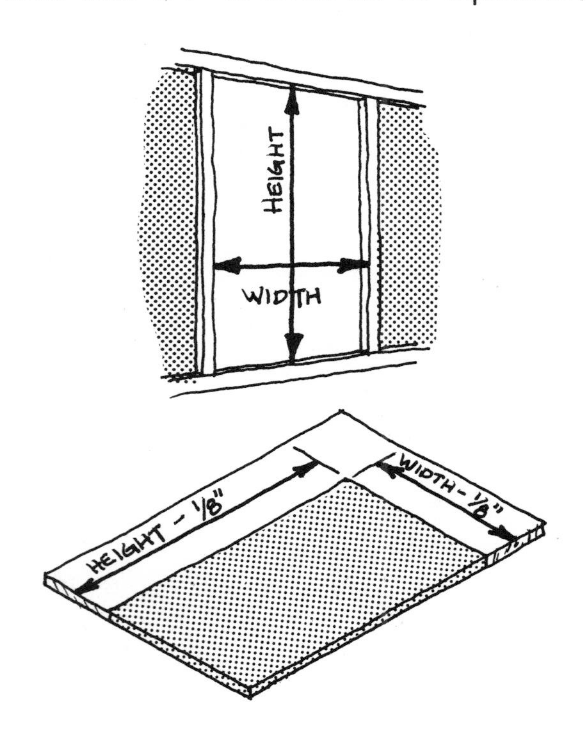

NOTE: 

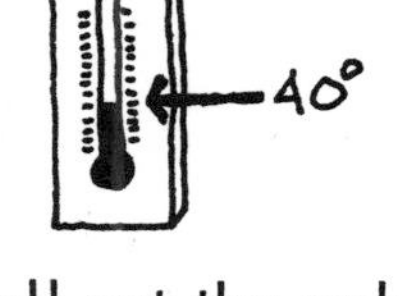

Do not re-glaze if the temperature is below 40° F. Putty should be stored in a warm place. If you must re-glaze, pull out the sash, install plastic sheet over the window opening, and work in a warm place.

2 Prior to placing the new piece of glass, apply a thin layer (1/8") of glazing compound to the inside of the rabbet around all sides of the pane. Besides cushioning and correcting for irregular-

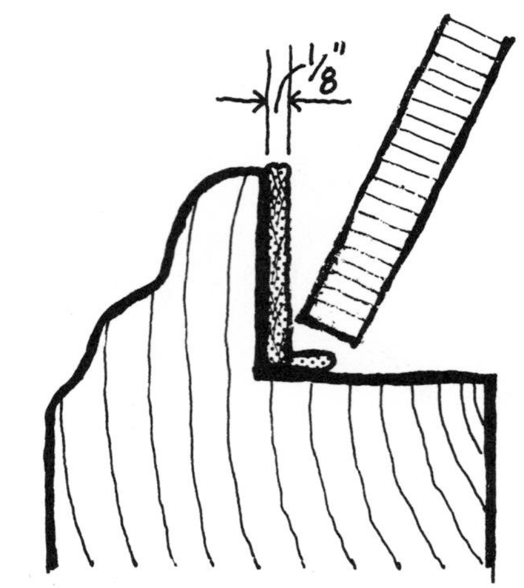

ities in the frame, this glazing compound will provide a water-tight seal once the glass is set in place. Press the glass firmly against the glazing compound.

3 Now for the glazier points. These small triangular shaped pieces of metal permanently hold the glass in place (don't rely on the glazing compound only). Glazier points are placed no more than 8" apart around the perimeter of the frame (a minimum of two glazier points per side for small panes). Push

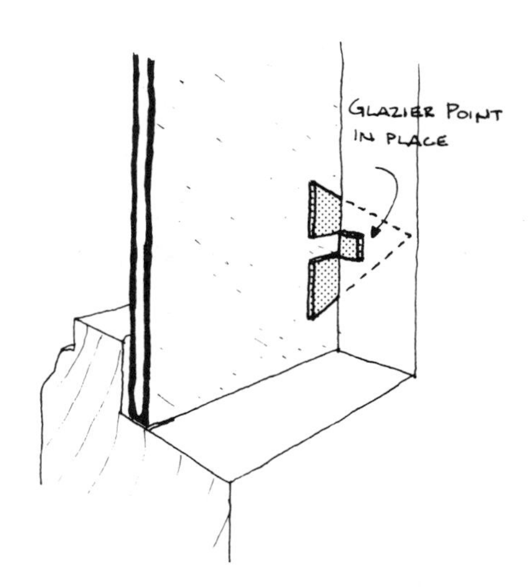

the glazier points in with the end of your putty knife or screwdriver. They're not hard to install.

4 Now place the glazing compound around the outside edge of the pane. An easy way of doing this is roll the glazing compound into strips about as thick as a pencil. Lay these strips on the recess around the edge of the pane with your hands. At this point, take your putty knife and form a bevel (triangular shaped). The important thing is to pack down the glazing compound firmly, leaving no cracks or open seams into which water can seep (or you may be referring to Chapter 8 in the future).

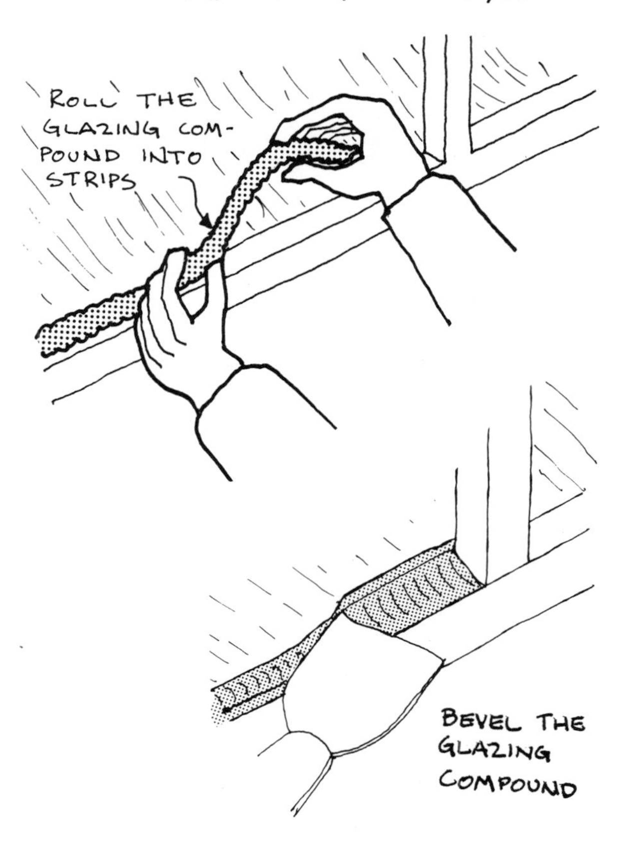

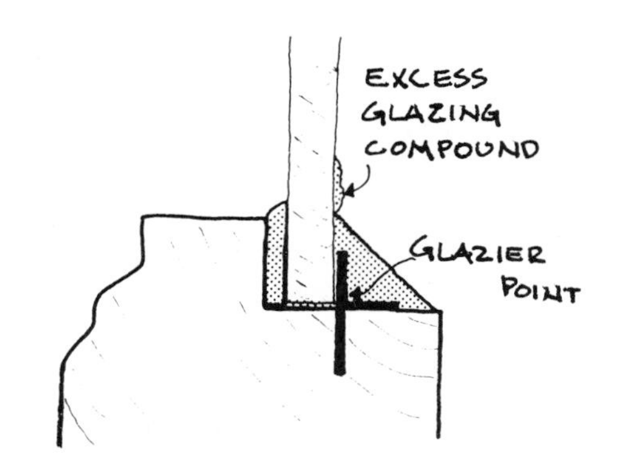

5 Wipe away the excess compound that you may leave on the glass.

6 Now you're ready to finish the job. The glazing compound has to be painted after it is applied. The time at which it can be painted after application depends on the type of glazing compound that you're using.

- wood sash putty (oil base)
  paint after it dries thoroughly
- metal sash putty
  paint as soon as the surface hardens

METAL SASH

The main difference between wood and metal sash frames in reglazing windows is that instead of using glazier points, a special "spring steel clip" is used to hold the glass in place. The glass should be placed on "setting blocks", so don't misplace the setting blocks you find along the bottom of the frame, but reinstall them before placing new glazing in the sash. You can use glazing compound, assuming it's not an oil base putty, to "seal" the glazing within the sash.

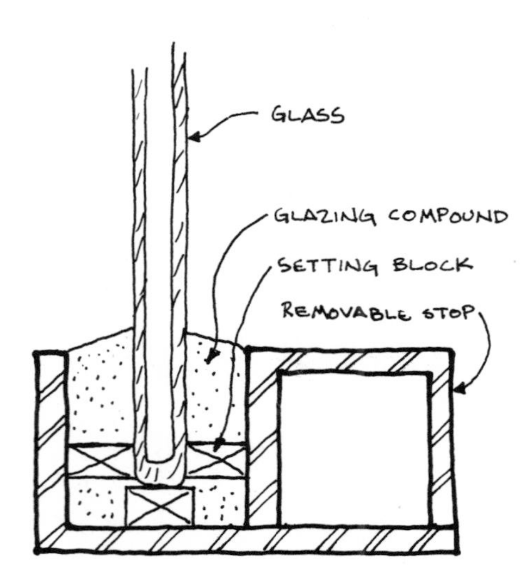

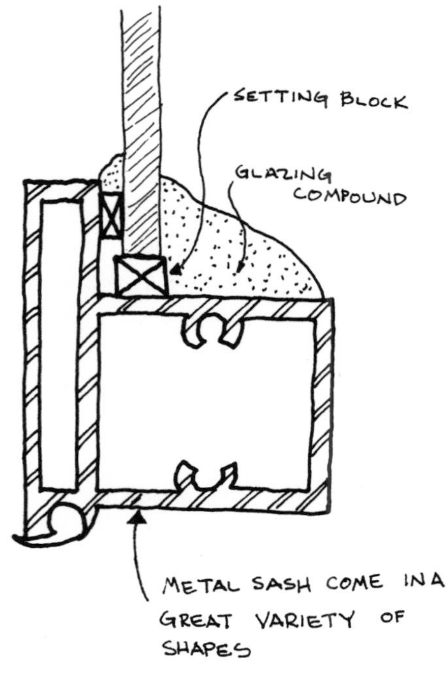

Some metal sash have a strip which holds the glazing in place (the strip, or molding, is, in turn, held in place by screws). In this situation, simply unscrew the strip, remove the broken glass, and replace the strip over the new glass.

NOTE:

The glass in some metal storm doors and windows is held in place by a plastic gasket. You may need the services of a glazing company or hardware store glass department for removal and reinstallation.

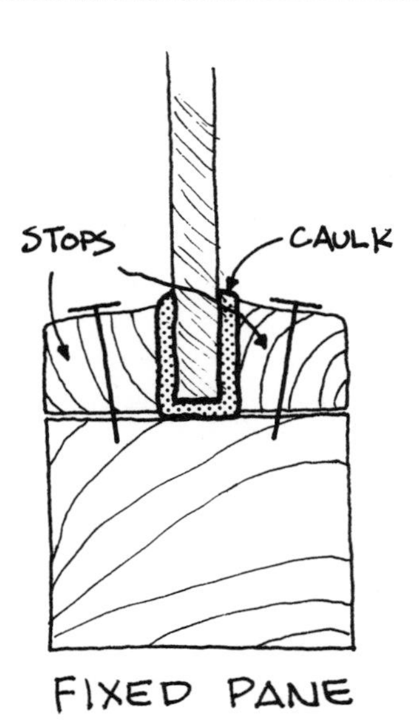

FIXED PANE

NOTE:

In some instances, especially with fixed windows, the glass may be set against wood or metal stops. Remove the stops and clean as described under "Preparation". Apply a continuous bead of caulk to the outside face of the pane and then force the stops against the pane. Secure the stops with nails and clean the excess caulk from the pane.

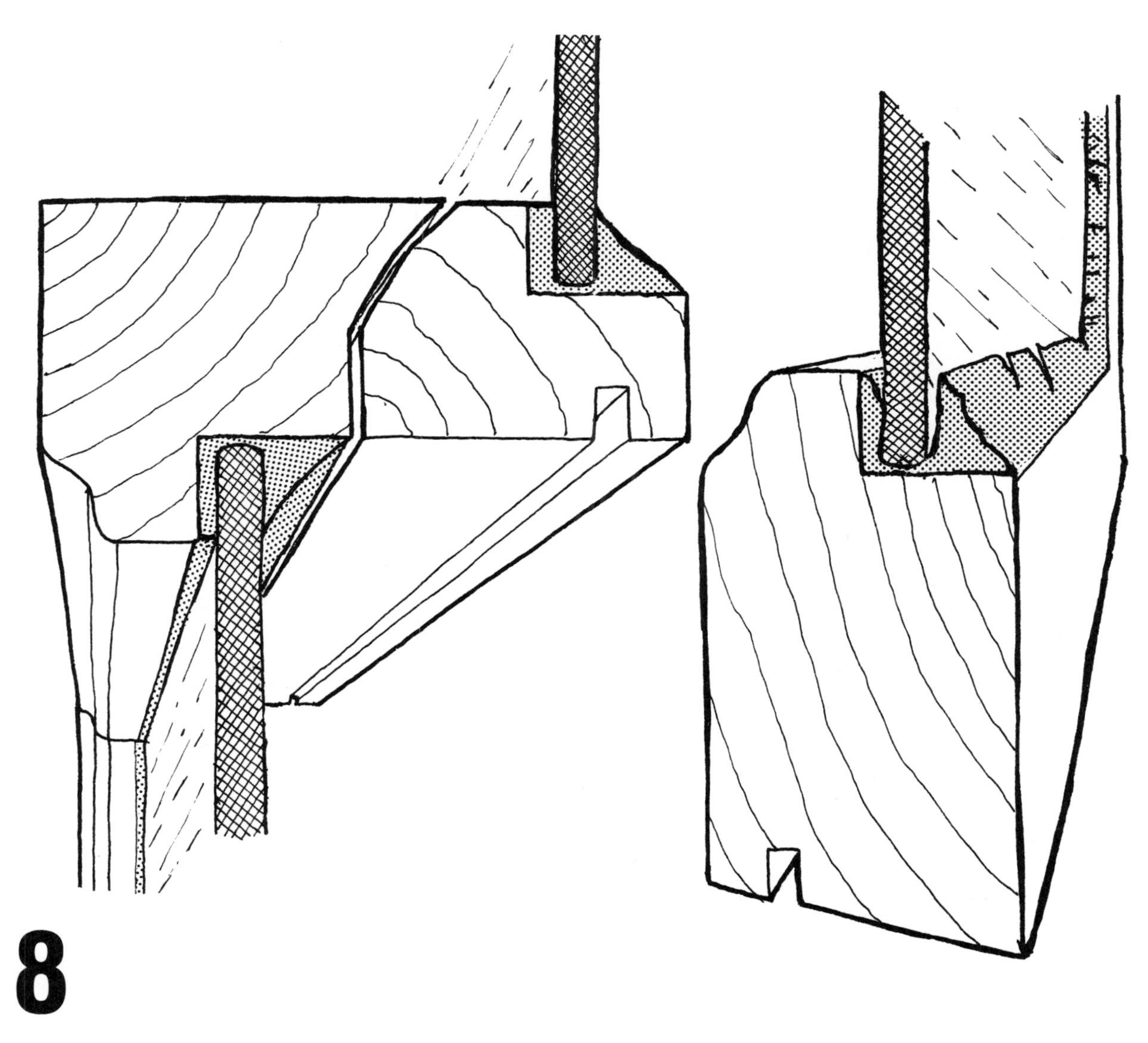

8

re-set glass

description

When glass is set within a sash, it is sealed to the sash with putty or glazing compound. This compound keeps water and air from getting into the sash, causing a wood sash to rot and a metal sash to rust. When the compound dries out, it becomes brittle and starts to crack. At this point, the deteriorated putty is no longer providing the proper seal and needs to be replaced.

If most of the putty has deteriorated, the glass has to be pulled out and re-set in new glazing compound. If there is "spot" deterioration of the putty, those spots can be re-puttied without having to remove the glass.

You can check the integrity of the putty yourself. Try pushing the glass away from the putty with your fingers; if a gap appears, the putty is not providing the proper seal. Pay carefull attention to the bottom of panes because it is here that water does the most damage as it condenses on the glass and runs down.

materials

putty

- wood sash putty (oil base)

- glazing compound - generally superior application and service performance than oil base putty

- glazier points - galvanized metal (triangle or diamond shaped), spring steel clips (for metal sash)

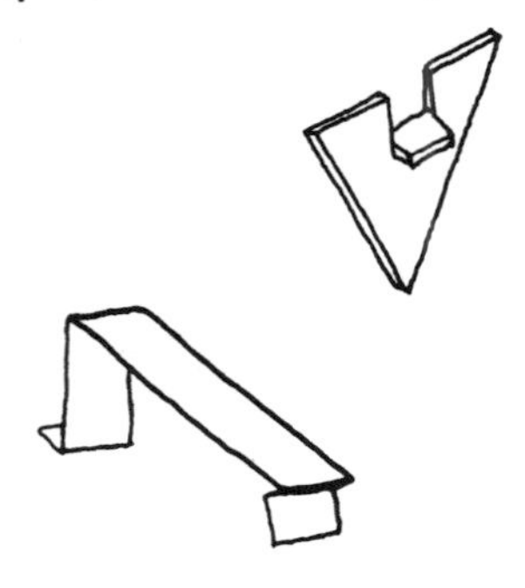

- setting blocks - rubber, neoprene, vinyl, or teflon

- wood or metal stops

preparation

1 Scrape out the old putty with a chisel or screwdriver. If the putty is extra hard, tap your chisel or screwdriver with a hammer to loosen the putty. Or, use a small electric heater that is designed for softening old putty.

2 Next, remove the old glazier points with your hand (tough ones may have to be pried out).

3 Now you're ready to remove the glass. The glass may be attached to a thin bed

of putty on the other side. If so, simply pry the glass away from it. Once the glass is out, place it in a safe place.

NOTE: 

If the glass breaks while you're removing it, see Chapter 7, "Replace Broken Glass".

4 Once the glass is out, scrape off any remaining putty from the rabbet. Be extra cautious if you use a torch to loosen the putty.

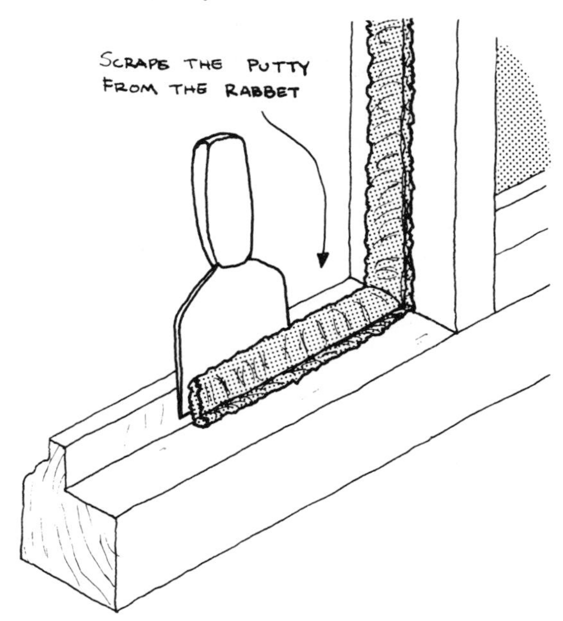

5 Clean the wood with a wire brush or steel wool and mineral spirits.

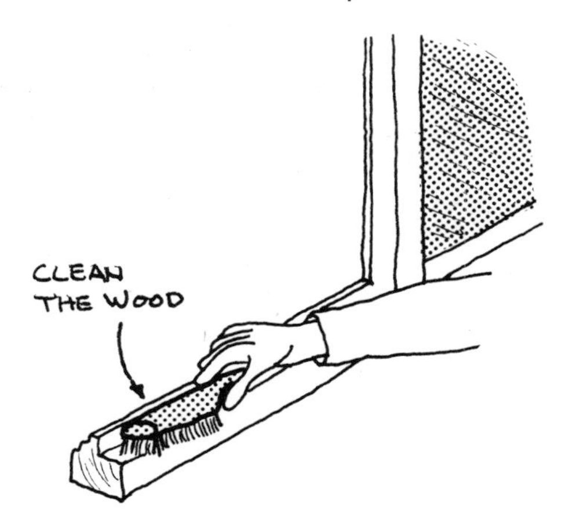

6 Seal the wood by painting the rabbet with linseed oil or primer. Do not use shellac or varnish for priming.

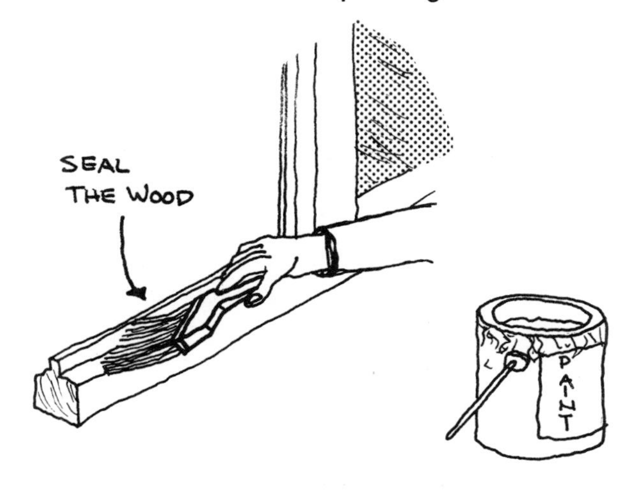

AN IMPORTANT NOTE: 

There will be instances where most of the putty on a pane is good while the remainder of the putty has deteriorated.

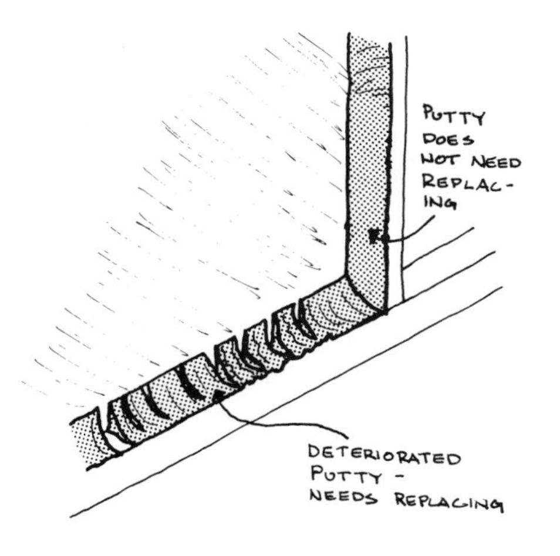

In this case, you can "spot patch" the deteriorated putty without removing the glass. The following steps are necessary;

- PREPARATION
  1, 5, and 6
- INSTALLATION PROCEDURES
  3, 4, and 5

installation procedures

1 Apply a thin layer (1/8") of glazing compound to the inside of the rabbet around all sides of the pane. Besides cushioning and correcting for irregularities in the frame, the glazing compound will provide a water-tight seal once the glass is set in place. Press the glass firmly against the glazing compound.

2 Insert glazier points around the perimeter of the frame, spacing them approximately 8" apart (a minimum of two glazier points per side for small panes). Push the glazier points in with the end of your putty knife or screwdriver. For metal frames, use **spring metal clips rather than glazier points.**

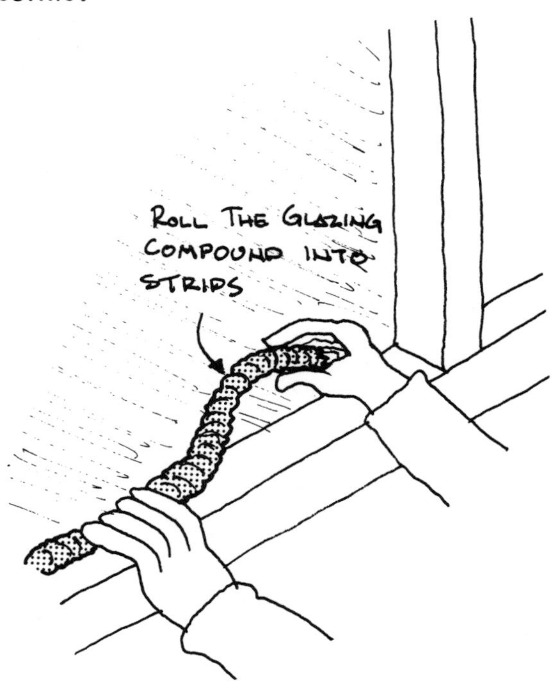

3 Seal the outside edge by rolling strips of glazing compound (about as thick as

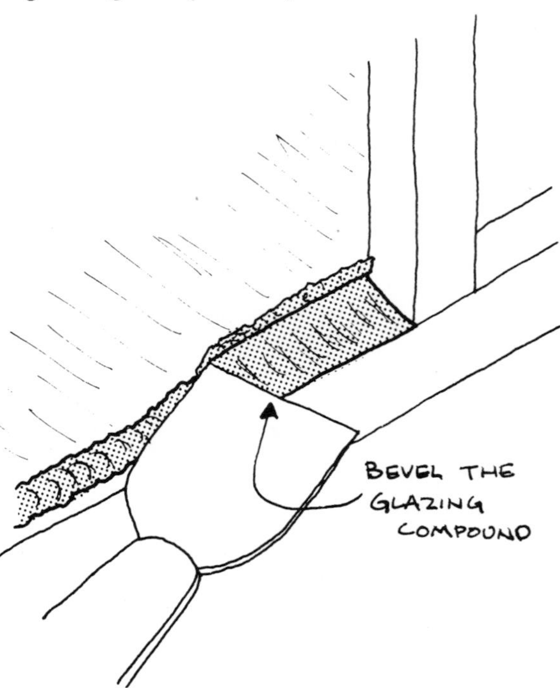

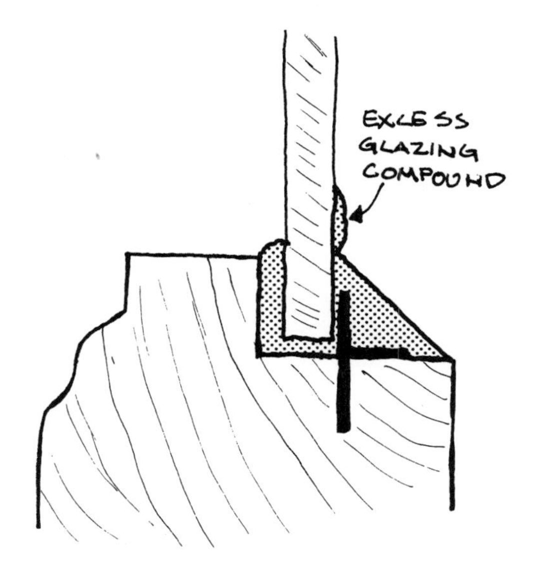

a pencil) and placing them on the recess around the edge of the sash with your hands. Then take your putty knife and form a bevel (triangular shaped). The important thing is to pack down the glazing compound, leaving no cracks or open seams into which water can seep. (Remember, it was water that caused the putty to deteriorate in the first place.)

4 Wipe away the excess compound that you may leave on the glass.

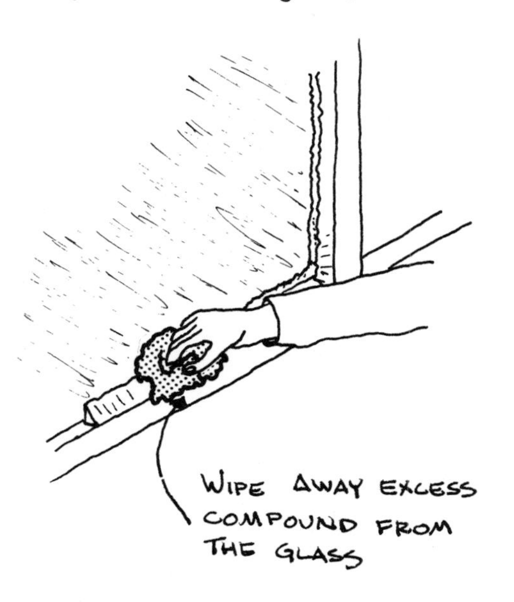

The glazing compound has to be painted after it is applied. The time at which it can be painted after application depends on the type of glazing compound that you're using.

- wood sash putty (oil base)
  paint after it dries thoroughly
- glazing compound
  generally superior in application and service performance to oil base putty; should be painted after a skin has formed

---

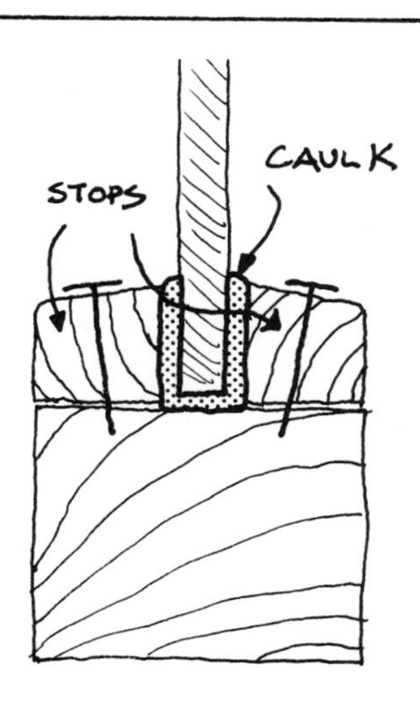

FIXED PANE

NOTE: 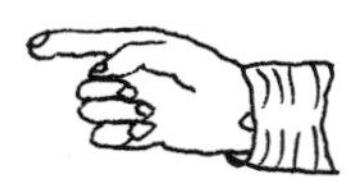

In some instances, especially with fixed windows, the pane may be set against wood or metal stops. Remove and clean the stops as described under "Preparation". Apply a continuous bead of caulk to the outside face of the pane and then force the stops against the pane. Secure the stops with nails or screws and clean the excess caulk from the pane.

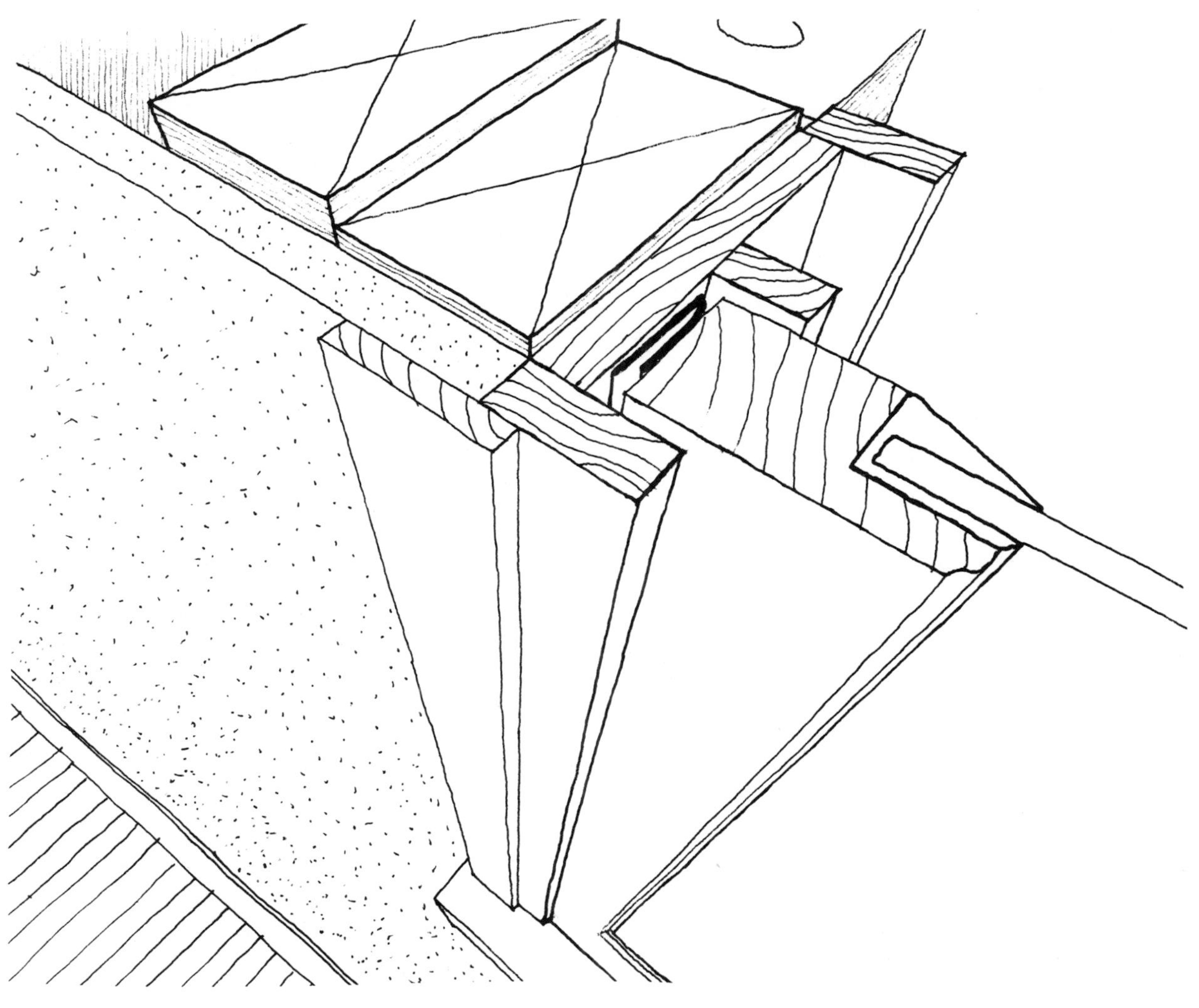

9

weatherstrip windows

description

Weatherstripping is a material used to prevent air leakage between the window sash and frame. Air leakage is caused by pressure pushing air through this joint. Weatherstripping makes a pressure fit across this joint and prevents air from being pushed through. Existing weatherstrips not providing this pressure fit should be removed and replaced with new weatherstripping.

Weatherstrips must have a flexible material to make a seal, but they also must stay in place. Either they are placed directly between rigid surfaces (such as foam tape) or they must come with the needed rigidity (such as wood backed rubber tubing). Select weatherstrips on how long they will remain flexible and how well they provide the needed rigidity.

Most weatherstrip material you will find is designed and packaged for "do-it-yourself" installation. Reviewing this chapter will help you determine which types you can work with the best. Make sure you examine products carefully to insure you have the right type. You may need more than one type, even for use on the same window.

materials

spring metal strips

(Bronze, Brass, Aluminum)

- sold as flat "tension strips" in rolls or packaged strips (wire brads for installation are usually included)
- for use between sliding and closing surfaces which bear against each other
- floor edge strips is often "passed off" as spring metal strip, but does not perform satisfactorily as a weatherstrip

rubber, PVC, EPDM, silicone or vinyl plastic tube

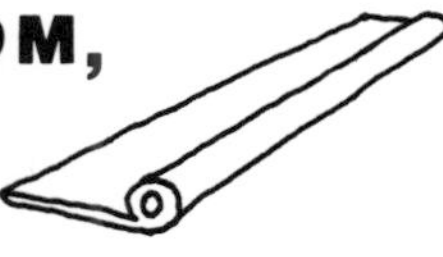

(Tube Weatherstrip)

- comes with flexible or rigid flange for fastening
- usually applied on the outside
- should not be painted
- EPDM and silicone are especially desirable for flexibility and long life

felt strips

(Metal or Wood Reinforced or Plain with Adhesive Backing)

- used where little abrasion or weathering is expected
- not very durable; do not use outside
- use in narrow, uniform gaps
- comes in a variety of widths, thicknesses, qualities, and colors
- painting around windows often renders felt ineffective if the felt itself is painted

foam strips

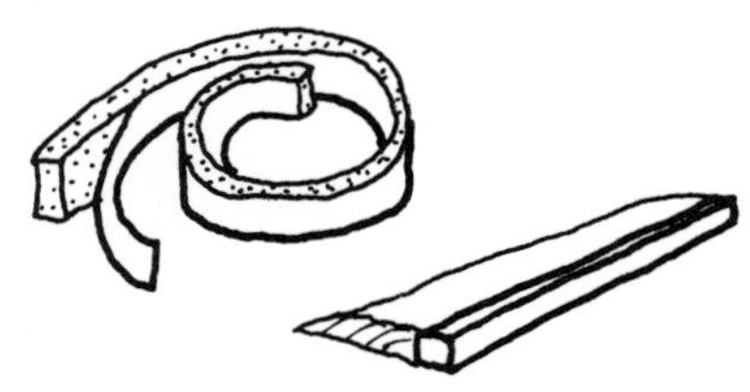

(Vinyl, Sponge Rubber, Neoprene, Polyurethane, Expanded EPDM Rubber)

- comes with adhesive backing or wood or metal flange backing for fastening
- use in same location as felt strips where the gap is wider and non-uniform

plastic strips

- comes with adhesive backing
- similar in shape to spring metal strips
- can be used between sliding surfaces that bear directly against each other

specialized weatherstrips

(Tube, Lip Type, and Slotted Fastening)

- special grooves are cut into the window jamb, sash, or stops to accept special weatherstrip shapes of PVC, EPDM rubber, silicone, or aluminum interlocking channels

- this is to be installed by a skilled carpenter or licensed installer and is therefore not covered here

other materials

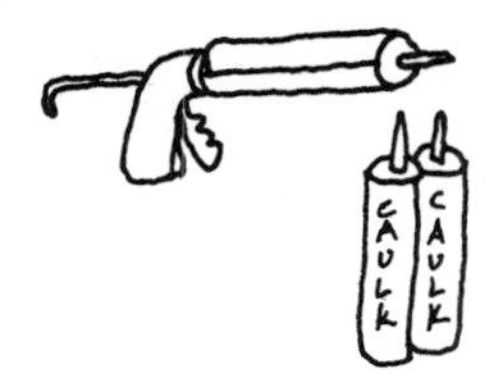

- caulk
- staple gun
- nails, nailset, or screws

NOTE:

This is only a brief survey of weatherstrip materials, which are actually available in numerous shapes, types, materials, flexibilities, and life expectancies.

preparation

Preparation procedures are the same, regardless of the window operation type.

1 The surfaces that you're weatherstripping are to be smooth, level, and cleaned of old weatherstrips, dirt, and old paint. Scrape or brush clean and wipe with clear mineral spirits using a clean rag. If you're using adhesive backed weatherstripping, wait for the surfaces to dry before pressing in place.

2 If the surface to which you're applying the weatherstripping is uneven, apply a bead of caulk (or adhesive) to the surface. In this manner, you'll seal any gap between the weatherstripping and framing.

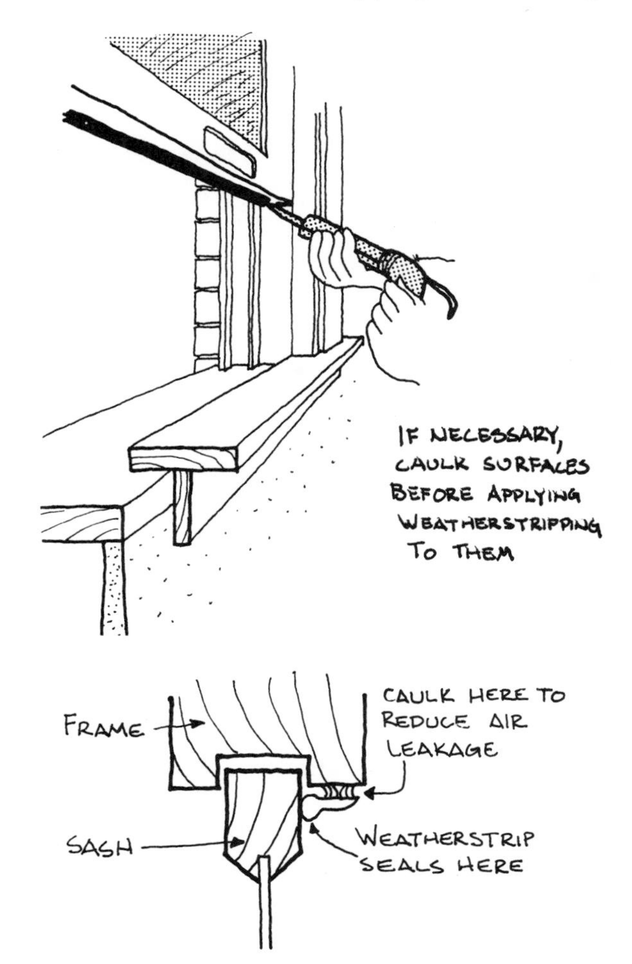

installation procedures

Installation procedures for the various weatherstrips are described for the following window operation types;

- Double hung
- Horizontal sliding
- Casement
- Tilting
- Jalousie

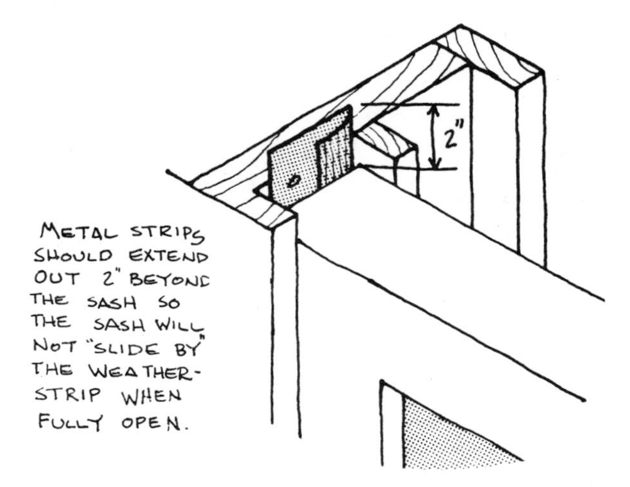

Installed weatherstripping is not to hinder the normal operation of the window. Although, by its nature, properly installed weatherstripping will increase the closing and/or locking pressure of the window.

WEATHERSTRIPPING A DOUBLE HUNG WINDOW...

...with spring metal strips

1 Begin by weatherstripping the jambs. Open the sash and slide a strip into the channels (the strips should extend out about 2" longer than the sash so that the strip is exposed when the window is closed). Cut the strips to avoid interference with any hardware that might be located within the jamb (pulleys, locks, etc.). The flared flange faces out, towards the sash.

2 Tack a wire brad at the top of the strip and continue tacking down the strip against the jamb, spacing the brads about 4" apart. While you're tacking the brads, make sure that the strip stays properly aligned. To avoid damaging the strips, drive the brads about half way in, then use a nailset to drive them flush to the strip. (If the strips do not have prepunctured holes, you can make holes by using an awl.) Strips may also be stapled in place with a medium weight staple gun.

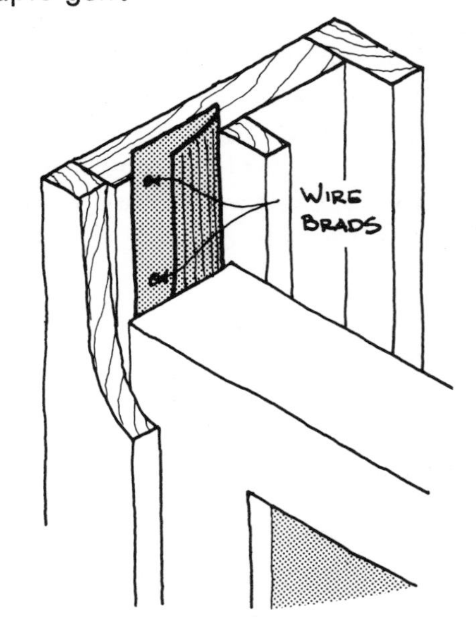

3 Cut and attach a strip along the bottom rail of the lower sash. (If the top sash is operable, also attach a strip on the top rail of the upper sash.) When the window is locked, these strips should press firmly against the window sill and head. Instead of attaching these strips to the sash, they may be attached to the sill and the head jamb.

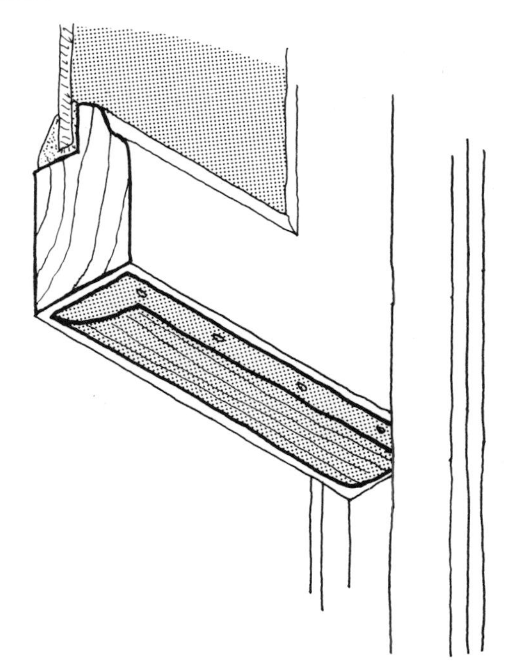

4 Now attach a strip to the inside face of the bottom rail of the upper sash. (Open the bottom sash and lower the upper sash so you can get at this rail.)

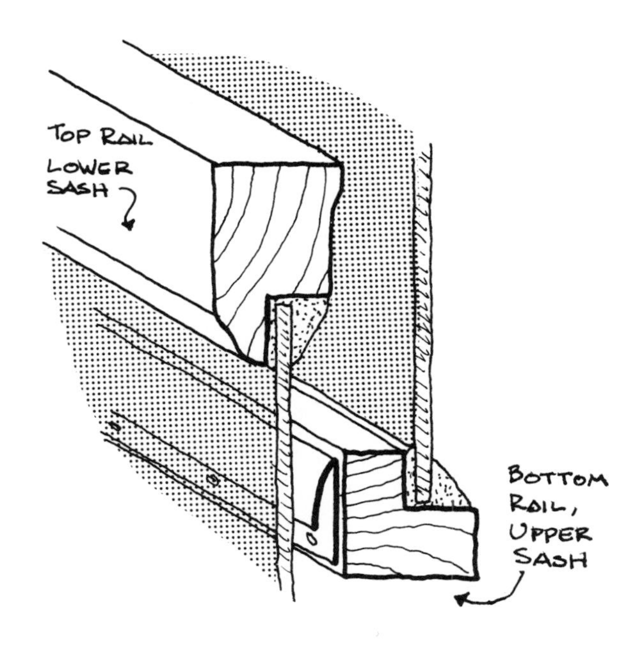

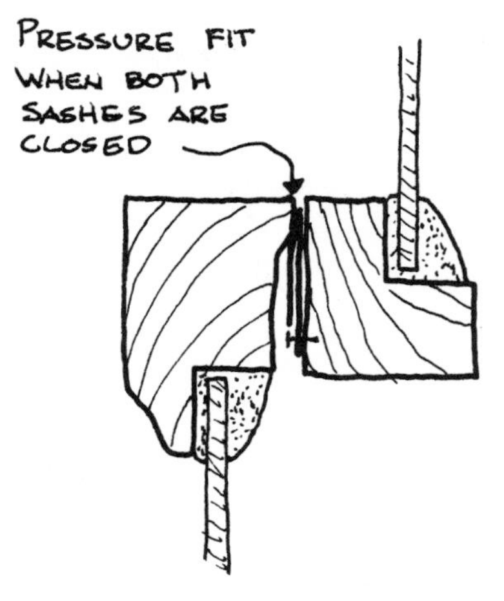

5 Depending on how the rails fit together, spring metal may not be installed unless one rail is planed down. Or, instead of planing, (or if the top sash is fixed in position), you can attach rolled vinyl to one of these rails to seal this opening (see "With rubber, PVC, EPDM, silicone, or vinyl plastic tube", step 4, in this chapter).

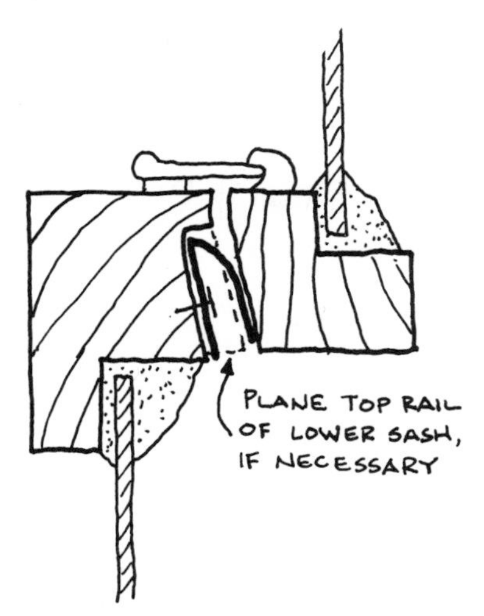

NOTE:

Spring metal is sold in widths of 1-3/8" and 1-3/4", each fitting that respective jamb dimension. To be safe, carry around 1-3/8" wide spring metal which will also work well in 1-3/4" jambs. Inspect spring metal weatherstrip. Flooring edge trim is sometimes "passed off" as weatherstrip.

6 Finally, flare out the free edges of all the strips with your screwdriver. The strip within the jamb should press firmly against the sash edge when the window is closed. The window should feel "tighter", but should operate as well as before weather-stripping.

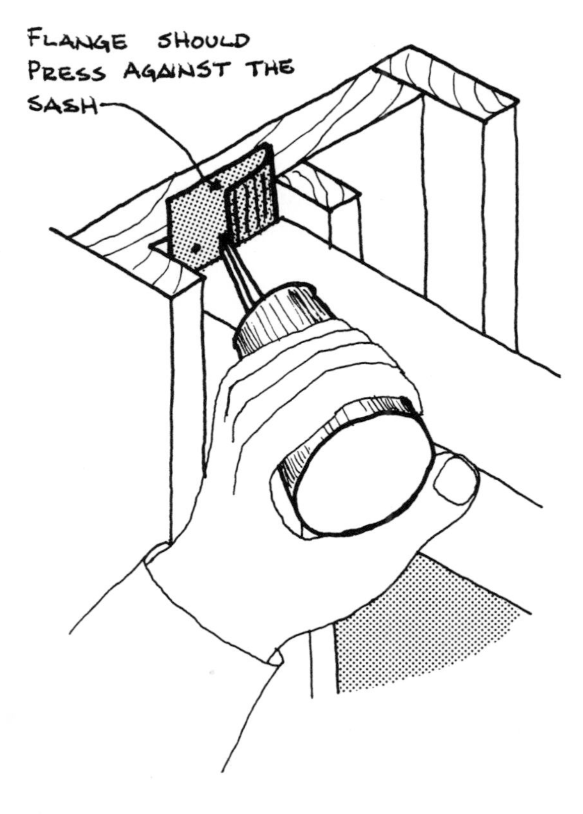

...with rubber, PVC, EPDM, silicone or vinyl plastic tube

Tubular weatherstripping is conspicuous and is therefore usually installed on the outside of the window. Also, installation on the interior of the upper sash may interfere with the opening and closing of the window. Measure the strips and tack in place with wire brads spaced every 4".

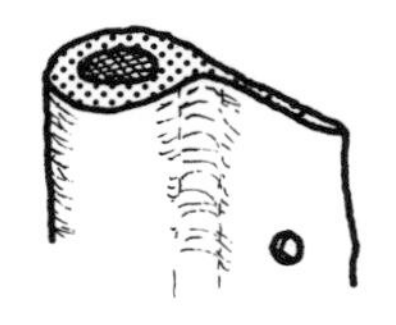

AN IMPORTANT NOTE :

If the top sash is operable, you can not insulate the lower sash in the manner described above. Otherwise, you won't be able to lower the top sash . In this situation, tubing has to be installed on the interior.

---

1 To weatherstrip the stiles of the bottom and top sash, attach the flange of the tubes

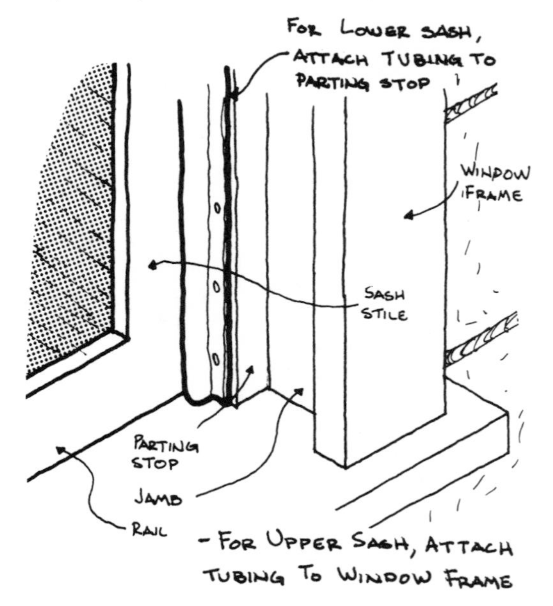

to the side face of the parting stop (bottom sash-only, if the top sash is non-operable) and window frame (top sash) with the tube extending over the edge, forming a seal against the sash stile.

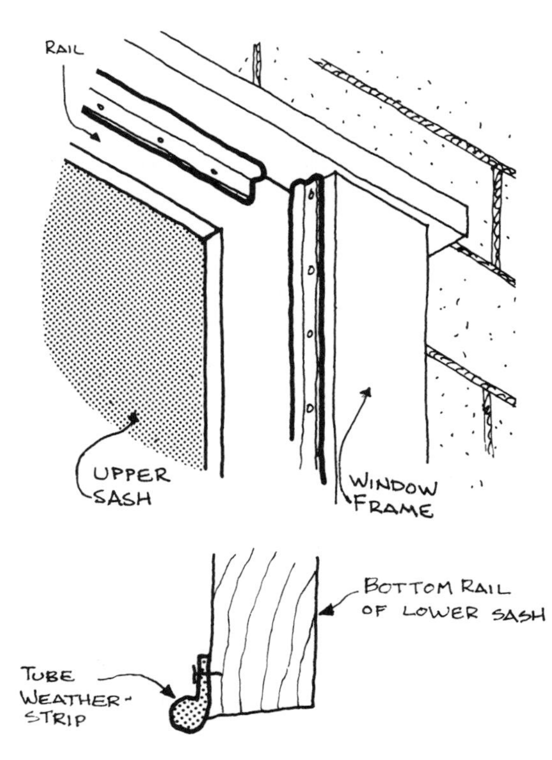

2 To weatherstrip the bottom rail of the lower sash, attach the flange to the rail with the tube facing down. Attach it in such a manner that the tube butts against the window sill firmly when the window is closed.

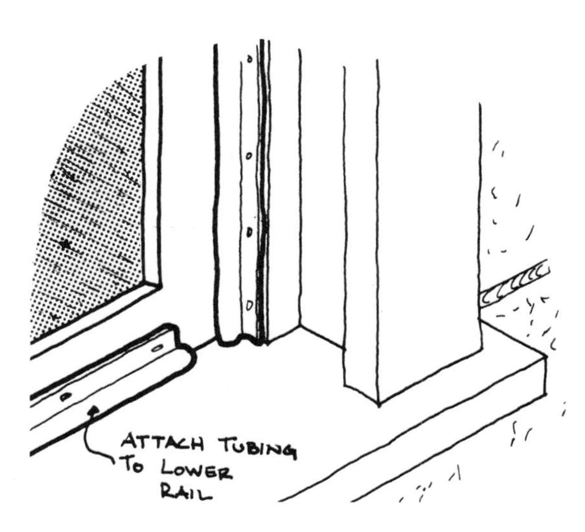

3 If the top sash is operable, attach the tubes to the top rail of the upper sash in the same manner as the bottom rail of the lower sash, but with the tube facing up.

4 Weatherstrip the meeting rails by attaching the flange to the top of the upper rail on the lower sash. The tube should extend over the edge and butt against the lower rail of the upper sash.

5 Alternately, you can weatherstrip the meeting rails on the outside by attaching the flange to the bottom of the lower rail of the top sash. But if the window does not lock so that the rails even up, the tube may not come in contact with the lower sash.

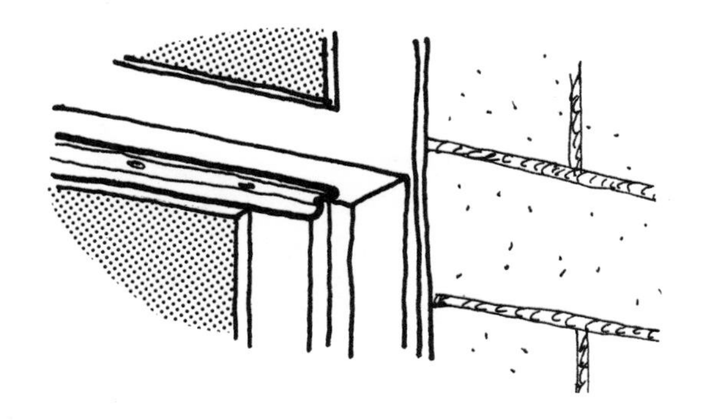

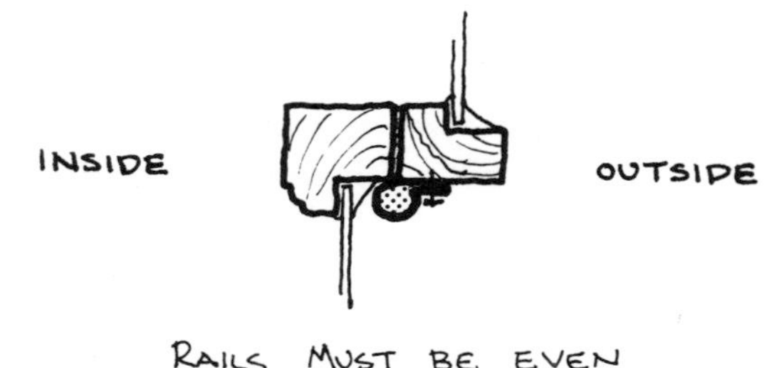

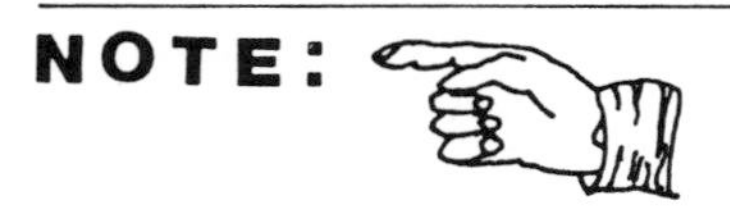

If this weatherstrip is installed in the usual fashion, the sash will not move in and out as freely as before, but it may still slide back and forth sideways.

If the window is mostly loose against the jambs instead of the stops, you may want to install this weatherstrip along the sides of the sash pressing against the stops.

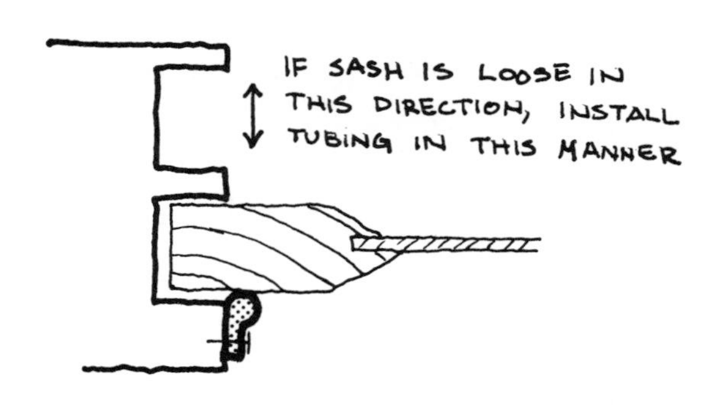

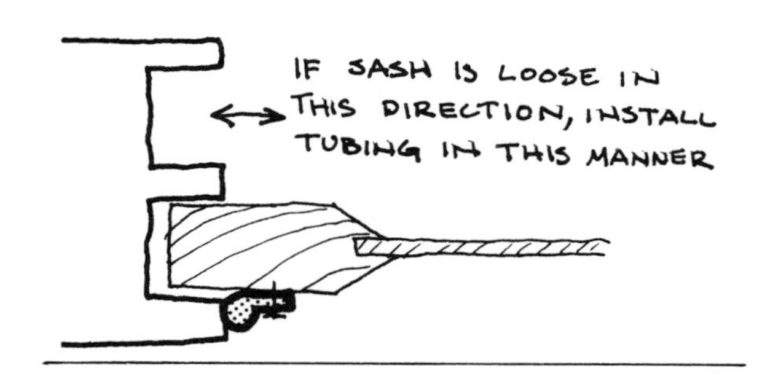

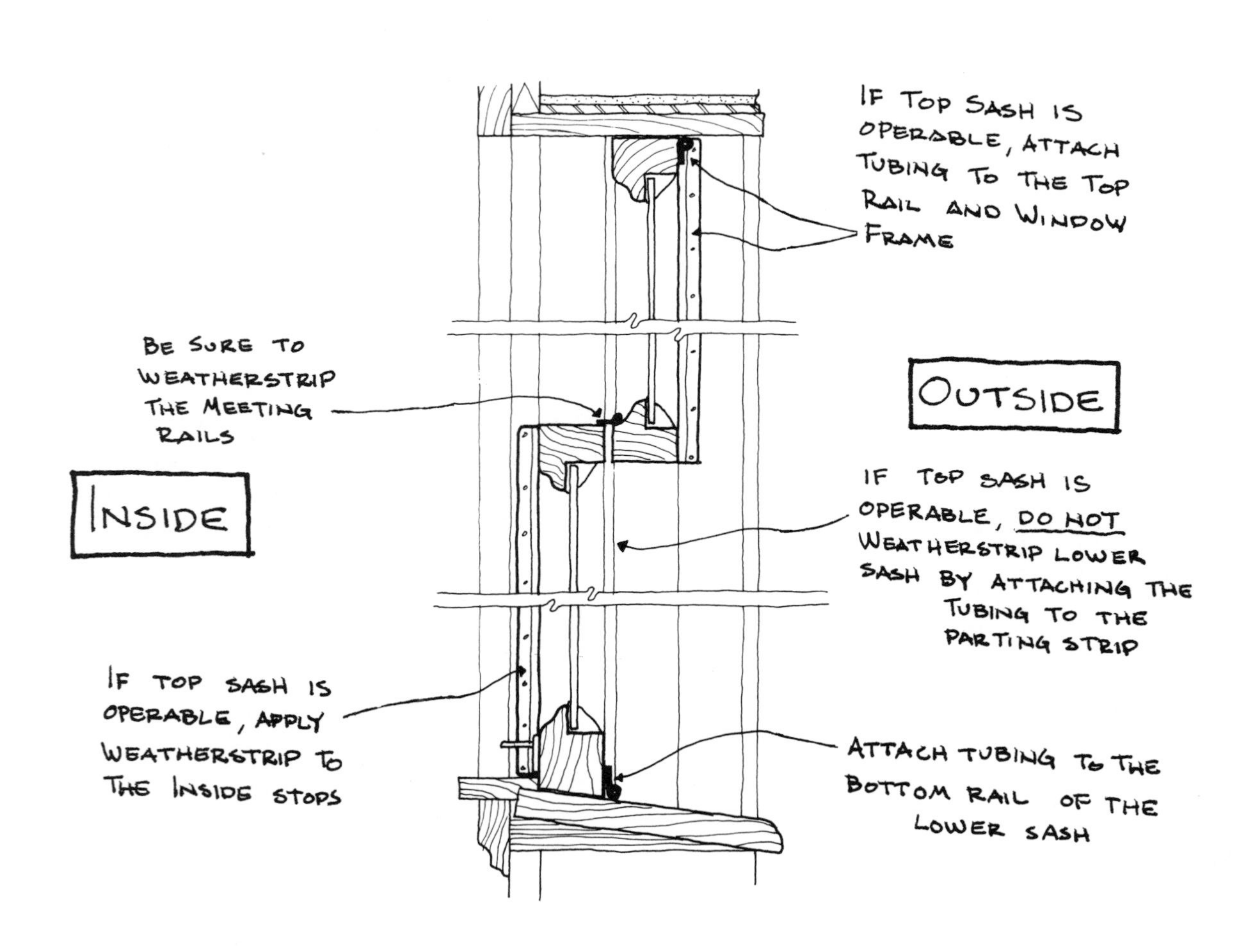

...with felt strips

Felt strips that are reinforced with wood or metal are installed as per "tube" weatherstripping, but only on the inside. (Steps 1 and 2 here are for adhesive backed felt.)

Adhesive backed felt should only be used on surfaces that close against each other (a compression seal). On a double hung window, this is the bottom rail of the lower sash and the top rail of the upper sash.

Some adhesives don't stick well unless they are at close to room temperature. Be sure to check the product and, unless the minimum temperature is marked on the package, test the adhesive bond at temperatures below 60°F before proceeding.

1 Cut the felt to match the width of the sash. Begin by attaching the felt at one end of the sash by removing the backing and tacking this end up with a wire brad.

2 Continue attaching the felt by pressing it firmly in place and driving a wire brad in or stapling every 4". Pull the felt tight as you go. Trim off any excess felt at the other end.

...with foam strips

Foam strips that are reinforced with wood or metal are not to be installed on double hung windows where surfaces slide against each other.

Felt strips and adhesive backed foam without backing strips or stiffeners are used only on surfaces that close against each other (the bottom rail of the lower sash and the top rail of the upper sash).

Again, check the package for temperature tolerance information, or test the adhesive bond at temperatures below 60°F before proceeding.

Foam is either "open cell" or "closed cell", and different types have different firmnesses. Neoprene, EPDM, or other rubber type foams are a better choice where the weatherstrip may be exposed to the weather. Choose a softer foam (open cell is ordinarily softer) when your surfaces are rough or uneven.

1 Cut the foam strips to match the width of the sash.

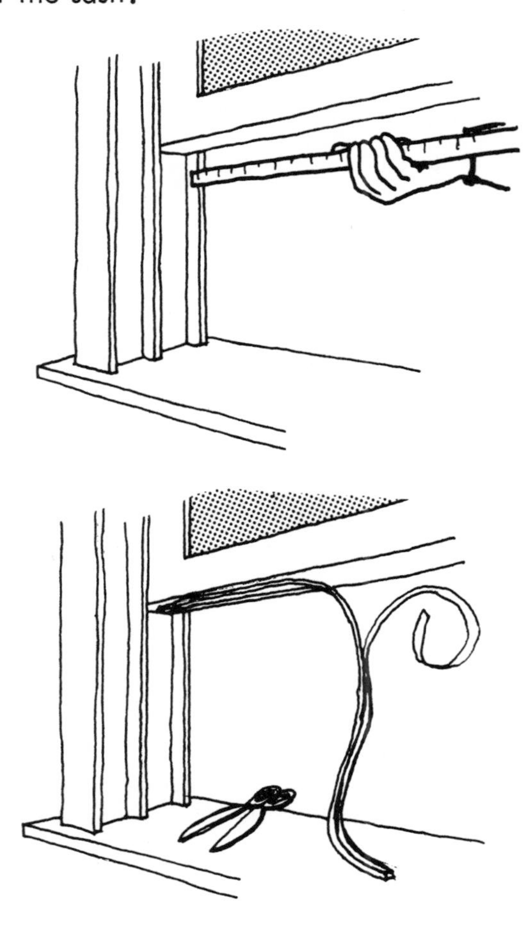

2 Start at one end of the sash and peel off the paper backing. Push only a short length of foam in place at a time so as to maintain control. Alternately, peel and push as you go. Do not stretch the foam or it may not stick properly.

...plastic strips

Install as per spring metal strips. You may wish to attach polyethylene strips more securely between jamb and sliding sash by stapling in place.

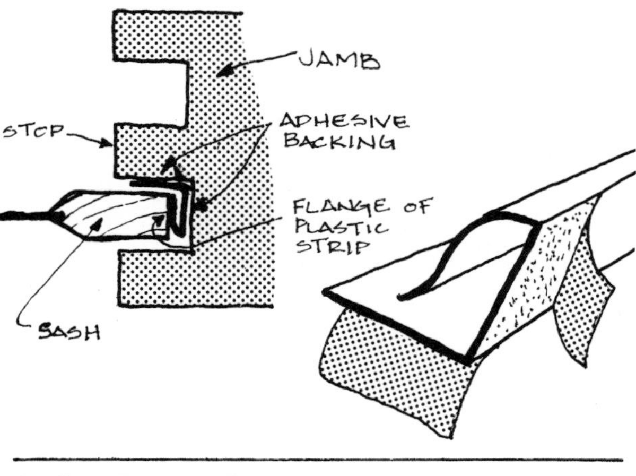

NOTE:

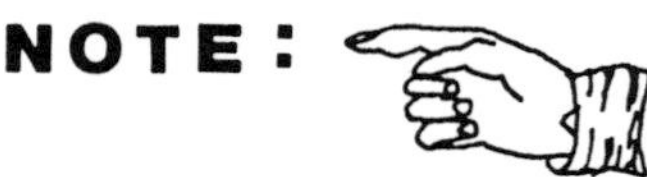

Rope caulk and weatherstrip tape are also available for weatherstripping windows. Simply cover the opening between the sash and the stop with the caulk or tape. When spring arrives, pull the material off (save the caulk; it can be used next year).

WEATHERSTRIPPING A HORIZONTAL SLIDING WINDOW...

Think of a horizontal sliding window as a double hung window turned on its side. It's weatherstripped in the same fashion. Think of the outer sash as being the upper sash of a double hung window. Pay particular attention to the top of the sash. Sash may rest on the jambs at the sill and be very leaky along the top of the window.

...with spring metal strips

Tack the strips to the face of the top and bottom channels, the side of the sash that closes against the jamb and where the two sash meet (meeting rails).

ATTACH SPRING IN THE FOLLOWING PLACES;

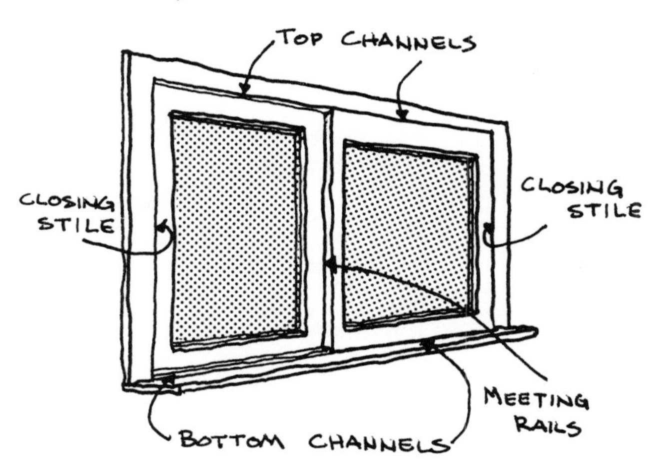

...with rubber, PVC, EPDM, silicone or vinyl plastic tube

Attach the flanges to the stops along the sill and upper frame and to the closing sashes. In all cases, the tube itself should extend over the edge to form a tight seal.

ATTACH TUBING IN THE FOLLOWING PLACES;

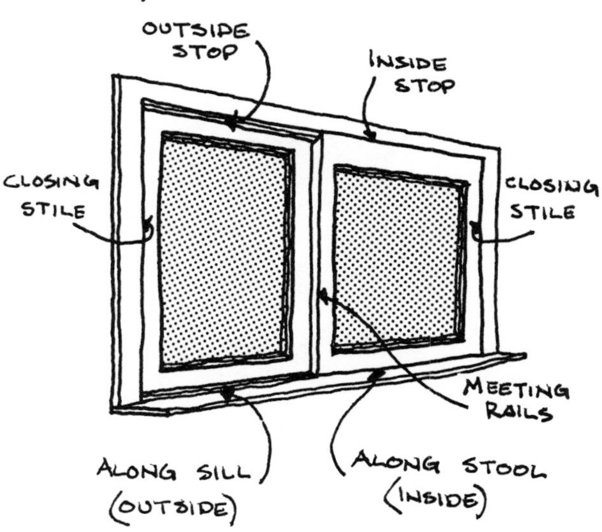

...with felt, foam, and polyethylene strips

Felt and foam should only be attached to the closing stiles. Install polyethylene strips as per spring metal strips.

ATTACH THESE WEATHERSTIPS IN THE FOLLOWING PLACES;

metal windows...

Metal windows are usually grooved around the edges so that that the flanges interlock. However, over the years, gaps will appear and weatherstripping should be applied.

One type of weatherstripping that will work here is the adhesive backed felt or foam (wire brads or screws that you need for the other type weatherstrips may interfere with the operation of the window). Look for weatherstripping with a strong adhesive (such as an acrylic type) that will hold the weatherstrips to the metal.

Brushing the metal with a wire brush, dusting, and wiping clean with solvent will usually make a great difference in getting adhesive backed weatherstrips to stick. Don't be surprised if preparing to weatherstrip a window takes more time than applying the weatherstrip.

Other metal windows (the newer ones) are fitted with neoprene gaskets or "piles" made of hair felt on a fabric backing. These windows are made so that if the neoprene or "pile" wears out, it can be pulled out and a new one slipped into place. However, these weatherstrips may be special for windows of a particular manufacturer.

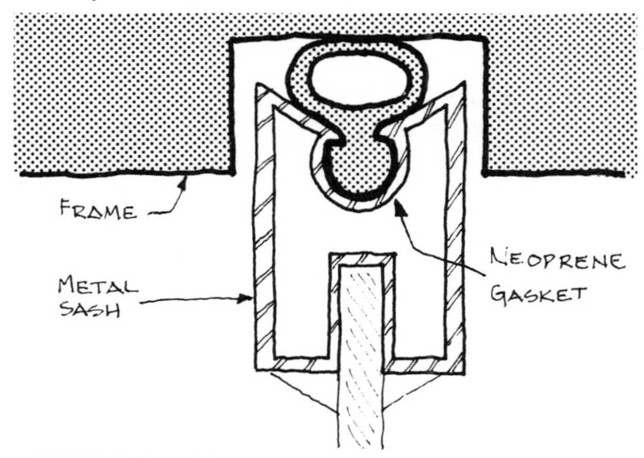

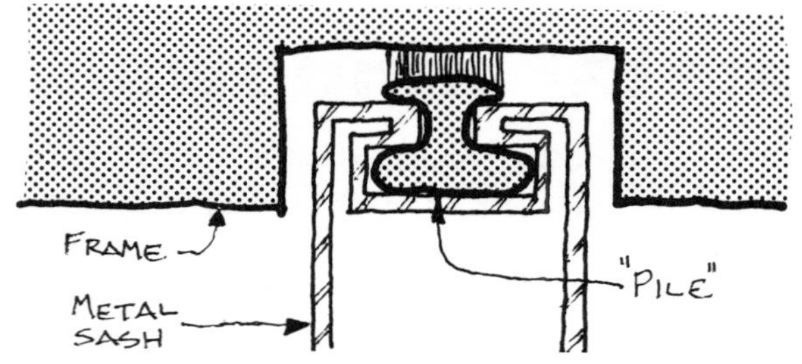

WEATHERSTRIPPING A CASEMENT WINDOW...

Weatherstripping a casement window is very similar to weatherstripping a door.

...with spring metal or polyethylene strips

Install the spring metal or plastic strips to both jambs, the head jamb, and along the sill of the window frame. The flange opening is to face out towards the sash. After you're through installing it, take your screwdriver and lift up the edge to form a better seal when the window is closed.

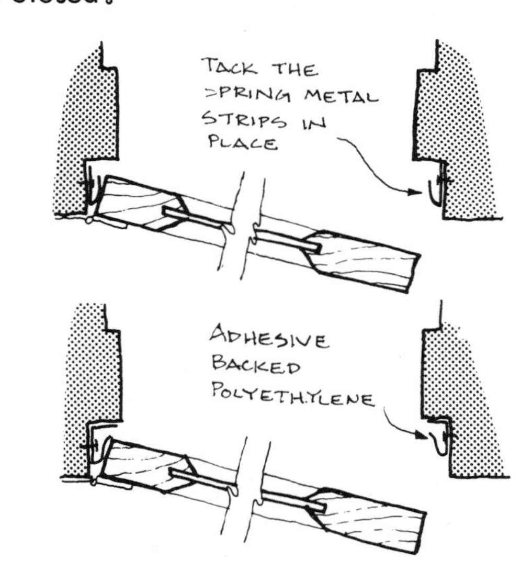

...with rubber, PVC, EPDM, silicone or vinyl plastic tube

Tack the flange of the weatherstripping to the stops along the jambs, the head jamb, and the sill. The tube should extend over the edge of the stop, so that

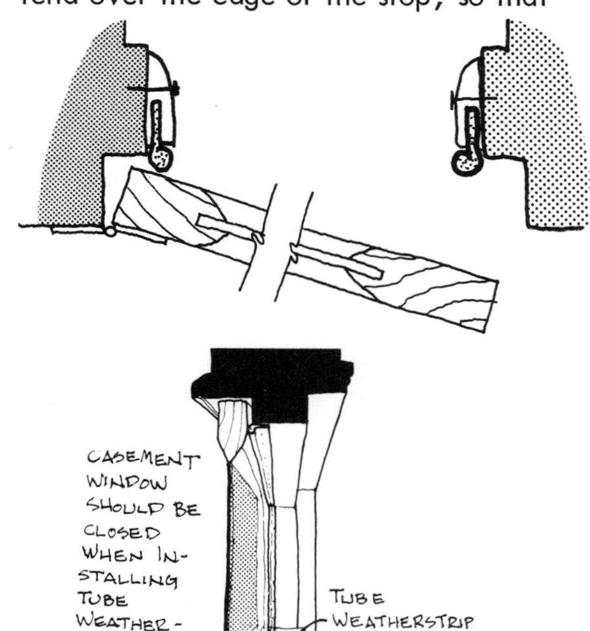

it forms a tight seal with the sash (the window should be closed when this weatherstripping is installed).

...with adhesive backed felt or foam

Attach the weatherstripping to the inside face of all the stops. Be careful to select foam that is thin and flexible enough so that the window can be closed!

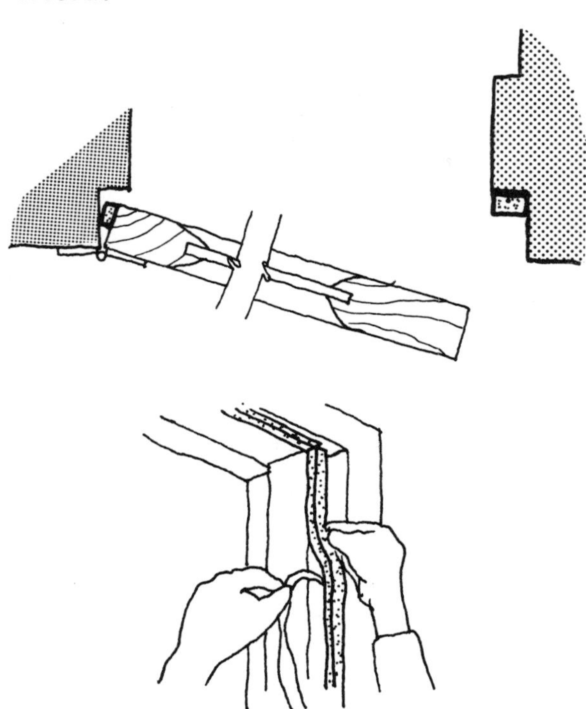

...with wood or metal reinforced felt or foam

Install in the same fashion as with rubber or vinyl plastic tubing. Close the window when you're installing this weatherstripping to insure a good seal.

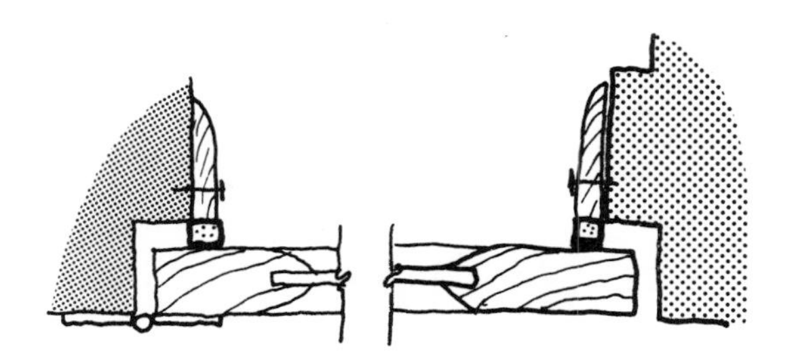

WEATHERSTRIPPING A TILTING WINDOW

Weatherstripping a tilting window is very similar to weatherstripping a casement window, only on its side. Tilting windows are hinged at either the top or bottom of the sash (casements are hinged on the side). Attach spring metal and adhesive backed felt or foam to the jambs, head jamb, and sill. Attach rubber or vinyl plastic tubing and wood or foam to the stops around the window. In all cases, the weatherstripping should form a pressure closure with the window when it is closed.

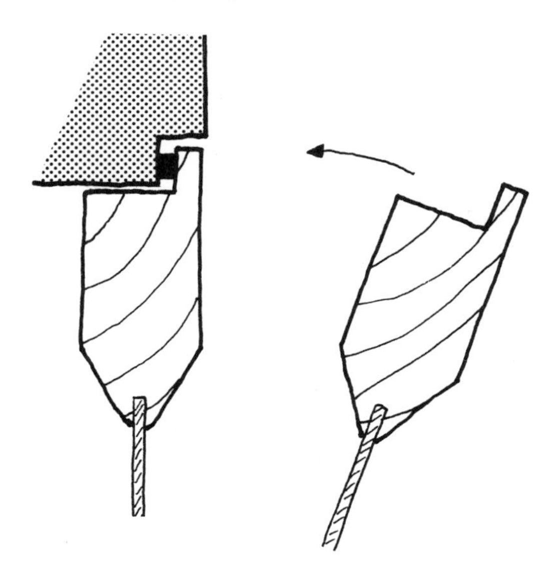

WEATHERSTRIPPING A JALOUSIE WINDOW

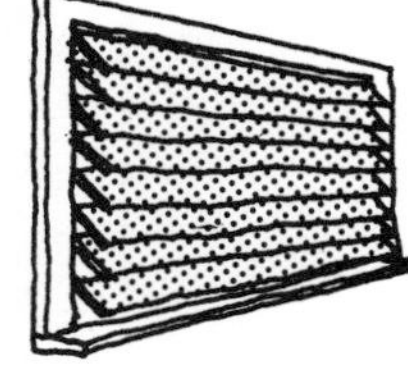

Foam with adhesive backing may be suitable for weatherstripping jalousie windows. Special "edge gaskets" are made, however, which can be cut with scissors and then snapped in place along the edges of the glass. Larger gaskets are made to snap over casement and tilting window sash edges.

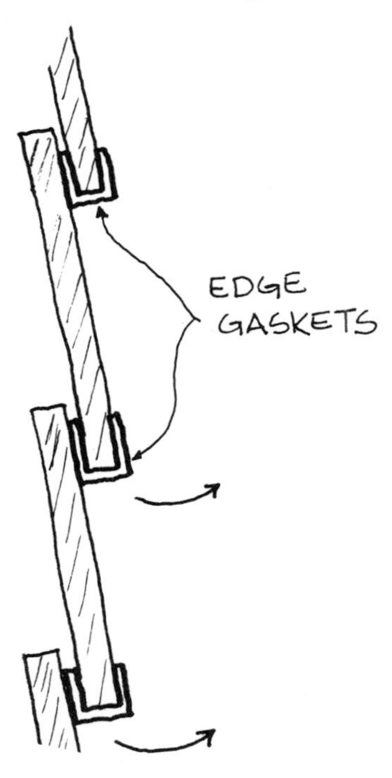

NOTE:

Fixed windows don't need weatherstripping. If air is getting in through cracks around a fixed window, the cracks should be caulked (see Chapter 10, "Pack and/or Caulk Windows and Doors"). Caulk is not only more durable than most weatherstrips, it's a lot cheaper.

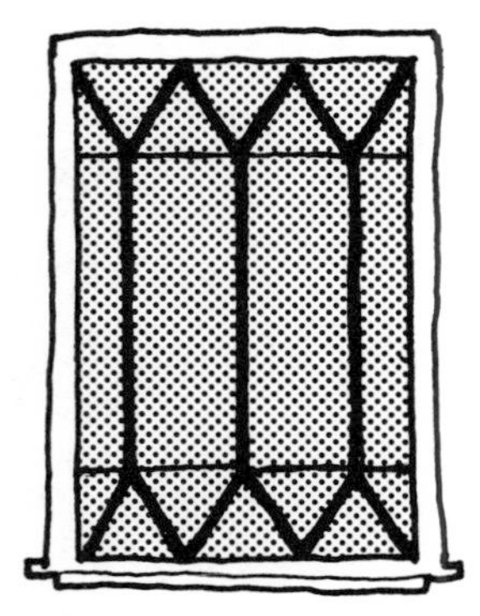

let's summarize...

With certain window conditions, you may want to use more than one type of weatherstrip. For example, you may use spring metal on the bottom rail of the lower sash (double hung window) while weatherstripping the stiles with felt that's reinforced with wood.

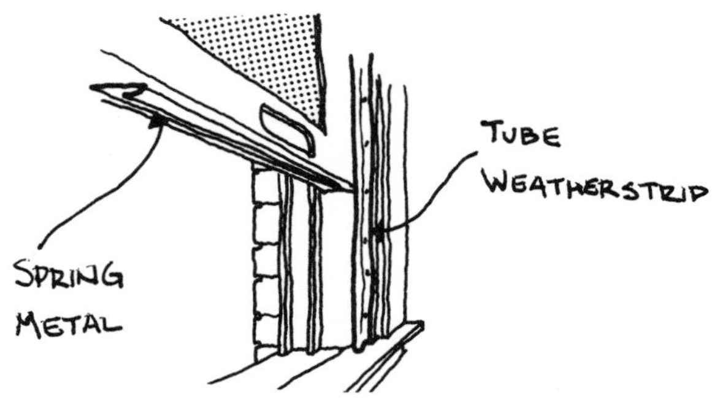

Or, you may get to a window and discover that it is already weatherstripped with spring metal . The spring metal is old and is no longer providing the proper seal. Instead of pulling it out and replacing it, you may be better off to leave it on and install, for example, rolled vinyl.

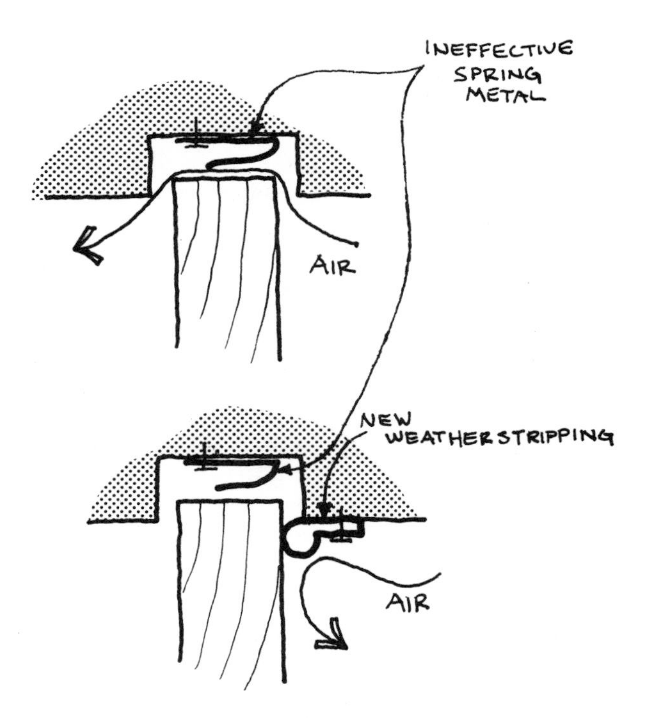

These sketches illustrate different window operation types with common weatherstrips applied. Of course, you won't be using all of the weatherstrips on the same window. However, if you find a window with existing weatherstripping that does not form a good seal, refer to these sketches for weatherstrips that can be applied without removing the existing weatherstrip.

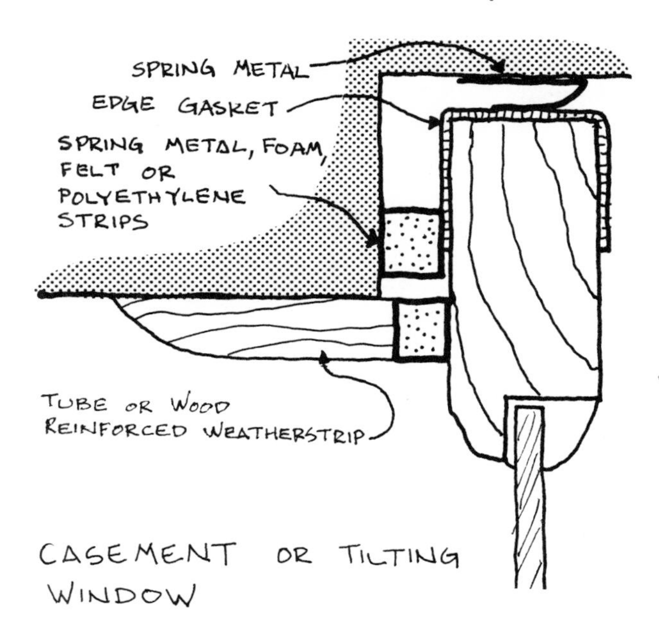

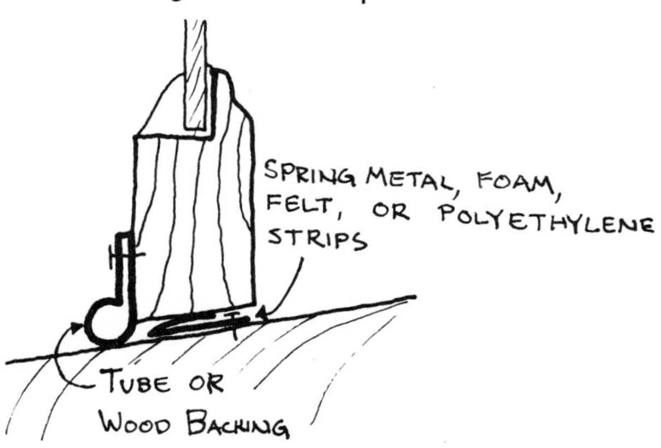

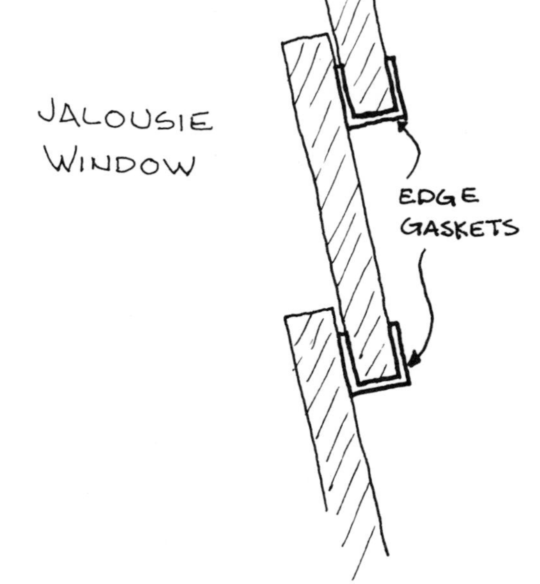

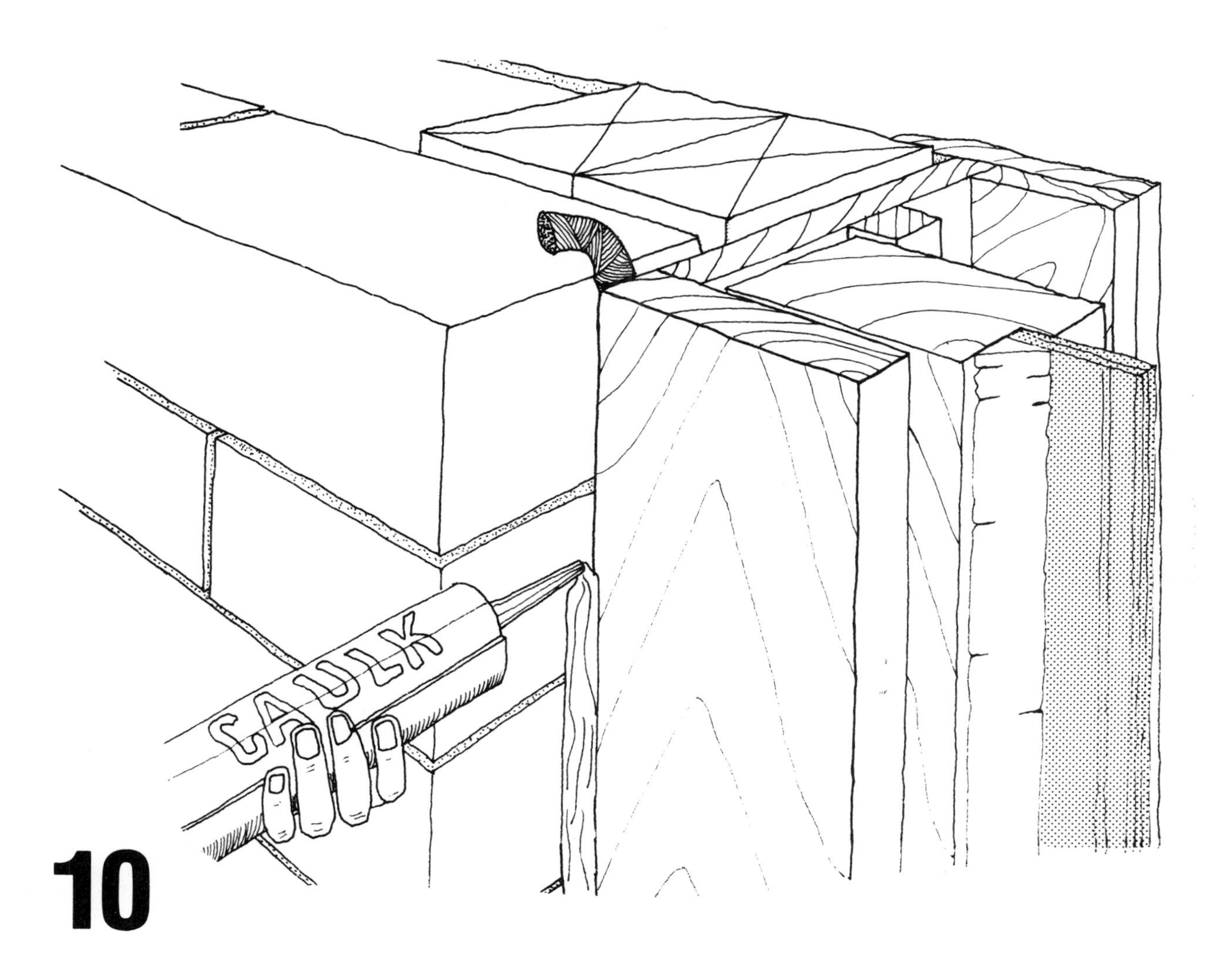

10

pack and/or caulk windows and doors

description

Just as you "sealed" an operating window and door panel to the frame with weatherstrip, the frame should be sealed to the wall with caulk. Examine the joint between the window frame and the wall on the exterior. Caulk, in good condition, is flexible when you push on it with your finger. It should be continuous and it should stick (no cracks) to both the frame and the wall. The joint is not adequately sealed if there are cracks. Where there has not been any recent caulking and if the caulk is dried, cracked, or missing entirely, the joint should be cleaned of all old caulk and dirt and re-caulked. Where the joint width is greater than 1/4", packing material should be used to provide a backing for the caulk.

Caulking is easy to learn and inexpensive. Besides, it's fun! There are a variety of caulks on the market. Before selecting a caulk, refer to the table of caulk properties in this chapter. You'll find that this table summarizes the more common types of caulk available. Using the right type of caulk is just as important as applying it properly.

materials

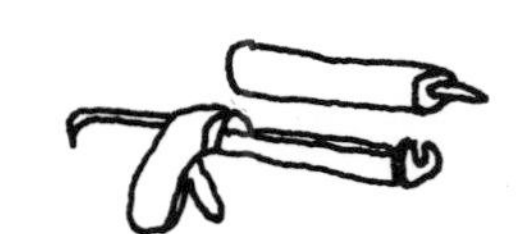

caulks

(For a detailed description of the following caulks, see "A Word about Caulks" in the first section of this guide.)

- oil based caulks
- latex caulks
- solvent based acrylic sealants
- butyl caulks
- polysulfide sealants
- urethane sealants
- silicone sealants
- nitrile rubber
- neoprene
- chlorosulfonated polyethylene

packing

- polyethylene (closed cell rod)
- polyurethane (closed cell rod-vertical joints only)
- oakum (oil saturated hemp)
- polyethylene bond breaker-tape (for joints too shallow for foam rod)

cleaning solvents

- turpentine
- mineral spirits
- lacquer thinner

preparation

The importance of preparing the surfaces for caulking must be emphasized. Good adhesion cannot be obtained without proper surface preparation. Preparing the surfaces for some caulks is a tedious process, but is necessary to obtain an effective and long-lasting seal.

1 Begin by removing all the old caulk. Use a putty knife or screwdriver along with a stiff brush to remove it (some caulks require "dusting" after wire brushing). You'll need a gap that is at least 1/4" deep for an adequate amount of caulk.

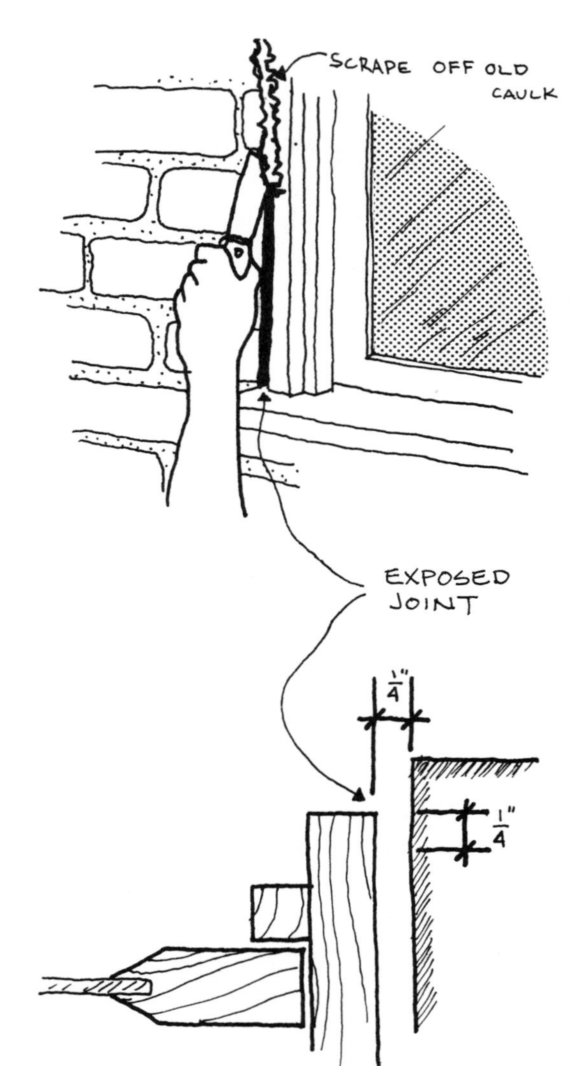

2 Clean wood surfaces of all dirt and loose paint with a stiff brush. Wipe clean with mineral spirits. Aluminum windows with protective lacquer should have the lacquer removed with a solvent stronger than mineral spirits, such as xylol, so that the caulk will adhere directly to the metal. Concrete and masonry surfaces that must be cleaned can be done with a wire brush, or ground down to the original surface.

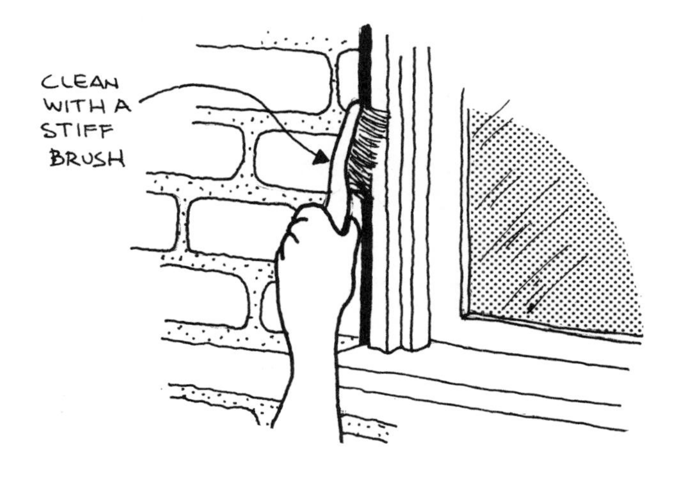

3 Some caulks adhere best to surfaces that have been primed. Paint wood surfaces with primer. Metal surfaces, other than aluminum, should be primed with rust inhibitive primer. Check with the manufactorer's recommendations concerning the necessity of priming.

installation procedures

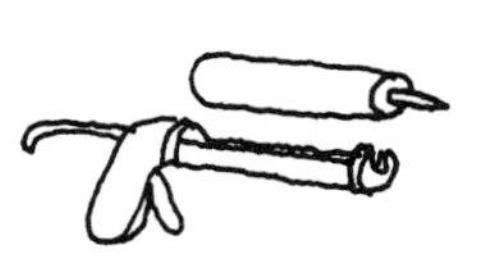

CAULKING

Before applying caulk, the surfaces should be clean, dry, and frost free. Caulking should not be done if it's raining or threatening to rain.

If the joint width is greater than 1/4", pack the joint within 1/4" of the edge with packing material prior to caulking (see the section titled "Packing" in this chapter).

NOTE:

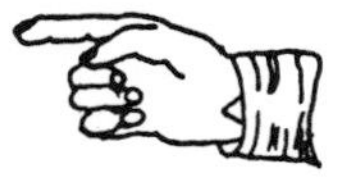

Check manufactorer's recommendations for minimum outdoor air temperature at which the caulk can be applied. Also, check the recommended caulk temperature at time of application.

---

1 Cut the nozzle of the cartridge at a 45° angle. Some cartridges have an inner seal that will need to be punctured with a long nail. If applying a caulk that will stain adjacent surfaces, mask the surfaces with tape to assure neat sealant lines.

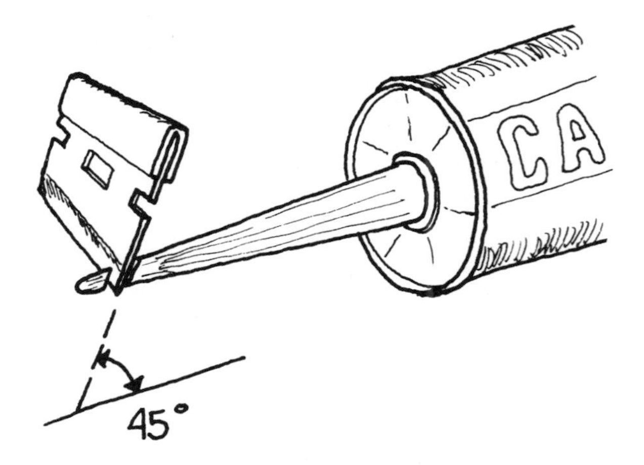

2 When caulking, hold the gun parallel to the joint at a 45° angle. Push, rather than pull, the gun while "drawing a bead" (applying the caulk). Pushing the gun forces the caulk out ahead of the nozzle doing a better job of filling the joint. Pulling the gun will give a neater appearance but may leave voids in the caulk unless it is tooled (see Step 5).

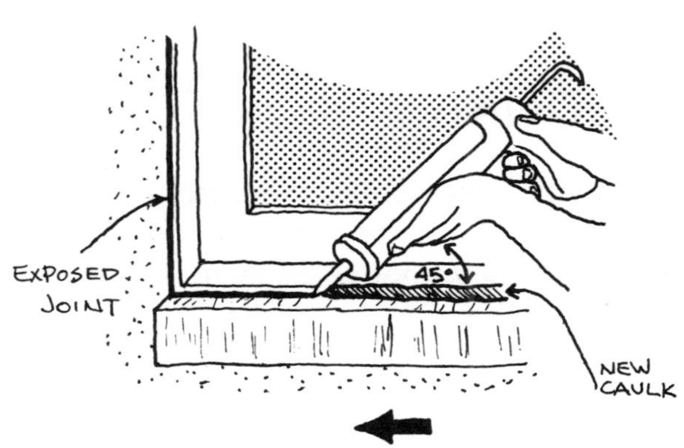

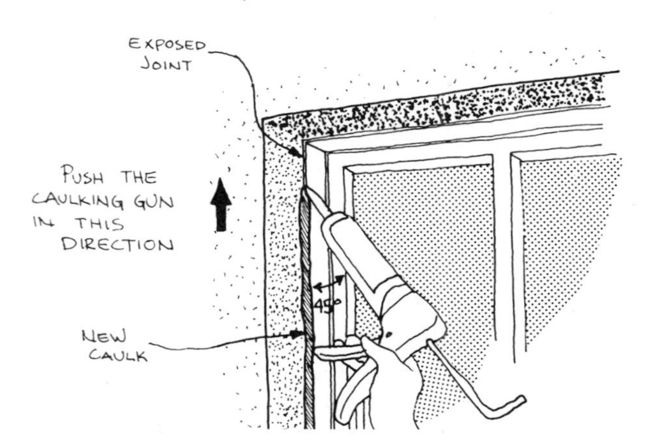

NOTE:

Caulking around a window and door frame may be ineffective with lap siding as any air flow between siding pieces will bypass this seal. To obtain a "tight" seal around these frames, remove the interior casing and caulk between the frame and rough opening.

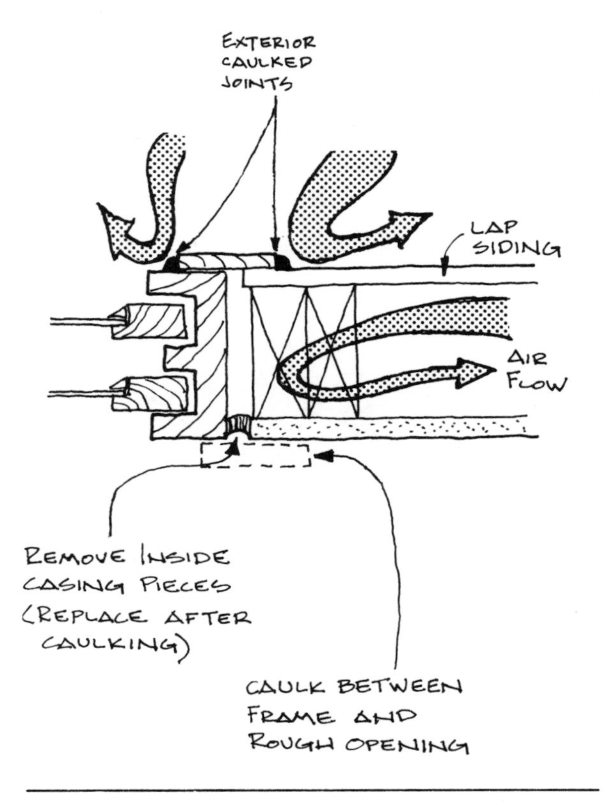

3 If possible, the bead should be concave. The depth of the caulk should be greater than or equal to the width of the crack. When the width is greater than 1/4", packing should also be used. The bead of caulk should fill the joint completely.

good

PACKING IF NECESSARY

MINIMUM DEPTH 1/4"

PACKING IF NECESSARY

IF THE JOINT IS TOO NARROW FOR CAULK, PACK PRIOR TO CAULKING.

PACKING

bad

4 When you're through caulking, disengage the rod to stop the flow of caulk. When storing, seal the nozzle with a nail, machine screw, or masking tape.

5 It is recommended that some caulks be "tooled" (smoothed and/or pushed in place to seal the joint) after application. Check the label on the caulk for this information.

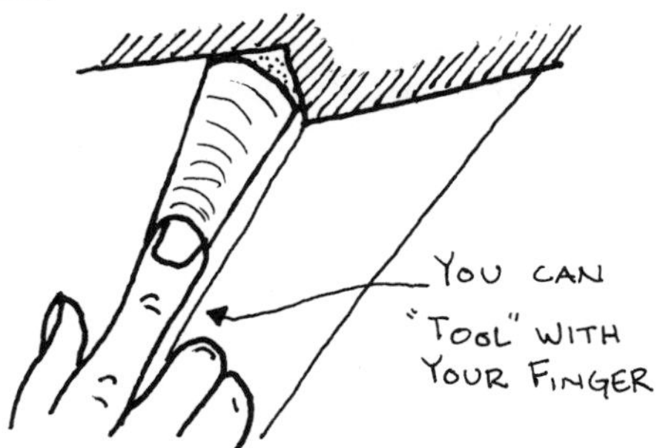

	How Long Will It Last?	Elongation	Min. Temp. For Use	Particular Good Features	Particular Problems	Relative Cost ($/11 oz. Tube)	Staining
Oil Base Caulk	◦ High quality last 5-10 yrs ◦ Low quality last 3-5 yrs	• Little elongation and no recovery		◦ Good adhesion to most surfaces while there is little or no movement	◦ Frequently used for glazing; with disappointing results	◦ 0.75	◦ Staining & bleeding of most substrates
Latex Caulks	◦ 8-10 yrs. Perhaps longer	◦ Approximate 60% when formed into ½" x ½" specimens	40°F	◦ May be applied to absorptive surfaces - brick, wood, good adhesion	◦ Adhesive property good w/little or no joint movement	◦ 1.50	◦ Non-staining to any type of substrate
Solvent Based Acrylic Sealants	◦ 10-20 years	◦ Below average (25-60%)		◦ Excellent adhesion, can be used for corner beads, & other odd shaped joints	◦ Strong odors - must ventilate interiors, vapors are contaminent to foods	◦ 3.25	◦ Non-staining to most materials
Butyl Caulks	◦ 10 years	• Average (75-125%)	40°F	◦ Particularily good adhesion metal to masonry & for sidings - alum, vinyl, etc.	◦ Only for joints with moderate movement	◦ 1.25	◦ Non-staining
Poly-Sulfide Caulks	• One component has good durability ◦ Two component 20-30 yrs.	◦ Superior (150-200%) ◦ Superior (150-350%)	• Cannot be applied below 40°F • Will not cure below 40°F	◦ Good ◦ Good performance on wide joints & working joints		◦ 7.30	◦ Non-staining to masonry
Urethane Sealants	◦ 20-30 years	◦ Excellent (300-450%)	32°F	◦ Excellent adhesion to wide variety of materials ◦ Abrasion resistant	◦ May adhere so strongly to weak masonry; spalling occurs	◦ 7.50	◦ Non-staining
Silicone Sealants	◦ 30 years	◦ (100-200%)	◦ Can be applied down to -35°F	◦ Good adhesion when applied to metal & glass	◦ Poor adhesion when applied to concrete; cannot be painted	◦ 4.50	◦ Non-staining on most materials
Nitrile Rubber	◦ 15-20 years	◦ (75-125%)		◦ Adheres well to metal, masonry & high moisture areas	◦ Poor performance on moving joints or wide cracks	◦ 3.25	
Neoprene	◦ 15-20 years	◦ (20-40%)	◦ Difficult to apply at low temps.	◦ Good with asphalt or concrete (foundations)	◦ Difficult to apply		
Chlorosulfonated Poly-Ethylene	◦ 15-20 years	◦ (15-20%)	◦ Difficult to apply at low temps.	◦ Working joints	◦ Very slow cure (3-4 mths.)	◦ 3.50	

Actual performance of caulks and adhesives depends upon many factors, including application technique, condition of crack or joint, and climate, and may vary significantly from those figures listed in this table. Refer to manufacturers' literature for more specific information concerning the product that you use.

	VERTICAL SAGGING	TEMPERATURE RESISTANCE LOW-HIGH	SURFACE PREPARATION	EASE OF APPLICATION	SHELF LIFE	EASE OF PREPARATION	SHRINKAGE
OIL BASE CAULK		° POOR LOW TEMPERATURE PERFORMANCE -10° → 180°F	° QUICK DUSTING OF JOINT	° APPLY AND TOOL EASILY	° GOOD	° EASILY MIXED	° LOW INITIAL SHRINKAGE 5-25%
LATEX CAULKS	° SUITABLE FOR SMALLER JOINTS (3/8" MAX.)	-20°F → 150°F	° CLEAN THE JOINT, BUT NOT AS THOROUGHLY AS HIGH GRADE SEALANTS	° EASY APPLICATION CLEAN UP W/ DAMP CLOTH	° GOOD	° GOOD, AVAILABLE IN 1/10 GALLON SQUEEZE TUBES.	° HIGH DEGREE OF SHRINKAGE CAUSES UNSIGHTLY WRINKLED APPEARANCE.
SOLVENT BASED ACRYLIC SEALANTS	° CAN BE OBTAINED IN NON-SAGGING FORM	° POOR LOW TEMPERATURE ELASTICITY -20° → 150°F	° MINIMUM CLEANING REQUIRED	° VENTILATE WHEN APPLYING IN INTERIORS - VAPORS CAN CONTAMINATE FOOD	° GOOD	° RECOMMENDED TO BE HEATED TO ABOUT 120°F BEFORE APPLICATION	° NEGLIGIBLE - HOWEVER LOW RECOVERY CAUSES WRINKLED APPEARANCE
BUTYL CAULKS	° AVAILABLE IN LOW-SAG FORM	-25° → 210°F	° CLEAN CONCRETE WIPE: METAL W/ OIL FREE CLOTH & SOLVENT, WOOD W/ CLOTH OR SOFT BRUSH, GLASS W/ CLOTH	° CLEAN UP W/ SOLVENT - BELOW 40°F, CAULK STIFFENS & IS DIFFICULT TO APPLY	° VERY LONG	° NO MIXING OR HEATING	° HIGH
POLY-SULFIDE CAULKS		-60° → 250°F			° LOW, POOR	° MESSY MIXING	° LOW, NEGLIGIBLE
URETHANE SEALANTS		-60° → 275°F	° MUCH: REQUIRES VERY CLEAN SURFACE JOINTS TO OBTAIN ADHESION	° ABOVE 70% RH CAN REDUCE EFFECTIVENESS OF SEAL	° VERY SHORT	° MIXING IS EXTREMELY CRITICAL	° NEGLIGIBLE
SILICONE SEALANTS	° NON-SAGGING ON VERTICAL WALLS	° HIGH - -90° → 400°F	° SURFACE REQUIRES CAREFUL PREPARATION		° VERY SHORT		° NONE
NITRILE RUBBER							° HIGH
NEOPRENE							° HIGH
CHLOROSULFONATED POLY-ETHYLENE			° POROUS SURFACES REQUIRE PRIMING		° EXCELLENT		° HIGH

Information from Sealants, edited by Adolfas Damusis, Reinhold '67, and personal communication with manufacturers.

PACKING

Packing is a material used to control the depth of caulk in deep joints (where the width-to-depth ratio is greater than 2 to 1). The packing should be inert, compressible, cellular, compatible with the caulk, and non-staining to the surfaces.

For working joints the caulk must be kept from adhering to the packing. This can be done with polyethylene tape (or strips), special wax-backed tape, or even masking tape. When used to prevent caulk from adhering to the packing, these tapes are known as "bond breakers".

1 Packing rod material should be approximately 1/8" wider than the joint to provide a compression (pressure) fit.

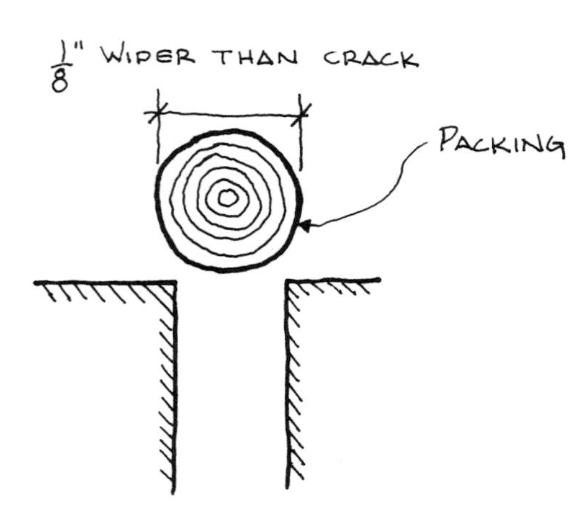

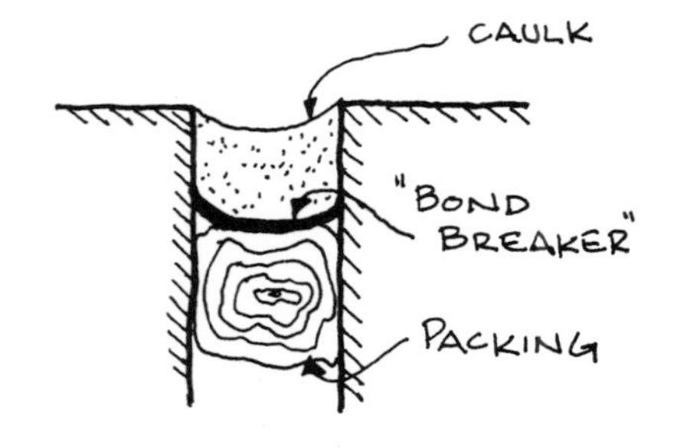

2 Use a blunt, rounded tool when installing packing rod to avoid puncturing the surface of it. Excessive stretching and compressing should be avoided.

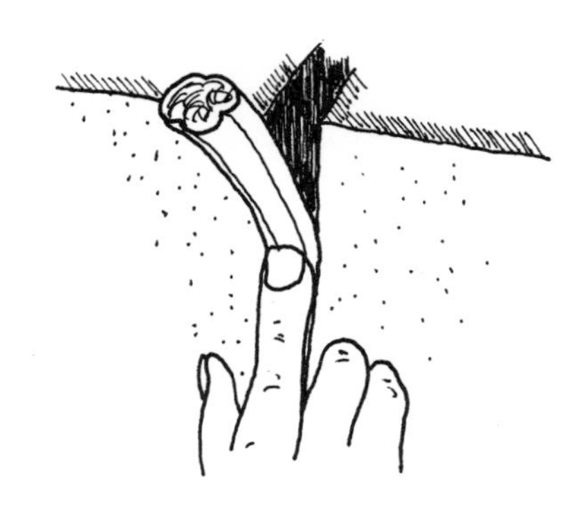

YOU CAN INSTALL PACKING WITH YOUR HAND

FOAM SEALANT

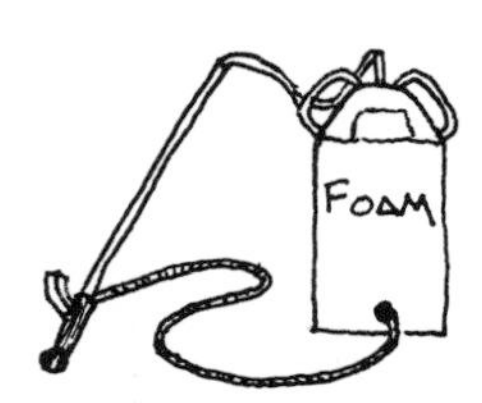

Polyurethane foam, either 1 or 2 part, can be used for filling large interior wall cavities that are opened to the outside by way of cracks in the exterior face of the wall or joints formed where the wall meets the window or door frame.

Two part foam is difficult to apply in the exact amount needed due to its rapid expansion after the foaming agents mix together. To get acceptable control, use a foaming kit with a "trigger" that has an adjustable flow valve. It can be used as packing on the exterior if caulk is applied over it. On the interior, it can be used as packing with no caulk applied over it. But since exposed foam is combustible, it must be installed behind or be covered with cement plaster or gypsum board.

11
change window operation

description

This retrofit involves making windows fixed or half operable. For sash that are never opened, you need not weatherstrip if sealing the sash closed would be acceptable.

Only windows that are not opened for ventilation purposes and are not needed for emergency exit can be fixed shut.

Windows like this might be found in basements, attics, stairways, entry-ways, and pantries. Instead of weather-stripping the windows to reduce infiltra-tion, it may make more sense to seal the window shut with caulk and/or wood nailing strips.

Making a window half operable involves sealing shut the top sash of double hung and double operable awning windows and the outer sash of sliding windows. This may be more acceptable since the other sash remains operable for ventilation and emergency exit. In this situation, the operable sash may be weatherstripped.

AN IMPORTANT NOTE:

Changing the window operation must meet local building code requirements for light, ventilation, and egress. You should thoroughly understand these codes prior to making windows fixed and/or half operable. Since the required ventilation opening is usually half as large as the required glazed area for natural lighting, making windows half operable will rarely present a conflict with building codes.

materials

wood nailing strips

fastening nails

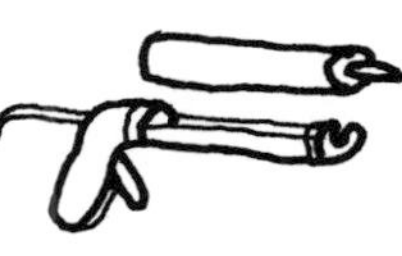

caulk

plastic wood

- nail finish putty

sealant tape

- vinyl, teflon, or polyethylene

packing

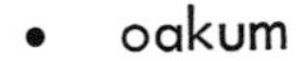

- polyethylene
- oakum
- polyurethane

cleaning solvents

- mineral spirits
- turpentine
- lacquer thinner

preparation

To assure that the caulk adheres to the window sash and frame, clean both these surfaces thoroughly with a stiff brush to remove all dirt, loose paint, and rust. Wipe clean with a cleaning solvent.

installation procedures

To fix a window shut, seal the jambs, head jamb, sill, and parting stiles.

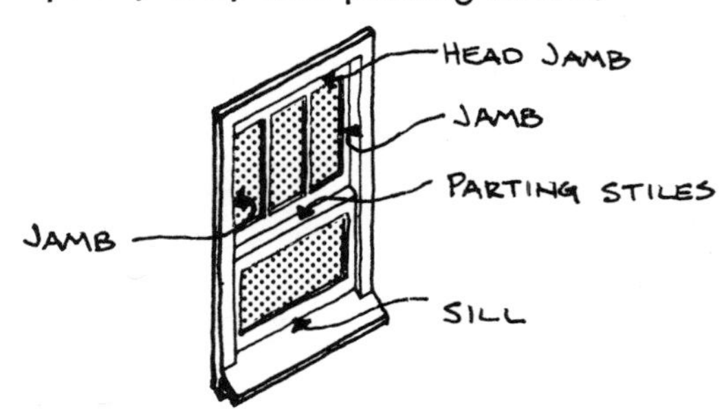

To make a double hung window half operable, seal the head jamb and the upper jambs of the top sash. The parting stile, sill, and the lower jambs may be weatherstripped.

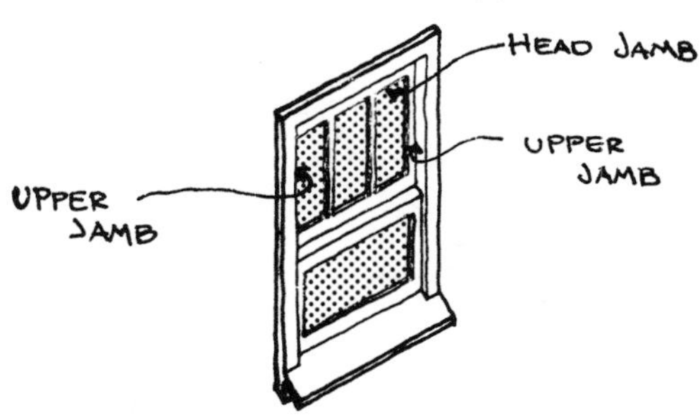

Use the same procedure for horizontal sliding windows, sealing the outermost sash, upper sash of double operable awning windows, and either of two operable sash of double operable casement windows. If one sash of a double operable awning or casement window cannot be opened, seal this sash.

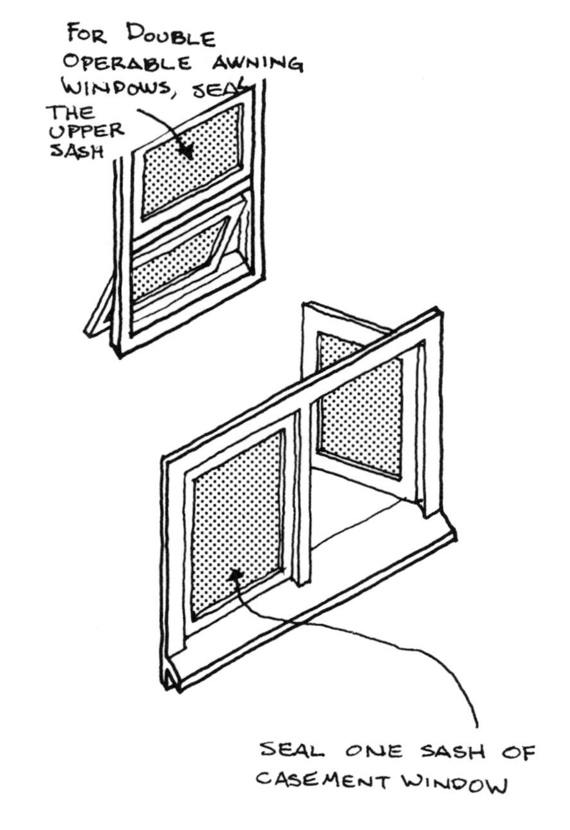

The two methods outlined here for changing a window operation are;

- Caulking the sash and frame
- Using wood stops in combination with caulk

CAULKING THE SASH AND FRAME

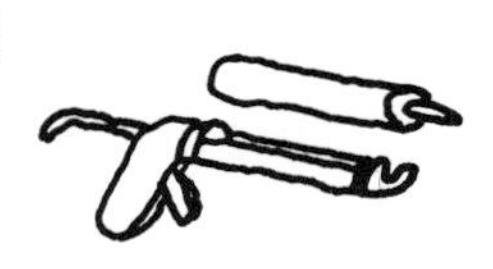

1 Carefully check the juncture between the sash and frame to determine if packing material is needed. After application, the bead of caulk should not be greater than 1/2" wide (to assure proper adhesion) and 1/4" deep (to avoid wasting caulk). If the width between the sash and frame is greater than 1/2", use wood stops in combination with caulk (see next section). If the bead of caulk will be greater than a 1/4" deep, use packing material.

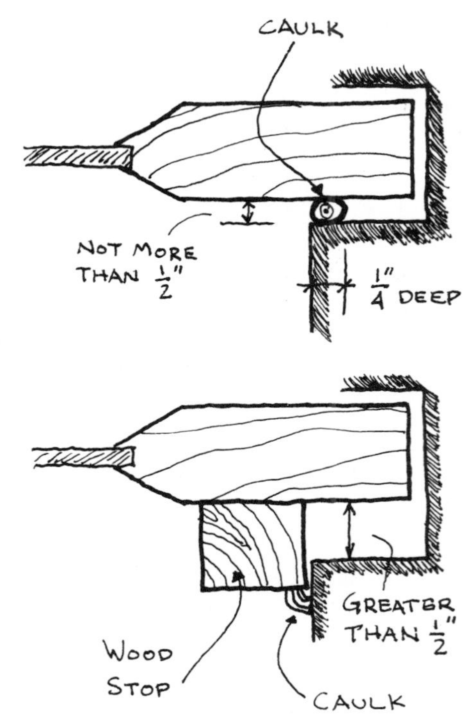

2 If packing material is needed, the packing rod should be approximately 1/8" thicker than the juncture to obtain a compression (pressure) fit. Use a blunt, rounded tool (like your fingers) when installing it to avoid puncturing the surface.

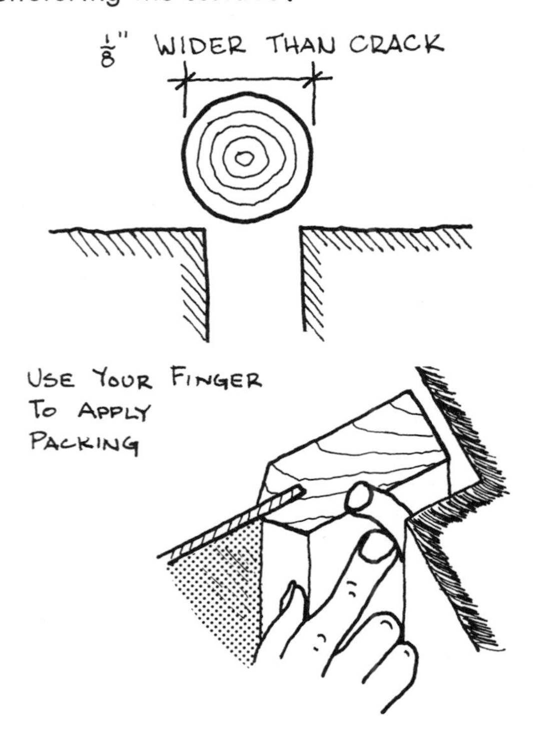

3 At this point, apply a continuous bead of caulk to the juncture. You can apply the caulk to the inside or outside of the window, depending on what's easier. For details on installing caulk, see Installation Procedures in Chapter 10, "Pack and/or Caulk Windows and Doors".

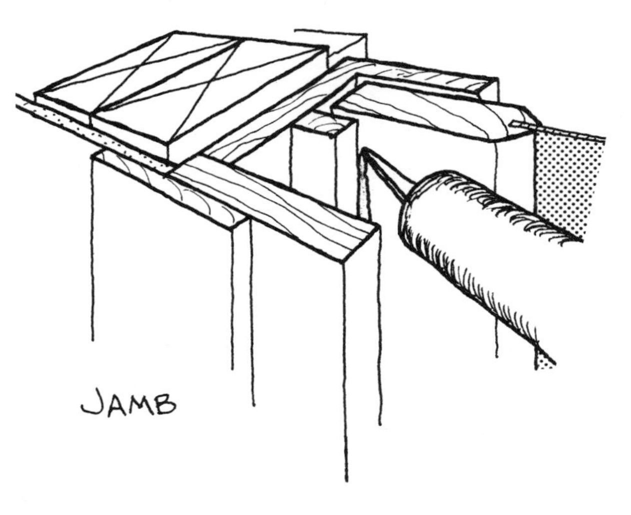

NOTE:

Do not apply caulk on the outside if it's raining or threatening to rain. Also, check the manufacturer's recommendation for proper installation temperature.

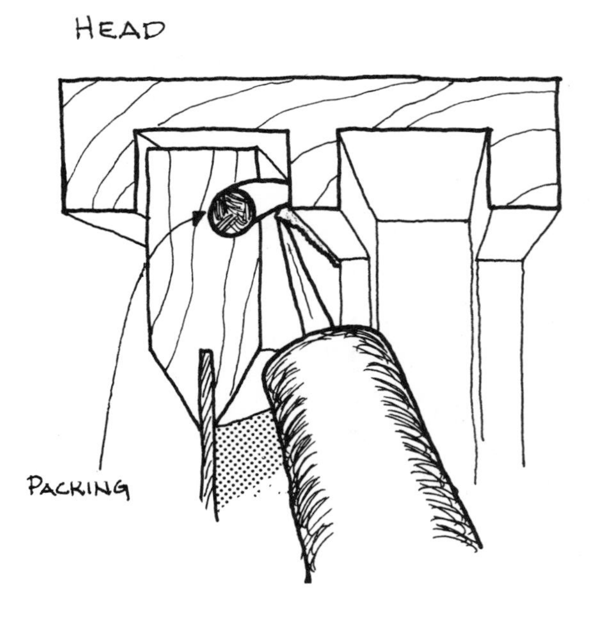

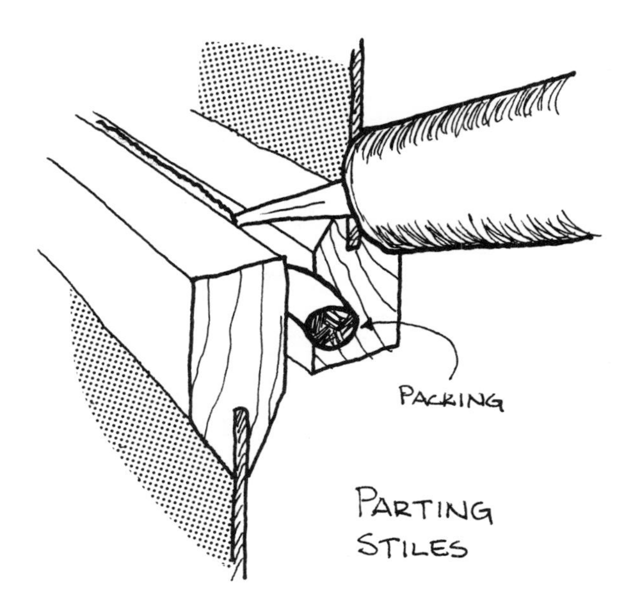

4 Additionally, you can toenail finishing nails through the sash stile into the jamb to secure the window. Use a nail set punch to set the nail heads back into the wood. Tap the nails in to keep the glass from breaking. Cover the nail heads with plastic wood or nail finish putty.

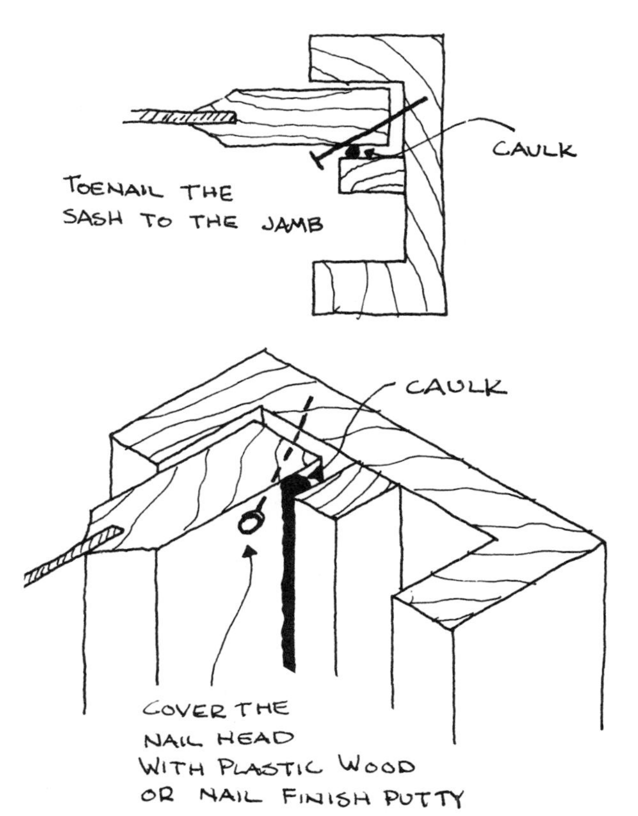

WOOD STOPS IN COMBINATION WITH CAULK

If the gap between the sash and frame is greater than 1/2" or very uneven, secure the window with wood nailing strips.

1 Apply a bed of caulk to the sash and the window sash.

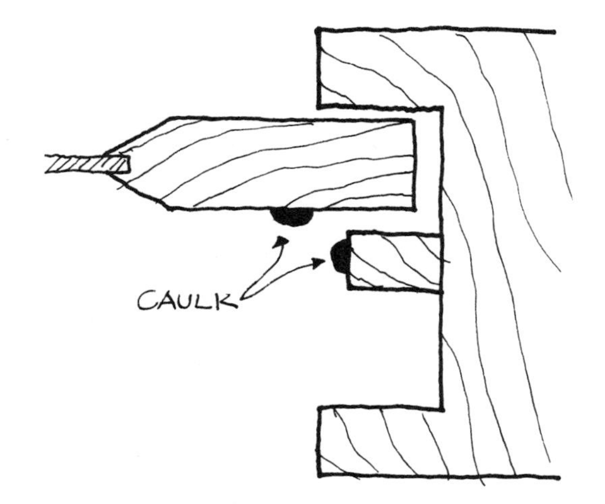

2 Nail the strips (use finishing nails) so that the strip butts against the sash and the window stop.

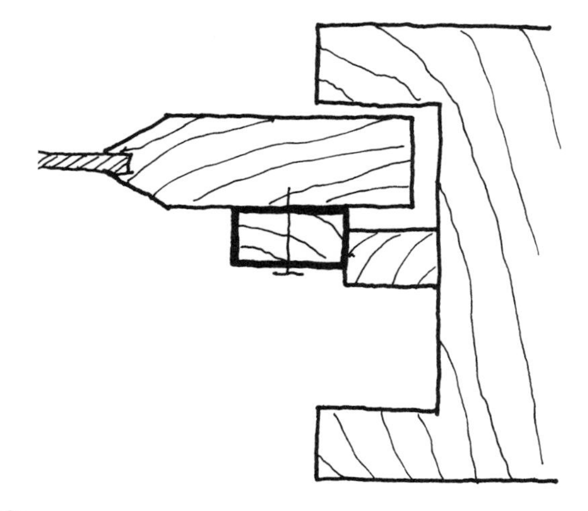

3 Use a nailset punch to set the nail heads back into the wood then cover the heads with plastic wood or nail finish putty.

The strips should be painted, especially if they're installed on the exterior.

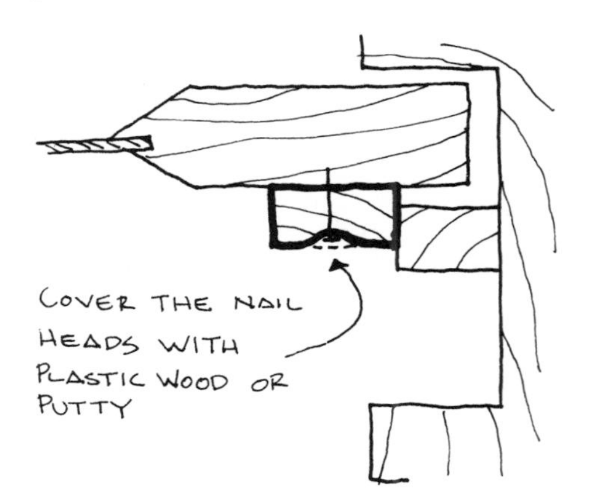

AN IMPORTANT NOTE:

When making a double hung or horizontal sliding window half operable, be certain that the nailing strip does not extend over the window stop. If it does, it will prevent the other sash from opening at all. In this situation, you may be better off installing the wood strips on the exterior.

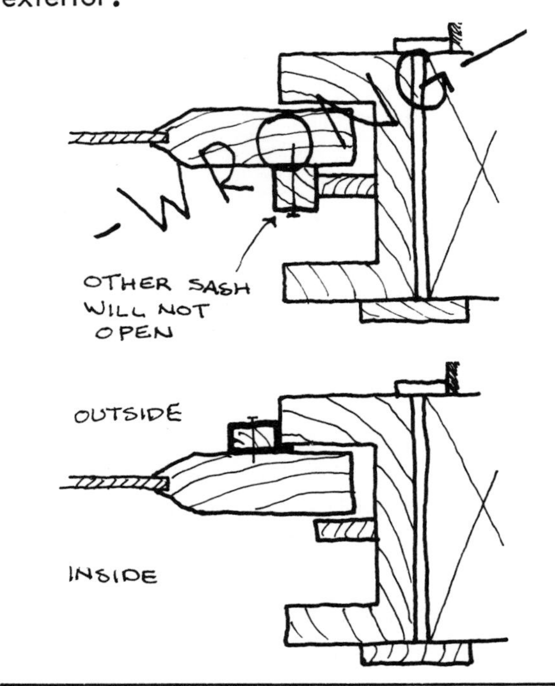

12

install plastic storm windows

description

Plastic storm windows* are fairly easy to assemble and install. Some plastic storms are available in kits. Others can be built at home with a few simple tools and readily available materials. We'll discuss both types in this section.

The primary reason that storm windows are installed these days is to make the home more energy efficient. Storm windows reduce conduction of heat by providing

* See table for properties of glass and plastic glazing materials in this chapter.

These products do not represent a comprehensive list of glazing products on the market for each generic type listed.

TYPE	GLASS — SINGLE				GLASS — DOUBLE			
BRAND NAME	ASG	SUNADEX	PPG	CLEAR VISION GLASS	ASG	SUNADEX	PPG	TWINDOW
THICKNESS (INCHES)	.125	.187	.125	.187	.5	.75	.5	1.
AIR SPACE (INCHES)	—	—	—	—	.25	.5	.25	.5
% TRANSMITTANCE — TOTAL SOLAR RADIATION	91.6	91.3			83.9	83.9		
% TRANSMITTANCE — VISIBLE LIGHT	91.	91.	90.	89.	83.	83.	82.	80.
TRANSPARENT	—	—	X	X	—	—	X	X
TRANSLUCENT	X	X	—	—	X	X	—	—
U-VALUE — SUMMER	1.04	1.03	1.04	1.04	.58	.53	.62	.58
U-VALUE — WINTER	1.14	1.13	1.16	1.14	.61	.55	.58	.49
WEATHERABILITY — DISCOLORATION	NONE							
WEATHERABILITY — IMPACT & ABRASION	ABRASION RESISTANT							
WEATHERABILITY — LIFETIME	200+ YEARS							
STRUCTURAL CONDITIONS — WEIGHT (LBS/FT²)	1.60	2.41	1.63	2.45	3.20	3.60	3.25	6.5
MAX. SIZE AVAILABLE (INCHES)	34 x 102	46 x 120	72 x 120	120 x 120	34 x 76	34 x 76	48 x 110	48 x 110
TYPES OF CAULKING								
FIRE RESISTANCE	NON-COMBUSTIBLE							
EXPANSION (INCH/INCH/°F)	$.50 \times 10^{-5}$							
COST ($/FT²)	.85-2.00	.90-2.30	1.50-3.00	1.60-3.00	2.00-4.00	2.00-4.50		
EVALUATION AND COMMENTS	PROS: • EXCELLENT SELECTIVE TRANSMISSION, WEATHERAB. • RESISTANT TO UV, AIR POLLUTION • FIRE RESISTANT • GD. INSULATION (DBL. ONLY)				CONS: • BREAKS EASILY • HEAVY • INSTALLATION RELATIVELY DIFFICULT			

TYPE	ACRYLIC — SINGLE				ACRYLIC — DOUBLE
BRAND NAME	ROHM & HAAS PLEXIGLASS K		CY/RO ACRYLITE GP		CY/RO ACRYLITE SDP
THICKNESS (INCHES)	.125	.187	.125	.187	.630
AIR SPACE (INCHES)	—	—	—	—	.51
% TRANSMITTANCE — TOTAL SOLAR RADIATION	90.	90.			83.
% TRANSMITTANCE — VISIBLE LIGHT	92.	92.	92.	92.	83.
TRANSPARENT			X	X	X
TRANSLUCENT	—	—	—	—	—
U-VALUE — SUMMER	.98	.93	.98	.94	.55
U-VALUE — WINTER	1.06	1.01	1.06	1.03	.58
WEATHERABILITY — DISCOLORATION	NONE				
WEATHERABILITY — IMPACT & ABRASION	STRONGER THAN GLASS, 1200 IMPACT STRENGTH .4 SCRATCHES, CLEANABLE				
WEATHERABILITY — LIFETIME	20+ YEARS				
STRUCTURAL CONDITIONS — WEIGHT (LBS/FT²)	.75	1.10	.75	1.10	1.00
MAX. SIZE AVAILABLE (INCHES)	120 x 144	120 x 144	120 x 144	120 x 144	47¼ x 144
TYPES OF CAULKING	GE SILICONE 1200 SILGLAZE, POLYSULFIDE LASTOMERIC, BUTYL TAPES				
FIRE RESISTANCE	SLOW BURNING FLASH PT. 560°-860°F WILL DRIP				
EXPANSION (INCH/INCH/°F)	4.1×10^{-5}				
COST ($/FT²)	2.48-1.33	3.13-1.67	1.20*		2.80-3.00*
EVALUATION AND COMMENTS	PROS: • EXCELLENT OPTICAL CLARITY, WEATHERABILITY • GOOD IMPACT STRENGTH • LIGHTWEIGHT • GOOD INSULATION (TWINWALL ONLY)				CONS: • SUSCEPTABLE TO ABRASION • HEAT SAG AT HIGH TEMPS.

TYPE	POLYCARBONATE — SINGLE				POLYCARBONATE — DOUBLE	
BRAND NAME	ROHM & HAAS TUFFAK		GE LEXAN MR4000		ROHM & HAAS TUFFAK-TWINWAL	CY/RO CYROLON SDP
THICKNESS (INCHES)	.125	.187	.125	.187	.220	.630
AIR SPACE (INCHES)	—	—	—	—	.187	.55
% TRANSMITTANCE — TOTAL SOLAR RADIATION	89.	85.	89.	88.	77.	74.
% TRANSMITTANCE — VISIBLE LIGHT	85.	83.	80.	85.	80.	73.
TRANSPARENT	X	X	X	X	X	X
TRANSLUCENT	—	—	—	—	—	—
U-VALUE — SUMMER	.98	.93	.97	.93	.62	.55
U-VALUE — WINTER	1.06	1.01	1.05	1.00	.62	.58
WEATHERABILITY — DISCOLORATION	YELLOWS AFTER 5-7 YRS.		YELLOWS CAN BE CLEANED		YELLOWS AFTER 5-7 YRS.	YELLOWS AFTER 2-3 YRS.
WEATHERABILITY — IMPACT & ABRASION	300X STRONGER THAN GLASS, 1200 IMPACT STRENGTH 16. SCRATCHES EASILY					
WEATHERABILITY — LIFETIME	TURNS MILKY WHITE 5-7 YRS. USABLE LIFE SPAN 5-7 YRS. **					
STRUCTURAL CONDITIONS — WEIGHT (LBS/FT²)	.78	1.17	.78	1.17	.25	.92
MAX. SIZE AVAILABLE (INCHES)	72 x 96	72 x 96	96 x 96	96 x 96	48 x 96	47¼ x 144
TYPES OF CAULKING	GE SILICONE 1200, SILGLAZE POLYSULFIDE LASTOMERIC, DOW 781, WELDON #35					
FIRE RESISTANCE	SELF-EXTINGUISHING FLASH PT. 800°-1090°F WILL DRIP					
EXPANSION (INCH/INCH/°F)	$3.3\text{-}4 \times 10^{-5}$					
COST ($/FT²)			3.23-4.40	4.30-5.93	1.25*	3.80-4.00*
EVALUATION AND COMMENTS	PROS: • VERY HIGH IMPACT RESISTANCE • GOOD SOLAR TRANS. • TRANSPARENT • GOOD INSULATION (TWINWALL ONLY)				CONS: • SCRATCHES EASILY • NON-RIGID • TURNS MILKY AFTER 5-7 YRS.	

TYPE	FIBERGLASS — SINGLE		FIBERGLASS — DOUBLE	
BRAND NAME			KALWALL	SUNLITE
THICKNESS (INCHES)	.030	.037	.50	1.50
AIR SPACE (INCHES)	—	—	.42	1.42
% TRANSMITTANCE — TOTAL SOLAR RADIATION	82.	79.	77.	65.
% TRANSMITTANCE — VISIBLE LIGHT	94.	94.	84.	84.
TRANSPARENT	—	—	—	—
TRANSLUCENT	X	X	X	X
U-VALUE — SUMMER	.76	.76	.55	.46
U-VALUE — WINTER	.81	.81	.55	.46
WEATHERABILITY — DISCOLORATION	STABLE, NON-FADING			
WEATHERABILITY — IMPACT & ABRASION	1200 IMPACT STRENGTH .4			
WEATHERABILITY — LIFETIME	GUARANTEED FOR 8, 10, OR 15 YEARS			
STRUCTURAL CONDITIONS — WEIGHT (LBS/FT²)	.25	.312		
MAX. SIZE AVAILABLE (INCHES)	26 x 144	26 x 144	33¾ x 77⅝	47⅞ x 167½
TYPES OF CAULKING	HORMOUNT SILGLAZE FOAM TAPE		GE SILICONE 1200 BUTYL, BEAD	
FIRE RESISTANCE	BURNS SLOW F.PT. 750°F		BURNS SLOW F.PT. 950°F	
EXPANSION (INCH/INCH/°F)	2.3×10^{-5}			
COST ($/FT²)	.59	.70	3.00	3.29
EVALUATION AND COMMENTS	PROS: • GOOD SOLAR TRANS. • HIGH DURABILITY		CONS: • TRANSPARENT NOT AVAILABLE	

* MANUFACTURERS' PRICES ** A NEW SILICONE COATED POLYCARBONATE IS BEING DEVELOPED BY GE TO EXPAND LIFE-SPAN OF LEXAN TO 15-20 YRS. ROHM & HAAS HAS ALSO JUST INTRODUCED CM-2 COATED TUFFAK, WHICH IS SUPPOSE TO WEATHER LONGER.

Reprinted with permission of "The Neighborhood Works Information Center", Center for Neighborhood Technology, Chicago, Illinois

another layer of resistance about equal to that of a sheet of glass. They also reduce window air leakage. However, don't be misled by fantastic claims of savings, as plastic storms perform only a little (if at all) better than glass storms. In this regard, placing the storm on the inside is as effective as placing it outside. However, inside storms will not prevent weathering of windows, which may also be of concern. Most plastics can be installed outside the window, but problems of fading and becoming brittle, which are common to most plastics, will occur sooner. On the other hand, plastics that are flammable should not be used on the inside (be sure to check with salespersons to obtain information on flammability before purchasing plastic material for inside use).

There are several cases where plastic storms provide a unique benefit. One is in apartments where you don't intend to stay or you don't pay directly for your heat. In this case, this temporary and inexpensive way to cut heating costs or increase comfort may be the most reasonable. Reducing air leakage of very drafty windows is another case. An exterior storm window, being toward the cold, will allow house water vapor to condense and become "fogged up" if the storm fits tighter than the prime window. You may be able, however, to use tightly fitted plastic interior storms to reduce air leakage in cases of tight fitting outer storms that do fog up.

Perhaps a final consideration when selecting a plastic storm is ease of operation. Plastic storms that are difficult to demount or move aside could block egress from a home in case of fire. No storm window that is difficult to open should be installed in a window intended to provide exit in your family fire escape plan. Likewise, keep one window that can be easily egressed in bedrooms of children, elderly, or handicapped.

NOTE: 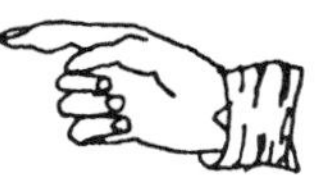

Temporary storm windows must meet local building code requirements for egress, ventilation, and flammability.

---

materials

plastic films (flexible)

- polyethylene film (6 mil is recommended)
- polyvinyl fluoride
- teflon
- polyvinyl acetate

rigid plastic sheets

- acrylic
- polycarbonate
- butgrate
- styrene (not for interior applications)

wood nailing strips

- approximately 1" x 2"

vinyl frames

fasteners

- galvanized nails (exterior)
- staples
- turn buttons
- wingnuts

adhesives

- liquid vinyl or acrylic based

weatherstrip tape

preparation

Clean the frames (or sash) of all rust, dirt, and loose paint. Wipe clean with mineral spirits or turpentine. The frame should be cleaned for either interior or exterior application.

installation procedures

For double hung and horizontal windows, install the storm window on the window frame. The same applies to casement and tilting windows where the sash swings away from the storm. Install the storm window on the sash of casement and tilting windows, where the sash would swing into the storm.

The installed storm windows should be straight, level, and without distortion. The frame of the storm should contact entire primary window frame or sash on which it is mounted. Remember, plastic storm windows should be installed in such a manner that the storms can be removed, but won't blow off.

storm windows made with plastic film

1 Cut the plastic film to the size of the window frame with an excess of 2"on each side. Use scissors or a utility knife for cutting.

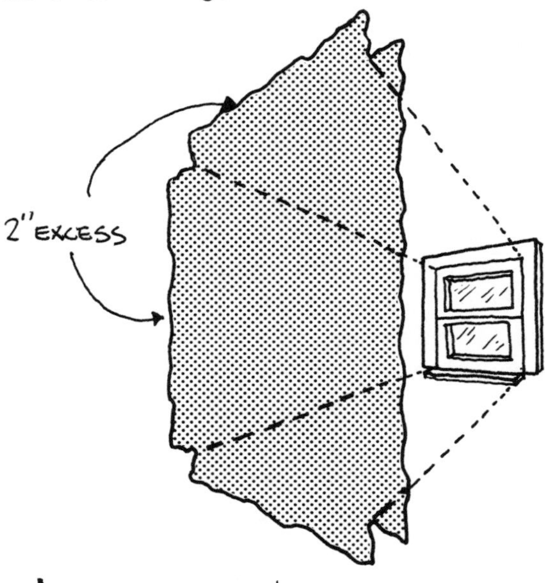

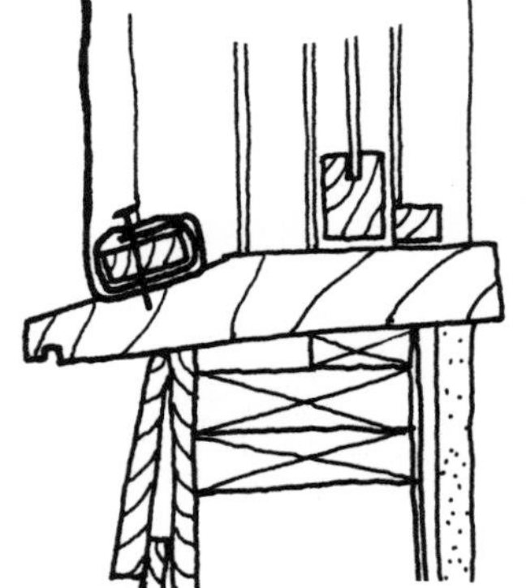

2 Roll the bottom "excess" edge around a wood slat (about 1/4" x 1-1/4"; old wood lath is about this size). Roll the plastic on the outside of the slat so water will run off and not collect in the wrapped film. Nail the bottom in place along the sill with wire brads spaced about 4" apart (staples can be used rather than brads).

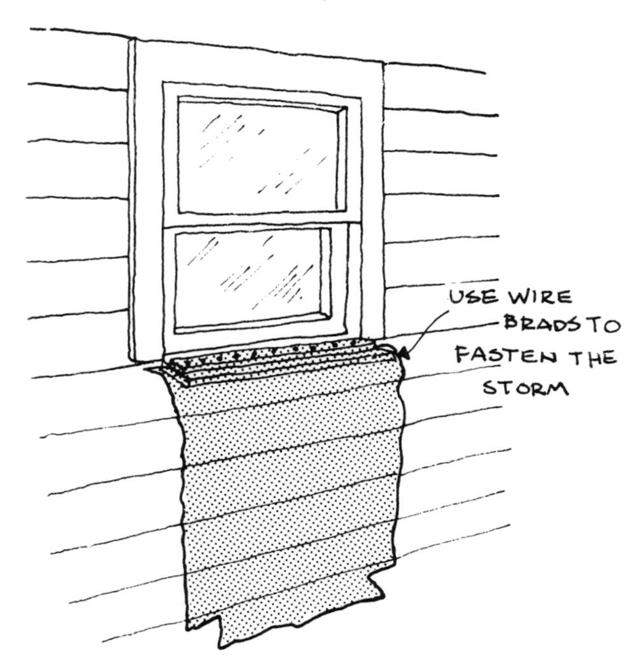

3 Now fasten the top edge of the plastic film to the window frame. Roll the plastic around a wood slat so that the plastic is taut when you attach it to the window frame.

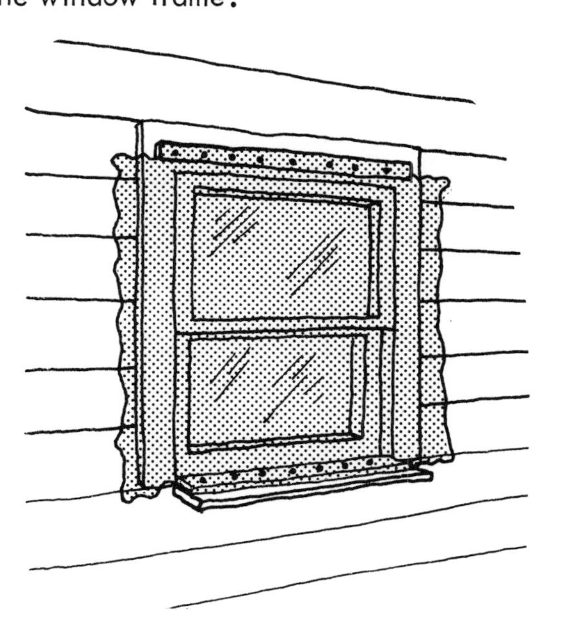

4 Roll the sides of the plastic and attach in the same manner.

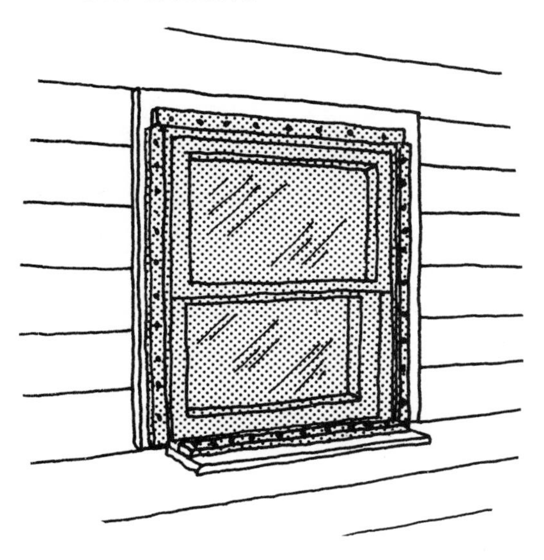

NOTE:

Rigid vinyl frames with channels for holding the polyethylene are available. For installation procedures, see "Storm windows made with plastic sheets" in this chapter.

5 When using plastic film on the interior, tape can be used in place of nails and wood slats. Roll the edges as before (you can roll them around cardboard strips), then tape them to the frame. Weatherstrip tape is less likely to damage paint when the storms are removed.

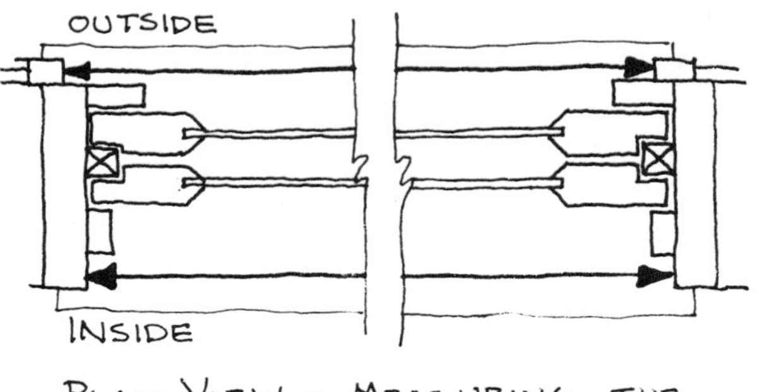

storm windows made with plastic film ...an alternative

An alternative to the previous method is framing the plastic with wood. In this manner, the storm is of a more permanent nature and is less likely to rip when removed in the spring.

1 Measure the interior of the window and cut the furring strips (1" x 2" or 2" x 2") to fit.

2 Put the furring strips together with brackets and screws. At this point, the frames should be painted (especially if they're going to be used on the outside) to seal the wood against moisture.

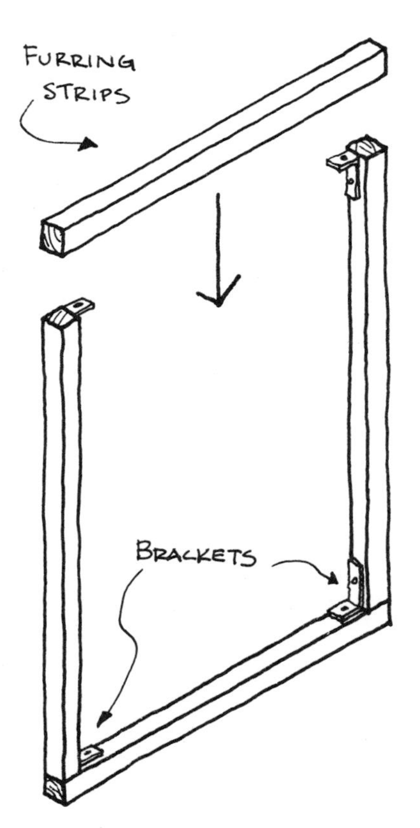

3 Roll the plastic around wood slats and nail or staple the plastic to the frames. Mount the storms to the window frames with turn button screws. At the end of winter, you can unscrew the storms and store them. It would be a good idea to label the storms so they can be remounted in the correct window frame the following year. An advantage of making frames for the plastic is that when the plastic wears out, it can be replaced over the frames that you built.

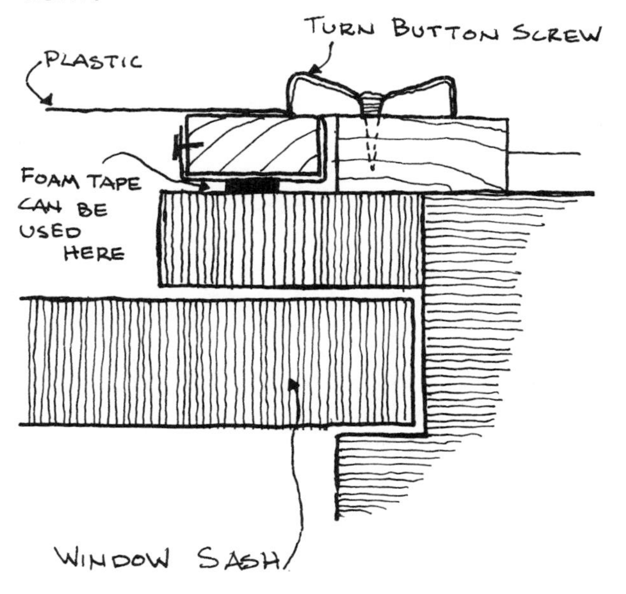

storm windows made with plastic sheets

Do-it-yourself storm window kits are available from lumber companies and hardware stores. Included in the kits are plastic sheets and plastic frames with adhesive backing or special turn buttons (plastic sheets are also available separately). Before purchasing the do-it-yourself kits, check to see if the storm is for both interior and exterior application or just interior. Most storm window kits are interior only. Also check flammability of materials.

Here are step-by-step instructions to install one type of plastic storm.

1 Measure the window to determine the size of the plastic sheet. Measure the width across the window to the outside of both frames and measure the height from the top of the frame to the sill.

Now, when the frame pieces are attached, the edges of the plastic frame should be flush with the edges of the window frame. Consequently, you have to subtract a little to get the plastic sheet size.

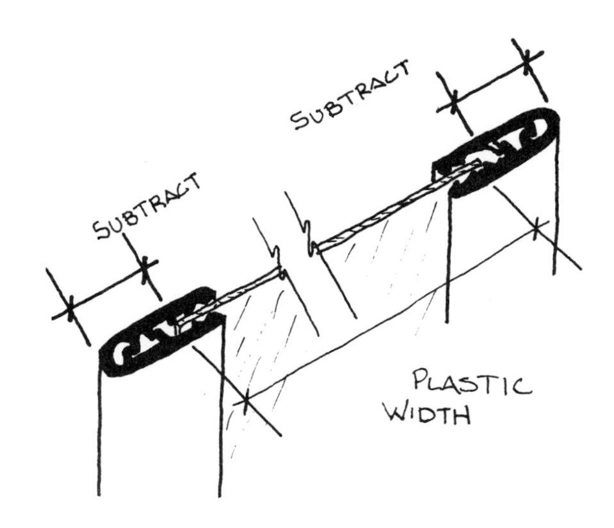

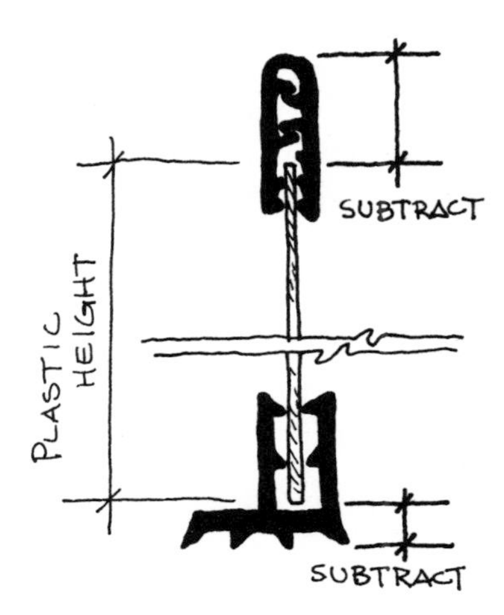

Now bend the sheet along the score and snap off the excess.

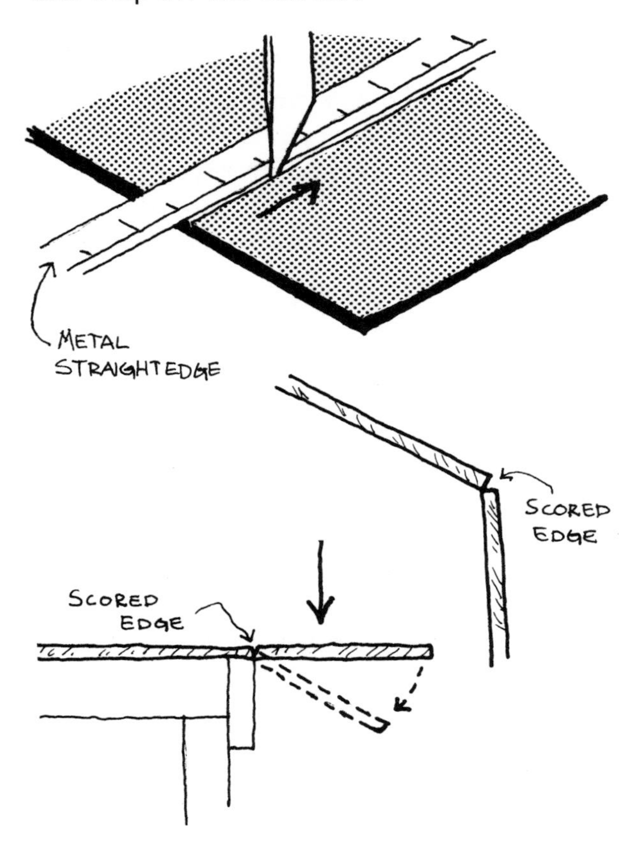

2 Lay the sheet of plastic on the floor and mark off your measured dimensions (to protect the floor, place a board under it). Put a straight edge along the dimension line and pull your utility knife across it 4 or 5 times. You're not trying to cut it all the way through - you're just cutting a groove in it. This is called "scoring". Be sure to score the plastic from edge to edge. Place the sheet of plastic on a table with the scored edge on the edge of the table.

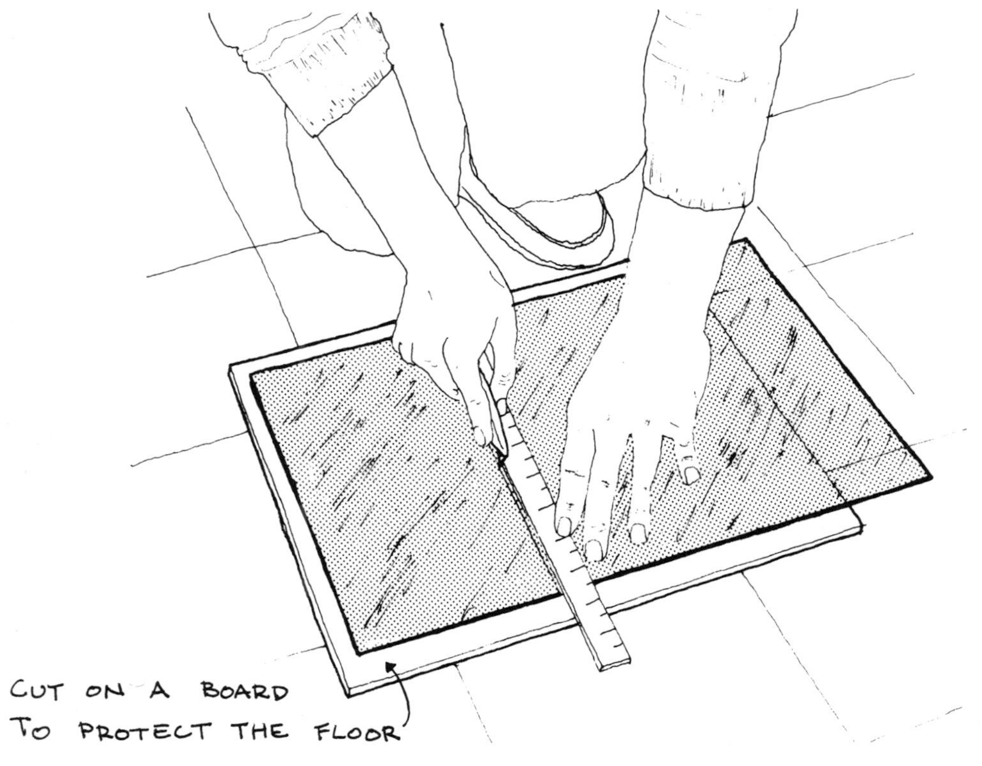

3 With a small hand saw (or "case cutting" knife), cut the framing pieces. The width of the top and sill pieces should correspond to the frame to frame dimension.

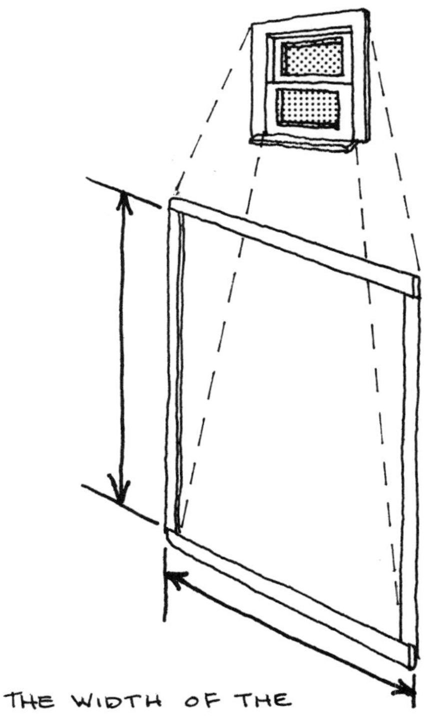

THE WIDTH OF THE TOP AND SILL PIECE SHOULD BE THE SAME AS YOUR FRAME TO FRAME MEASUREMENT (STEP #1)

NOTE:

It's easier to cut the strips if they're snapped closed.

4 Now you're ready to attach the frame strips to the sheet. Open the strips and attach the sill strip first, followed by the side pieces, and finally the top.

ATTACH THE SILL PIECE FIRST

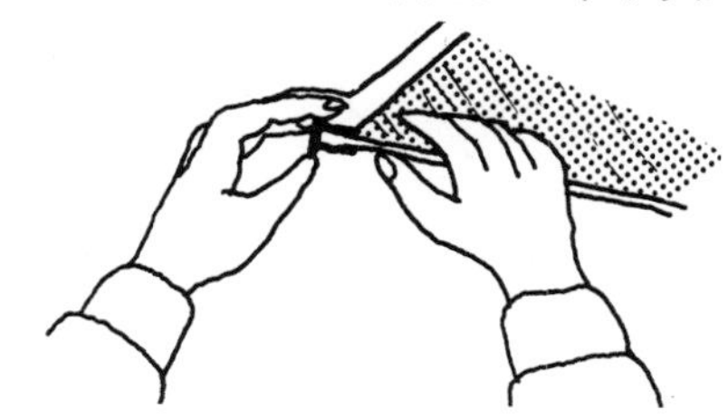

It's easier to close the strips on a hard surface such as the floor.

SNAP THE STRIPS SHUT

5 Some plastic frames have an adhesive backing. A frequent problem has been, however, that the frame will come loose from the window. To insure a good bond, apply a bead of a compatible adhesive to the window frame. On the exterior, apply a bead of caulk to the frame to strengthen the bond. Test any adhesive on a piece of frame before use as some adhesives don't work well on some plastics.

NOTE:

When plastic frames do not have an adhesive backing, apply a bead of vinyl or acrylic adhesive (interior) or caulk (exterior) to the window frame and fasten the plastic frame over the adhesive with nails or screws. Then snap in the glazing sheet.

6 Now you're ready to put up the storm window. Remove the covering from the adhesive tape and press firmly into place (begin at the sill and work your way up). After the storm is in place, make sure the perimeter of the storm is in full contact with the adhesive/window frame. That is, check for gaps or openings between the storm window and the window frame. Any openings should be closed with adhesive of caulk. To remove the storm, unsnap the frame and remove the glazing, leaving the frame attached to the window casing.

on the interior...

If frames are unavailable for plastic sheets, you can tape the sheets (with weatherstrip tape) to the window frames on the interior.

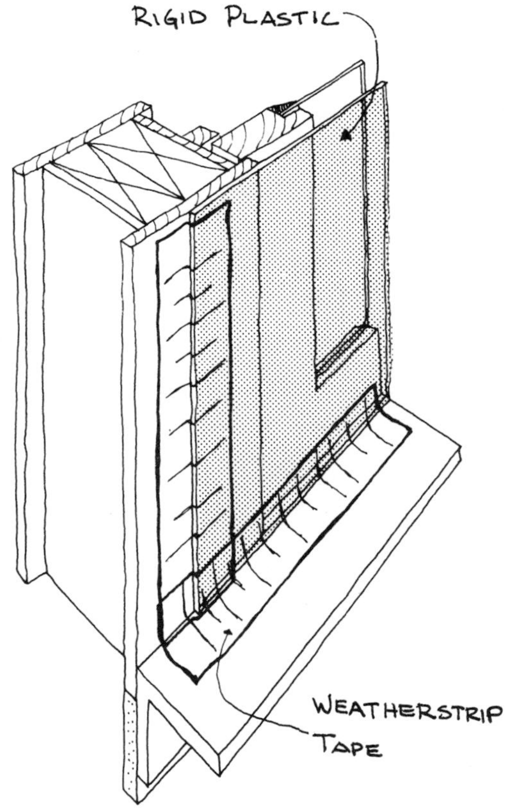

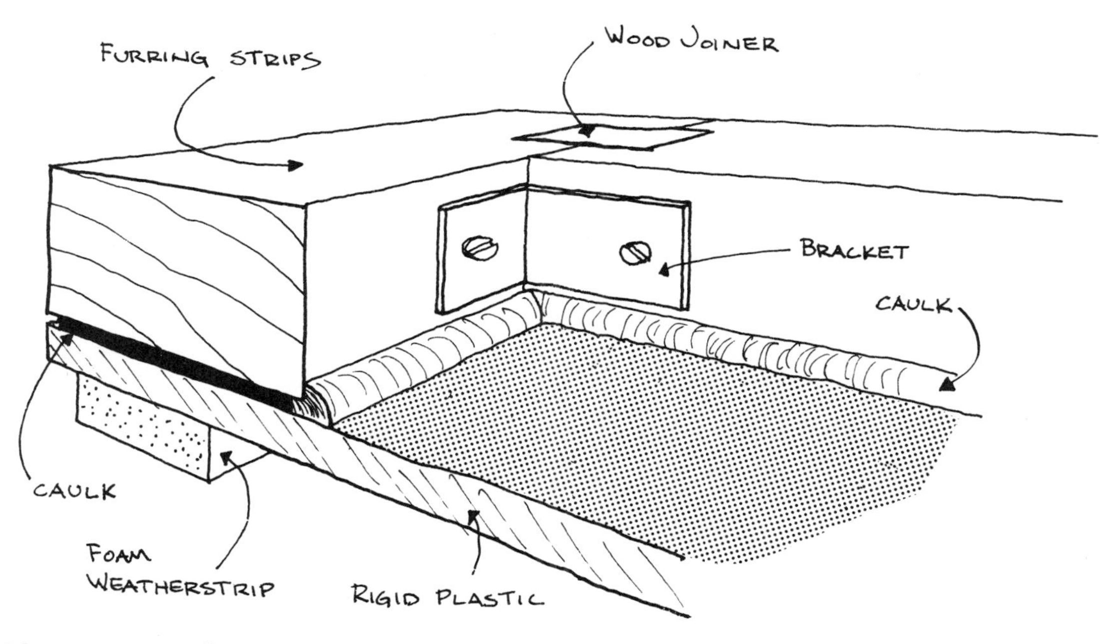

on the exterior...

1 Glue the sheet of plastic to wood slats or furring lumber. Apply a bead of caulk to the joint between the wood and the plastic.

2 Apply strips of adhesive backed foam weatherstripping to the back of the plastic to seal any air leaks. Mount the storm to the prime window frame with turn button screws or clips.

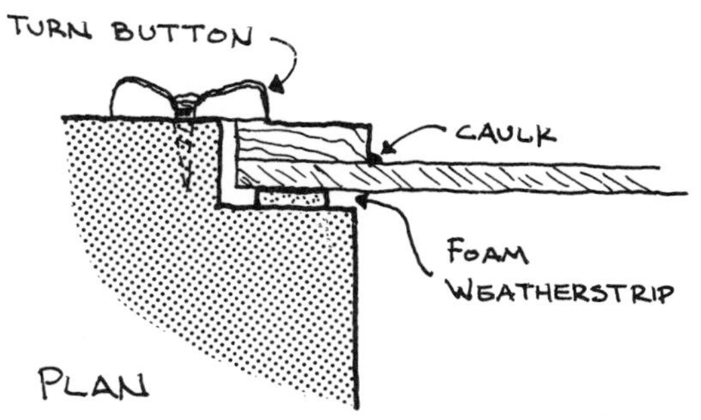

storms surface mounted to windows

Storms can be "surface mounted" to the window sash rather than the window frame. You may want to consider this for tilting and casement windows. Since it moves with the sash, the storm can be left on permanently without affecting the operation of the window. However, cold air can still leak in between the sash and frame. Consequently, if you "surface mount" storms, you may also need to weatherstrip the window.

Mount the storms using wood slats and/or tape (see "Storm windows made with plastic film...an alternative" in this chapter). If the sash is metal, you'll have to drill holes for screws rather than using wire brads.

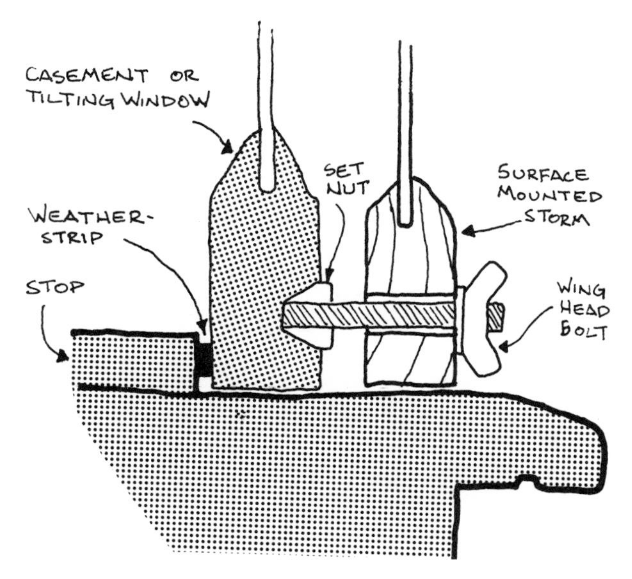

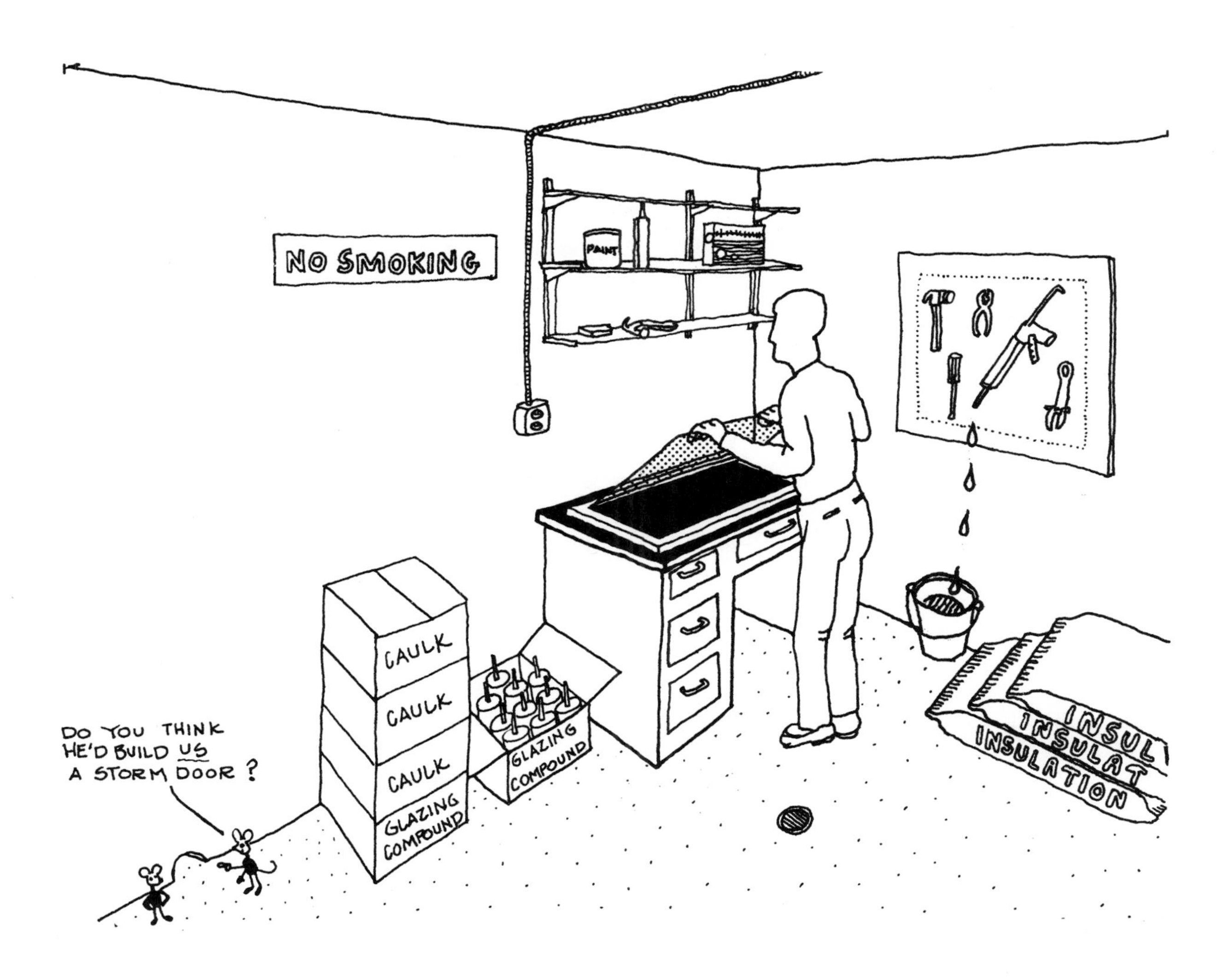

13

install glass storm windows

description

In Chapter 12, methods for constructing and installing plastic storm windows were presented. Glass can also be used for glazing storm windows (refer to the table for properties of glass and plastic glazing materials in Chapter 12). Glass storms are more permanent in terms of scratch and fade resistance than plastic storms.

Like the plastic storms, you can build your own glass storm windows. If you're not up to building your own, then it may be worthwhile for you to measure for and install pre-assembled storms, which would reduce your cost outlay. Build-it-yourself storms usually require seasonal operation (removed from the window for the summer, put back in place for the winter). Frames for this type of storm can be fabricated from aluminum, wood, rigid vinyl, or steel (rigid plastic can also be substituted for the glass). A method of fabricating storms using aluminum and wood frames is presented in this chapter.

Some storms, referred to as "double" or "triple" track storms, are installed on the exterior of double hung windows. During the summer, the storm merely slides up within the track and a screen is lowered in its place. These storms are also called "combination storm screens". This type of storm is relatively expensive and should be fabricated by a reputable company. This is known as an "operable" storm window. Information for installing the finished product is given in this chapter.

NOTE:

Storm windows must meet local building codes for egress, ventilation, and flammability. Review your codes before fabricating or purchasing storm windows. Particular concerns for maintaining egress are the same as for plastic storms.

---

materials

operable storm windows

- double or triple track (for single hung, double hung, and sliding window)
- hinged (top, side, or hopper)

non-operable storm windows (do-it-yourself storms)

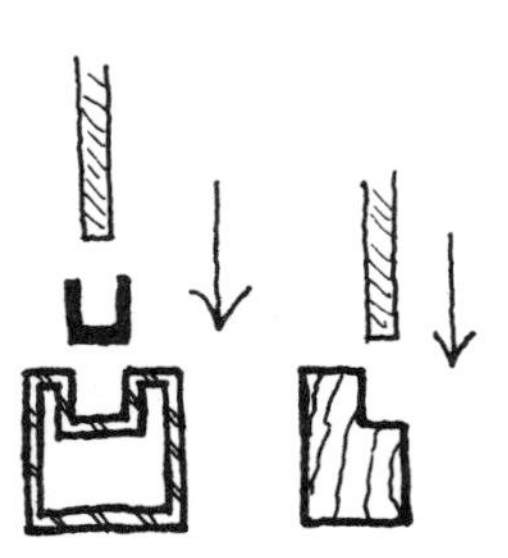

Framing
- aluminum
- wood sash stock

Glazing
- glass
- rigid plastic
  - acrylic
  - polycarbonate
  - butyrate
  - styrene (not for interior applications)

other materials

Fasteners
- wood or sheet metal screws
- galvanized nails and screws
- wing nuts
- clips
- turn buttons
- wood joiners

Cleaning Solvents
- mineral spirits
- turpentine
- lacquer thinner

Caulk

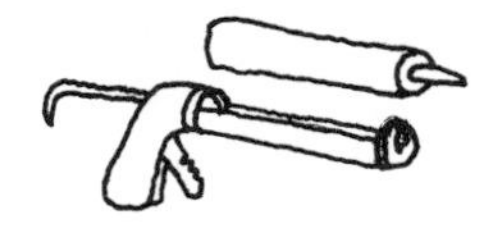

Non-metallic Spacers (1/4" thick minimum)
- wood
- vinyl rubber
- high density polyurethane

preparation

Prior to installing an operable or non-operable storm, clean the frame of all rust, dirt, and loose paint. Wipe clean with mineral spirits or turpentine. The frame is to be cleaned for either interior or exterior application. Frames found to be decayed, insect infested, or unsound should be corrected prior to installing storm windows.

installation procedures

Procedures for building non-operable storms with aluminum and wood sash are presented here. In addition, information is given for installing standard (pre-assembled) operable storms.

The installed storm windows should be straight, level, and without distortion. The frame of the storm should contact the entire primary window frame. Interior storm windows should be backed with foam tape weatherstrip.

Remember, non-operable storm windows should be installed so that they can be removed, but won't blow off.

The parts that you'll need to build your own storm windows are available at lumberyards, large hardware stores, or home centers.

DO-IT-YOURSELF aluminum storm windows

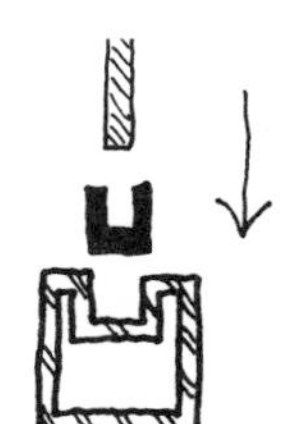

1 Begin by measuring the area where the storm window will go. It might be a good idea to measure the width and height in two or three different locations and use the smallest width and height measurements to insure a proper fitting storm window. Measure the width from reveal to reveal and the height from reveal to sill.

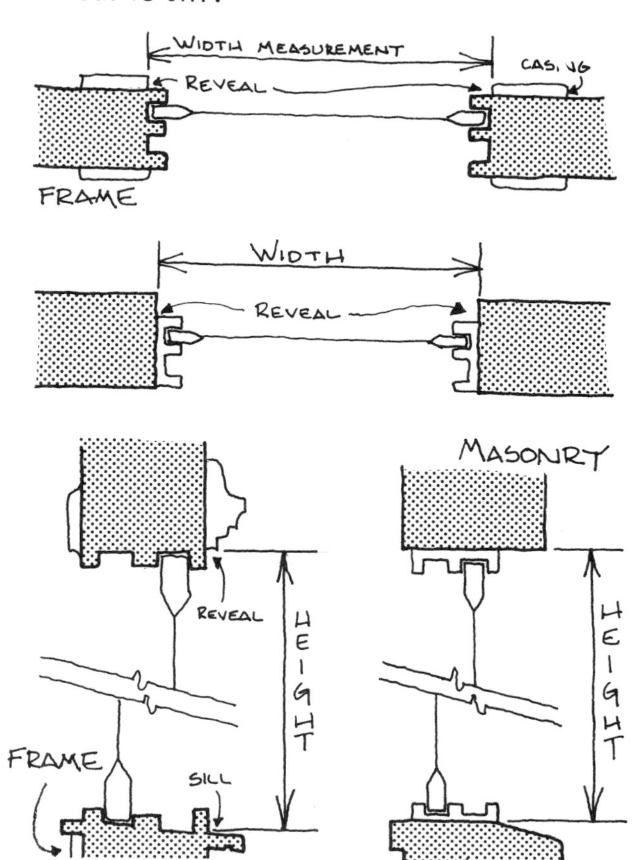

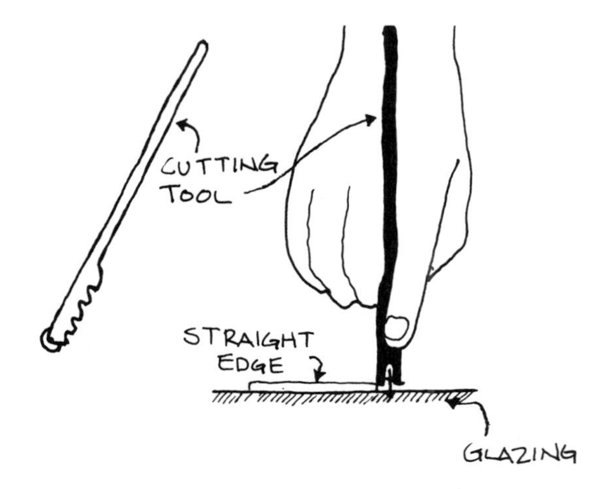

2 Cut, or have cut, the glass (or rigid plastic sheets). To fit properly, the glass must be square. Score the glass with a glass cutting tool and a metal straight edge. Bend the sheet along the score line and snap off. Remember, wear sturdy gloves when handling glass and dispose of (or recycle) the excess glass in a container marked "Broken Glass".

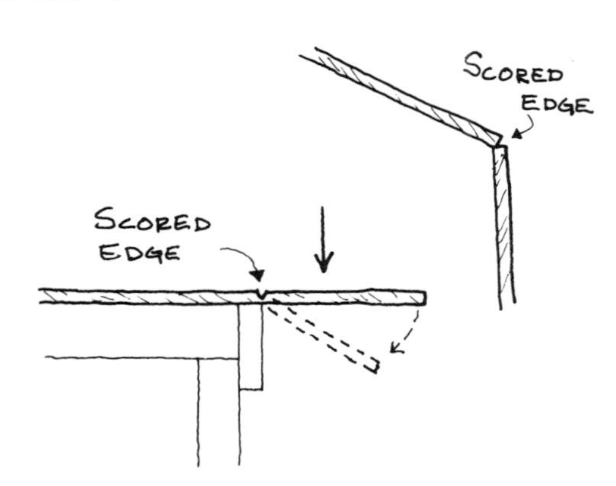

3 Remove the rubberized glazing channels that you'll find within the metal sash. Use metal snips or a utility knife to "bite out" a triangular piece of each side of the channel, cutting at a 45° angle on each side. Don't cut through the entire channel; merely cut out a triangle in the flanges. Replace the rubber glazing piece and place the channels around the edges of the glass. The channel should bend at the cuts to form a neat miter* joint. If needed, use tape to hold the channels in place. These channels should be made and attached soon after cutting the glass so as to cover sharp edges and to keep the glass from breaking.

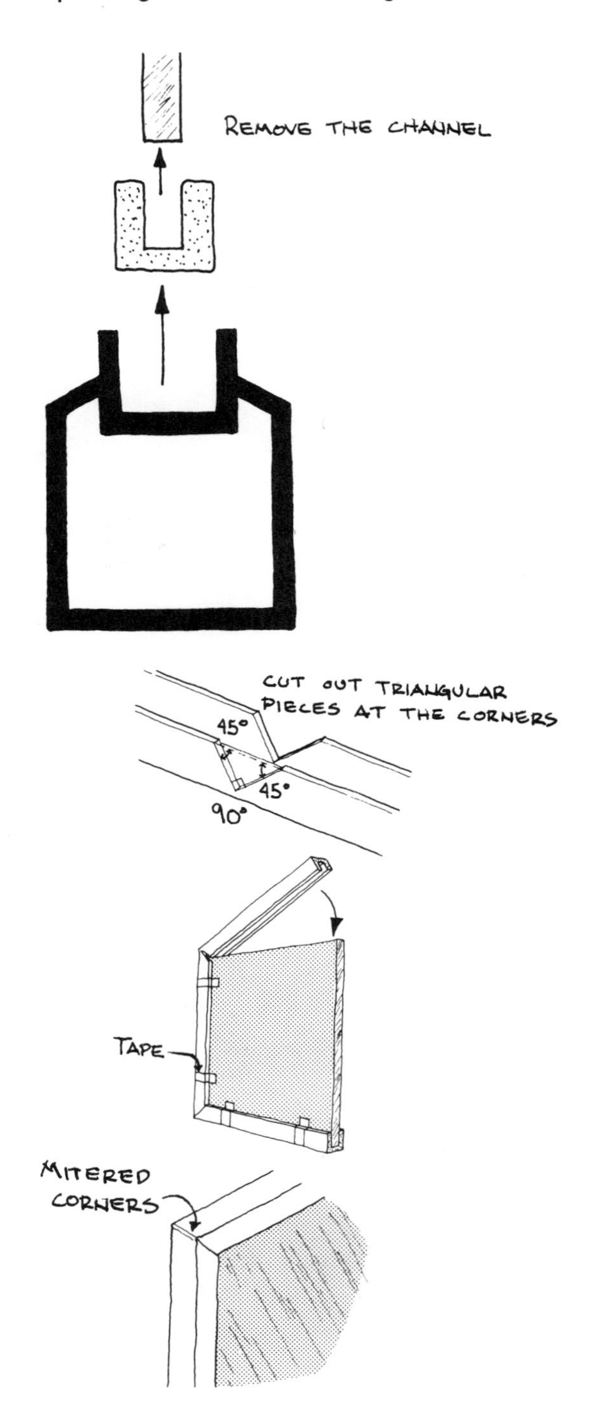

* The joining of two pieces of material that are beveled so that, when attached, they adjoin each other at an angle.

4 Sash pieces are used to make a rigid frame for the glass. Use a hacksaw and a miter box to cut the aluminum sash pieces.

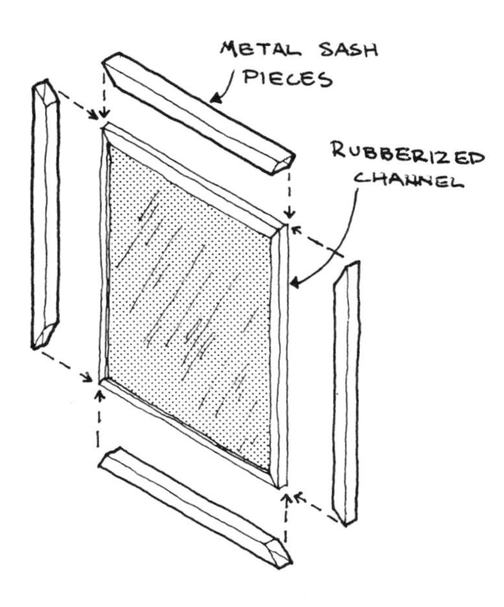

MITER BOX AND SAW

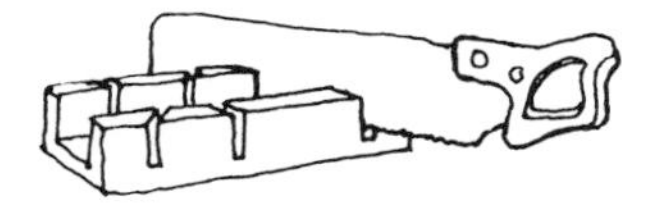

5 Attach the top and bottom pieces over the glass and glazing channel.

ATTACH TOP AND BOTTOM SASH PIECES

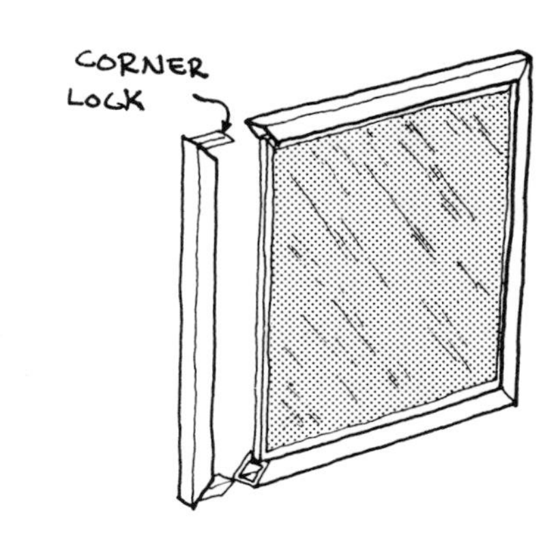

6 Install the corner lock hardware in the two pieces. Now attach these two side pieces to the glass. Hang the storm with turn button screws or clips.

DO-IT-YOURSELF wood storm windows

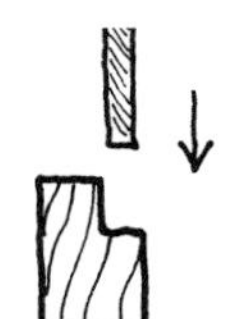

1 Once again, begin by measuring the window opening. Measure the width from reveal to reveal and the height from reveal to sill. Measure along the perimeter of the window, since these are the measurements you'll use to cut the sash stock.

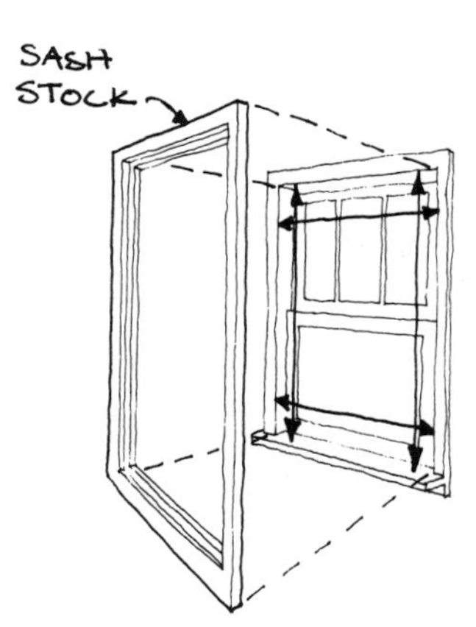

NOTE: 

Sash stock comes in different thicknesses.

2 With a miter box and saw, cut the sash pieces. Fasten the pieces together with wood joiners. If the storm is to be used on the outside, prime and paint the sash. Allow the sash to dry before setting the glass in place.

3 At this point, measure the opening of the storm and cut the glass (or plastic)

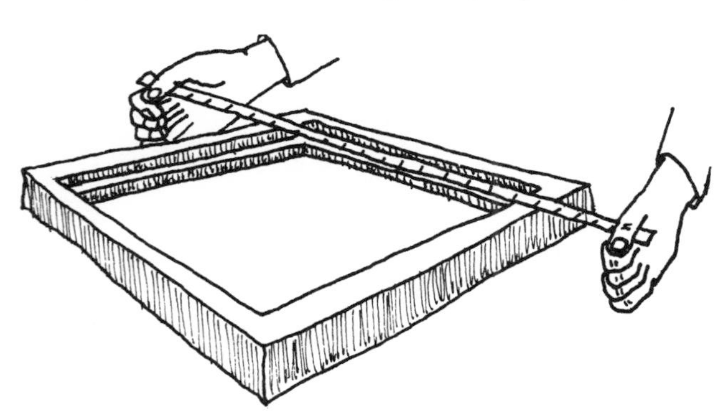

to fit. Remember, to cut glass or plastic, all you have to do is "score" it. Then "snap" the excess over the edge of a hard surface.

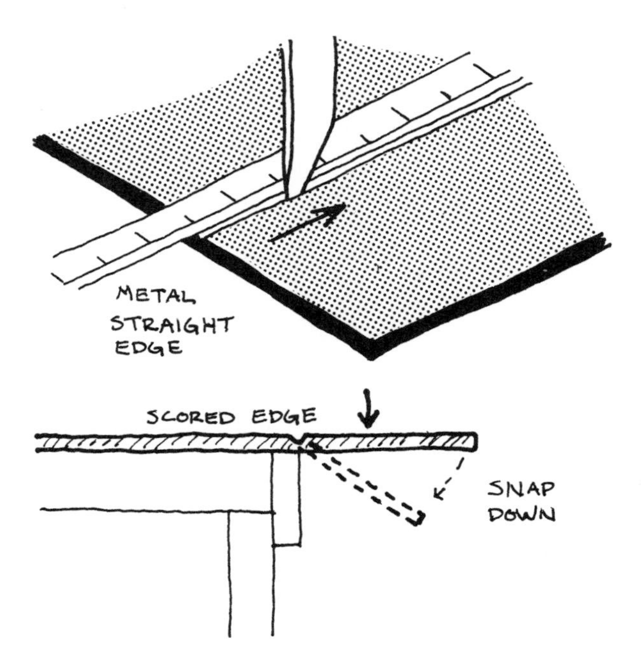

4 Apply a bead of caulk or glazing compound to the inside face of the rabbet. Now press the glass into place. Apply enough caulk so that excess caulk is pressed out when the glass is set in. This insures good adhesion and will keep moisture out.

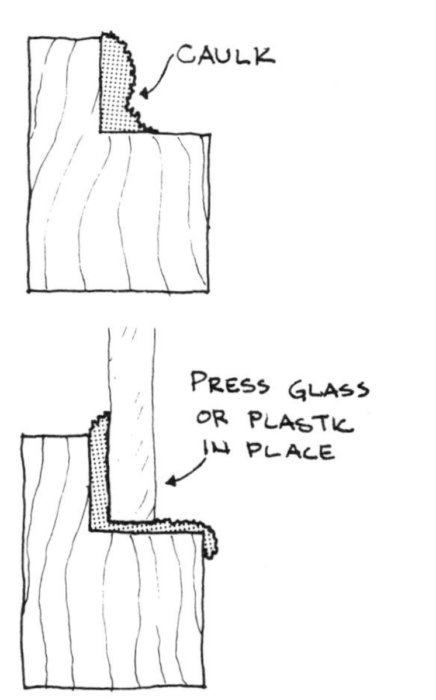

5 Now apply another bead of caulk to the glass and, using wire brads or finishing nails, tack wood strips or apply glazing compound to hold the glass in place. When you're through with this, take a rag and wipe up the excess caulk from the glass. Hang the storm with turn button screws or clips. Mount angle iron or a wood strip along the sill to stabilize the storm. Drill weep holes through the angle iron or wood to prevent water from collecting behind it. When you're through mounting the storm, make sure these weep holes are open and not clogged with dirt.

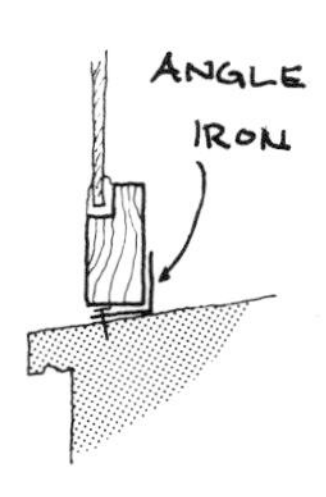

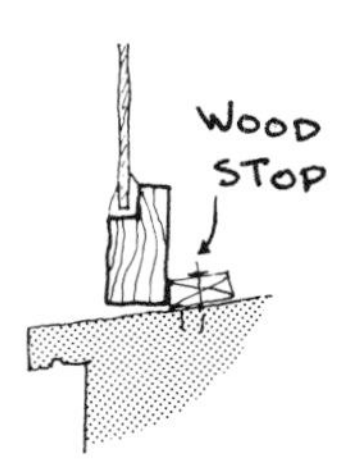

A COUPLE OF SUGGESTIONS

It would be a good idea to label the storms at some point while you're building them to avoid confusion as to where they belong. For example;

- north wall, dining room
- southeast wall, master bedroom

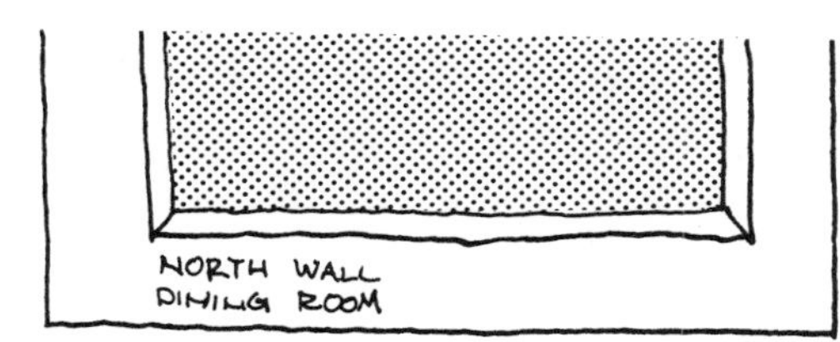

Your preliminary evaluation may include window location and designation. If it does, then mark the storms using this scheme.

Before hanging the storms, apply adhesive backed foam weatherstripping around the perimeter of the storm to seal out air leaks. Leave a gap in the weatherseal along the sill to let water escape to the outside - this is called a "weep hole" and it lets a little air between the window and the storm to keep the storm window from "fogging".

INSTALLING OPERABLE STORM WINDOWS

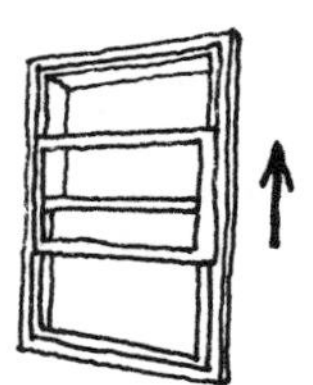

before ordering...

1 Check with the manufacturer as to how the prime windows should be measured.

- reveal to reveal ?
- frame to frame ?
- how much tolerance or "error" is allowed in measurement

Also, most storm window distributors will be pleased to supply a person who will handle installation.

2 Label the window dimensions with window locations to avoid confusion upon receiving the windows from the manufacturer.

Window
Location: kitchen, east wall,
first window

Window
Dimensions: 36-3/4" x 42-3/8"

after receiving...

1 Clean the prime window frames (see Preparation).

2 To seal air leaks, apply a continuous bead of caulk to the frame on both sides and at the top.

3 Fasten the storms with nails or screws. Screws are preferred. Use fasteners of the same metal type as the storm so as to avoid corrosion. Aluminum storms must be attached using aluminum fasteners. Steel storms must be attached with galvanized or exterior type fasteners. If the storm and prime window sash are made of different metals, install the storms so that they do not contact the prime window. Alternately, use non-metallic spacers and coated fasteners to prevent corrosion by galvanic action.

STEEL SCREWS + ALUMINUM FRAMES = RUST

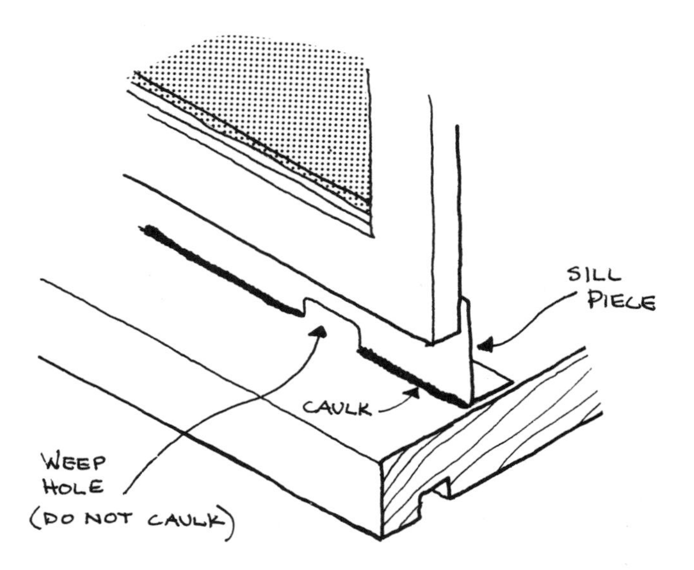

4 Install the sill piece to close the gap which would exist between the storm sash and the sill using fasteners as above. Caulk along the sill, but do not caulk continuously all the way across. Leave gaps in the bead where vents or weepage holes exist.

14

install window insulating shutters

description

Insulate a window? Essentially, that's what thermal shutters do. Even though the previous window retrofits reduce window heat loss, a significant amount of energy is still lost through windows. For example, the "R" factor* of a typical, uninsulated, wood frame wall is about 4.0 (with insulation, about 15.0). On the other hand, a single glazed window with a storm window has an "R" value of 2.6. If a thermal shutter is added to the window, the "R" value can increase to 9.0, depending on the type and thickness of insulation used, when the shutter is closed. A thermal shutter can increase the "R" value of a window well beyond that of an uninsulated wood wall --

*Resistance to heat flow.

almost to the point of an insulated wall.

A thermal shutter is most easily constructed of rigid insulation and, if located on the exterior, must be protected by a weather resistant material.

The shutter is operated (opened and closed) on a daily basis. It is opened during the day to let sunlight in and provide a view. At night, the shutter is closed to minimize heat loss. The shutters are mounted on the window frame with hinges or magnetic clips.

materials

rigid insulation

- polystyrene
- polyurethane
- glass fiber

weatherstrips

- spring metal
- adhesive backed foam
- adhesive backed felt

wood

- stops (as required)
- framing lumber

tape

- vinyl
- polyethylene
- aluminized edge tape (for exterior shutters)

magnetic clips* and strike plates

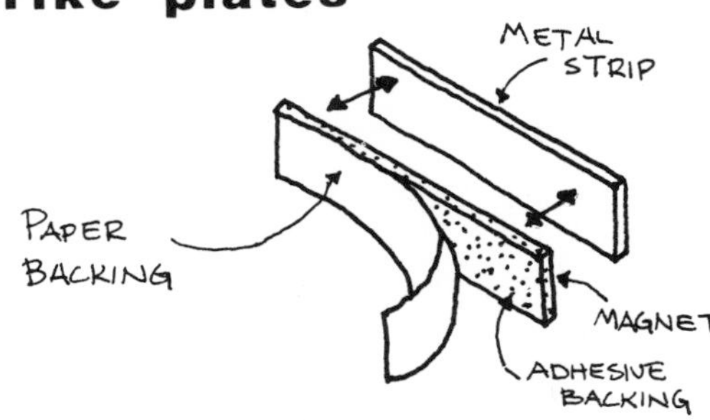

other materials

- heavy duty clothesline rope

- pulley with surface mounted housing

preparation

Before beginning, carefully look over the shutter designs presented in this chapter and compare them along with the installation techniques to the existing window condition. For example, is there a shade, drape, or blind that will have to be removed or taken off the window to accommodate the shutter? Or will wood blocking have to be added to the inside casing to assure a continuous contact surface around the window? One design may be more appropriate than another, or you may have to make certain modifications to the shutter for it to work on a certain window.

The three types of shutters discussed in this chapter are described on the following page.

---

* Zomeworks Corp.
P. O. Box 712
Albuquerque, N. M. 87103

FOLDING LEAF

- Shades and blinds must be taken off
- Curtain rods may have to be lengthened
- Window handles, if they protrude out from the window, will have to be removed and reinstalled
- Shutter is held in place by magnetic clips, therefore it can be taken down and stored
- Framing the shutter with wood is not necessary

SIDE HINGED INTERIOR SHUTTERS

single panel

- Drapes have to be moved to one side*

double panel

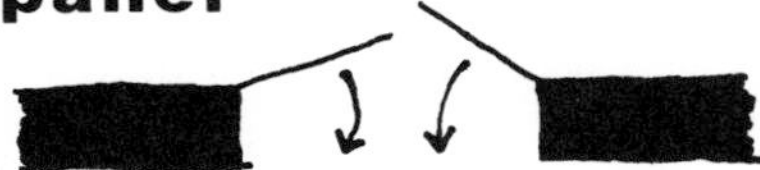

- Drapes have to be removed*
- Swing distance not as great as single panel

accordian type

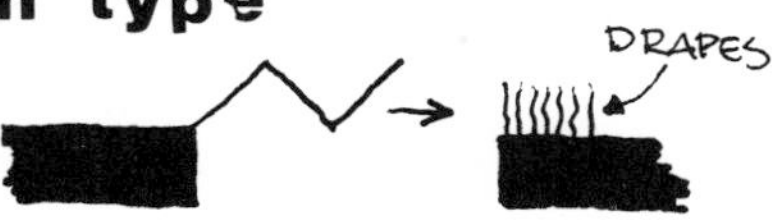

- Drapes have to be moved to one side*
- Good for large windows
- Requires relatively little wall space for storage

SIDE HINGED EXTERIOR SHUTTERS

- Should be built by a person skilled in carpentry
- Care must be taken to weatherproof the shutter

*Note: drapes may be mounted onto the shutter panel

- Construction techniques are similar to side hinged interior shutters
- Exterior shutters offer the advantage of shading during the summer
- Can be hinged at the top (south facing window) or on the side (east and west facing windows)

installation procedures

FOLDING LEAF SHUTTER

The "folding leaf" shutter is the shutter pictured on the first page of this chapter. It's recommended that glass fiber insulation board be used for this as well as for other interior shutters. Otherwise, the insulation should be protected with a fire resistant material.

The top leaf folds down over the bottom leaf, allowing a view out the window. On a bright, sunny day, the entire shutter can be taken down and stored (the shutter is held in place with magnetic strips, rather than hinges).

Insulation is wrapped with a fabric and sealed with vinyl or polyethylene tape. The tape is also used to "hinge" the two leaves together.

1 The shutter is attached to the lower sash of a double hung window. Thus if there is any hardware, such as handles, that protrude out from the sash, they may have to be removed (alternately, the insulation can often be notched to accommodate the hardware).

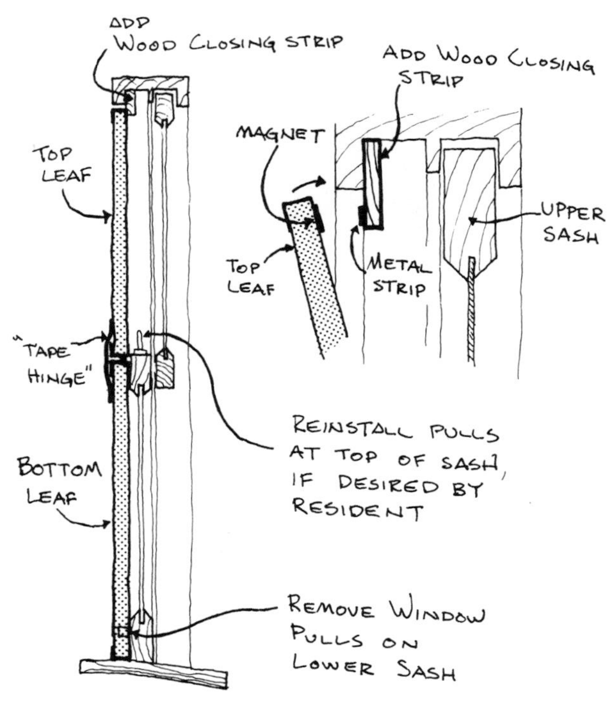

2 Measure the width and height of the window. Measure the width from the inside faces of the two stops. Measure the height from the top of the sill to the bottom face of the upper stop. You want to transfer the shape of the window opening to the insulation board, therefore, measure the width and height of the window at different spots to obtain a good profile of the window opening.

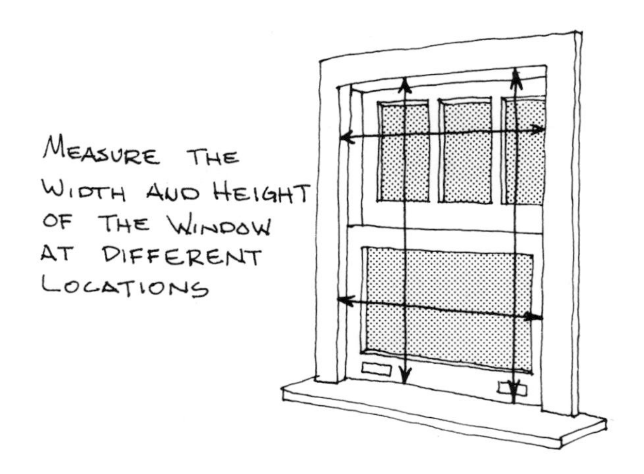

3 Locate the dimensions from Step 2 on a sheet of rigid insulation board. To allow for clearance, subtract 1/8" from the width and height dimensions. Cut the insulation with a utility knife and a straight edge. Run the blade into the cut with several strokes to cut through. Always keep the knife in the existing cut and perpendicular to the board to produce a clean square edge. Cut the board outside, in a garage, or in a basement to avoid getting the shavings over everything.

4 Check the fit of the insulation by setting it in place over the window

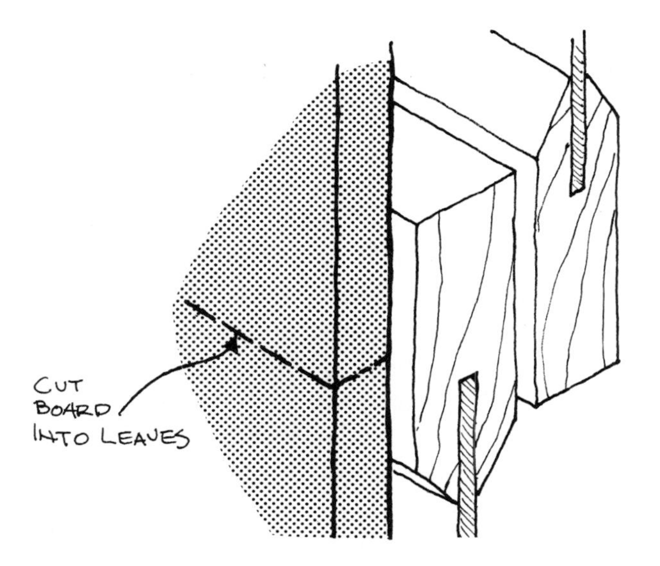

(trim if necessary). Now mark the top of the lower sash on both sides of the insulation. Cut across the insulation at these points to get your top and bottom leaves. Mark "top" and "bottom" on the appropriate leaf to avoid confusion later on. Depending on the insulation board dimensions, separate leafs may be cut from separate boards to minimize material usage.

AN IMPORTANT NOTE:

Rigid fiberglass insulation often has an asphalt coating on one side. Cut through the board from the opposite side and build the shutter so that the asphalt coating faces <u>in</u> towards the room.

5 For appearances sake, as well as to keep the edges of the insulation from fraying, the insulation should be wrapped with a fabric. You may have a fabric that matches (or complements) other interior finishes. Lay the top and bottom leaves over the fabric and cut two pieces of fabric for each leaf. Cut the fabric so that it does not drape over the edges of the leaves. A vinyl wall covering is easy to work with and is available in several colors.

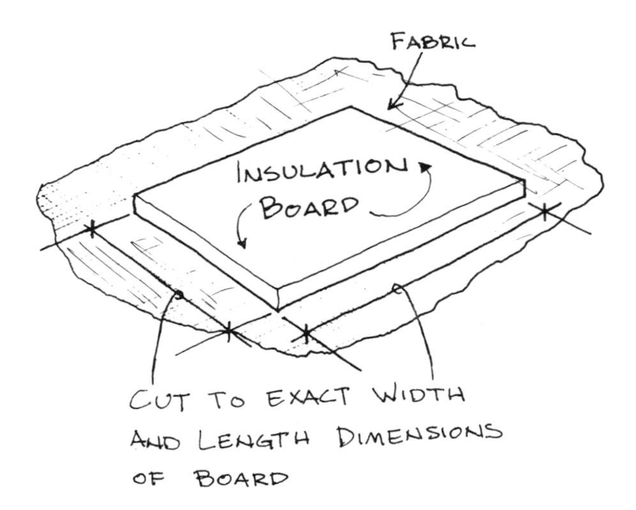

6 Tape the pieces of fabric together and to the edges of the leaves with a heavy vinyl tape (polyethylene tape, or weatherstrip tape, will also work). Lap the tape over the edges of the cover at least 3/4".

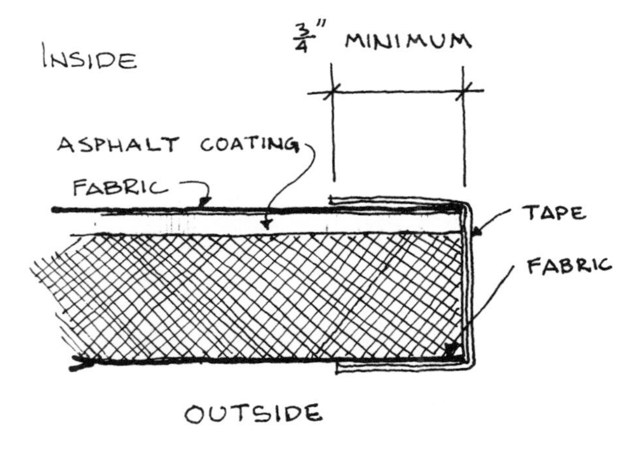

7 Now it's time to fasten the two leaves together. Use the same

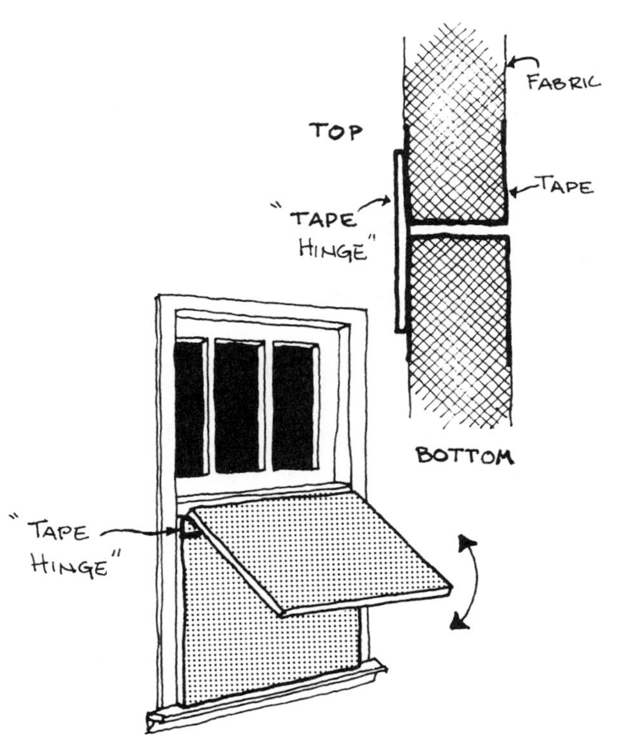

vinyl or polyethylene tape to tape the top and bottom leaves together. Apply the tape to the side that faces into the room. This "tape hinge" is quick and cheap and it seals the crack between leaves, but pay attention to how durable it is. You will probably have to reinforce the hinge with "crossover" strips so that the upper leaf can hang from the lower leaf when it is folded open.

NOTE:

For shutters in windows that face south or west, the surface of the shutter that faces outside should be light colored (preferably, white or silver). In this manner, the shutter can be used during the summer to reflect heat back outside and keep it from building up between the glass and the shutter. It would also be helpful to open the window prior to closing the shutter so the heat will vent out.

8 Magnetic clips* can be used to hold the shutter in place. The clips have an adhesive backing and come with metal strips. Attach the metal strips to the sash with silicone caulk. Place the clips over the metal and then remove the backing to expose the adhesive.

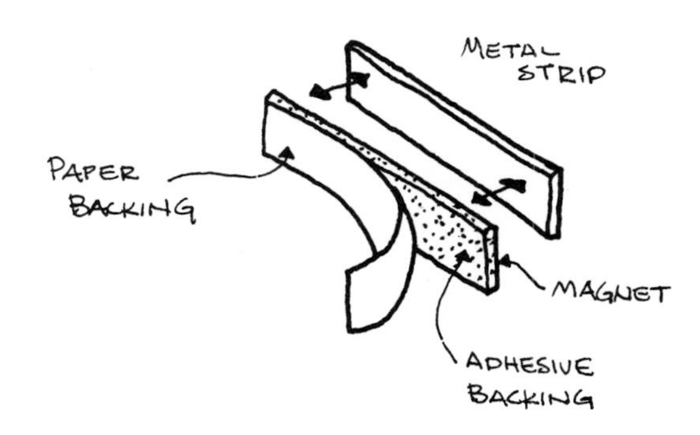

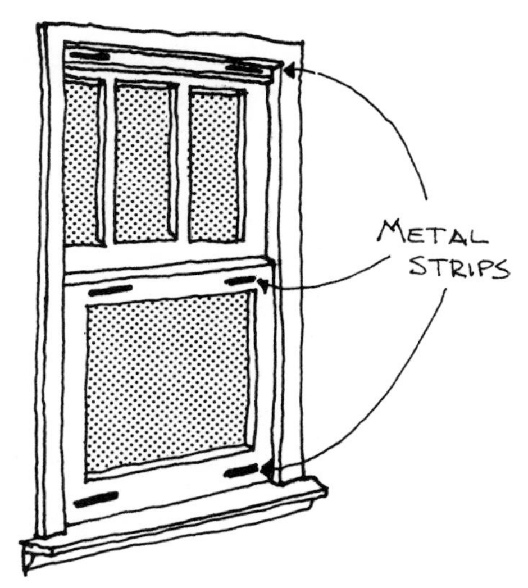

9 Place the shutter in position and, with your fist, strike the shutter at the points where you've located the clips. This is a quick way to locate the magnetic strip which is (after striking) adhered to the insulation, exactly over the metal strip attached to the sash.

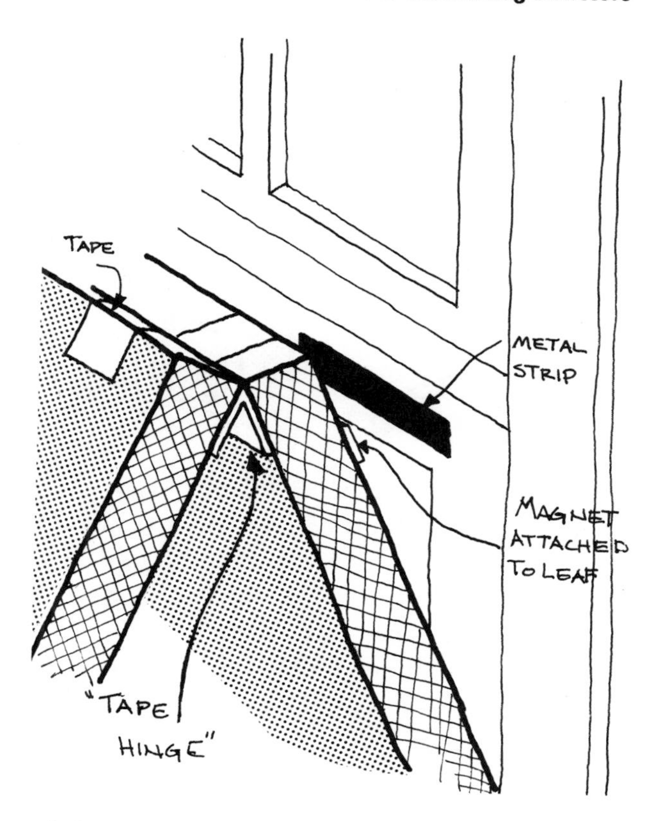

10 Use the shutter just as if there were a window shade which rolled up at the bottom of the window. That is, leave the bottom leaf in place for privacy. Open the top for natural light and view. If sash handles are left in place or re-attached to the top of the lower sash, the lower sash can be raised for ventilation or egress without necessarily having to remove the shutter. For this reason, you may find it easier to attach the lower leaf "semi-permanently" with bolts set

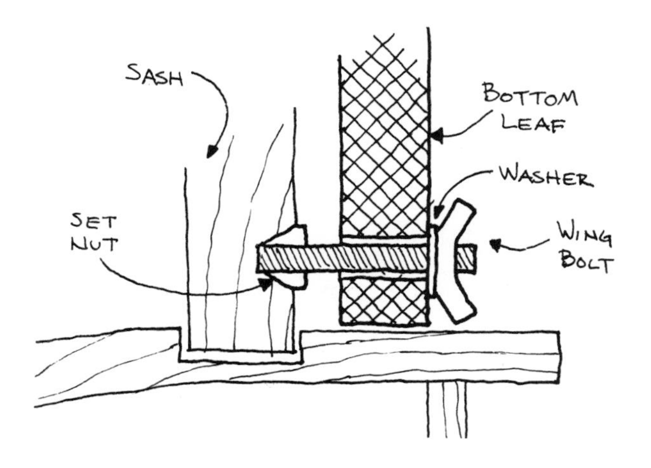

---

*Zomeworks Corp.
P. O. Box 712
Albuquerque, N. M. 87103

into the sash (at four corners of lower sash) and wing nuts to hold the insulation onto the bolts.

SIDE HINGED INTERIOR SHUTTERS

This shutter is basically a "door" (or set of doors) located over a window. Three variations of interior side hinged shutters - single panel, double panel, and accordian type - are "permanently" mounted with hinges (unlike the folding leaf shutter which is held in place with magnetic clips). However, magnetic clips can be used along the closing stile to hold the shutter closed. If the width of the window is such that a single panel shutter will create a clearance problem, a double panel or accordian type shutter can be used. All these shutters are framed with wood strips and faced with plywood or masonite. These shutters can be painted or covered with a fabric to match the interior decorations.

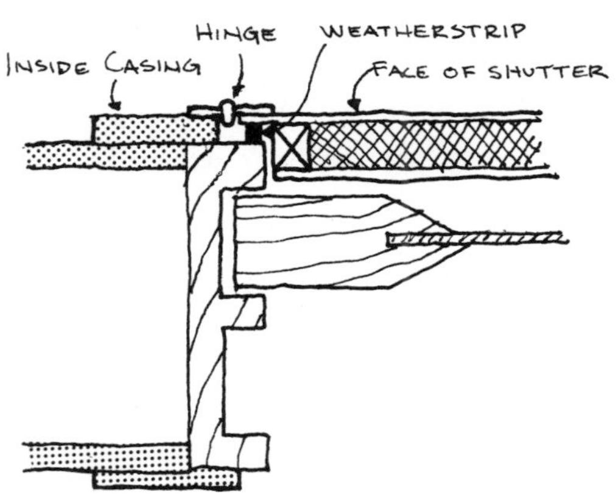

There are two methods for attaching hinged shutters. The first method, and the easiest, is a "flush" mounted shutter where the outside face of the shutter is flush with the inside casing.

The second method is a "block" mount where wood strips are attached to the inside casing around the entire window.

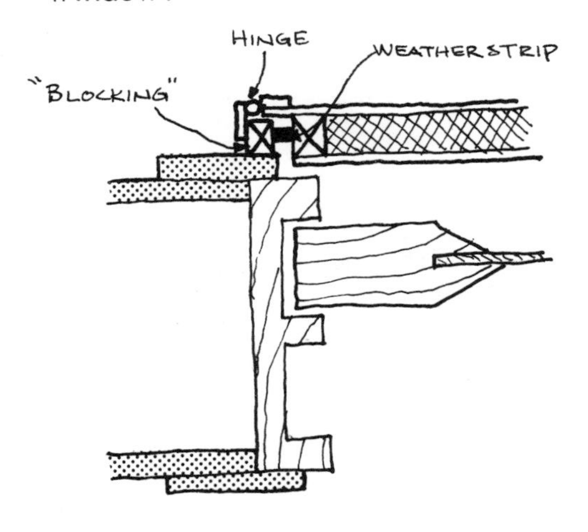

How do you know what type of mount to use? First, determine the distance between the top edge of the inside casing and the sash--if the thickness of your insulation is greater than this dimension, you'll need to add blocking.

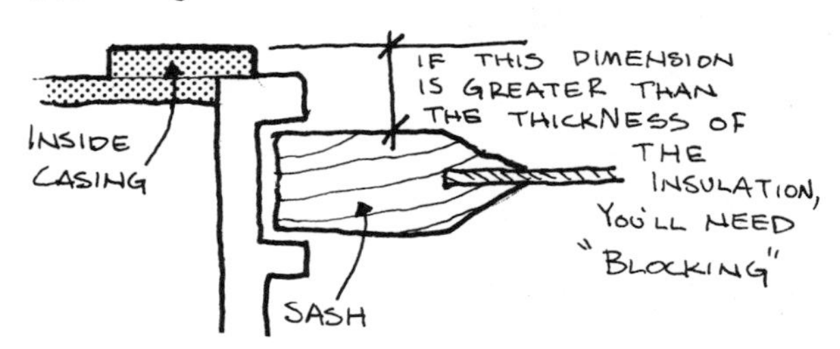

Second, if there's a shade, or some other window ornamentation, within the window frame dimension that you want to keep, you'll need to "block out" to get the shutter over it.

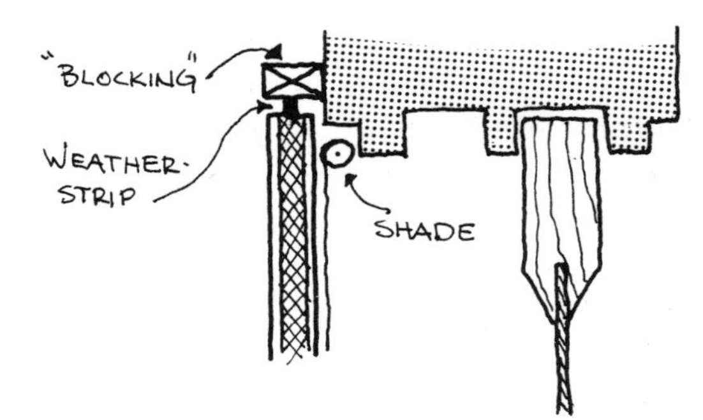

Finally, on most casement, awning, hopper, jalousie, and on some double hung and horizontal sliding windows, the operating mechanisms extend out beyond the window sash making blocking necessary.

If you're "flush" mounting shutters, begin with Step #3.

1 Strips of 2" x 2" wood attached to the inside window casing will probably provide you the necessary

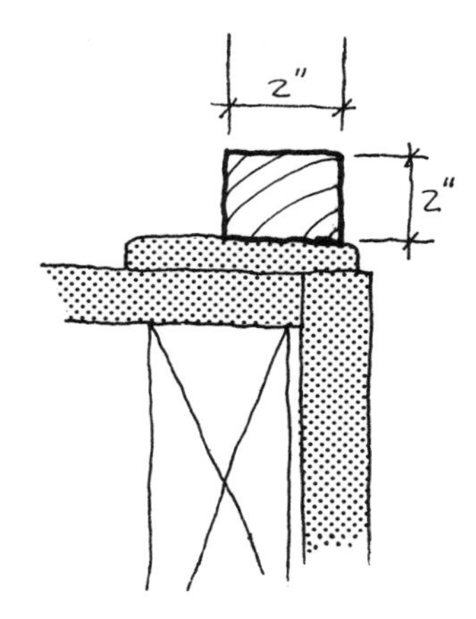

clearance for the shutter to close over the window. A thinner strip of wood will work along the sill. With

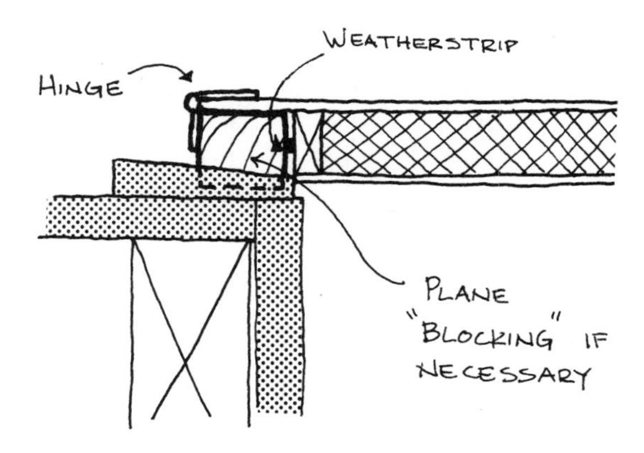

some windows, the casing and/or sill is tapered. In this situation, cut or plane the wood strips so that surfaces of the strips are parallel to the surface of the wall.

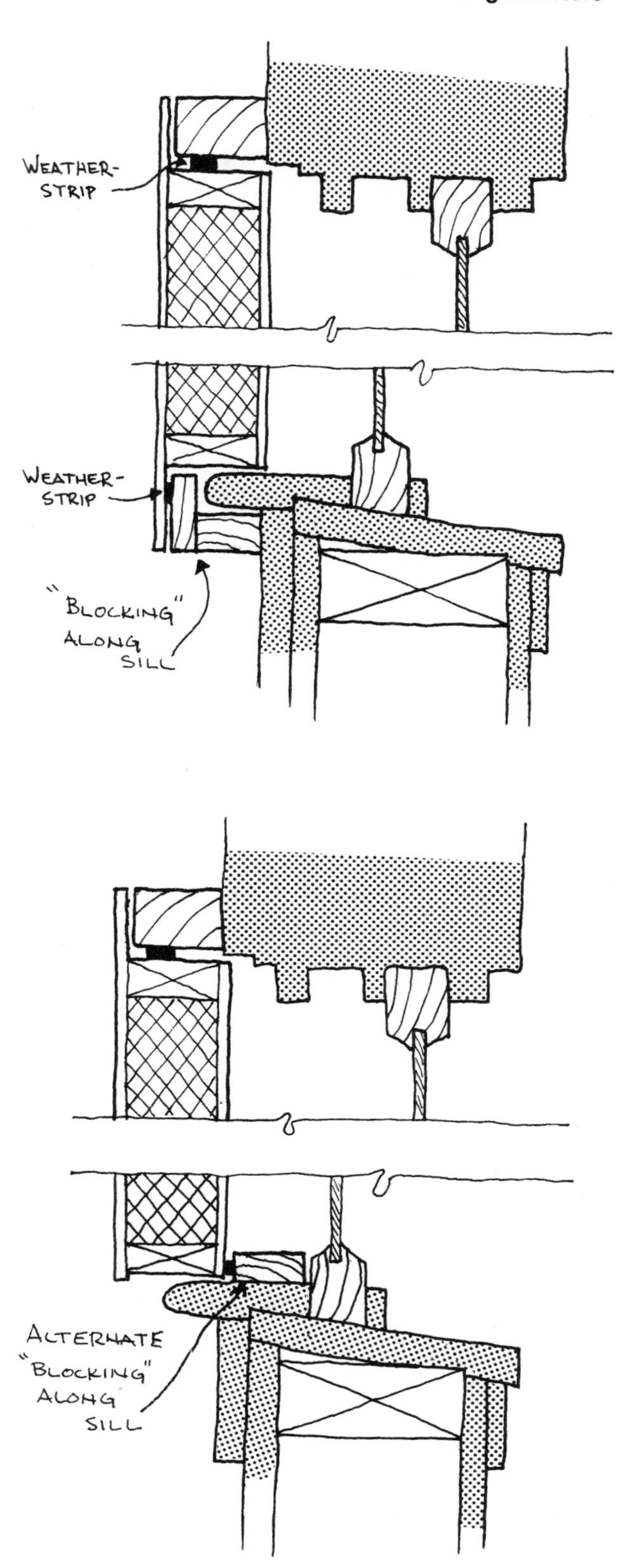

2 For a neater appearance, miter the corners of the wood strips. Attach the strips with finish nails.

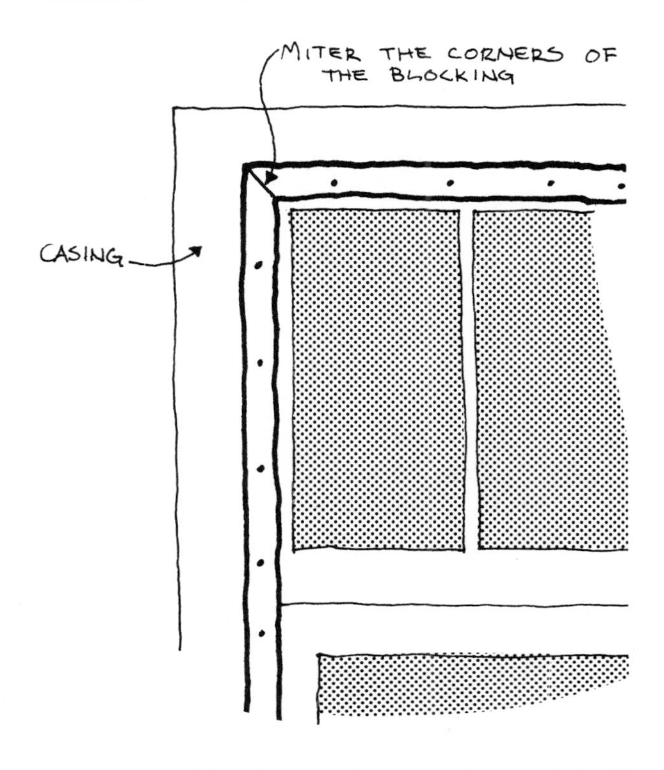

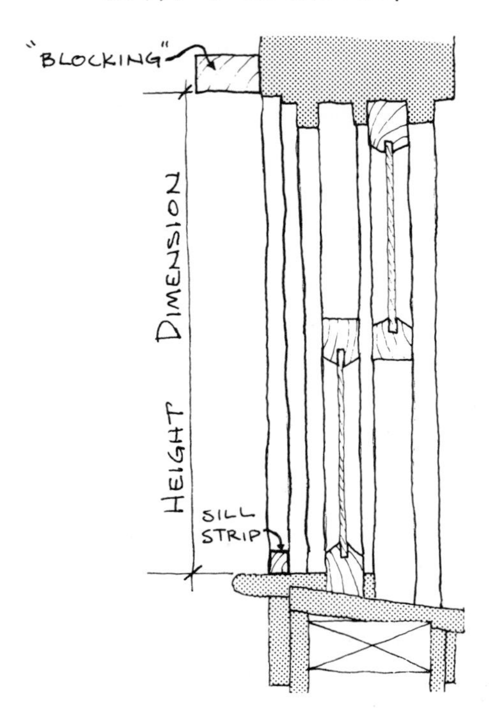

NOTE:

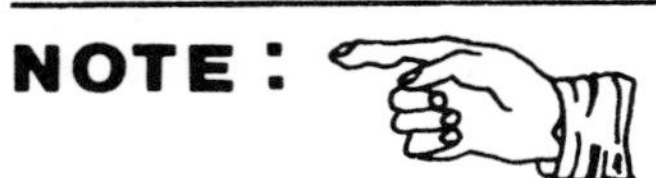

If you're going to paint the strips, do it before you nail them to the casing.

3 Measure the opening to determine the dimensions of the insulation. If you added blocking, measure the width from the inside face of the blocking and the height from the sill to the bottom face of the wood blocking along the head. If you're flush mounting the shutters, measure the width from the inside face of the stops and the height from the sill to the bottom face of the stop along the head.

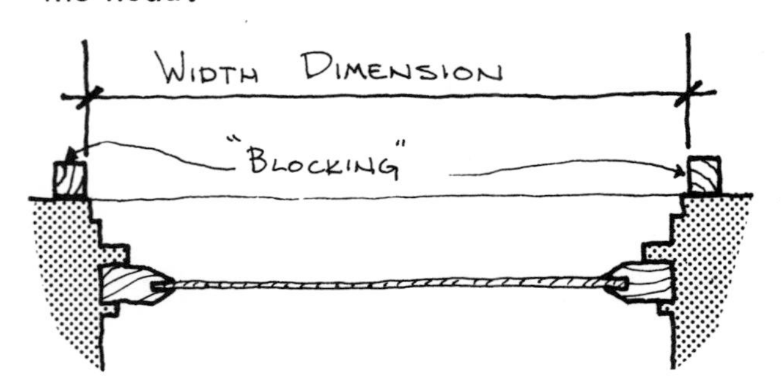

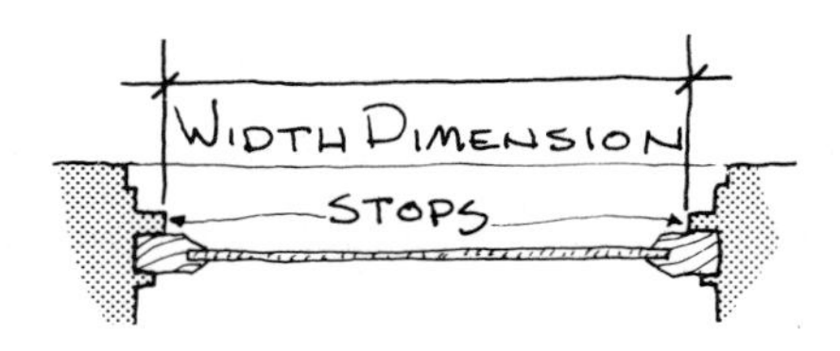

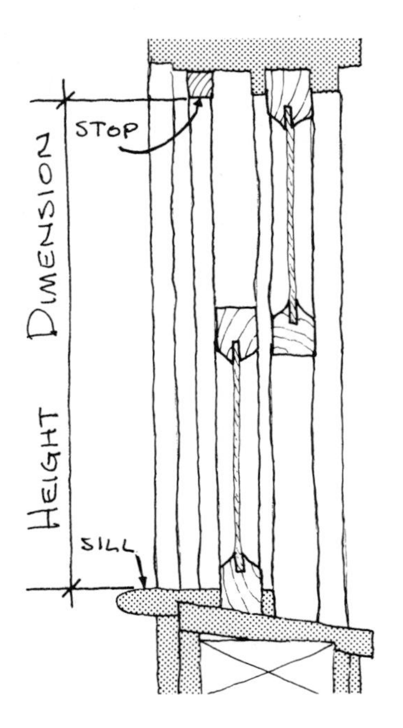

4 From these dimensions, you'll have to subtract the necessary clearances for the shutter frame plus an additional 1/4" on each edge. Furthermore, if you're building a double panel or an accordian type shutter, you have to subtract additional clearances for extra framing strips. Cut the insulation as described in step 3 under "Folding leaf shutter".

HEIGHT CLEARANCES
(applies to all hinged shutters)

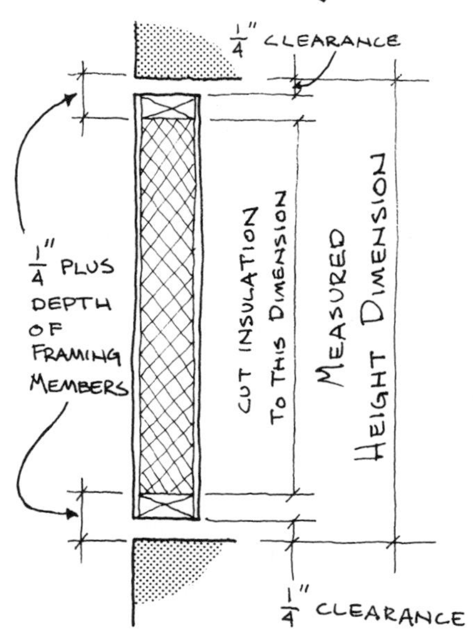

WIDTH
(single panel)

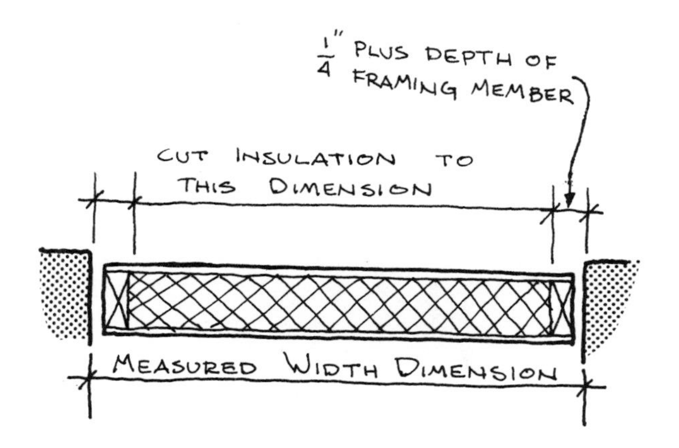

WIDTH
(double panel)

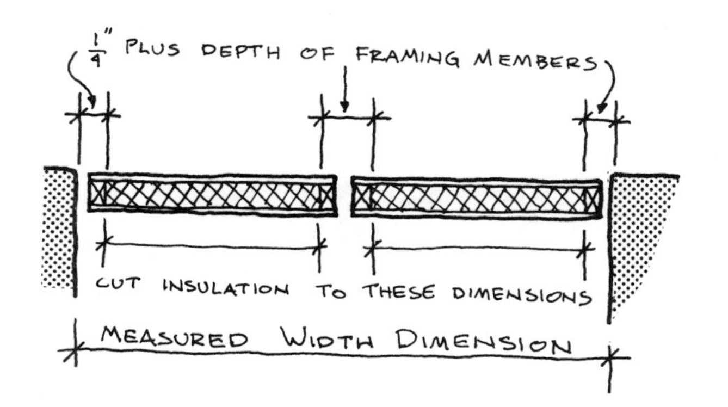

WIDTH
(accordian type)

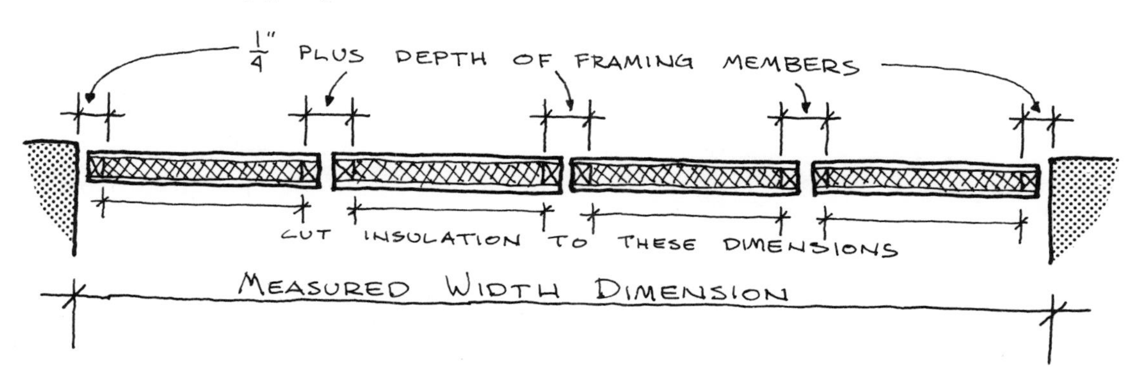

5 Cut and attach the wood framing members. Depending on the thickness of the insulation, use 1" x 1", 1" x 2", or 2" x 2" wood strips. Use nails and glue.

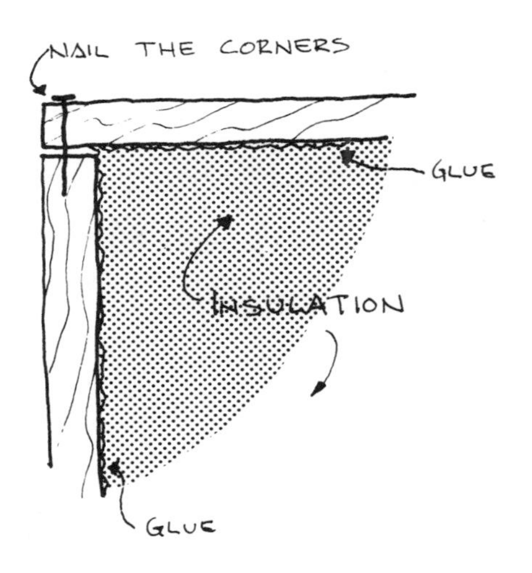

6 Cut 1/4" plywood or masonite sheets for the facing. The sheet used to cover the back should be cut to the same dimensions as the shutter. Cut the front sheet with enough excess around the edges to allow for the addition of weatherstrips that will close against the blocking and/or stops when the shutter is closed. Nail and/or glue the facing sheet to the framing members.

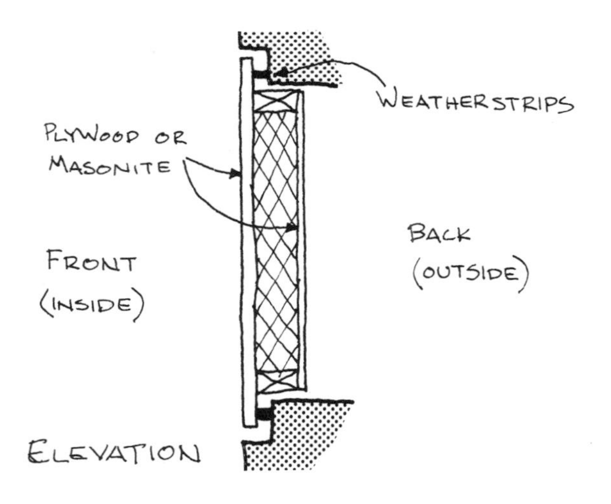

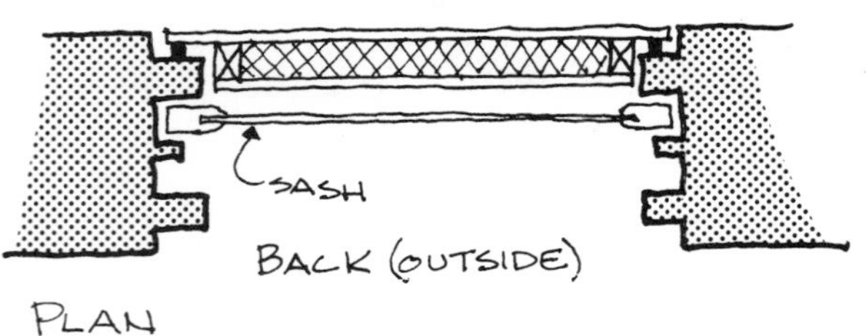

7 Screw the hinges to the shutter. If you used blocking, you'll need to bore recess holes in the blocking to accommodate the screws. Then attach the hinges (shutter) to the blocking or casing.

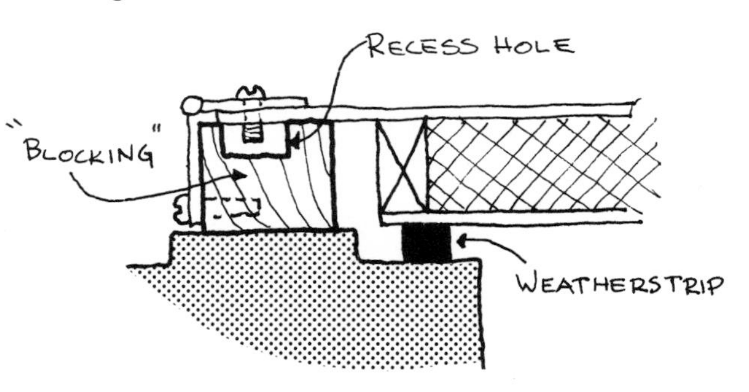

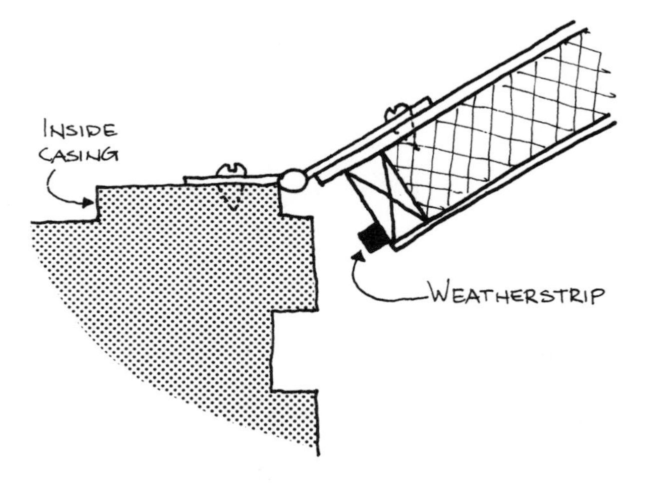

NOTE:

Note that the hinge must pivot from a flat surface away from the frame to allow the shutter to swing a full 180° back against the adjacent wall to "get it out of the way" when opened.

8 Use magnetic clips or latches to hold the shutter closed. "Turn buttons" may also be used but they should press against a screw head to protect the shutter wrap and finish.

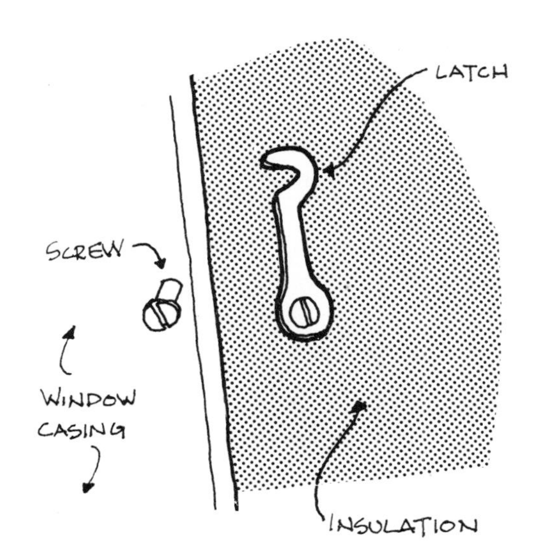

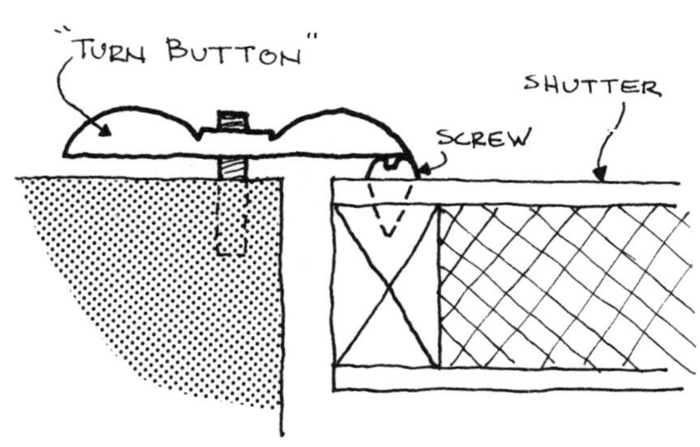

SIDE HINGED EXTERIOR SHUTTERS

Operable exterior shutters were common on many old buildings. The shutters were not used primarily for decoration, but rather to moderate the interior temperature (heating systems were not as developed as they are today and air conditioning had not yet been invented). By closing the shutters, the occupants reduced both winter heat loss and summer heat gain through the window, making the interior temperature more comfortable. Unfortunately, most (if not all) exterior shutters that you see today can only provide a decorative effect.

The exterior shutter described in this chapter lends itself well to both side and top hinging (side hinge is preferred for east and west facing windows, top hinge is preferred for south facing windows). This shutter should be built only by a person skilled in carpentry who understands what the construction and operation of this shutter entails.

1 Construct two shutter panels (each panel being half of the window area) following Steps 1 through 6 for "Side hinged

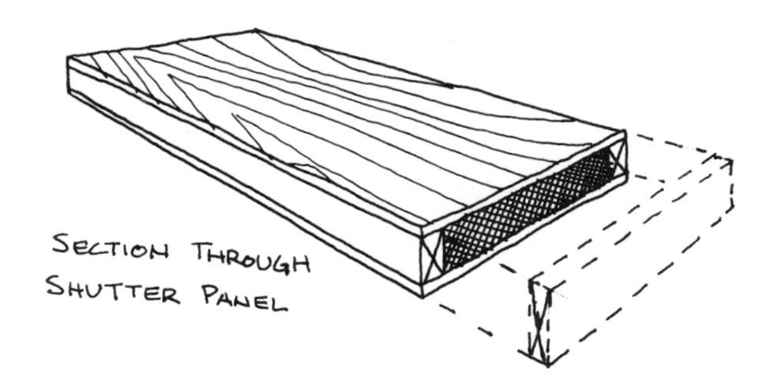
SECTION THROUGH SHUTTER PANEL

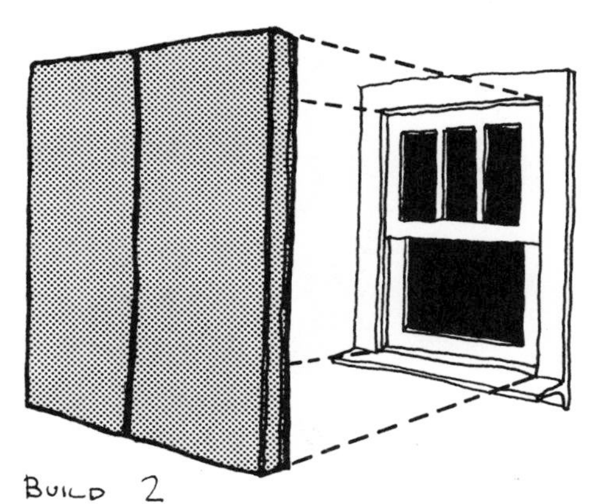
BUILD 2 SHUTTER PANELS TO FIT OVER THE WINDOW

interior shutters" in this chapter. The facing, however, should be flush with the sides of the shutter. Also, use a weather resistant material when facing the shutter. An exterior type oil or urethane base paint may be sufficient exterior protection.

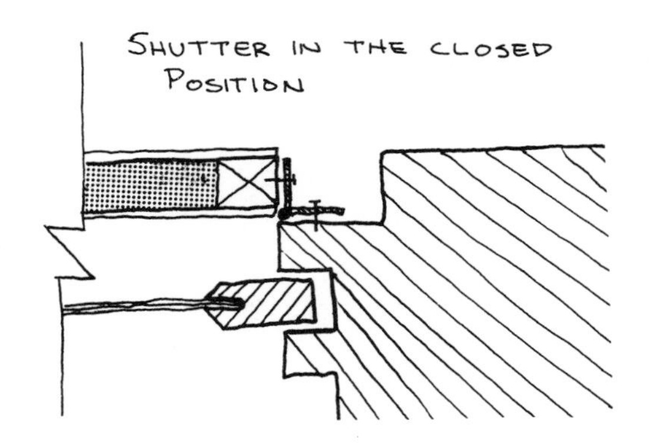

2 Join the two panels by mounting two hinges on the front of them (the side that faces away from the window).

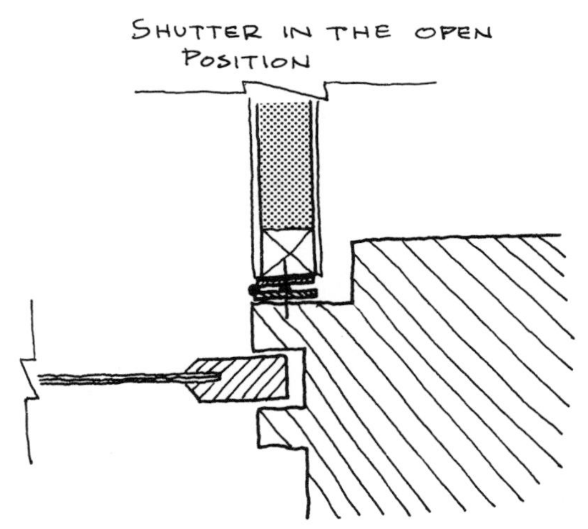

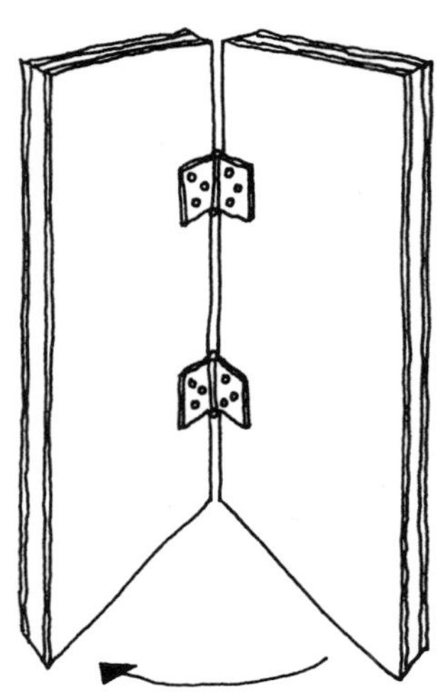

3 Attach the shutter to the window frame, again with two hinges. Allow enough clearance for the shutter to open and close without binding against the window casing. For west and east facing windows, mount these hinges on the north side of the window.

4 After checking to insure that the shutter opens and closes properly, attach a corner bracket (minimum 3" x 3") to the edge of the panel that is not mounted to the window casing. Attach the bracket near the middle of the panel.

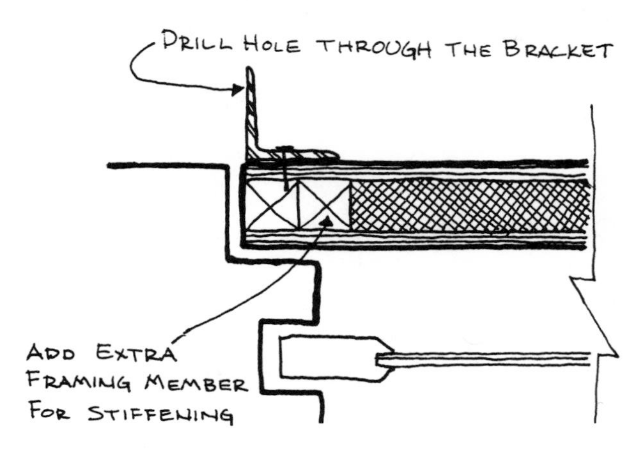

Drill a hole in the bracket so that a heavy, nylon clothesline rope will fit through it.

NOTE: 

To stiffen the shutter along this edge, you may want to add an additional framing member here.

5 At the same level of the bracket, drill two holes through the wall--one on each side of the window. Drill the holes to the back of the window casing. Cut two 1" diameter chrome pipes to the width of the wall section and insert them in the openings. Caulk around the pipes to seal them against the wall.

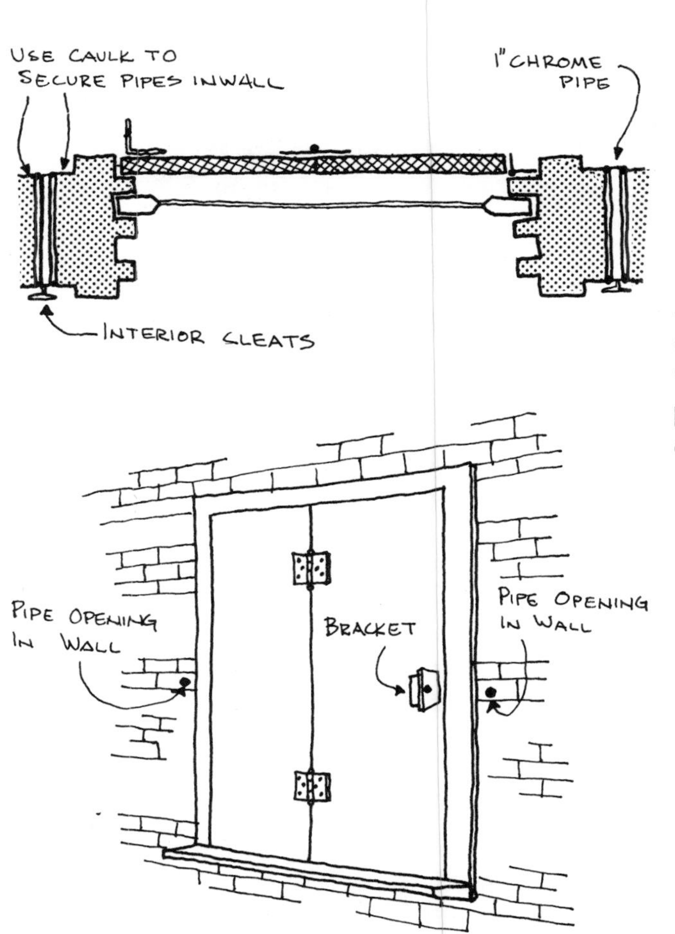

6 Locate two cleats on the interior beneath the pipes. Beginning with one

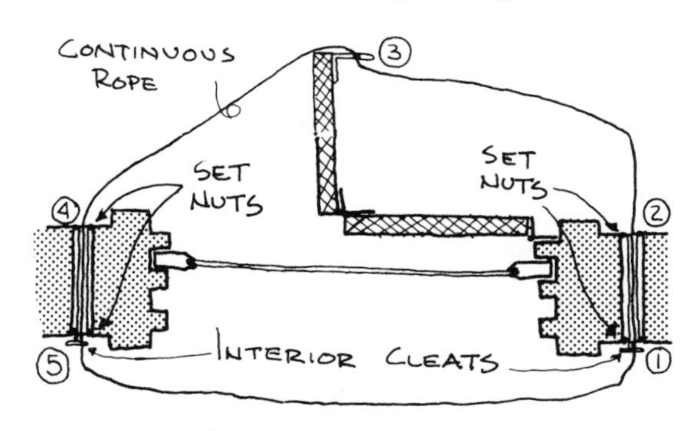

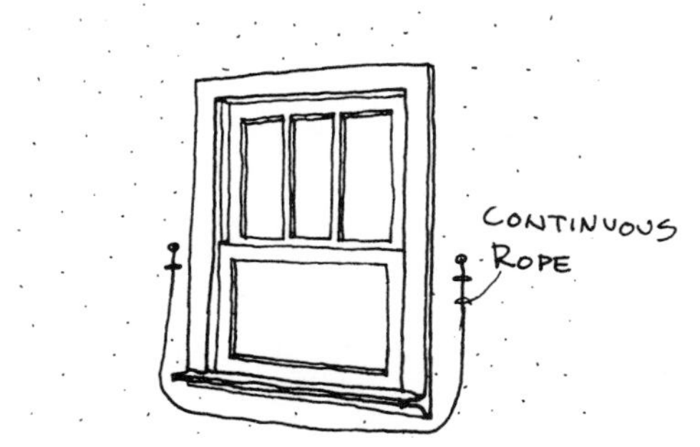

cleat (1), thread the clothesline through one pipe (2), the bracket (3), back through the other pipe (4), around the other cleat (5), and back to the cleat where you began (1). Tie the rope together here. A continuous rope like this will keep it from pulling outside. As you're doing the threading, slip set nuts over the clothesline and tap them in place over the opening of the pipes. The set nuts will prevent the clothesline from fraying.

7 With the shutter fully closed, tie off the rope on the cleat that is located on

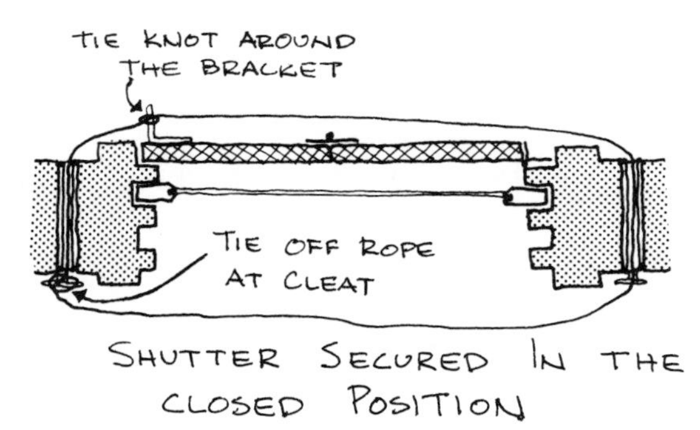

SHUTTER SECURED IN THE CLOSED POSITION

the same side as the bracket and then tie a knot around the bracket. This will keep the shutter tight against the window. Note that the knot around the bracket is permanent.

8 To open, untie the rope at the cleat and pull the shutter. Then tie off the clothesline on the other cleat to keep the shutter drawn against the wall to prevent the wind from banging it back and forth.

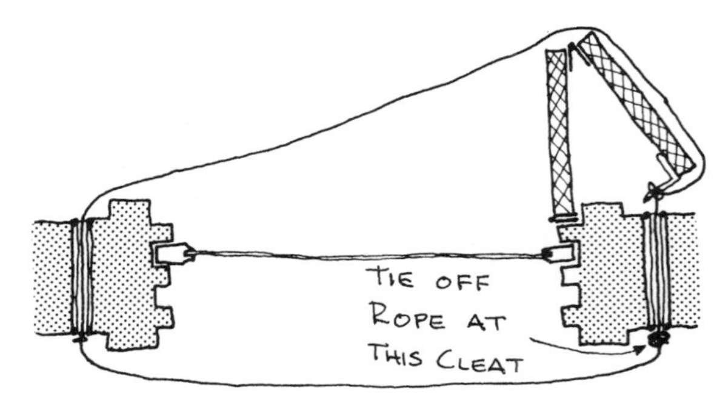

SHUTTER SECURED IN THE OPEN POSITION

9 Install rope guides or pulleys over the exterior opening of the pipes. This will keep the clothesline from binding against the edge of pipe when the

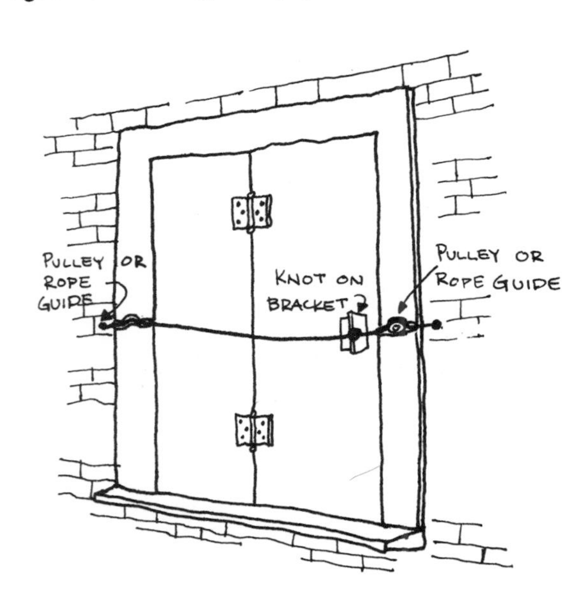

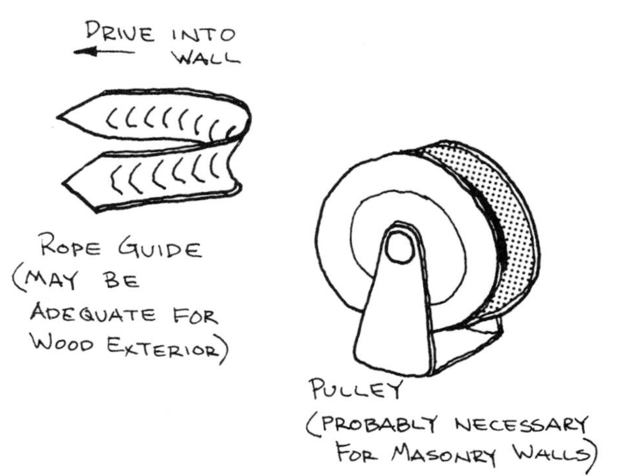

shutter is being operated. Also, to assure a continuous contact between the shutter and window frame, apply weatherstripping to either the frame or shutter.

top hinge shutter

Follow the same procedure for hinging the shutter on the side of the window,

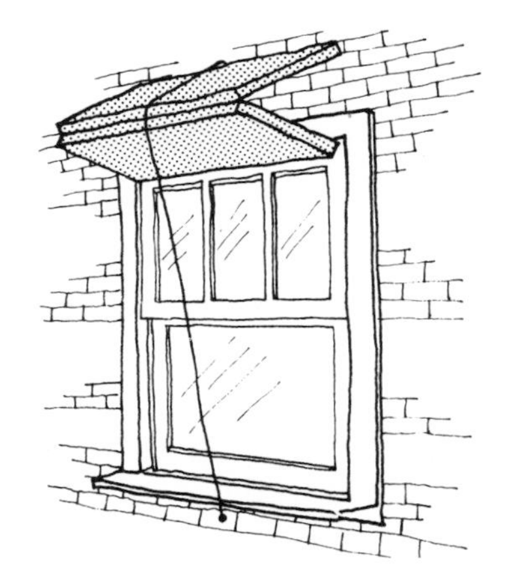

except mount the hinges at the window head. In this manner, the shutter will provide an "awning" to keep sunlight off of a south facing window in summer.

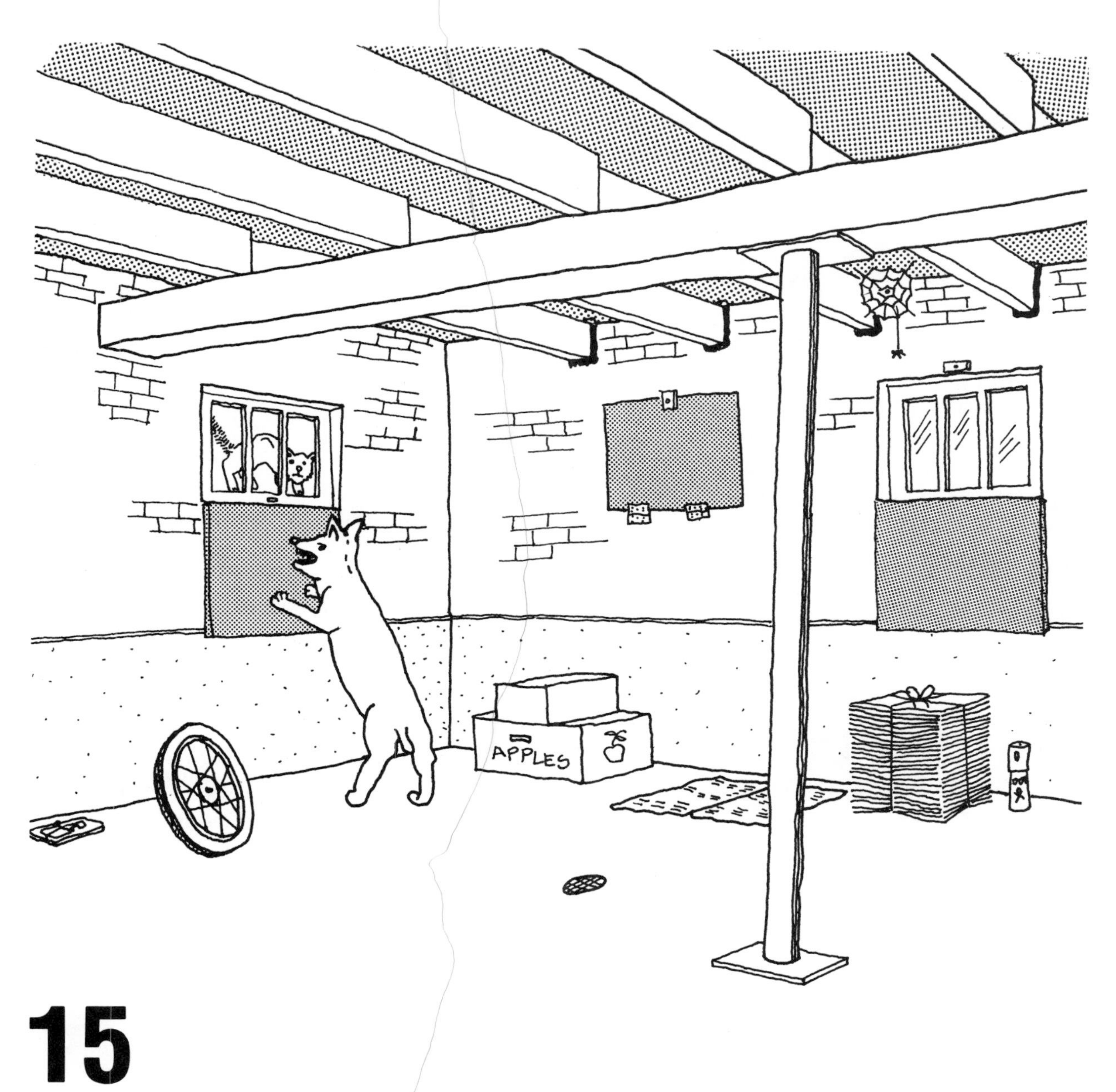

15

install window insulating panels

description

There may be windows in your home that don't warrant the time and effort in building insulating shutters. Insulating panels can be used in this case – they are just as effective as insulating shutters, but easier to make.

The difference between insulating shutters and insulating panels is that shutters can be opened and closed on a daily basis where as panels are used on a seasonal basis. That is, the panels are set in place over the windows at the beginning of winter and then are removed and stored over the summer.

Windows that are prime candidates for insulating panels are windows that are not needed for light, view, or emergency exit purposes. Also, insulating panels can be effective over windows that should in effect, be replaced, but cannot be. Likely locations of these windows are basements, garages, attics, stairwells, and pantries.

In its simplest form, an interior insulating panel can be nothing more than a piece of rigid insulation held in place by friction. That is, rigid insulation is cut to fit the window opening and pressed tightly into the opening. Alternatively, turn buttons, wing nut bolts, coverings, etc., can be used for a more thorough job. On the exterior, the panel is framed with weather resistent materials.

materials

rigid insulation

- polystyrene
- polyurethane
- glass fiber

weatherstrips

- spring metal
- adhesive backed foam
- adhesive backed felt

wood

- stops and framing lumber (as required)
- panel facing

tape

- vinyl
- polyethylene
- aluminized edge tape (for exterior panels)

other materials

- turn buttons
- set nuts
- wingnuts

preparation

The panel should fit snugly against the window sash. If there are any obstructions (window pulls, shade, etc.) that would prevent this, they should be removed. If these obstructions can not be removed, or if you do not want them removed, you'll have to "block out" from the window (see Chapter 14, "Install Window Insulating Shutters", for details).

Before you begin, compare the window with the various panel designs presented in this chapter and determine if any modifications have to be made. Be sure you understand the type and use of the panel that you're going to install.

installation procedures

1 Begin by cutting a piece of rigid insulation to fit within the window opening. Measure the width from the inside faces of the two stops. Measure the height from the top of the sill to the bottom face of the upper stop. You want to transfer the shape of the window opening to the insulation board, therefore, measure the width and height of the window at different spots to obtain a good profile of the window opening.

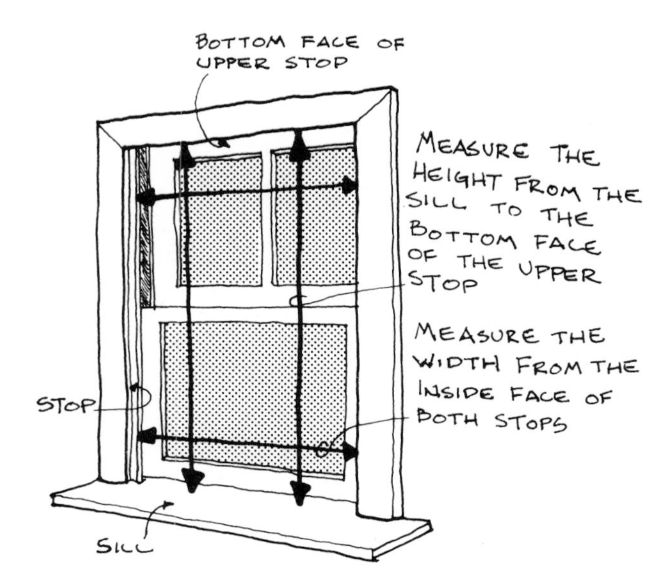

2 Locate the dimensions from step 1 on a sheet of rigid insulation board. Cut the insulation with a utility knife and a straight edge. Run the blade into the cut with several strokes to cut through. Always keep the knife in the existing cut and perpendicular to the board to produce

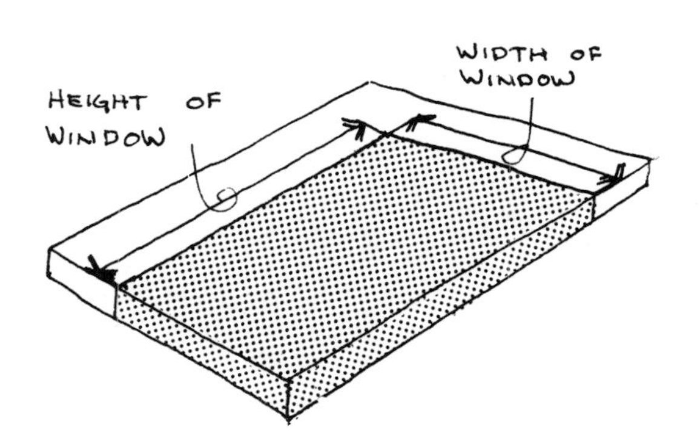

a clean, square edge. Cut the board outside, in a garage, or in a basement to avoid getting the shavings over everything.

3 When placing the panel on the interior of the window, wrap it with a vinyl wall covering or canvas (you may need to cut an additional 1/8" off the edges of the panel to allow for clearance). Cut two pieces of fabric to the exact dimensions of the panel. Tape the pieces of fabric together and to the edges of the panel

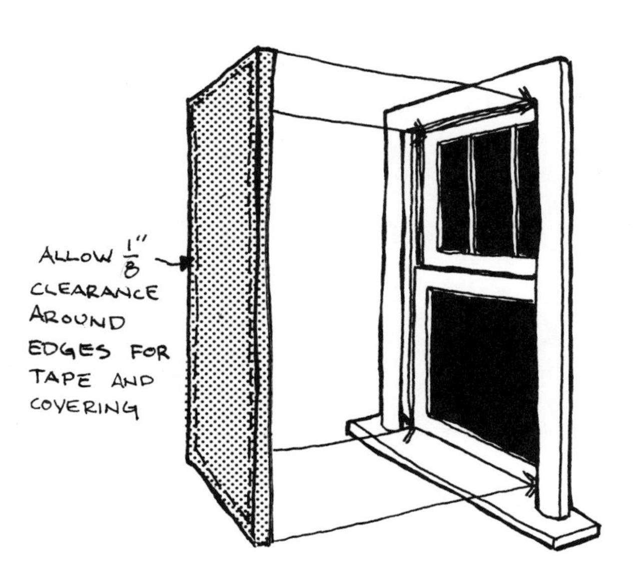

with a heavy vinyl tape (polyethylene tape, or weatherstrip tape, will also work). Lap the tape over the edges of the fabric at least 3/4". For additional details, see Chapter 14, "Install Window Insulating Shutters".

NOTE:

For panels in windows that face south or west, the surface of the panel that faces outside should be light colored (preferably, white or silver). This way, the panel can be used during the summer to reflect heat back outside and keep it from building up between the glass and panel. It would also be helpful to open the window prior to closing the panel so the heat will vent out.

---

4 When locating the panel on the exterior, frame the panel with wood strips (1" x 1", 1" x 2" - depends on the thickness of the insulation) and face it with plywood or masonite. Paint the finished panel with an exterior type oil or urethane base paint. Be sure to trim the edges of the insulation to accommodate the wood strips. Assemble the wood strips, facing, and insulation with nails and glue. For additional details, see Chapter 14, "Install Window Insulating Shutters".

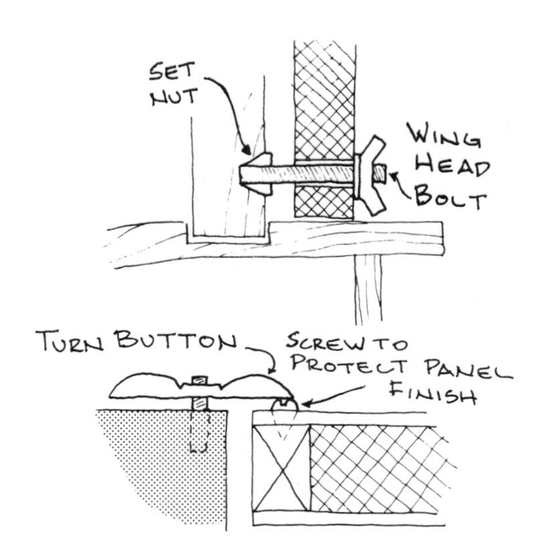

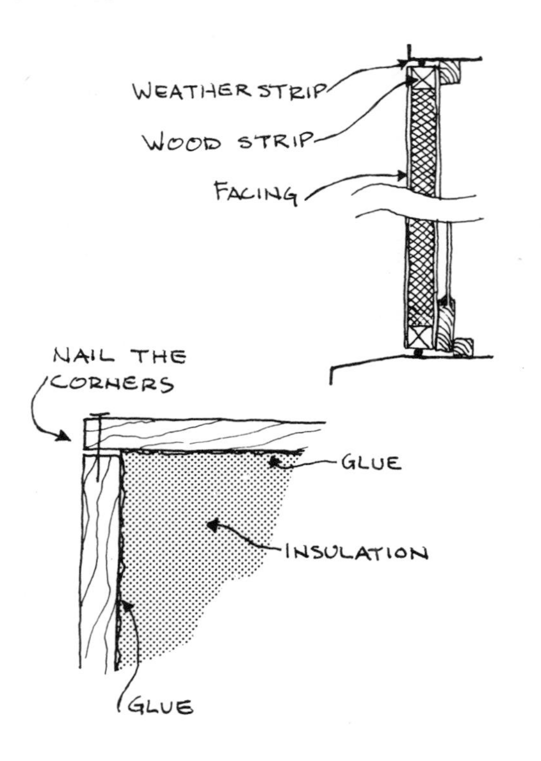

5 Hold the interior and exterior panels in place with turn buttons or set screws and wing head bolts. Turn buttons should press against a screw head to protect the panel covering and finish. If the panel extends out beyond the plane of the window frame, provide blocking for the turn buttons. Alternatively, place set nuts into the sash at the four corners. Place bolts through the insulation into the set nuts. The panel can be held in place with wing head bolts. Apply weatherstripping to the panel for a tight seal.

6 You can also build wood "clips" (as pictured on the first page of this chapter) to hold the panels in place. Build the "clip" from scrap wood, a screw, and a washer. Mount the panel in place with two hinges and locate the "clip" opposite the hinges. Hinging the panel in place has the advantage of "storing" the panel against the wall when not in use.

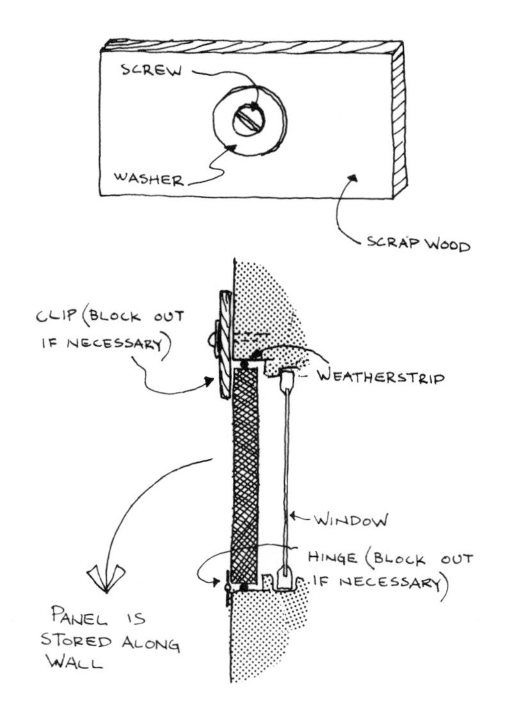

16
replace existing window

description

If the windows are in reasonably good shape, a combination of storm windows, insulating shutters (panels), weatherstripping, putty, and/or caulking will be the best answer for reducing heat loss through them. However, at the other end of the scale, if the windows are in such poor condition and have deteriorated beyond repair (see "The irreparable window" in this chapter), they will have to be replaced.

In some cases, only the window sash needs replacement (that is, the frame and sill are okay). In other cases, the entire window assembly, including frame, jambs, and sill, will have to be replaced. In this situation, the work may involve structural alteration. In any event, replacing windows should only be done by a person skilled in carpentry.

Replacement sash can be ordered, or the entire replacement window (sill, jambs, head, and sash) can be ordered. The windows are made of wood, wood clad with metal, steel, aluminum, bronze, or stainless steel. Thermalized metal windows have insulating materials (thermal breaks) installed in the sash and frame to resist heat conduction. Windows can be purchased unglazed, or with single, double, or triple glazing.

Replacement windows must meet local building code requirements for egress, ventilation, and lighting. This will not ordinarily present a problem if the replacement window is the same size as the window to be replaced. Most windows open for ventilation over half of the window area and most codes require only half as much area for ventilation as for natural light. Therefore, there is usually no problem unless an operable window is to be replaced by a fixed window.

materials

replacement windows

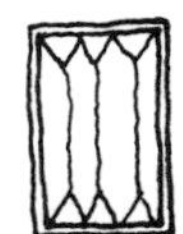

FIXED
- non-opening

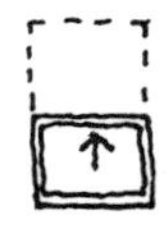

SINGLE HUNG
- only lower sash opens

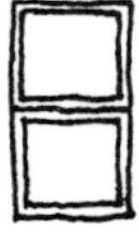

DOUBLE HUNG
- upper and lower sash open; some are built so that sash may be removed from the inside

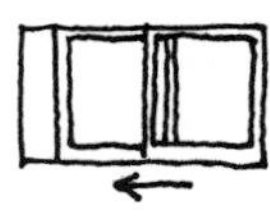

HORIZONTAL SLIDING
- one or both sash open; some are built so that sash may be removed from the inside

CASEMENT
- sash swings in or out on extension hinges attached to the hanging rail of the sash and the head/sill of the frame; some have a hand crank to open window

HOPPER or AWNING
- consists of one or more sash that swing horizontally in or out on extension hinges attached to the hanging-stile of the sash and jambs of the frame

HORIZONTAL or VERTICAL PIVOT
- sash rotates around pivot attached to the center of the top and bottom sash stile/rail and frame

caulk

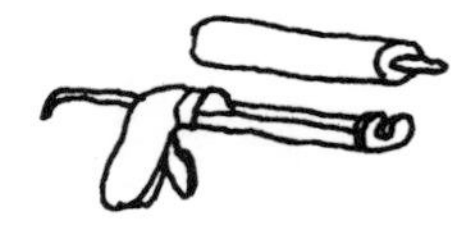

packing

- polyethylene rod
- polystyrene foam board
- oakum
- urethane board or 1 or 2
- part foam
- fibrous rigid foam board insulation

cleaning solvents

- turpentine
- mineral spirits
- lacquer thinner

framing lumber

- shims
- wedges
- blocks
- framing

THE IRREPARABLE WINDOW...

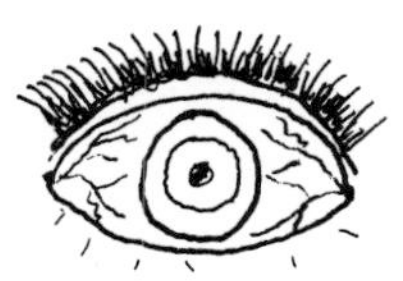

...what to look for

Described below are "symptoms" to look for in determining whether or not a window should be replaced.

1 The window may be "out of plumb" (not perpendicular; not vertical). The window will not open or close all the way or will not lock and it will probably not look straight to you.

NOTE :

Windows may not open or close because they are stuck to the stops (or frames) due to paint, dirt, or "swelling" (wood expanding due to moisture absorbtion). This problem can be corrected without replacing the window. See "Opening a stuck window" in this chapter.

2 In addition to being "out of plumb", look for wood rot. The wood is decayed, will flake apart, and in some cases, some of the wood is missing entirely. Wood rot is caused by water condensing on the glass and running down onto the wood. Therefore, look for wood rot along the parting rails, the bottom rail, and the sill.

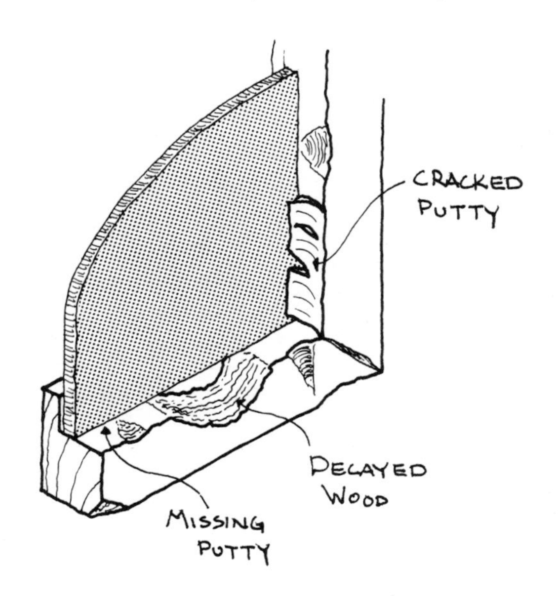

3 The window may be a candidate for replacement if the frame is decayed, insect infested, or damaged.

4 Finally, think of the other improvements which could be made to the window. Can these improvements be installed effectively? For example, the purpose of weatherstripping is to prevent the infiltration of air and moisture around windows and doors. What good is the weatherstripping if the wood is rotten or the window won't close all the way? Also, think of the window in terms of its location. If it's located in a room where it is not used (nor needed - check building codes), such as a pantry, stairwell, etc., it may make more sense to seal it (see Chapter 11, "Change Window Operation"), or install an insulating panel over it (see Chapter 15, "Install Window Insulating Panels") rather than replace it.

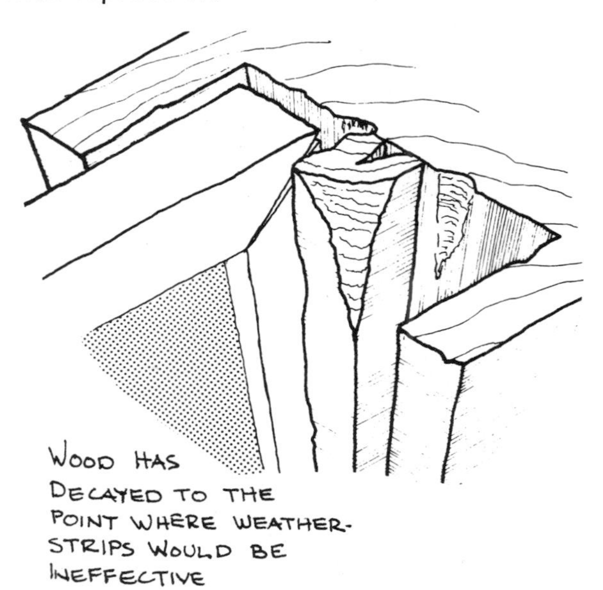

preparation

Material is presented here for both sash and window replacement (sash, head, jambs, sill and frame). If you're replacing the sash or the entire window assembly, the work should be done only by a person who is skilled in carpentry.

The most common residential window is the double-hung wood window. Consequently, the examples discussed are for a double-hung window. Basically, you'll be concerned with placing the frame of the "new" window into the structural rough opening, so window operation type makes little difference.

1 If you're just replacing the sash, carefully measure the opening for the new sash. (If you're replacing the entire

window, follow the manufacturer's specifications for the rough opening size required for the replacement window.) You may find a stock sash at a lumber company that will fit. If you don't find a sash that fits, you'll have to have one made. Custom sash must be ordered in advance and you will have to specify 1-3/4" or 1-3/8" thick sash.

2 Carefully pry out the inside stops with a putty knife or chisel.

3 Remove the ropes (chains), weights, and pulleys for the lower sash. Let the weights drop into the weight box if they can't be used. Remove the lower sash.

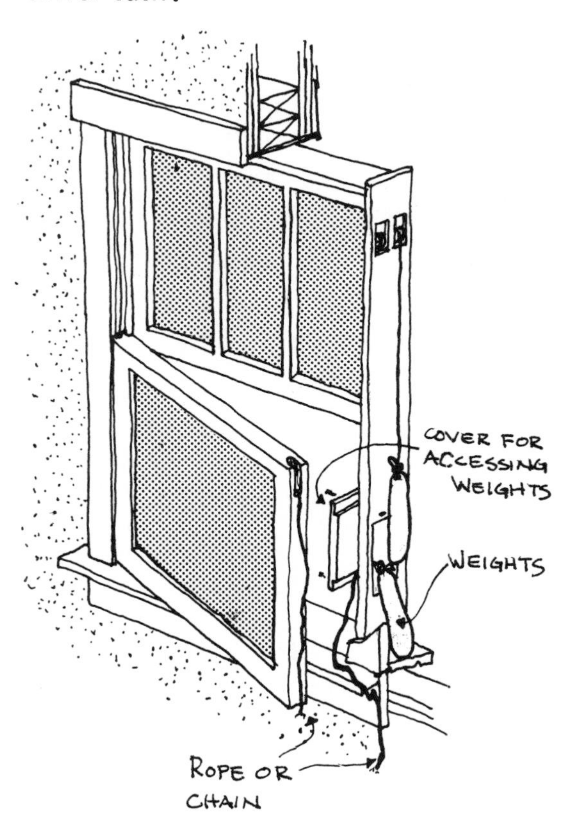

4 Remove the parting stop and upper sash in the same fashion as described in steps 2 and 3.

NOTE: 

Newer double-hung windows are held in place by springs or friction devices. They have no ropes, pulleys, or weights and can be taken out by pressing the sash into one jamb and pulling the sash edge past the opposite jamb.

5 If you're replacing just the sash, skip to "Installation procedures", step 7. If you're replacing the entire window assembly, remove the inside casing

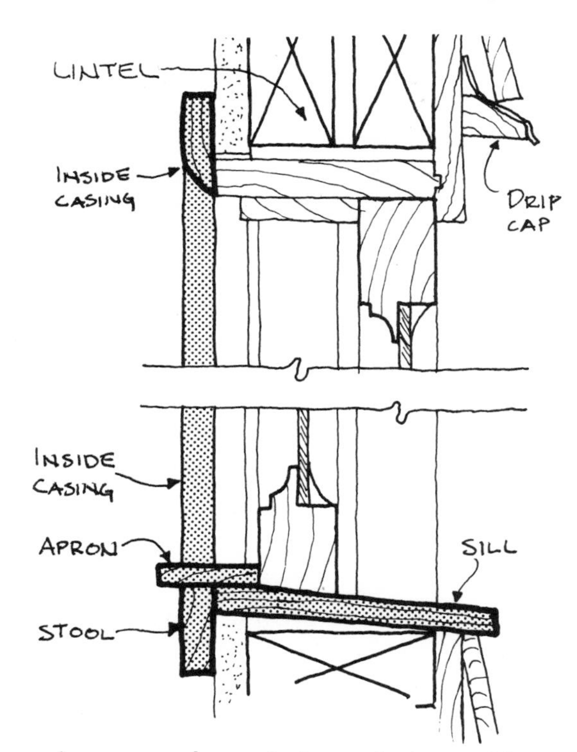

and remove the existing window. Also remove the stool and apron. Remove any interior and exterior wall finishes to expose the rough opening for the new window. Unless they are severely damaged or rotted, do not remove the drip cap, lintel, or sill. Do as little damage as possible as you'll have to repair these finishes once the new window is in.

AN IMPORTANT NOTE :

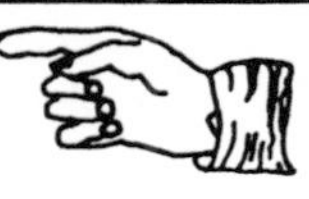

There is now a large opening in your wall. This opening is structurally sound. If, however, the header and/or studs need to be relocated to accommodate the replacement window, and the opening will be greater than 3'-6", you should temporarily support the ceiling before you relocate the header and/or studs. To support the ceiling, cut a 4" x 4" to a length slightly longer than the new window

opening. Then measure the floor to ceiling height and cut two 2" x 4" posts exactly 3-1/2" less than this height. Attach the 2" x 4" posts to the 4" x 4" with plywood. The assembly should fit tightly between the floor and ceiling. Use shims* if necessary. Once the headers and studs are replaced, the support can be removed.

*Shims are thin, wooden strips or wedges used to fill out or level surfaces.

6 Adjust the framing studs, head, and sill to provide the correct rough opening size -

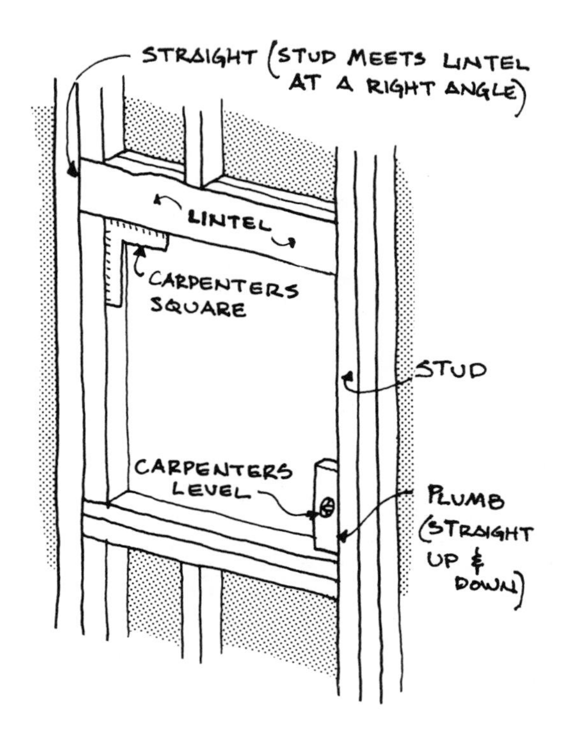

straight and plumb. Although rarely required, this may mean planing, chiseling out, or adding shims* to fill the opening in.

installation procedures

The specific problems and details that you'll encounter when installing replacement windows and sash depend on the window and wall construction type. Only general points and guidelines are covered here.

If you're replacing only the window sash, begin with step 7.

1 The rough opening should be square (plumb and level) with the header, studs, and sill. This may require moving framing members or bringing the framing members out by adding wood strips.

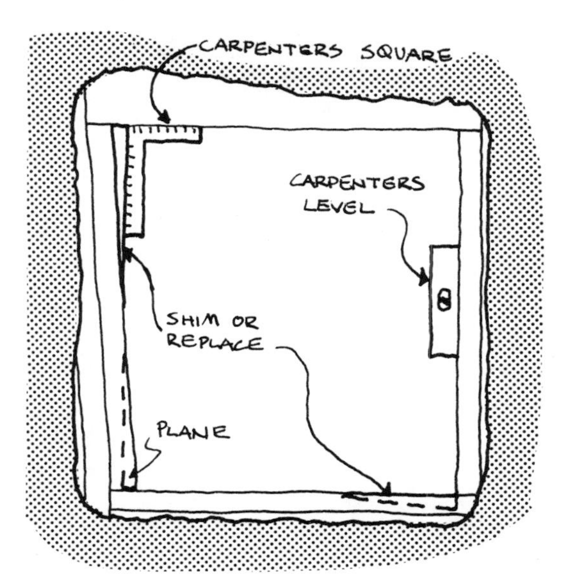

2 Set the replacement window in place (the sashes should be left in the window while doing this). The sill, jambs, and side casings are to be plumb, level, and square. If the frame is not square, the sash will not slide properly and a hinged sash will tend to bind when being opened or closed. If the rough opening is not exactly square and plumb, drive wedges in along the sill and jambs to lift and slide the frame within the

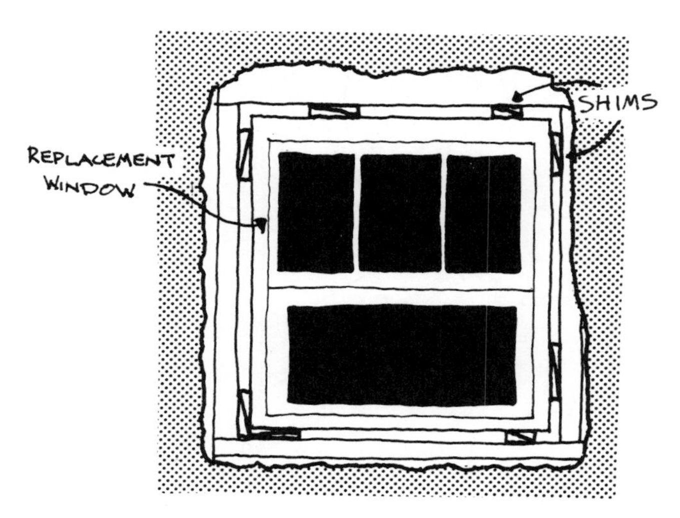

opening until it is square and plumb. Test sash operation and adjust frame position as you go.

NOTE: 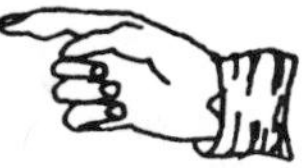

Do not paint the jambs (the track in which the sash slides). Wood jambs can be treated with a wax, however.

---

3 Once you are sure that the window is square and level, secure the window by nailing through the outside window casing into the framing members of the

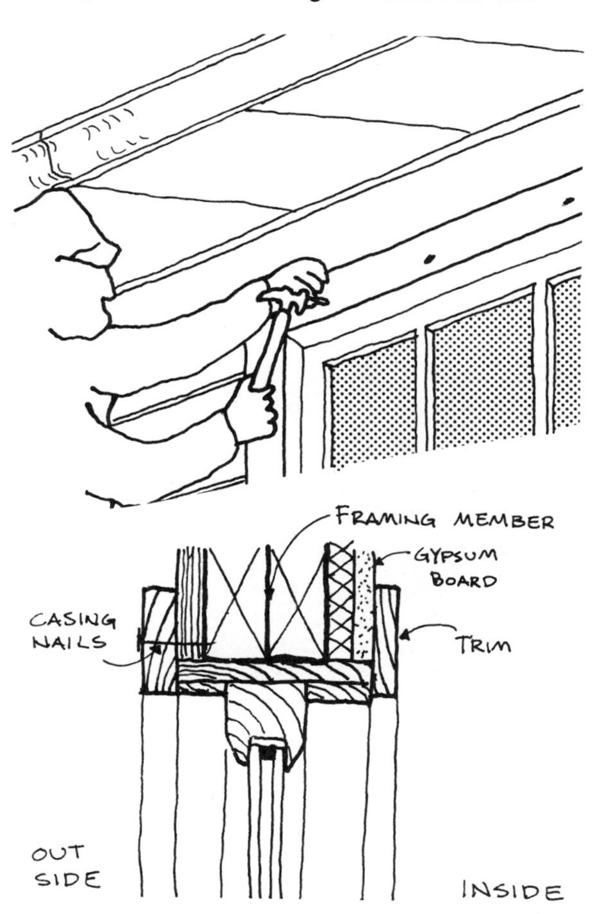

wall. Use casing nails spaced 16" o.c. and be sure that they're long enough to penetrate the framing members. Be sure and refer to the manufacturer's instructions for installation details.

4 Re-install, or patch, any exterior finish that was removed to get the original window out.

5 Pack the opening around the window with strips of batt insulation or spray with urethane foam. Re-install or patch

any interior finish. Caulk any remaining voids between the frame and rough opening before installing the inside casing.

NOTE:

It may be easier to add the insulation as you're installing the window frame (step 2).

---

6 Pack and/or caulk the exterior frame - see Chapter 10 for details.

SASH REPLACEMENT ONLY

7 Install the upper sash first. Check the operation of the sash after it's in, that is, it should slide freely without binding.

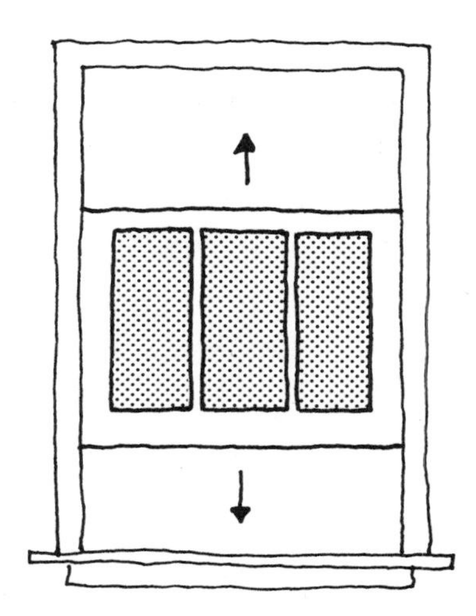

8 Replace the parting strip by applying a bead of caulk to the back of it then nailing it in place with finishing nails.

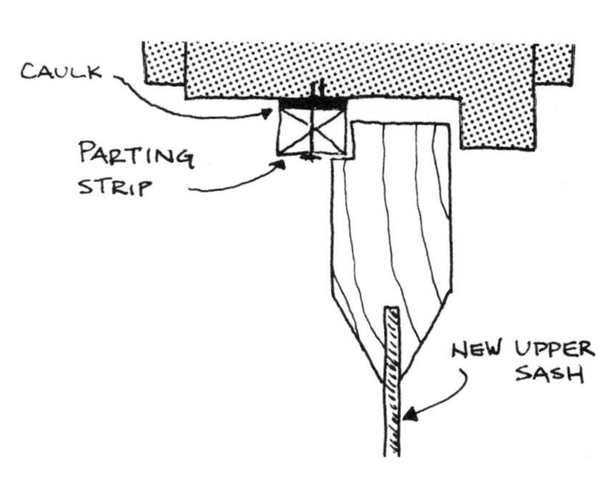

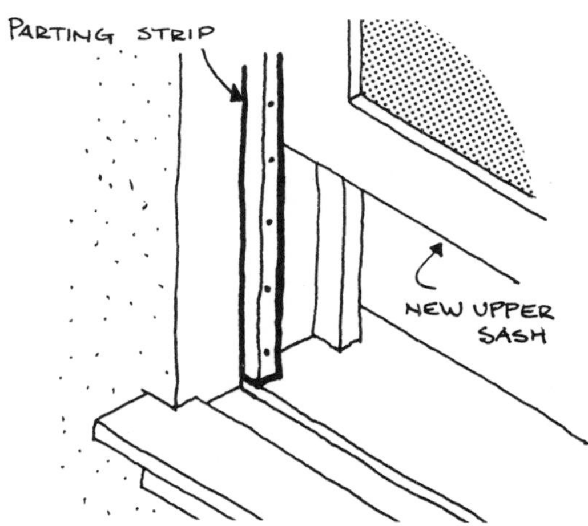

9 Install the lower sash and check it to make sure it opens and closes. Re-

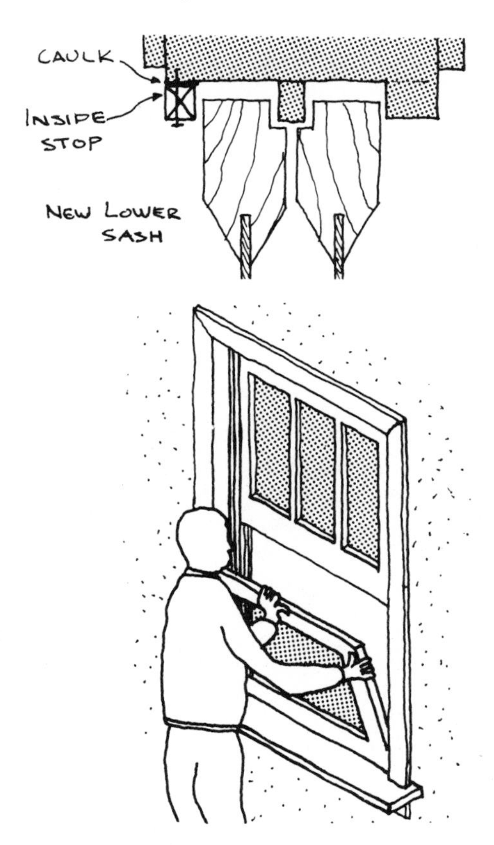

place the stop in the same fashion as described in step 8.

10 Patch interior finish and re-install existing casing, head, and interior apron or install new interior wood trim.

OPENING A STUCK WINDOW

1 Sometimes, paint will seal a window. Tap a putty knife between the sash and stop to pry the two apart. Move the putty knife up and down the length of the sash as well as along the width of the sill.

2 Clean the jamb with a rag, stiff brush, and/or sandpaper to remove excess paint, dirt, and other material. Scrape carefully to avoid chipping and gouging. Finally spray the jamb with a silicone lubricant or rub with soap or wax. Don't use oil.

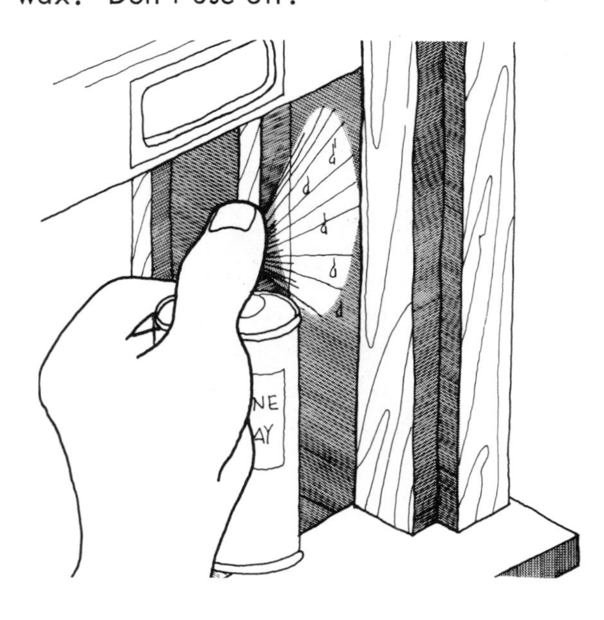

3 In some cases, the wood has swelled so that prying the sash and stop apart won't help (or, there is excess paint

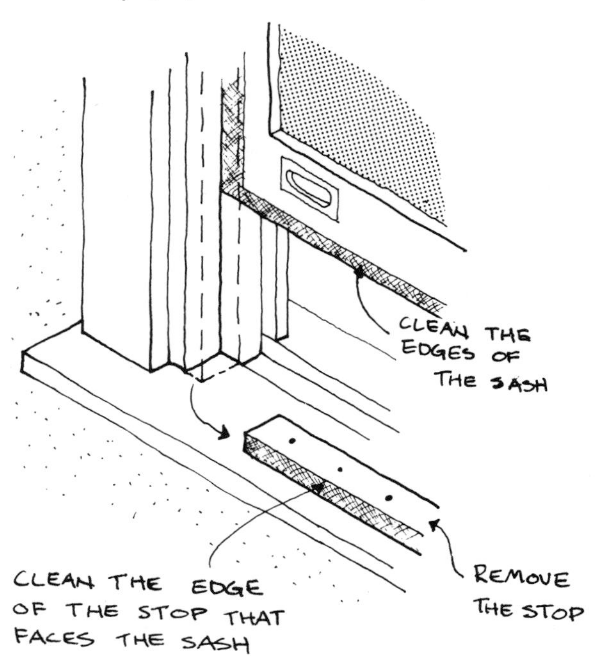

that you can't scrape off). Remove the stop (see "Preparation", step 2) and clean it along with the inside face of the sash (light sanding may work best).

4 If steps 1, 2 and 3 don't work, remove the sash by unscrewing the rope or

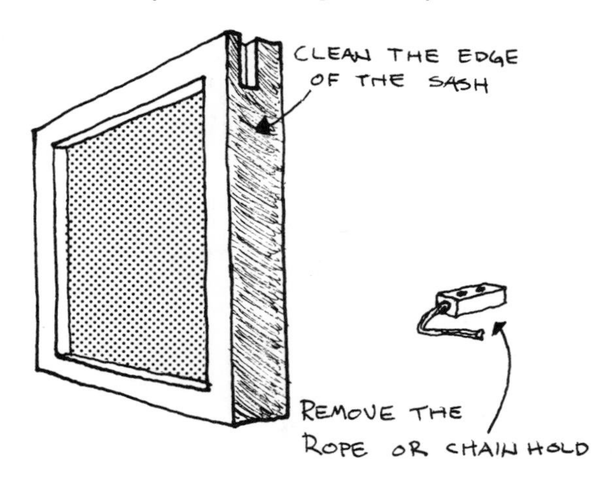

chain. Clean the edge of the sash, jamb, and parting strip as described in step 2. If the sash has swelled, plane or lightly sand the edge of it.

5 Windows that are difficult to move should be cleaned and sprayed with a silicone lubricant. Clean built-in metal weatherstripping with steel wool prior to lubrication.

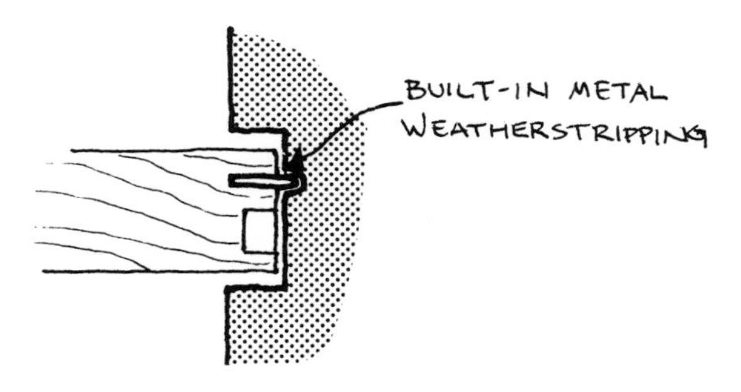

door retrofits

Install door threshold and/or bottom seal

Weatherstrip doors

Install storm door

Replace existing door

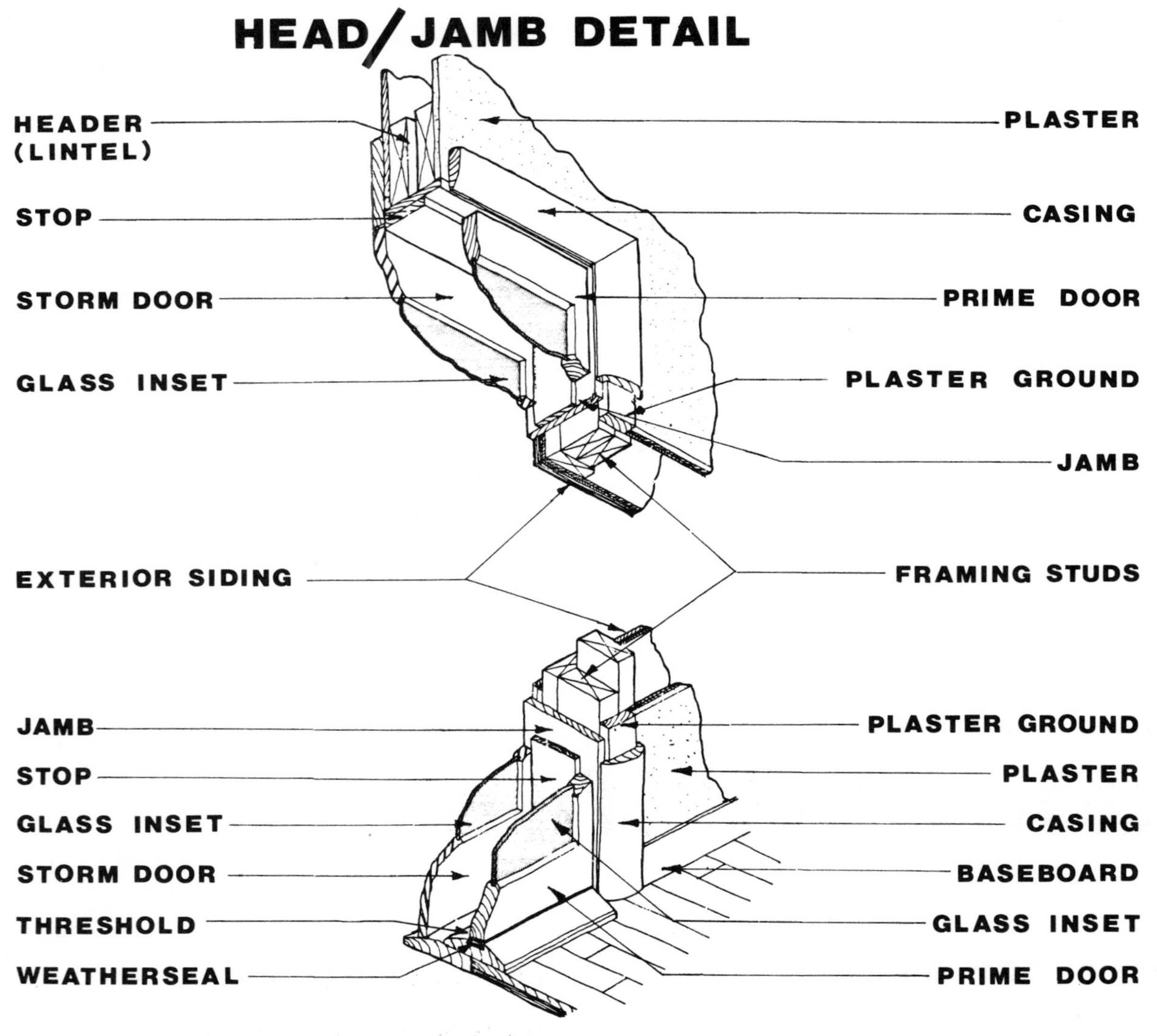

THRESHOLD/JAMB DETAIL

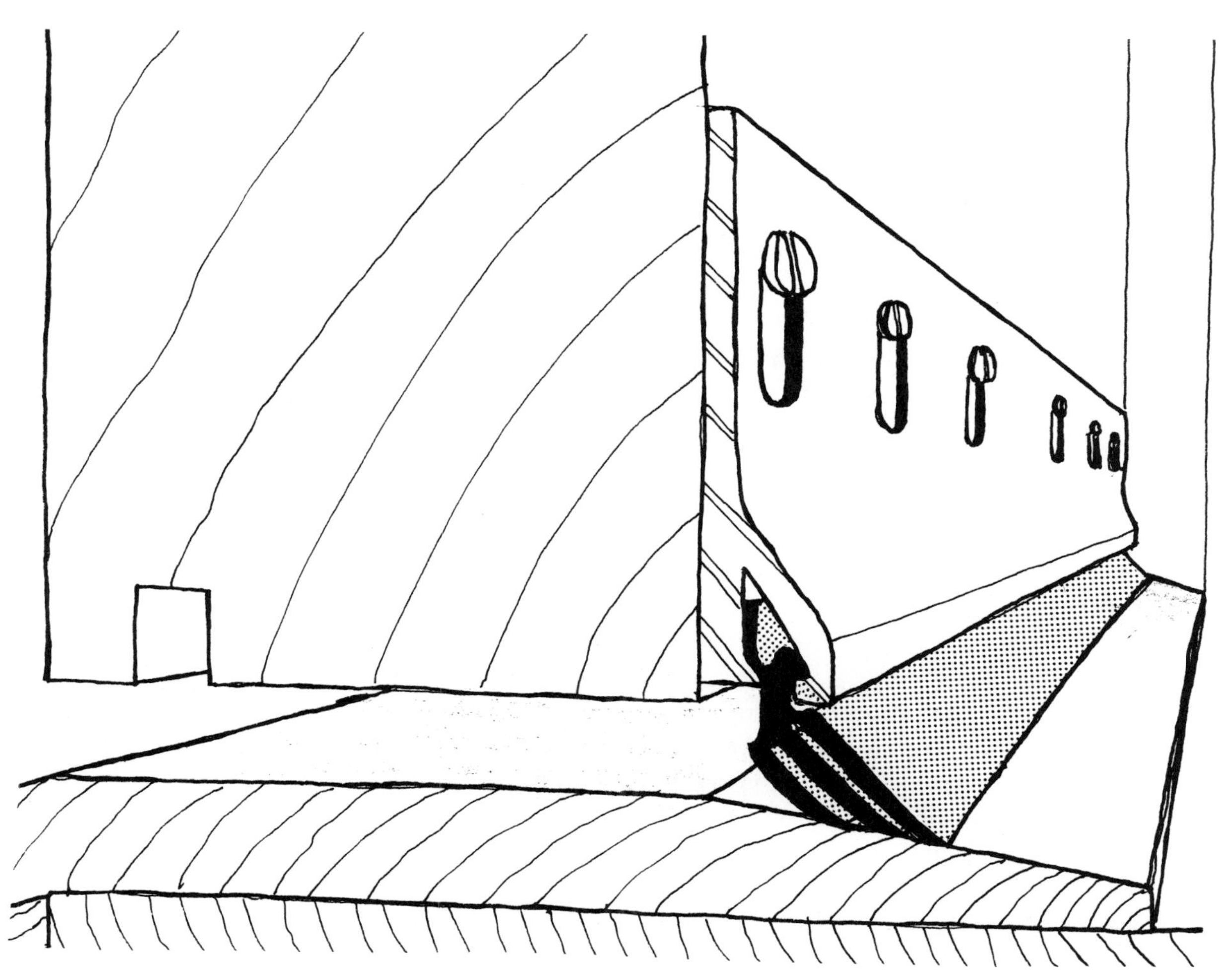

17

install door threshold and/or bottom seal

description

Doors have the same problem as windows; i.e., air leaks into the home through the crack around the door (window) and frame. Like windows, the crack along the top and sides of the door can be weatherstripped (see Chapter 18, "Weatherstrip Doors"). However, these same weatherstrips will usually not work along the threshold* because of the foot traffic here and because there is no stop or "door buck" at the bottom.

*A threshold is a strip of wood or metal with beveled edges, used over (sometimes it is set into) a finished floor, upon which the bottom of a door closes.

Consequently, the "gap" must be treated differently, either with door sweeps or door bottom seals (weatherseals).

Over a period of years, a wood threshold will wear down (or in some cases, a threshold was never installed or was removed). In this situation, the old threshold should be replaced with a wood or metal threshold before installing a weatherseal or door sweep. Some metal thresholds have integral weatherseals. If the seal wears out, the seal can be replaced with a new one without removing the threshold. However, the shape and size of the seal varies, so take the old seal with you when looking for a new one.

Not only is heat lost through the gap beneath a door, but it can also be the source of uncomfortable drafts. A good stopgap measure is stuffing this opening with a rag or rug. However, a proper fitting threshold and/or sweep will provide a more permanent and attractive seal.

materials

door sweeps

- Self Sticking
- Single Flange
- Triple Flange
- Automatic Door Bottom*

---

*An automatic door bottom, of which one type is a "flip seal", is essentially a sweep (usually triple flange) which is hinged so that the seal is normally up and out of the way of carpeting, but is forced down against the threshold or flooring as the door is closed.

- rigid or moveable
- easy to install
- good for doors without threshold
- may drag on carpet or rug
- usually installed on the interior

door weatherseal

(Door Shoe)

- durable
- difficult to install (you have to remove the door)
- door may have to be trimmed
- good for use with a wooden threshold

gasket threshold

- gasket can be replaced if it wears out
- useful where there is no threshold or on a wood threshold
- is usually part of a metal threshold

wood threshold

- easy to replace
- available at lumber companies
- prefinished or unfinished
- door may have to be trimmed

metal threshold

- seal pieces are replaceable
- many types of metal holds
- door may have to be trimmed

interlocking threshold

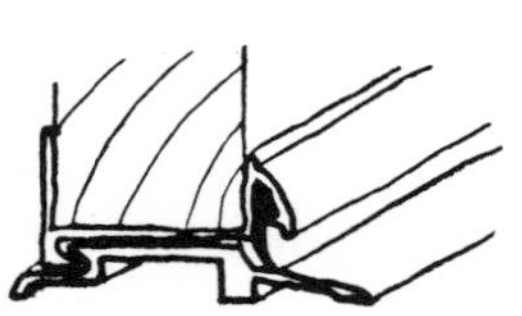

- two pieces
- should be installed by someone skilled in carpentry

preparation

All surfaces that are to receive door sweeps, seals, or thresholds should be smooth, level, and free of all dirt and old paint. Scrape or brush clean and wipe clear with mineral spirits using a clean rag.

1 Before installing a weatherseal and/or threshold, check the clearance at the bottom of the door. If the gap is not wide enough to accommodate the seal and/or threshold, remove the door and trim or plane the door bottom (see Chapter 20, "Replace Existing Door").

AN IMPORTANT NOTE:

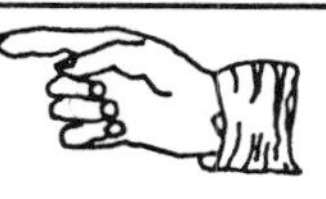

Some seals and thresholds require a beveled door bottom - check the manufacturer's recommendations for the specific seal and/or threshold that you're working with.

2 If you're installing an unfinished wood threshold, varnish the threshold and allow it to dry completely before installing.

installation procedures

DOOR SWEEPS

There are various types of door sweeps. Of the most common are;

- Self Sticking
- Single Flange
- Triple Flange
- Automatic Door Bottom

Sweeps are applied along the face of the door at the bottom. Install the sweeps on the outside face of doors that swing out and on the inside if the door swings in.

Before installing any seal which extends below the existing bottom edge of a door, open the door to see if installing a seal will cause the door to scrape the floor as it is opened.

...self sticking door sweep

1 Cut the sweep to fit approximately 1/16" in from the door edges.

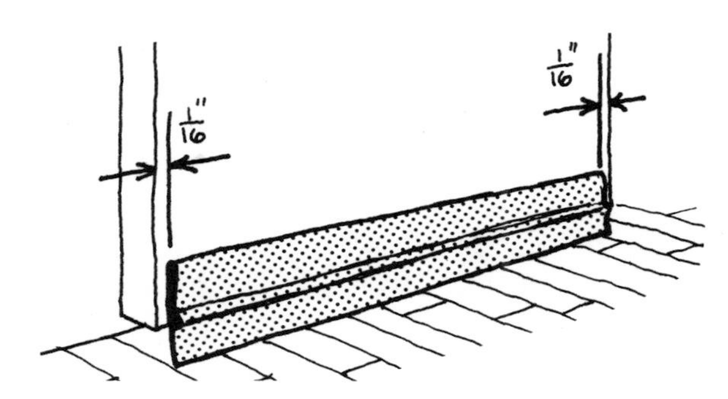

2 Peel off the back and then press firmly into place. This type of sweep is good for doors that you don't want to nail into.

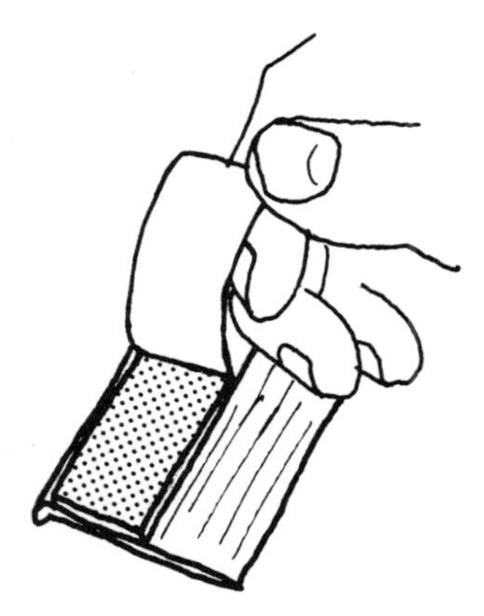

...single flange, triple flange

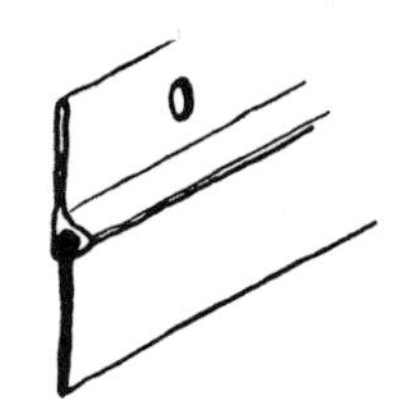

1 Cut the sweeps to fit across the base of the door, leaving 1/16" to the door edges.

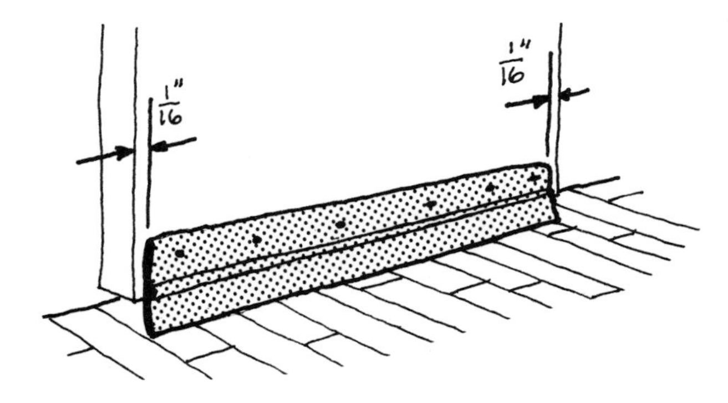

2 While holding the strip in place, mark the two end holes with a pencil. Be certain that the edge of the sweep rests firmly against the threshold. Use the slots in the seal to adjust its height.

3 Remove the strip and use an awl or wire brads to tack small starter holes.

4 Place the strip back into position and screw into place.

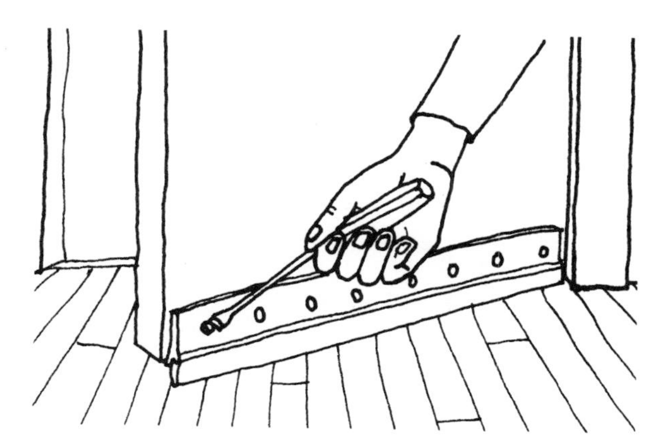

5 Now repeat the procedure for the remaining holes.

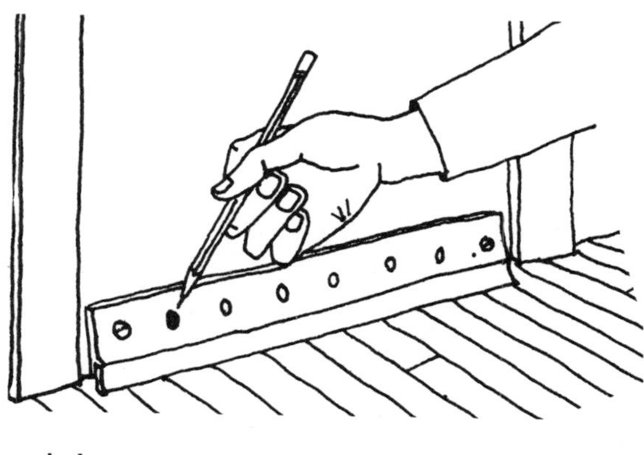

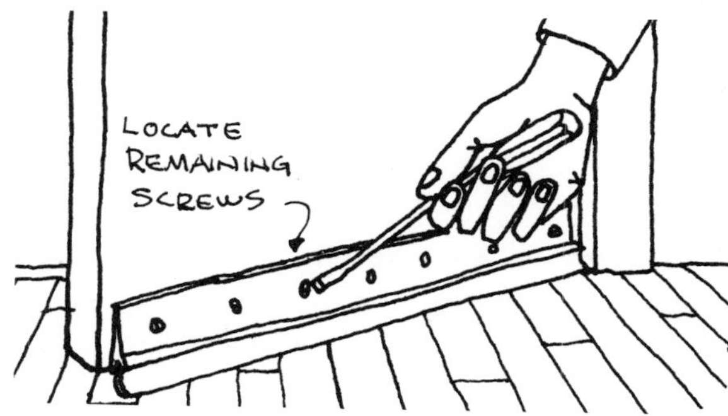

...automatic door sweep

Automatic door sweeps are especially desirable for doors that open across a rug or carpet where a plain sweep may interfere with the door operation by dragging across the carpet.

When the door is closed, a strike plate or screw on the door jamb forces the sweep (some sweeps are spring loaded) down against the floor or threshold. When the door is opened, the sweep opens or retracts allowing it to clear the carpet. Automatic door sweeps are usually installed on the outside of inward swinging doors.

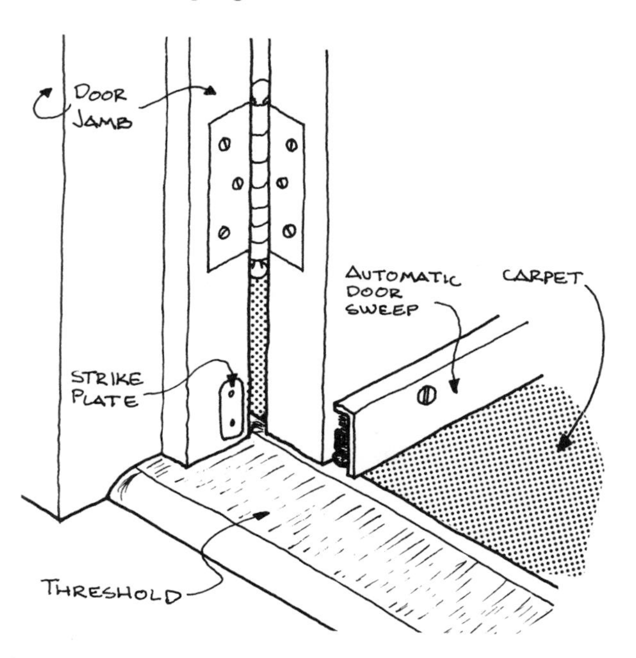

Install in the same fashion as single or triple flange door sweep, except carefully locate the special screw in the door frame to push the seal down firmly as the door closes.

DOOR WEATHERSEALS

Weatherseals, sometimes referred to as "door shoes", are installed along the bottom of the door. As they're somewhat difficult to install, it is important that you understand the clearance required and the amount the door may have to be trimmed or planed.

1 Check the existing clearance at the bottom of the door. Determine if the

door has to be planed for the weather-seal to fit; if so, mark this dimension on the door.

2 Remove the door. Do this by removing the "pin" in the butt hinges. It's best to have help to take the door off the

hinges and carry it to where the trimming will be done.

3 Trim or plane the door the required dimension. It's a good idea to mark this dimension on the door prior to removing it, otherwise, you may accidentally trim the top of the door rather than the bottom of it. For details on planing, see Chapter 20, "Replace Existing Door".

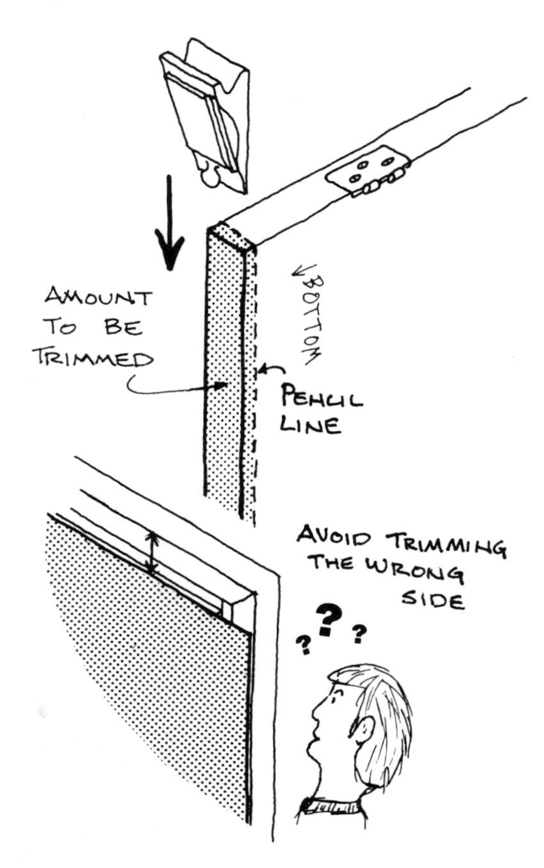

4 Cut the seal to match the door width. Install by removing the seal and fastening the metal channel with screws.

5 Slide the seal back into place before hanging the door.

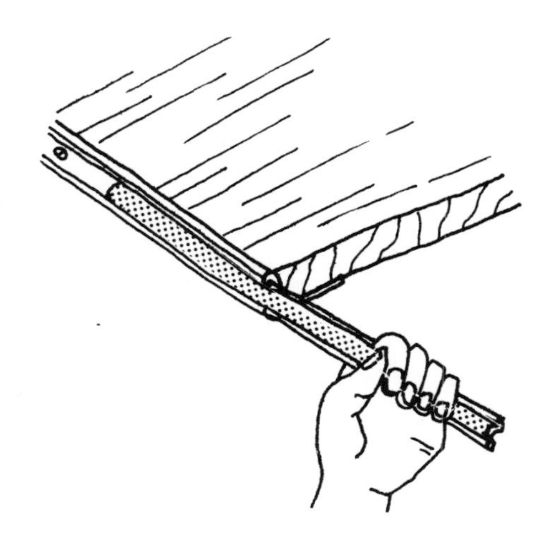

NOTE: 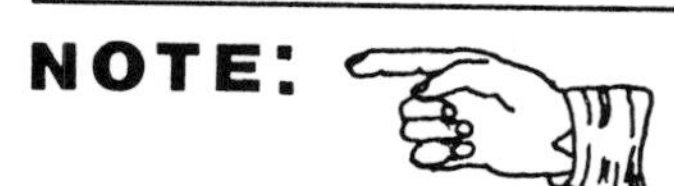

Some seals are installed in combination with thresholds.

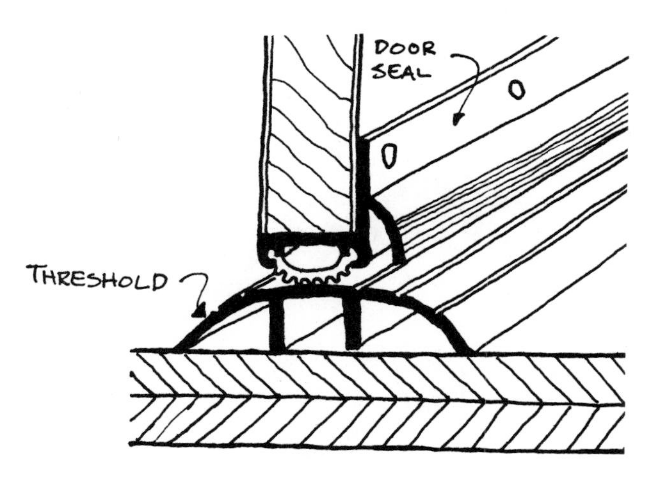

GASKET THRESHOLDS

Install over wood thresholds or beneath doors without thresholds. Cut the threshold to door width, remove the seal piece, fasten with screws, and replace the seal piece. This type of seal is easy to install and requires door removal only if the door has to be planed for the seal to fit. People, however, may trip over the seal.

Alternately, the seal can be installed on the bottom of the door (this requires removal of the door), eliminating the hazard of people tripping over it. The seal, however, may drag across carpeting or hang up on throw rugs over which the door opens.

Replacement seals are available for existing gasket thresholds with worn seals.

WOOD THRESHOLDS

Prefinished or unfinished thresholds are available at lumber companies. The bottom of the threshold may have to be planed to fit across the existing floor (the door bottom may have to be planed also). The threshold should be one continuous piece that spans the width of the door. Install with finishing nails. Set the nails with a nail set punch and cover them with plastic wood.

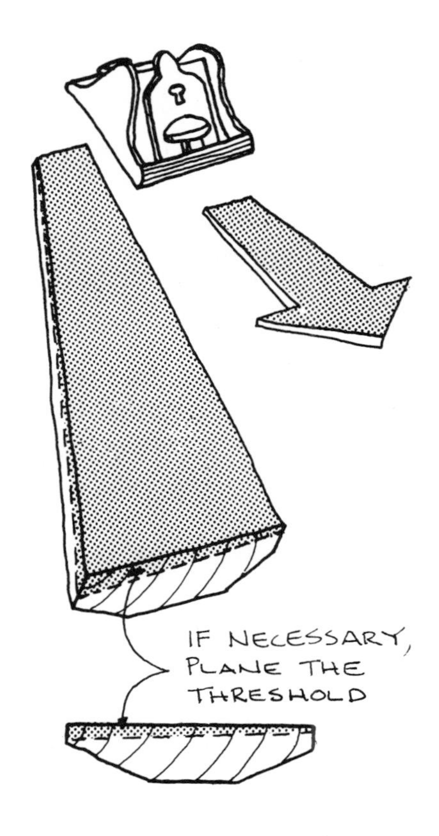

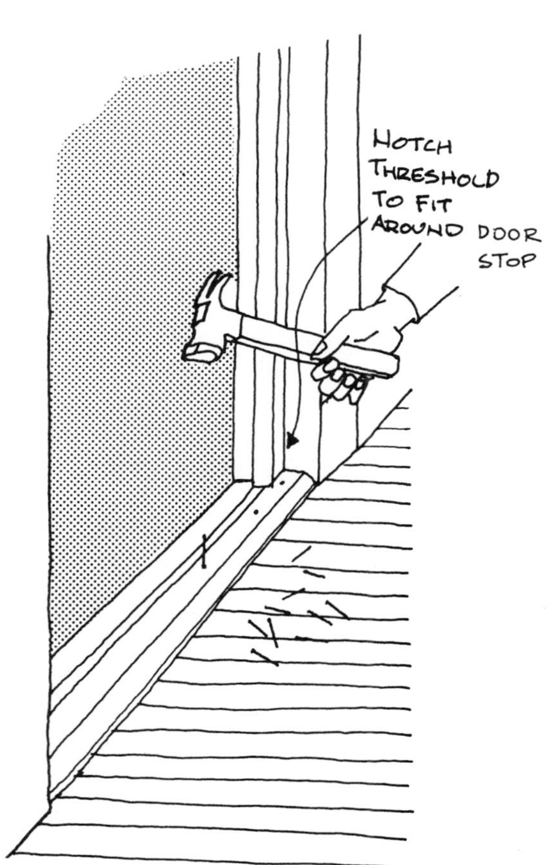

NOTE:

A door sweep, seal, or gasket threshold should always be installed along with a wood threshold on doors which have no seal.

METAL THRESHOLDS

A variety of metal thresholds are available, most of which have seal insert pieces that can be replaced if they wear out. Remove the seal piece to fasten the threshold to the floor. The door bottom will probably have to be planed.

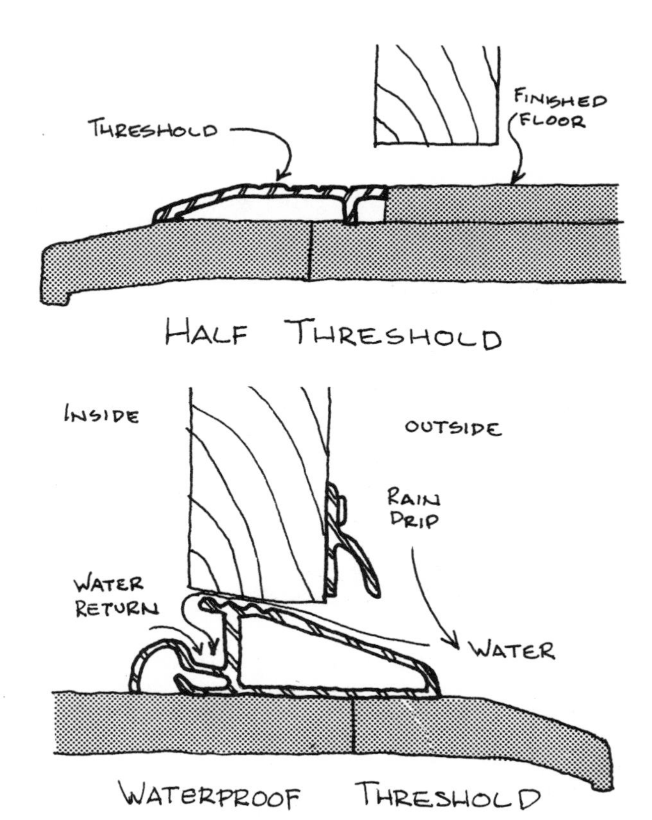

It has probably occurred to you by now that door sweeps, weatherseals, and thresholds have been discussed for hinged doors only. The problem of air infiltration across the threshold also applies to sliding doors. An air seal is obtained by the use of neoprene gaskets or "piles". If they wear out, they can be replaced. However, there may be a special weatherseal for your particular sliding door.

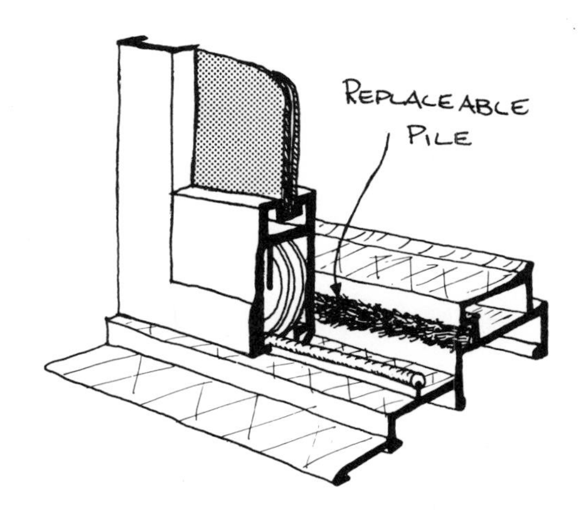

let's summarize...

Remember, the object of installing thresholds and seals is to create a smaller crack opening between the door and the floor beneath, and to assure that a flexible seal makes a firm continuous contact to close this opening whenever the door is closed.

If a seal piece is present, but the threshold is missing, then perhaps installing a wood threshold will provide a surface for firm (and continuous) contact with the seal. Likewise, a threshold may be present but the seal may be missing. Installing a replacement seal (for a bulb type threshold) or a sweep or gasket which closes firmly against the existing threshold may be all that's needed. If neither threshold or seal exists, then make sure that installing a seal will provide firm continuous contact and will be serviceable, or install both threshold and seal.

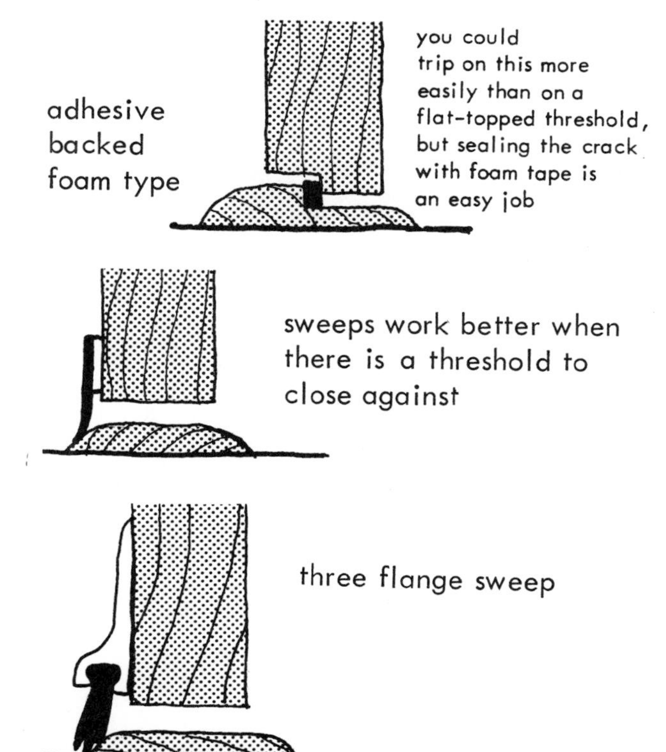

like sweeps, door shoes work best if they close against a threshold

thin bulb in aluminum strip may be mounted onto wood threshold

flip seal may sometimes be adequate

outside drip cap

bulbs are hard to replace because they have a special shape on the edges to keep them in the channel

gate threshold with felt seal-lifts vertically when door is opened-may be enough without use of a threshold on the floor

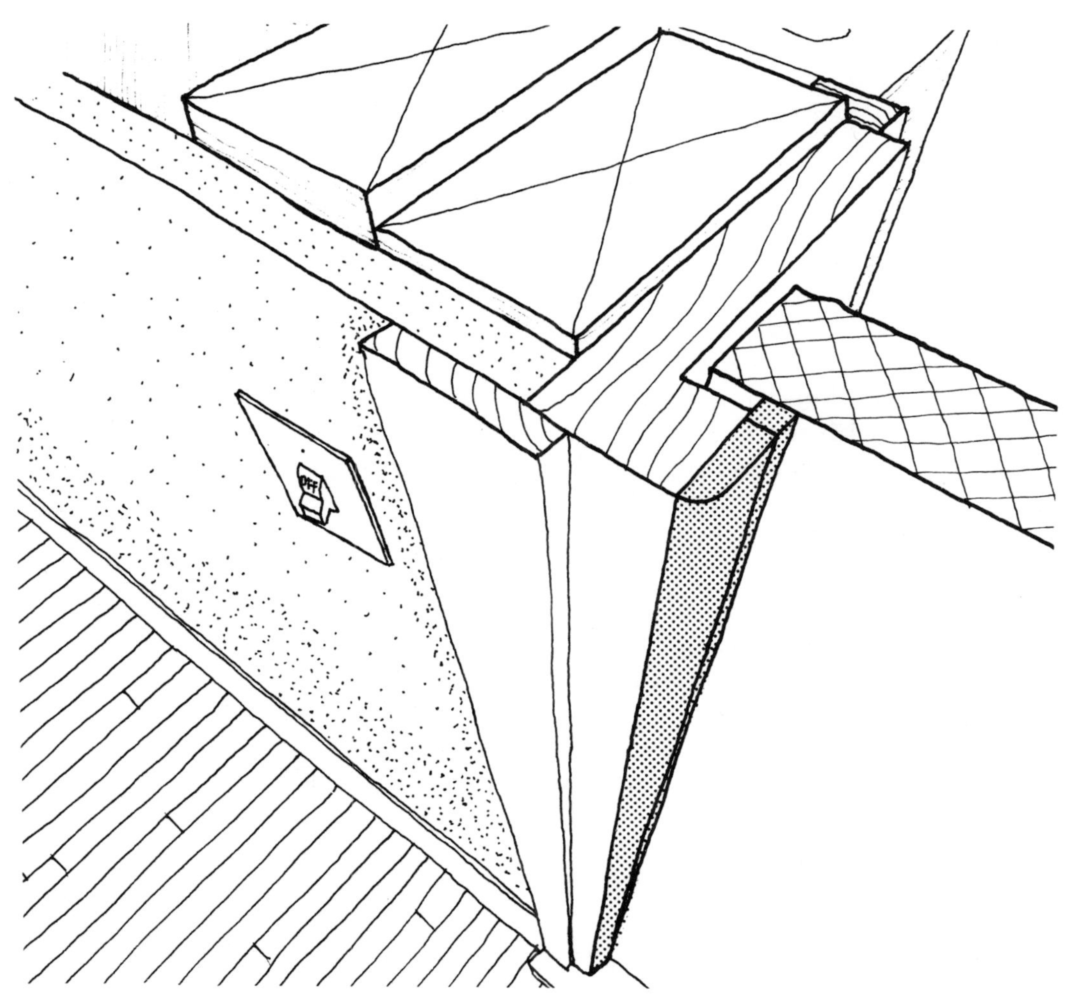

18

weatherstrip doors

description

To open and close, all doors need a certain amount of space between the door panel and the jambs. For doors that open to the outside (or doors that open onto unheated areas), this crack can be the source of drafts (air infiltration) into the home.

Air infiltration under a door was reduced in Chapter 17 by installing a threshold and/or a weatherseal. In this chapter, infiltration around the other three door edges is reduced by the installation of weatherstripping. Existing weatherstrips not providing a proper pressure fit should be removed and replaced with new weatherstripping.

Weatherstrips must have a flexible material to make a seal. They also must stay in place. Either they are placed directly between rigid surfaces (such as foam tape) or they must come with the needed rigidity (such as wood backed rubber tubing). Select weatherstrips on how long they will remain flexible and how well they provide needed rigidity.

Most weatherstrip material you will find is designed and packaged for "do-it-yourself" installation. Reviewing this chapter will help you determine which types you can work with the best. Make sure you examine products carefully to make sure you have the right type. You may need more than one type, even for use on the same door.

The same types of weatherstripping that you used for windows (Chapter 9) will also work for doors.

materials

spring metal strips

(Bronze, Brass, or Aluminum)

- sold as flat "tension strips" in rolls or packaged strips (wire brads for installation are usually included)
- for use between sliding and closing surfaces which bear against each other
- floor edge strip is often "passed off" as spring metal strip, but does not perform satisfactorily as a weatherstrip

rubber, PVC, EPDM, silicone or vinyl plastic tube

(Tube Weatherstrip)

- comes with flexible or rigid flange for fastening
- usually applied on the outside
- should not be painted
- EPDM and silicone are especially desirable for flexibility and long life

felt strips

(Metal or Wood Reinforced or Plain with Adhesive Backing)

- use where little abrasion or weathering is expected
- not very durable, will not hold up well when applied on the exterior
- use in narrow, uniform gaps
- comes in a variety of widths, thicknesses, qualities, and colors
- painting around doors often renders felt ineffective if the felt itself is painted
- felt inserts are available for sliding doors

foam strips

(Vinyl, Sponge Rubber, Neoprene, Polyurethane)

- comes with adhesive backing or wood or metal flange backing for fastening
- use in same location as felt strips where the gap is wider and non-uniform

---

NOTE:

This is only a brief review of weatherstrip materials, which are actually available in numerous shapes, types, materials, flexibility, and life expectancies.

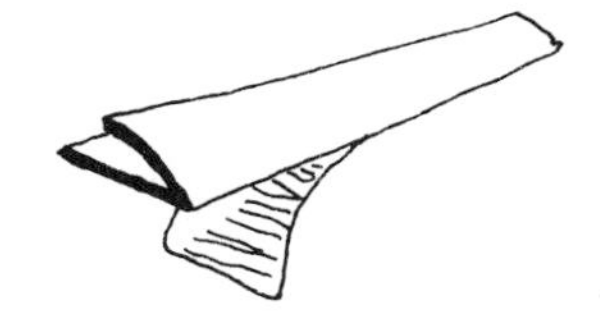

plastic strips

- comes with adhesive backing
- similar in shape to spring metal strips
- can be used between sliding surfaces that bear directly against each other

specialized weatherstrips

- special grooves are cut into the door panel, jamb, or stops to accept special weatherstrip shapes of PVC, EPDM rubber, silicone, or aluminum interlocking channels
- this is to be installed by a skilled carpenter or licensed installer and is therefore not covered here

other materials

- caulk

- staple gun
- nails, nailset, or screws
- tin snips

preparation

Preparation procedures are the same, regardless of door-operation type.

Check the operation of the door. Doors that do not close tightly within the frame may require shimming, planing, or rehanging. Refer to Chapter 20, "Replace Existing Door", for details.

1 The surfaces that you're weatherstripping are to be smooth, level, and cleaned of old weatherstripping, dirt, and old paint. Scrape or brush clean and wipe with clear mineral spirits using a clean rag. If you're using adhesive backed weatherstripping, wait for the surfaces to dry before pressing in place.

2 If the surface to which you're applying the weatherstripping is uneven, apply a bed of caulk (or adhesive) to the surface. In this manner, you'll seal any gaps between the weatherstrip and frame.

installation procedures

Installation procedures for the various type weatherstripping is described by the following door operation types

- Hinged
- Horizontal sliding
- Hatched (attic)
- Garage (overhead)
- Fixed

NOTE:

Installed weatherstripping is not to hinder the normal operation of the door, though, by it's nature, properly installed weatherstripping will increase the closing and/or locking pressure of the door.

WEATHERSTRIPPING A HINGED DOOR...

...with spring metal strips

1 Spring metal is applied to both jambs and the head jambs. Position the strips on the jambs so that the flared flange almost touches the stop. Trim the metal to accommodate hinges, locks, and other hardware (tin snips work well for trimming metal strips).

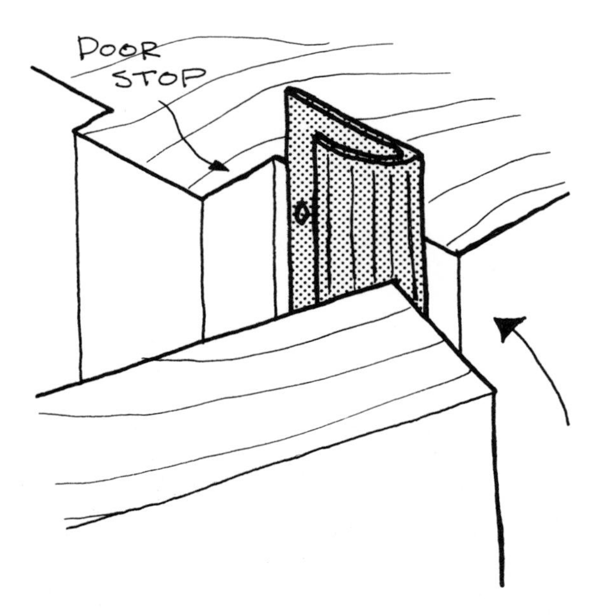

2 Tack a wire brad at one end of the strip. Continue tacking down the strip, spacing the brads about 4" apart. While you're

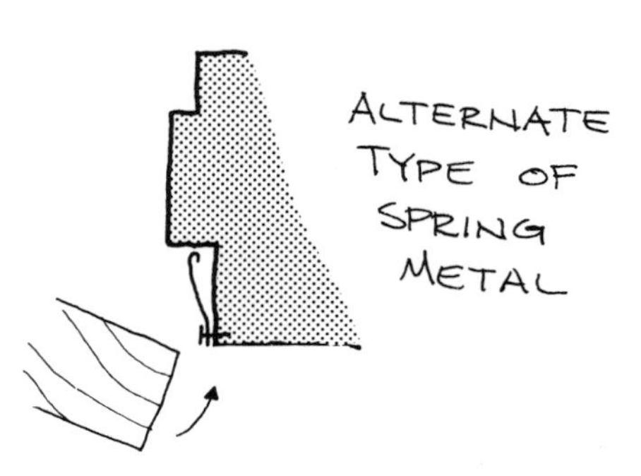

tacking the brads, make sure the strip stays properly aligned. To avoid damaging the strips, drive the brads about half way in, then use a nailset to drive them flush to the strip. (If the strips do not have pre-punched holes, you can make holes by using an awl). Strip may also be stapled in place with a medium weight staple gun.

3 Finally, flare out the edges of all the strips with your screwdriver to insure a tight fit.

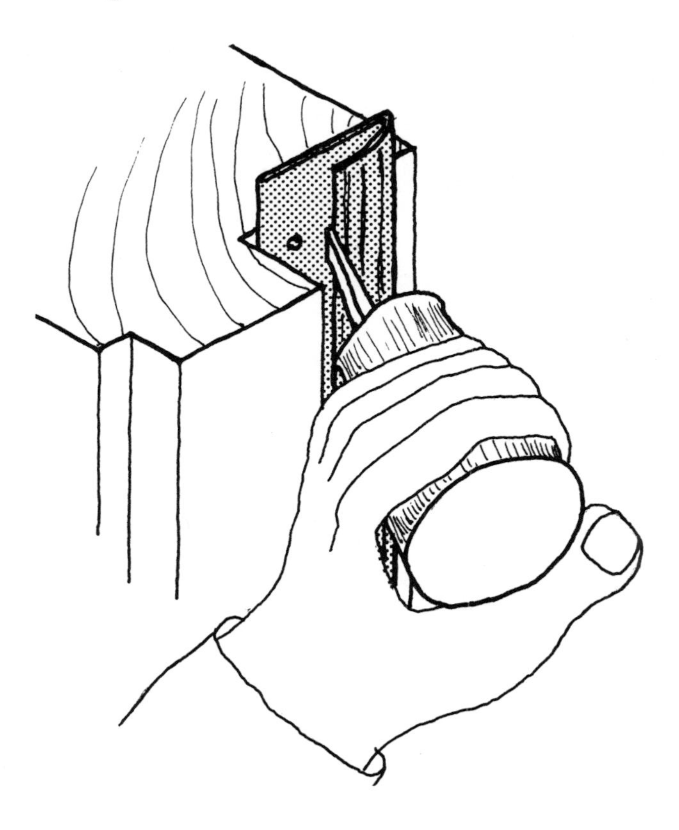

...with rubber, PVC, EPDM, silicone or vinyl plastic tube

1 Nail the first strip to the stop along the head jamb (space the brads about 4" apart). Then nail the strips along the side jambs. Be sure the tube extends over the edge of the stop forming a tight seal when the door is closed.

2 This is a good type of weatherstrip for warped doors which are leaky because the door does not close evenly against

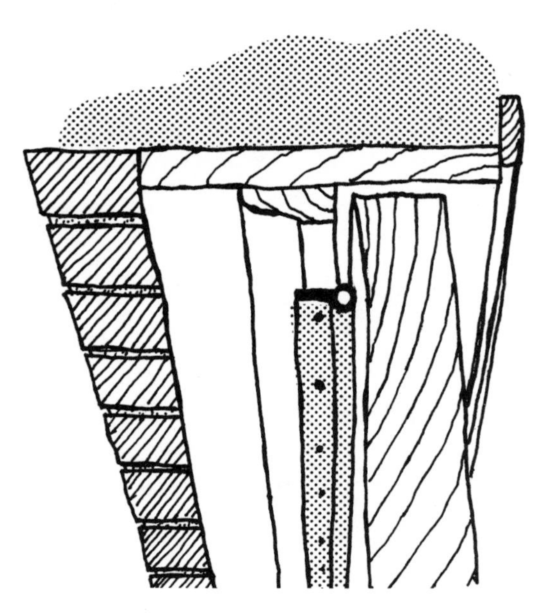

the stop. Nail the weatherstrip in place with the door closed and adjust it as you go to compensate for door warpage. This means the tube will project over the stop further in some places than others.

NOTE: 

Tube weatherstrips with metal flanges are available for metal hinged doors.

...with felt strips

Felt strips that are reinforced with wood or metal are installed as per "tube" weatherstripping.

With adhesive backed felt, peel off the backing and adhere it to the inside face of the three stops. Keep in mind that adhesive backed felt strips are relatively thin and should be used only where the joint between the stop and the closed door is narrow. Also, felt should not be used on any exterior door where weathering is expected. In other words, where it may get wet.

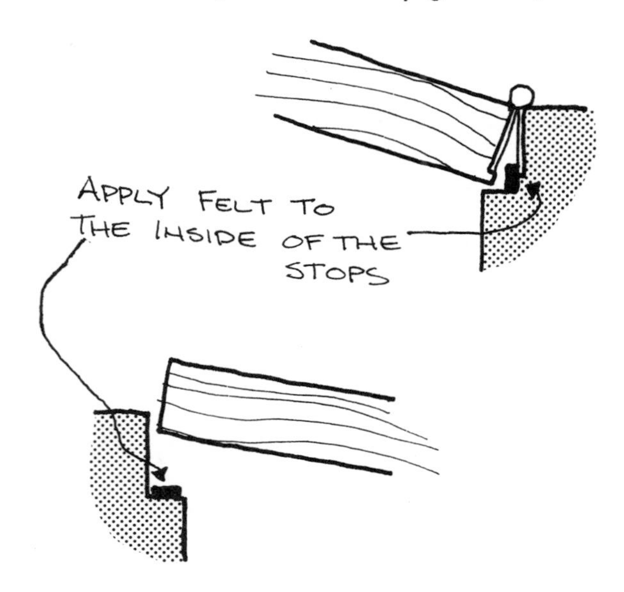

...with foam strips

Foam strips that are reinforced with wood or metal are installed as per "tube" weatherstripping.

Foam is either "open cell" or "closed cell", and different types have a different firmness. Neoprene, EPDM, or other rubber type foams are a better choice where the weatherstrip may be exposed to the weather. Choose a softer foam (open cell is ordinarily softer) when your surfaces are rough or uneven.

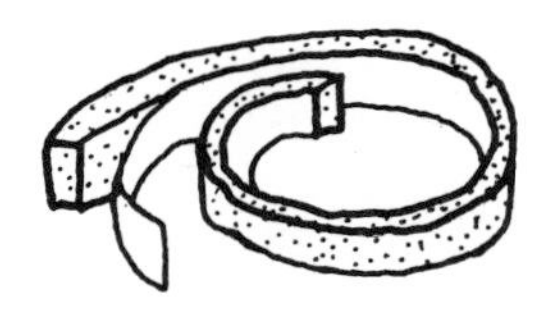

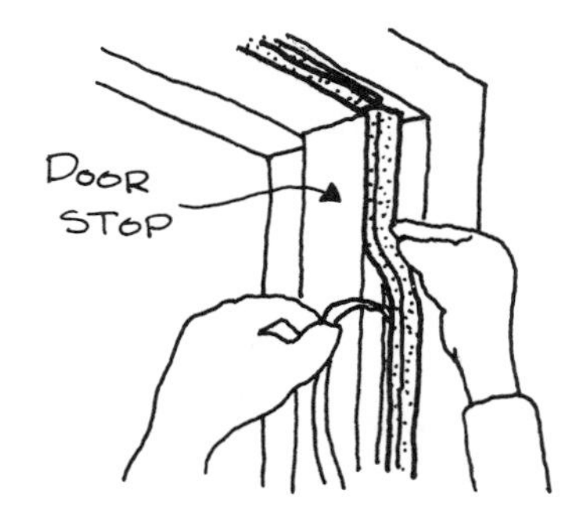

NOTE:

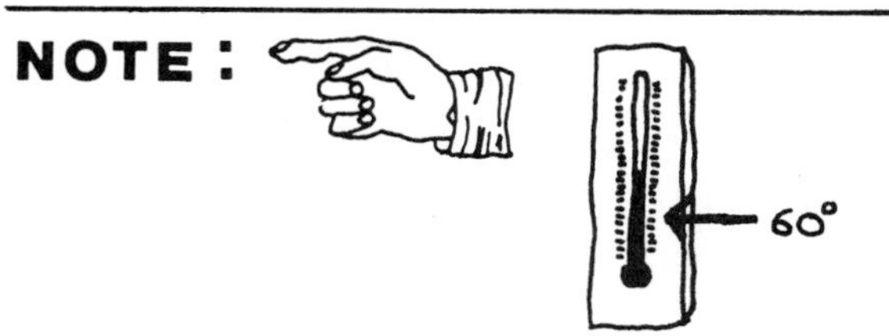

Some adhesives don't stick well unless they are close to room temperature. Be sure to check the product and, unless the minimum temperature is marked on the package, test the adhesive bond at temperatures below 60° F before proceeding.

...plastic strips

Peel the backing from the plastic strips and apply to the jambs so that the flared flange almost touches the stop. Unlike spring metal strips, wire brads are not needed to hold the plastic strips in place.

METAL HINGED DOORS

A special weatherstrip, similar to rolled vinyl, is available with a magnetic core. The flange of the strip is tacked to the jamb such that the core butts firmly against the closed door. Although the magnetic core assures weatherstrip contact with the door, most metal doors have metal frames and the weatherstrip must be screwed to the frame.

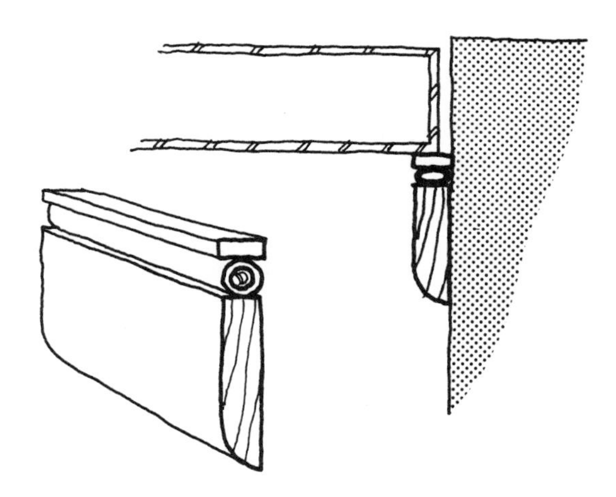

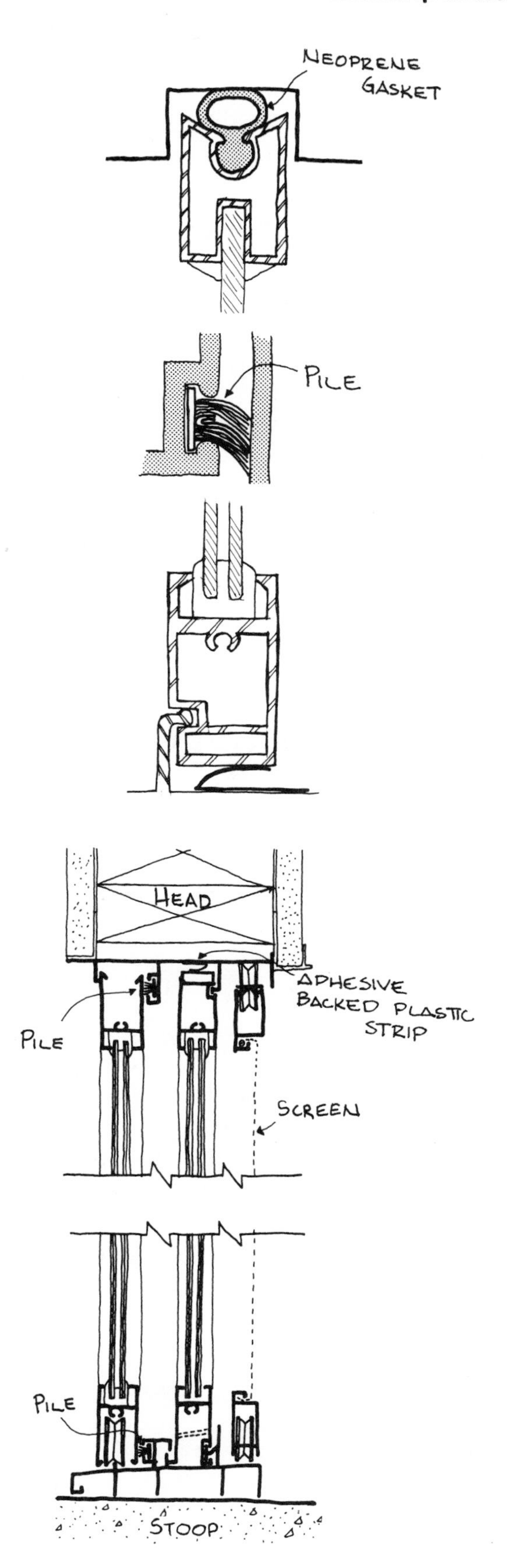

WEATHERSTRIPPING A HORIZONTAL SLIDING DOOR

Most horizontal sliding doors have metal frames and doors. They may be fitted with neoprene gaskets or "piles" that can be pulled out and replaced if they wear out. Adhesive backed plastic strips will also work here.

Other than the gasket, pile, and plastic spring strips, the stiles along both jambs can be weatherstripped with adhesive backed felt or foam.

If the sliding door has a wood frame and door, the door can be thought of as a sliding window and can be weatherstripped as such (except for the threshold). See "Weatherstripping a horizontal sliding window" in Chapter 9.

WEATHERSTRIPPING A HATCH DOOR

Hatch doors that open to the attic or crawl space from a heated living area should also be weatherstripped. Your best bet for weatherstripping hatch doors is adhesive backed foam strips applied to the inside face of the stops or along the edges of the door (adhesive backed felt may be too thin for application here).

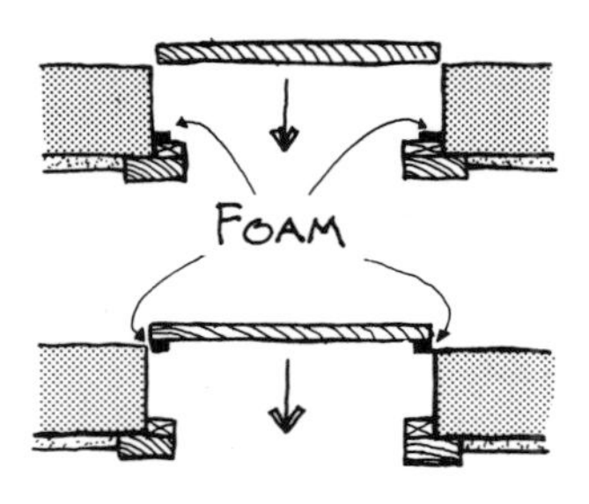

WEATHERSTRIPPING A GARAGE DOOR

1 Garage doors are usually not weatherstripped. Most garage doors, however, have a piece of heavy rubber at the base that cushions the door against the concrete. If this piece wears out, it can be the source of drafts into the garage and can be replaced.

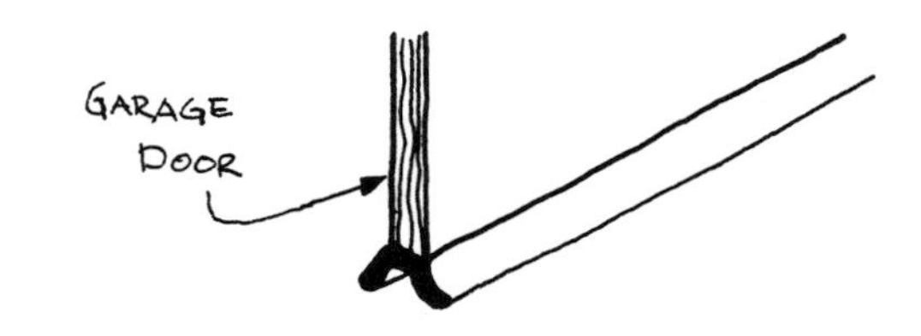

2 Top and side weatherstrips, made of vinyl, are available.

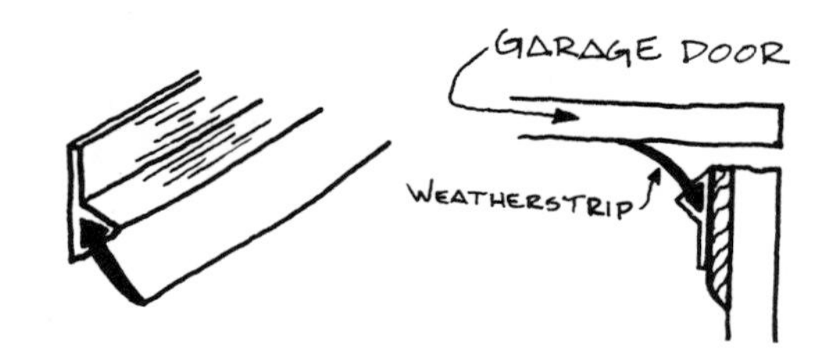

NOTE:

In some cases, a garage may have been converted into a heated living area of the home. In this case the garage door is probably fixed shut and should be treated as a fixed door in terms of sealing the edges.

---

3 Cracks between panels of overhead doors can be sealed with polyurethane weatherstrip tape. Leave a crease in the tape so it won't tear or split when the door is operated (or, apply the tape when the door is in a position where the cracks are the widest).

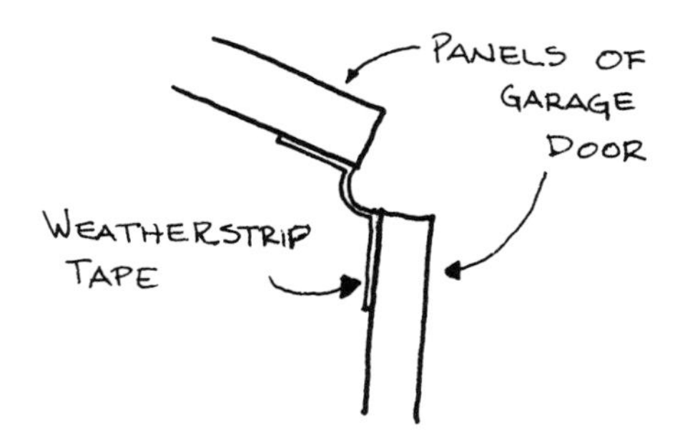

WEATHERSTRIPPING A FIXED DOOR

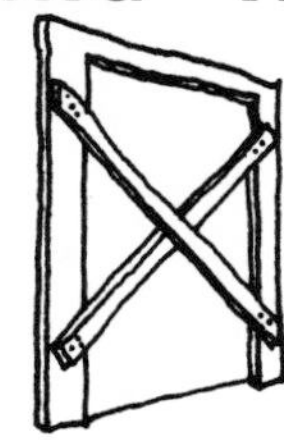

Fixed doors don't need weatherstrip. If air is getting in through cracks around a fixed door, the cracks should be caulked shut (see Chapter 10). Caulk is not only more durable than most weatherstrips, it's a lot cheaper.

19

install storm door

description

Although storm doors do not save as much energy as storm windows, there are advantages provided by storm doors in addition to reducing heat loss by conduction and infiltration. For example, an exterior storm door protects the primary door from weathering and reduces the amount of noise and pollution that may enter a home. Also, heat loss through the door can be further reduced if the storm door is equipped with an automatic closer. This is good if there are "forgetful" children because the storm door is never totally opened for long, even though the prime door may have been left open.

Storm doors are available for standard widths and heights of exterior door openings. Some suppliers will come out to your home and measure the doors for storms. If you do the measuring for the supplier, first check as to how the doors should be measured. Installing storm doors should be done only by someone who is skilled in carpentry (some suppliers will also install storm doors).

Storm doors are made of wood, rigid vinyl, or aluminum. They are available with a frame which is attached to the casing of the existing door. Plain (mill finish) aluminum will corrode. Use a door with an anodized or baked enamel finish. Use sturdy frames because children sometimes swing on the storm door while playing.

Storm doors are glazed with glass or plastic (acrylic, polycarbonate, butgrate). Combination storms (windows and screens) are also available. Since people can easily fall against a storm door, use only tempered glass which breaks into pellets which aren't likely to cut someone if an accident occurs. Like prime doors, storm doors should be weatherstripped (this can be done by the supplier when the storms are being fabricated).

Local building codes may require safety glass or a plastic glazing material. In certain buildings, exterior doors, including storms, must swing toward the outside to allow people to quickly exit the building. This is usually not required on small residential buildings. Don't use any storm door which has a narrower opening than the prime door.

materials

storm doors

- wood
- rigid vinyl
- aluminum
- steel

weatherstrip

- adhesive backed foam
- vinyl or rubber tube with fastening flange

cleaning solvents

- mineral spirits
- lacquer thinner
- turpentine

caulk

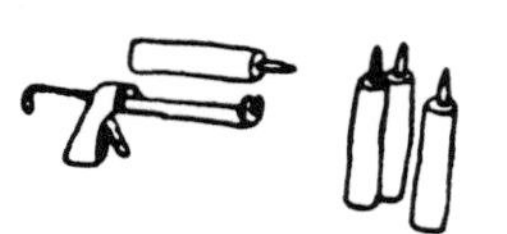

preparation

1 With a stiff brush and rag, remove all loose paint, dirt, and rust from the prime door frame. Wipe clean with a cleaning solvent.

2 If the storm door comes with a metal frame there may be a protective lacquer coating. Remove the lacquer coating on the back of the frame with lacquer thinner. This will assure adhesion between the metal and caulk.

installation procedures

Remember, storm doors reduce heat loss by creating a dead air space between the storm and prime door when the doors are closed. Therefore, you want the storm to be as air-tight as possible. Check the quality of the storm door construction; if you can see through the joints at the corner, it's going to leak air. Storms which come with a ready made and pre-weatherstripped frame are much easier to install. Hinges, latch, and closer are usually attached to this frame so once the frame is attached and the bottom seal is adjusted you're finished with installation.

Once the storm door is in place, regardless if you or a contractor install it, make sure the door closes tightly and that there are no cracks between the storm door frame and the prime door frame in or around the storm door.

1 Apply a continuous bead of caulk to the prime door frame along the line at which the storm door frame will be applied.

NOTE:

Wood storm doors should have the edges finished with a sealer to reduce swelling and warping. If door adjustments must be made by cutting or planing the door, the edges should be resealed.

---

2 Fasten the storm door frame in place with galvanized or aluminum wood screws (use galvanized screws for steel frames and aluminum screws for aluminum frames). If you're installing an unframed storm door, a wood strip 3/4" x 1-1/2" or so may be needed to provide a "stop" for the door. Install the stop as shown, or trim the door to fit into a "reveal" in the door trim. Take account of where hinges, latch, and bottom seal must be installed.

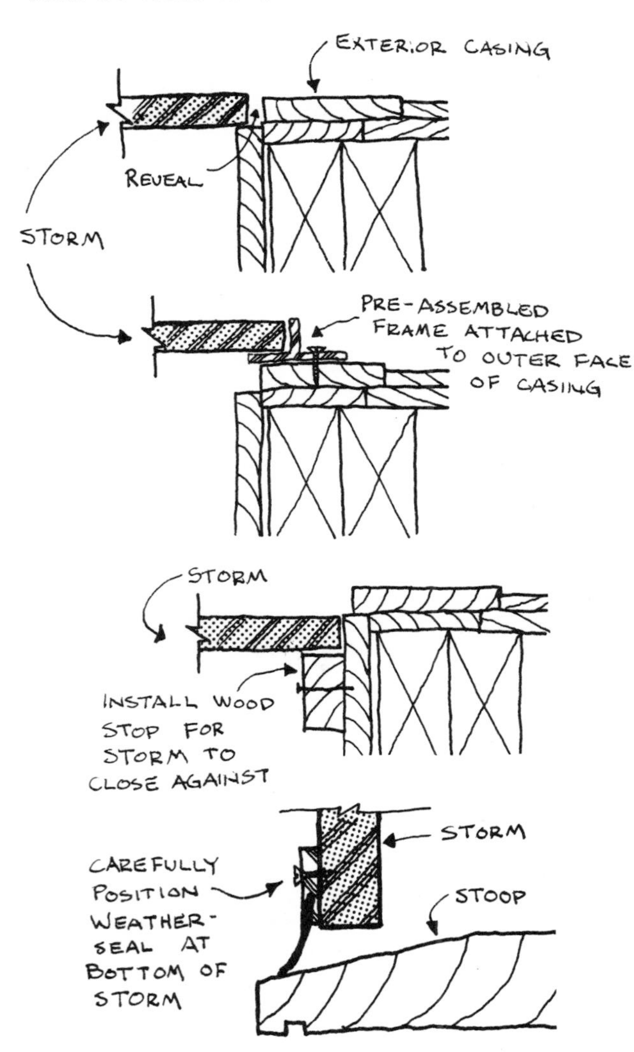

3 Make sure the storm door is installed in place straight, plumb, level, and without distortion. If the hinges are not aligned vertically, the door may tend to swing open if the latch or closer is broken.

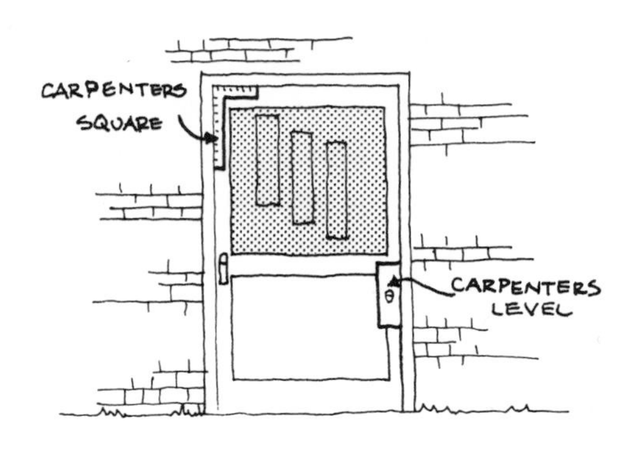

AN IMPORTANT NOTE:

The storm door must not block the movement of the prime door. As a general rule, storms swing in the opposite direction of the prime door and they should be hinged on the same side of the frame as the prime door.

4 Once the storm door is in place, check that it does not bind against the frame when opened or closed. If it's a combination storm, check to see if the screen insert fits into the storm tightly.

5 Storm doors are fairly light. Be sure that the storm latches easily but securely so that it won't be blown about in the wind and broken.

20 replace existing door

description

Doors are subject to wear and tear by people as well as by the weather. Thus, doors may stick, bind, rattle, or not close all the way. Doors must be operating for other improvements to be worthwhile. However, a door may be so warped, cracked, and beyond repair to make other improvements fruitless. In this case, replacing the door may be a better way to reduce heat loss.

These procedures are for replacing the door, the door and door frame, or just the door frame. As with replacing a window, this work should be done only by someone who is skilled in carpentry. Two people should work together when replacing a door.

Replacement doors can be purchased with or without frames at lumber companies. Stock sizes are available or custom size doors can be made. When purchasing a door, the size and type is usually printed on it. For example, if "2/6 6/8 H.C." is printed on the door, the dimensions are 2'-6" x 6'-8" and the door is hollow core. The "Hands of doors" is also important when ordering hardware for the door and is discussed later in this chapter.

Local building codes may require safety glass or plastic glazing material if glazing is used in the door. In certain buildings, exterior doors, including storms, must swing toward the outside to allow people to quickly exit the building. This is usually not required on small residential buildings.

materials

replacement doors

PANEL DOOR

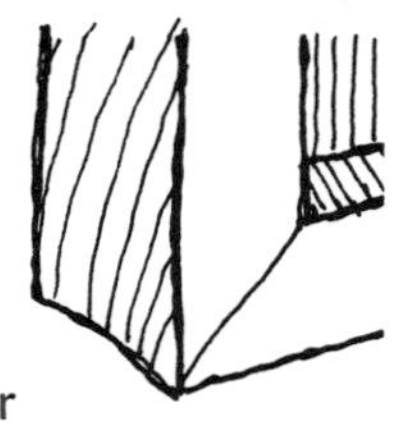

- consists of wood panels held in place by wood verticals (stiles) and horizontals (rails); panels can be made from solid wood, plywood, metal, or glass

SOLID CORE DOOR

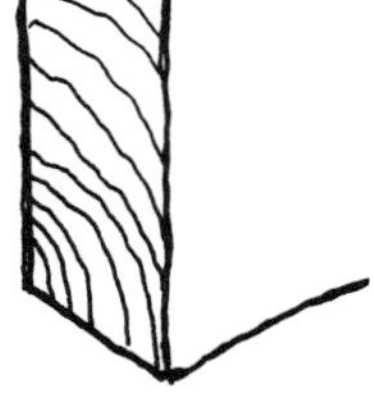

- consists of a solid lumber core, compressed wood fiber core, or rigid foam board (insulated door) laminated to plywood or metal facing

HOLLOW METAL DOOR

- consists of shop fabricated welded, sheet metal; mineral or asbestos board (fire resistant) or rigid foam board (insulation) can be added as a core

HOLLOW CORE WOOD DOOR

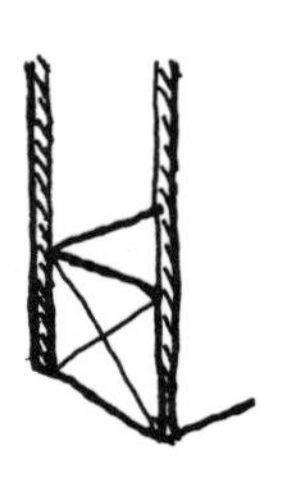

- consists of plywood or press-board face and a honeycomb wood or paper interior core; hollow core wood doors are usually for interior use only

caulk

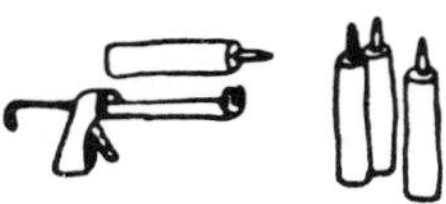

packing

- polyethylene strips
- polystyrene foam board
- oakum
- urethane, 1 or 2 part foam
- fibrous or rigid foam
- board insulation

cleaning solvents

- turpentine
- mineral spirits
- lacquer thinner

framing lumber

- shims
- wedges
- blocks
- framing

millwork & trim

- door jamb
- head piece
- stops
- casing

thresholds

- wood
- aluminum

HANDS OF DOORS

To determine the hand of a door, stand on the outside. If the door swings away from you, and the door knob is on your right (hinges on your left), you are looking at a "left-hand regular bevel" door.

LEFT-HAND REGULAR BEVEL

Other hands of doors are determined in a similar fashion and are illustrated here.

RIGHT-HAND REGULAR BEVEL

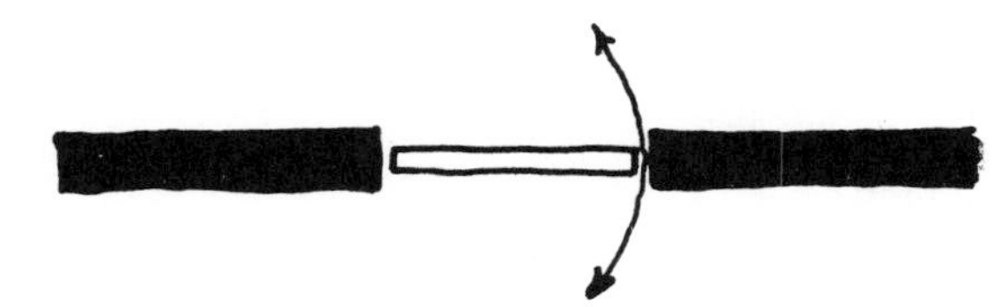

DOUBLE ACTING

LEFT-HAND REVERSE BEVEL

RIGHT-HAND REVERSE BEVEL

A NOTE ABOUT HINGES

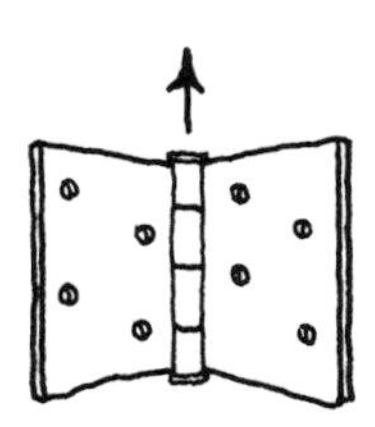

Hinges are made up of two "leaves", and the leaves are joined by a pin. When fastened to the edge of a door, only the butt end of the hinge can be seen. Thus, hinges applied to the edges of doors have come to be known as butt hinges ("gate hinges", which attach to the door and the door casing, are not commonly found). Butts are usually mortised* into the edge of a door and are the most common type of hinge used probably because they're concealed when the door is closed. The size of butt hinges vary from 2" to 6". Hinges are usually mounted on a door 5" from the head and 10" from the floor

*A slot cut into wood, usually edgewise.

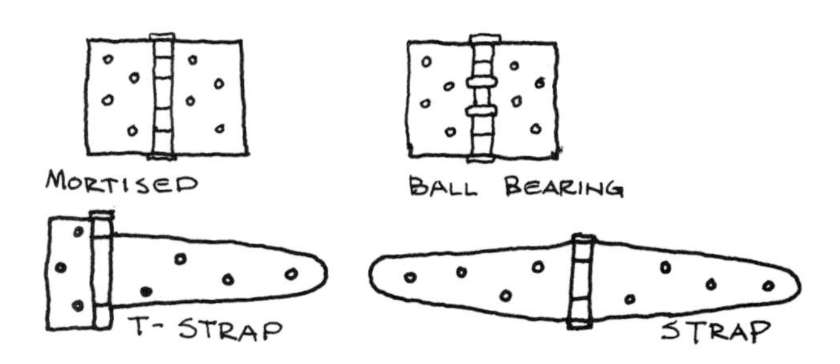

(when a third hinge is required, it should be mounted equidistant between the top and bottom hinges).

preparation

1 Begin by removing the old door. If the hinges are good (not rusted or bent) and are attached firmly to the jamb (not loose), remove the pins from the hinges to get the door out (it's best to have help in removing the door). If the hinges, as well as the door, have to be replaced, open and support the door and unscrew the hinges from the jamb to get the door out.

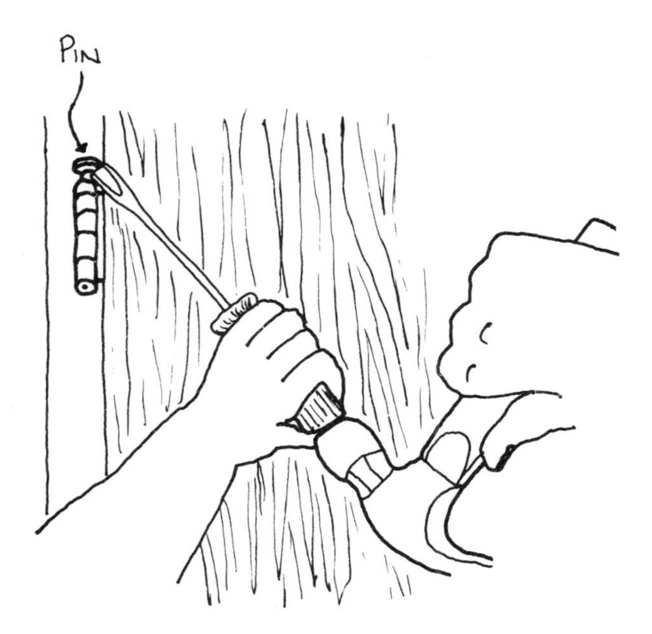

2 If you're replacing the door frame, remove the stops, jamb, outside and inside casing and threshold. Also, remove any interior and exterior wall finishes to expose the framing studs, headers, and sill. Do this work carefully as you'll have to replace these materials later. Unless severely damaged or rotted, do not remove the drip cap, lintel, or sill.

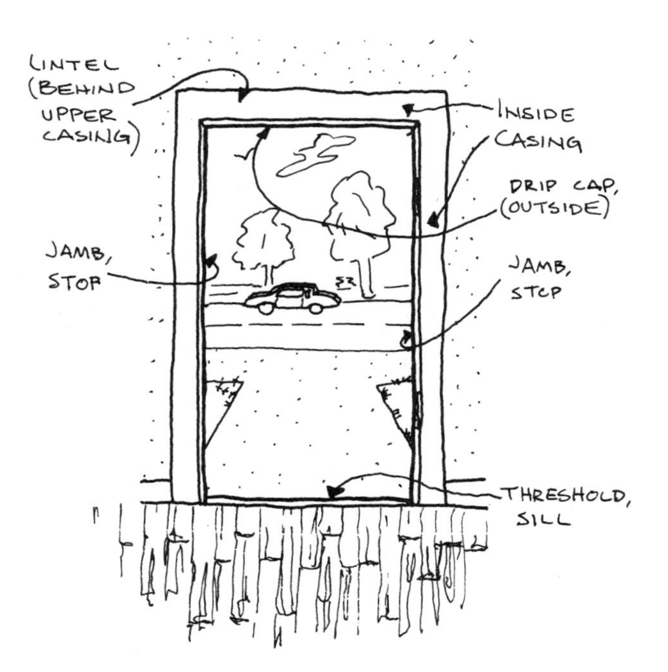

3 Adjust the framing studs, headers, and sill to provide the correct rough opening size. This may require planing or adding shims. Check the replacement door manufacturer's specifications for rough opening size.

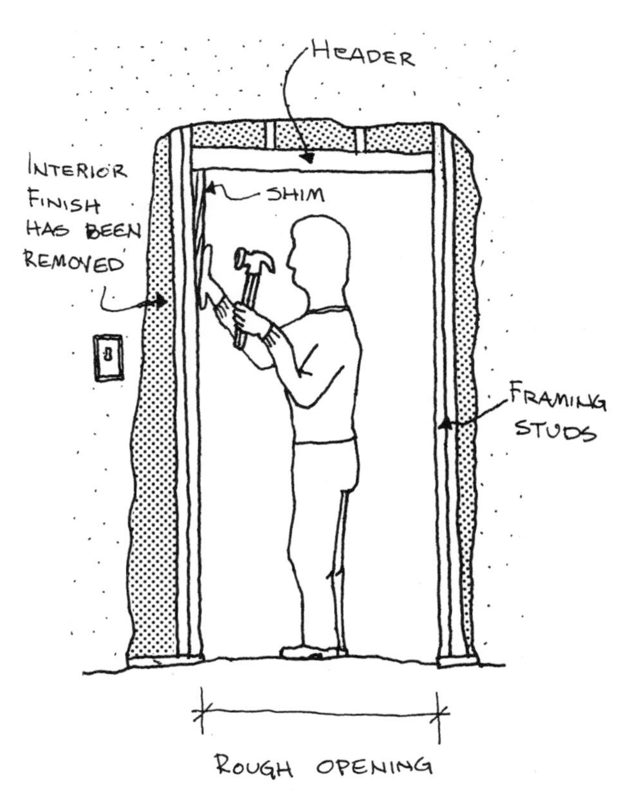

AN IMPORTANT NOTE:

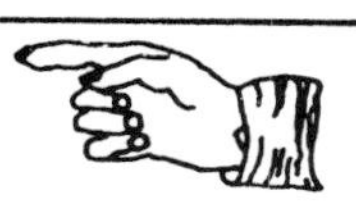

The rough opening is an opening in the structural support of the building. Therefore, if the header and/or studs need to be relocated to accommodate the replacement door, you must temporarily support the ceiling first. See Chapter 16, "Replace Existing Window", for details on temporary supports.

installation procedures

If you're installing a door frame complete with a prehung door, installation procedures I through 6 apply. If you're replacing only the door, begin with step 7.

The specific problems and details that you'll encounter when installing replacement doors are dependent upon the door and wall construction. Therefore, only general points and guidelines are covered here.

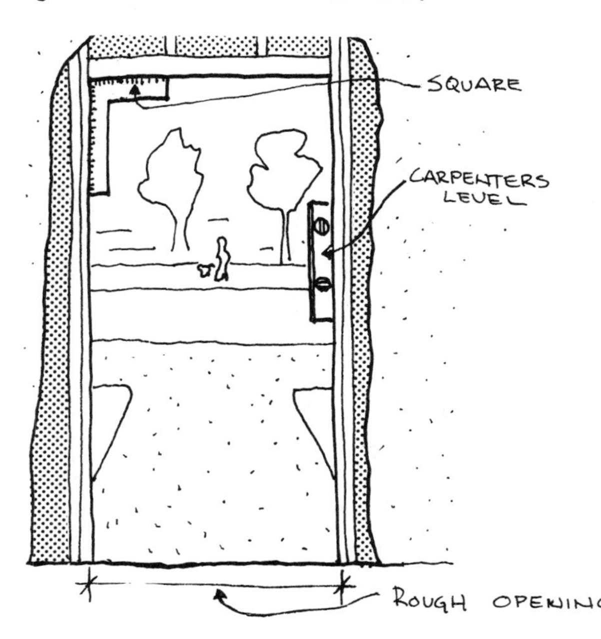

1 The rough opening is to be square. The header, studs, and sill are to be plumb and level. This may require moving framing members or bringing the framing members out by adding wood strips.

2 Install the framing with the door in place. The sill, jambs, and side casings are to be plumb, level, and square. Shim, or plane, if necessary.

The objective of these first two steps is to insure level surfaces for the hinges and to assure that the door closes tightly against the stops. After step 1, if the rough opening is still out of plumb and not square, shim or plane the frame to accommodate the rough opening.

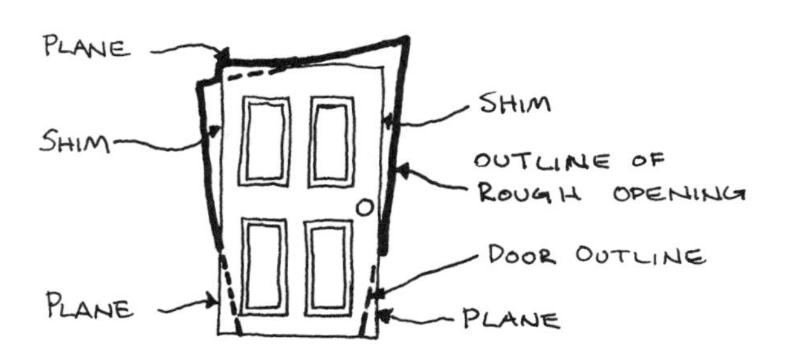

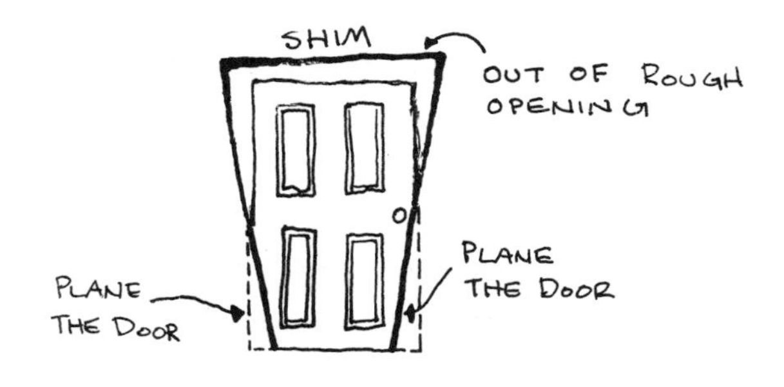

3 Set the door frame in place and secure it with a temporary brace. Insert blocks or shims between the frame and rough opening so that the frame is plumb and level. Secure the shims by driving a finishing nail through the jamb and shim into the framing.

Finally, nail through the outer door casing into the framing with casing nails spaced at 16" o.c. Block the frame solidly on both sides of the jamb at the height of the door knob/door lock.

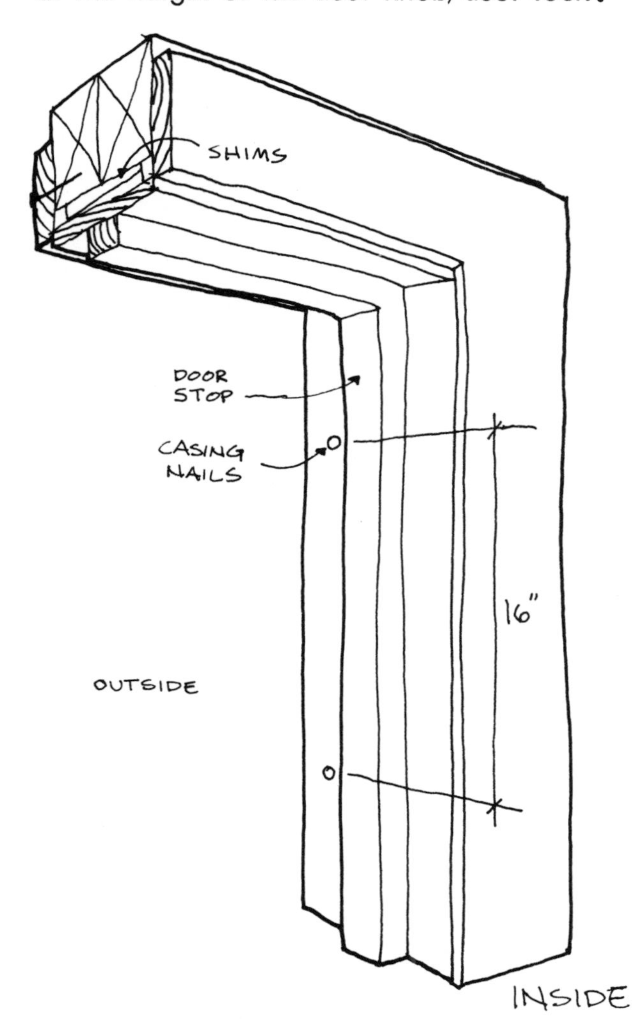

4 Re-install, or patch, any exterior finish that was removed to get the original door out.

5 Pack any voids between the structural wall members and the door framing with strips of insulation or urethane spray foam. Re-install or patch any interior wall finish. Caulk any remaining voids between the frame and rough opening before installing the inside casing.

NOTE:

It may be easier to add the insulation as you're installing the door frame (step 2).

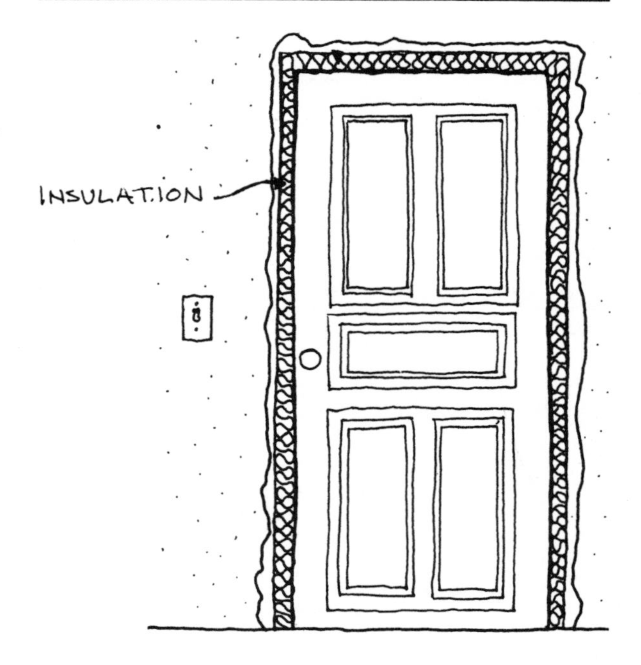

6 Pack and/or caulk the exterior frame - see Chapter 10 for details.

door replacement only

7 If needed, replace the threshold prior to hanging the door. For details on thresholds and door weatherseals, see Chapter 17, "Install Door Threshold and/or Bottom Seal".

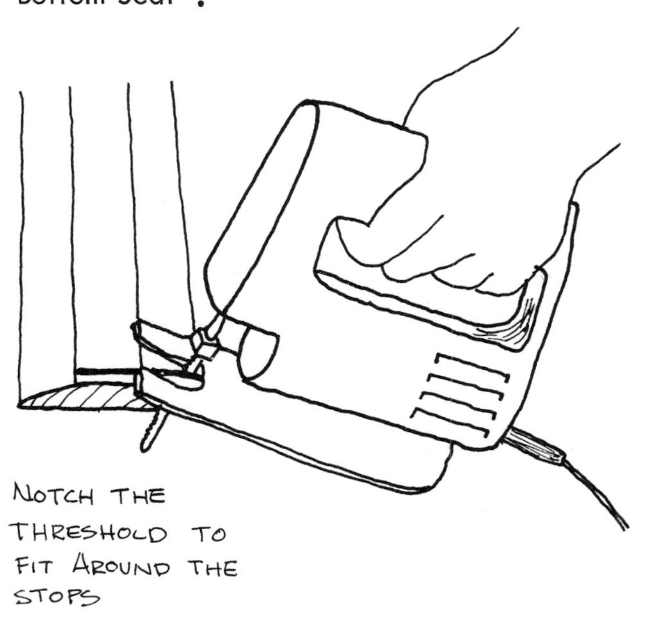

NOTCH THE THRESHOLD TO FIT AROUND THE STOPS

NOTE:

Thresholds which are built up so that the door closes against them are not advisable in homes of the elderly. It's easy to trip and fall over this type of threshold.

8 Trim the door to conform to the shape of the door opening. Any irregularities in the jambs should be transferred to the door. The door should be trimmed, providing a clearance of 1/16" between the door and the jambs/threshold. Mark these dimensions on the door. Where possible, limit your trimming to the hinge stile, since there is usually a slight bevel on the lock side.

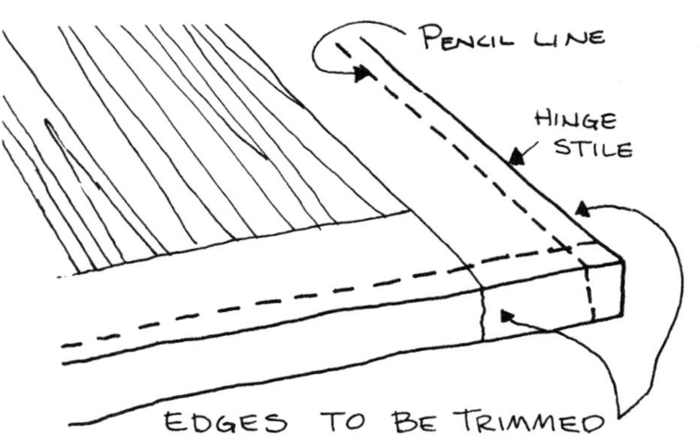

some tips on planing

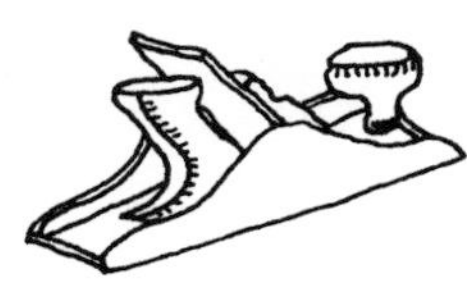

The amount to be trimmed is usually too small for a saw and can be done with a plane.

Place the door on the floor with the long edge down. There are a variety of ways of steadying the door for planing;

- prop the door into a corner
- clamp it to a door that is wedged opened (use pieces of cardboard to avoid damaging both doors)

- have a friend hold one end while you straddle the other end

In any event, you'll find planing easier if you straddle the door and brace it with your legs.

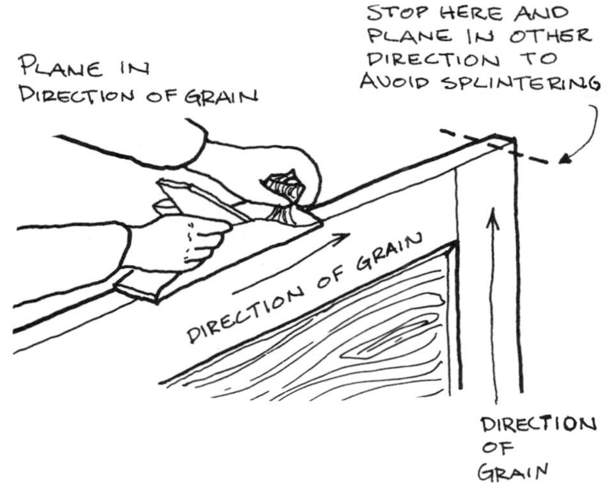

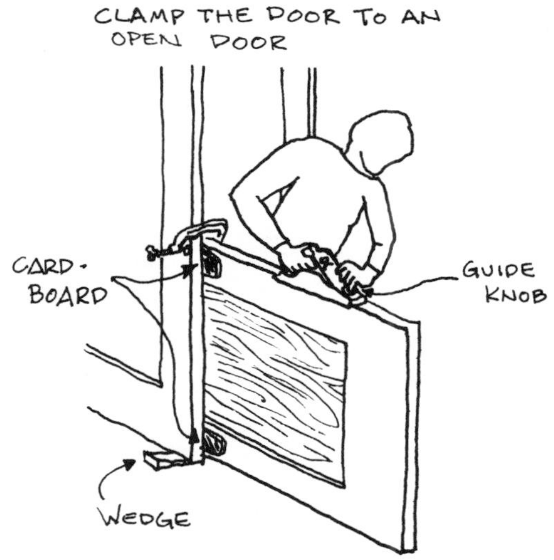

When planing;

- use a plane that has a blade wider than the edge of the door
- at the start of the stroke, keep pressure on the guide knob at the front of the plane
- at the end of the stroke keep pressure on the rear of the plane - not on the guide knob
- keep the plane parallel and even to the door edge
- when planing across the grain, do not plane across the edge as this may cause the door to splinter - instead, plane in from the edge towards the center

9 Next, position the door in the frame to locate the hinges on the door. This involves wedging the hinge stile and the top rail against the jambs. Place wedges, or shims, against the lock stile and the threshold to hold the door in this position. Place a 4-penny finish nail between the top rail and head jamb prior to wedging the door to assure proper clearance along the top of the door.

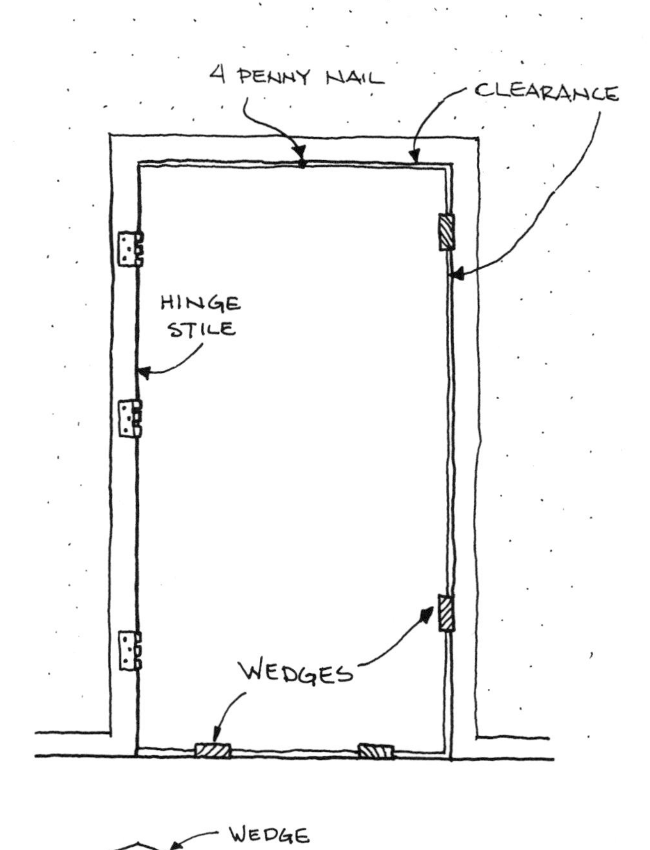

10 Now that the door is wedged into position, mark the location of the hinges along the hinge stile of the door.

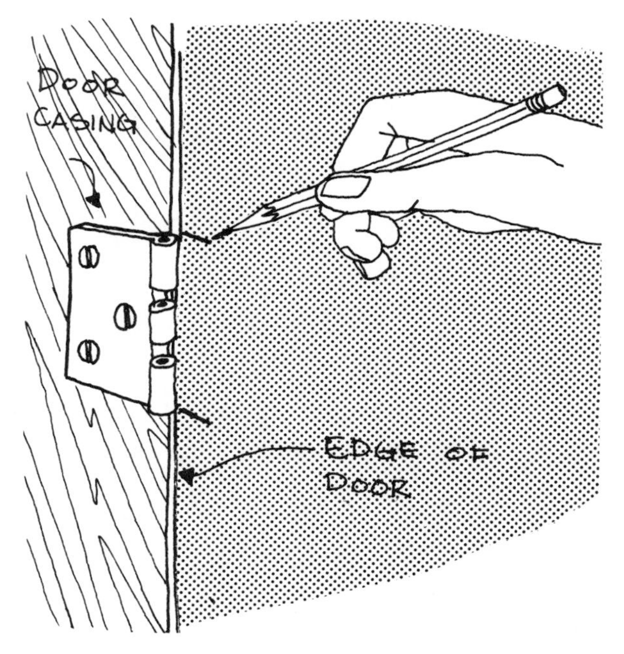

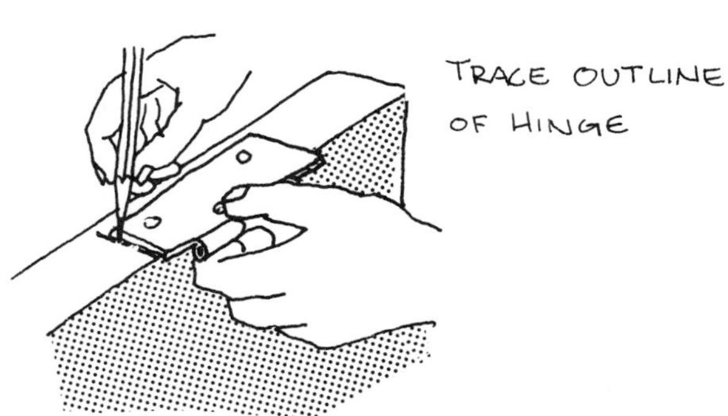

11 Remove the door and, once again, prop it up on the floor with the hinge stile facing up. Tap a chisel around the outline of the hinge location. Now make a series of ridge-cuts by tapping the chisel across the width of the outline. Finally, scrape off the excess wood by moving the chisel sideways against the wood. This recess is known as a mortise. A "router" * may also be used to mortise a door for hinge attachment.

* A router is a machine with a revolving vertical spindle and cutter for milling out the surface of metal or wood.

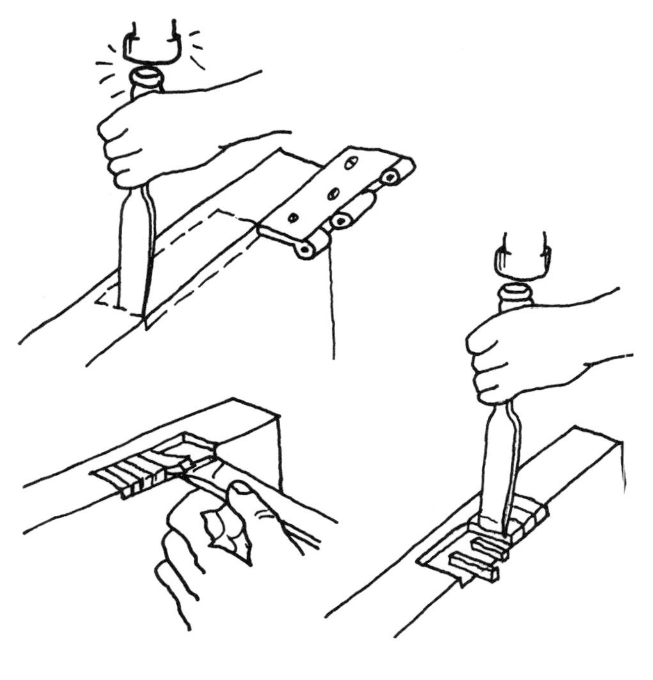

NOTE:

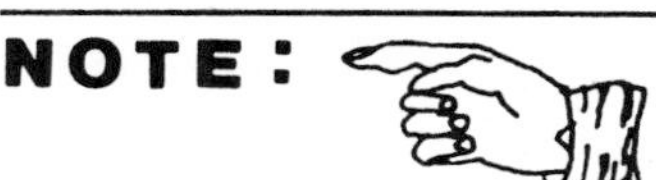

The width of the hinge leaf is not to be greater than the edge of the door.

12 The mortise should be deep enough so that the hinge leaf, when screwed in place, is flush with the edge of the door (this is usually not more than 1/8").

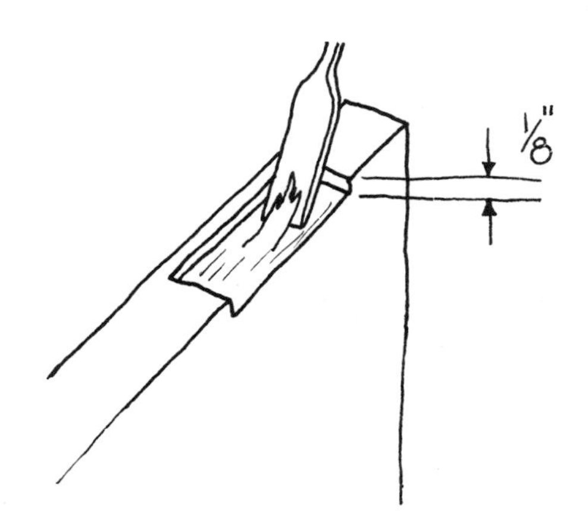

If the screws holding the hinges to the jamb are loose and you can't tighten them, remove the screws and force toothpicks into the holes.

Break off the ends and then replace the screws.

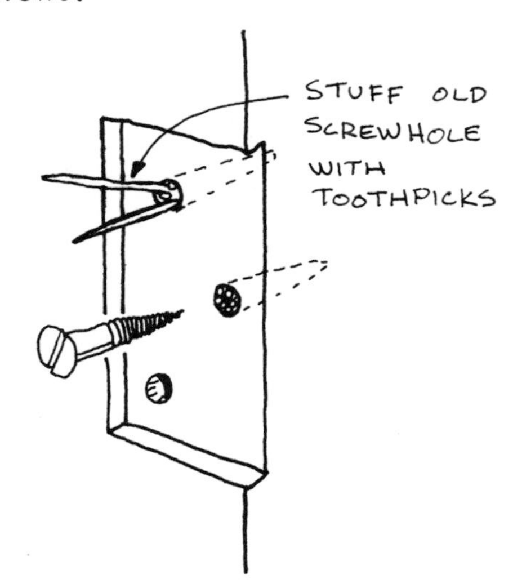

13 Mount the hinge leaves to the door with screws (be sure to mount them so that the hinge pin is up when the door is hung). Pre-drill holes for the hinge screws.

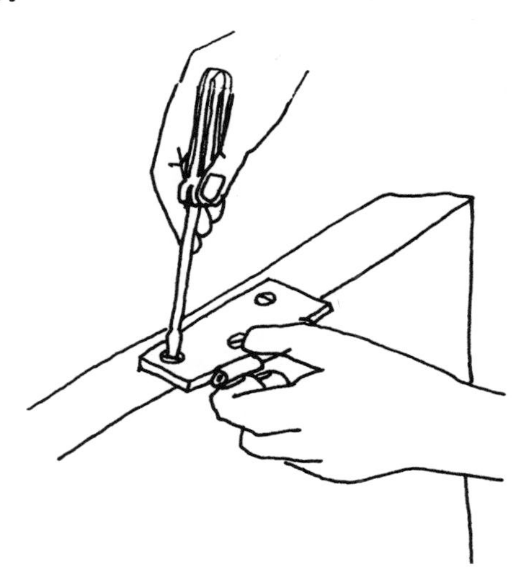

NOTE:

If the existing door frame is not plumb (straight up and down), then make sure you plumb the hinge side of the frame even if you must trim from the lock side. This will assure that the door holds its position and doesn't swing open or shut on its own.

14 Now you're ready to hang the door. It's best to have help while doing this - one person to hold the door while the other person slides the pins in place. Once the door is in place, check and make sure it opens and closes freely and locks.

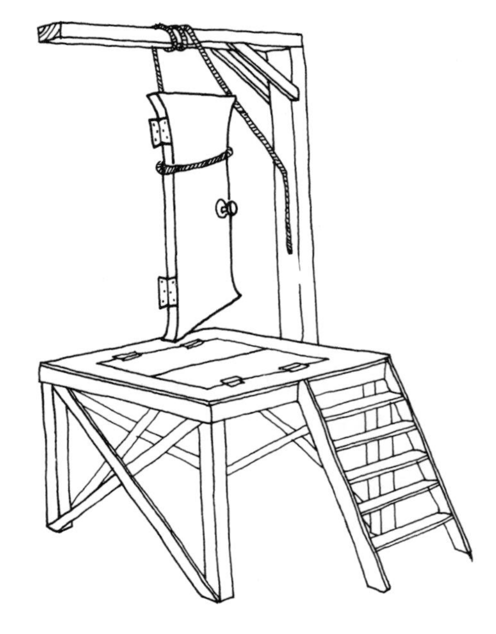

"HANGING THE DOOR"

IV

insulation retrofits

Insulate basement walls on the interior

Insulate basement walls on the exterior

Insulate crawl space walls on the interior

Insulate crawl space walls on the exterior

Insulate floor above crawl space

But what about accessing the crawl space

Insulate slab on grade

Insulate masonry wall on the interior

Insulate masonry wall on the exterior

Insulate open frame wall

Insulate closed frame wall

Insulate unfinished attic

Insulate finished attic

Insulate finished attic outer sections

But what about accessing the attic

But what about attic ventilation

Insulate finished ceiling against roof

where to insulate?

21
insulate basement walls on the interior

description

Many basements, though intentionally not heated, are relatively warm during the winter because of heat loss from the heating plant and upper floor. You may use the basement in the winter for storage, washing clothes, workshop, etc., and therefore wouldn't mind keeping it warm. In addition, the heat in basements keeps pipes from freezing. However, heat is still lost from basements to the ground and outdoor air through the basement walls. To reduce this loss, basement walls can be insulated (this includes the bandjoist and headers). After installation, the insulation must be covered with a fire rated material that is nailed to wood framing members.

If the basement walls are unfinished, roll/batt insulation (with a 2" x 4" framing system) or rigid board insulation (with furring strips) can be used. If the basement walls have already been finished, chances are that there is a furred air space behind the finish where blown in place insulation can be used.

NOTE:

In extreme northern climates, such as Minnesota and Maine, insulating the interior of basement walls may cause them to "heave" because of deep frost penetration. Insulating the exterior of basement walls (see Chapter 22) may be recommended. Check with your building department or local HUD/FHA office.

---

materials

roll/batt insulation

- glass fiber
- rock wool
  (foil faced, kraft paper faced, unfaced, or reverse flange)

rigid insulation

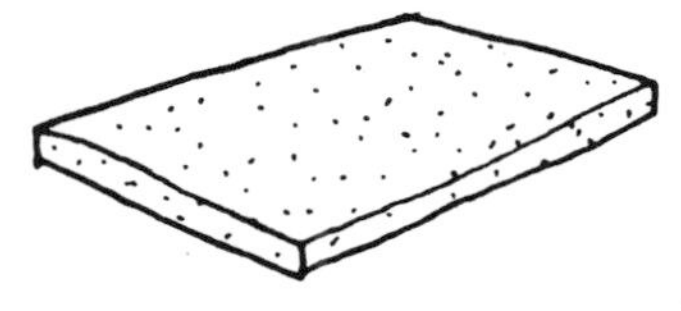

- polystyrene
- polyurethane
- polyisocyanurate
- mineral or wood fiber
  (sheathing board)

blown in place

- cellulose
- rock wool
- glass fiber
- vermiculite
- perlite

vapor barrier

- aluminized or oil based enamel
- latex vapor barrier paint
- polyethylene film
- foil faced gypsum board

waterproofing-adhesives

- fibrated mastic
- asphalt emulsion
- liquid epoxy
- panel adhesive

furring / framing

- 2" x 4" framing lumber
- furring channels
- metal channels/studs
- gypsum board or other fire rated
  material – check local codes

other materials

- masonry cleaners

- caulk
  (or patching cement)

- packing

- fasteners

- hole plugs

preparation

Before adding insulation to the basement wall, any water seepage through the wall must be stopped by waterproofing on the exterior of the wall. Basement walls which show evidence of leakage should not be insulated before waterproofing (and other needed repairs) are made.

Excavating the perimeter of the basement is necessary for waterproofing. For detailed excavating procedures, see Chapter 22, "Insulate Basement Walls on the Exterior". Waterproofing should be applied down to the footing.

If waterproofing is not necessary, skip to the Installation Procedures.

1 Clean all dirt and loose material from the wall with a stiff brush and water. A mild acidic masonry cleaner can also be used, but avoid skin contact with the cleaner by wearing gloves and a long sleeve shirt. After the wall has dried, seal all cracks and joints with caulk and, if necessary, packing. Cracks should be routed* 1/4" or more in width and depth. Working cracks (use an elastomeric sealant here) should be routed 3/4" to 1" in width and depth. Refer to Chapter 5, "A Word about Caulks", for details regarding "dormant" and "working" cracks. Large dormant cracks can be sealed with patching cement.

* Rout - the removal of material by cutting, milling, or gouging to form a groove.

2 Apply waterproofing to the wall from just above the grade line down to the footing. For backfilling procedures, see Chapter 22.

APPLY WATERPROOFING DOWN TO THE FOOTING

installation procedures

roll/batt insulation

Roll/batt insulation, whether it's faced or unfaced, is installed between wood framing. Essentially, this means building a frame wall over the masonry with 2" x 4" framing lumber or metal studs. The insulation, as well as the finish material, is then attached to the studs and plates*.

1 Begin by building the framework. Space the studs to correspond to the width of the insulation. Typically, this is 16" or 24". Nail the bottom plate to the floor and the top plate to the ceiling joists. Then toenail the studs in place.

Alternatively, framing can be "tilted up" after being built on the floor, just as a housing contractor does. If this is done, be sure to find the tightest dimension between the floor and ceiling (joists) and space the plates 1/4" or so less than this dimension. Frame around windows, pipes, and other obstructions on the wall (notch the framework if necessary).

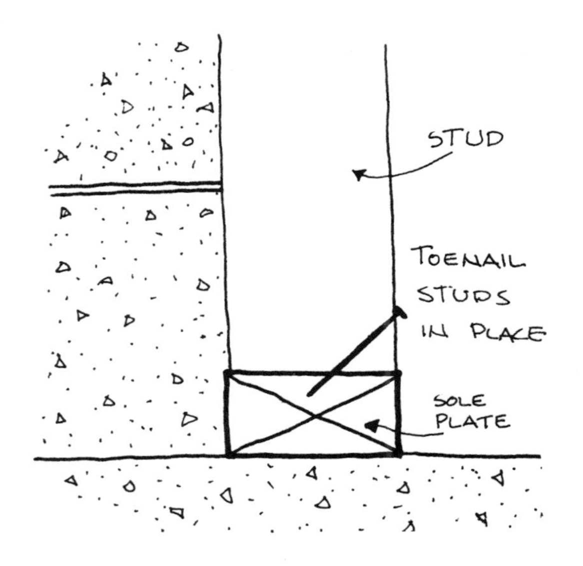

NOTE:

If wood is to contact the basement floor, floor drains should be present and in working order to drain any water accumulation from cleaning or possible plumbing leaks.

* Top and bottom nailing strips

SPACE THE STUDS TO CORRESPOND TO THE WIDTH OF THE INSULATION

FRAME AROUND WINDOWS, PIPES AND CONDUIT.

2 Next, cut small pieces of insulation to fit around the bandjoist and header*. Simply push the insulation between the joists against the bandjoist and header. Faced insulation should be installed with the facing side out (towards you).

* The edge joist that is perpendicular to the floor joists is the "header". The edge joist that is parallel to the floor joists is the "bandjoist". Oftentimes, though, these are collectively referred to as the "bandjoist" or "rim joist".

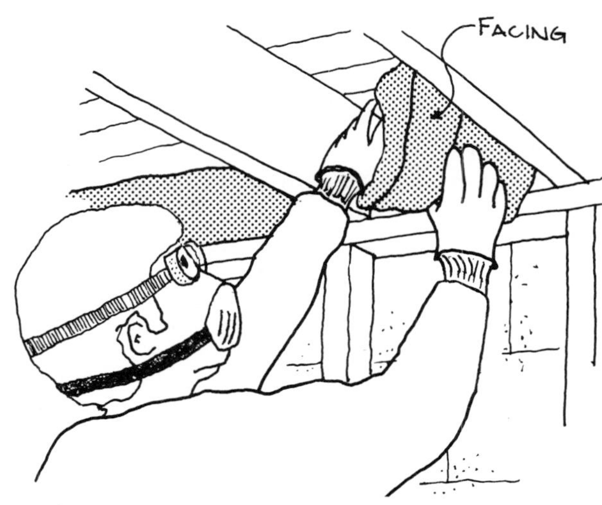

INSULATE THE HEADERS AND BAND JOIST WITH PIECES OF BATT INSULATION

3 Install the insulation between the studs from the top plate down to the sole plate. If the insulation has facing, install it so that it faces out towards you. Cut the insulation to the proper length with a utility knife and straight edge (it's easier to cut with the facing placed up).

NOTE:

If the basement is prone to flooding, extend the insulation down the wall to a point above the floor at which it won't get wet.

4 Place the insulation behind (on the cold side of) pipes, wiring, or electrical conduit. Keep insulation a minimum of 3" away from wiring that is frayed or not armored in conduit. Pack narrow spaces with scraps of insulation.

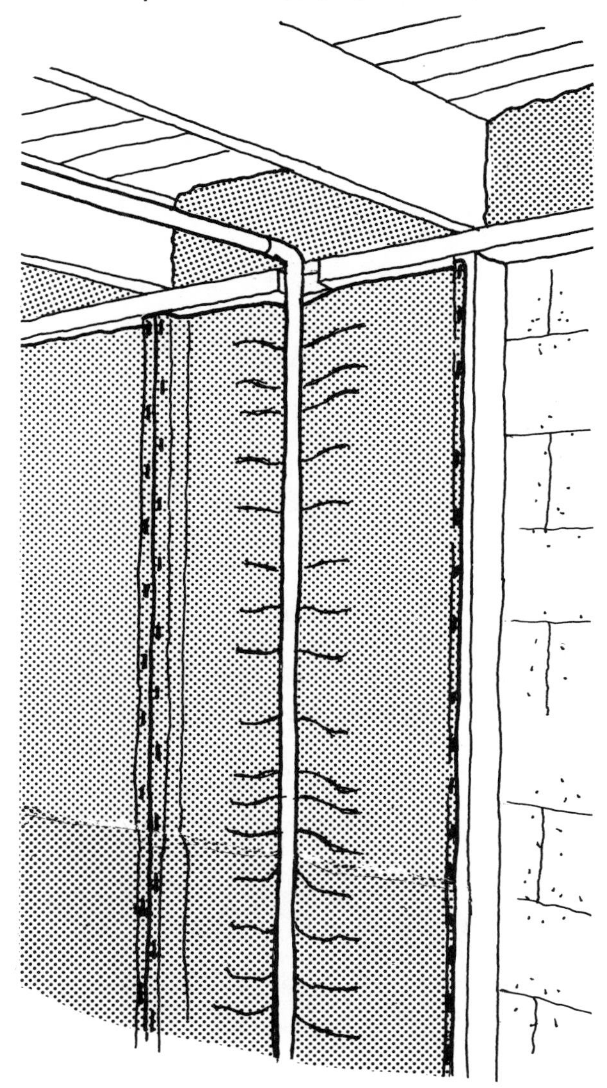

5 Staple, or tape, the insulation to the studs. Attach the flanges of the insulation to the edge or inside face of the studs. Avoid creases, folds, or "fishmouths" in the insulation. If staples are used, space them approximately 6" apart and cover them with polyethylene tape. Be sure that there are no voids between pieces of insulation that butt together.

THE INSULATION CAN BE STAPLED IN PLACE

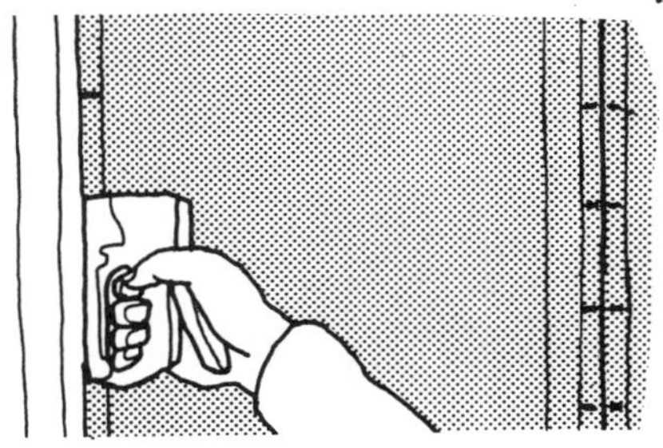

6 Wedge unfaced insulation between the studs. To install a vapor barrier (usually a polyethylene sheet) over the unfaced insulation, staple it to the studs. You'll need to go back and tape over the staples with polyethylene tape to keep moisture out. If unfaced insulation was used along the bandjoist, you'll need to cover that also with a vapor barrier. The foil or kraft paper backing on faced insulation serves as a vapor barrier.

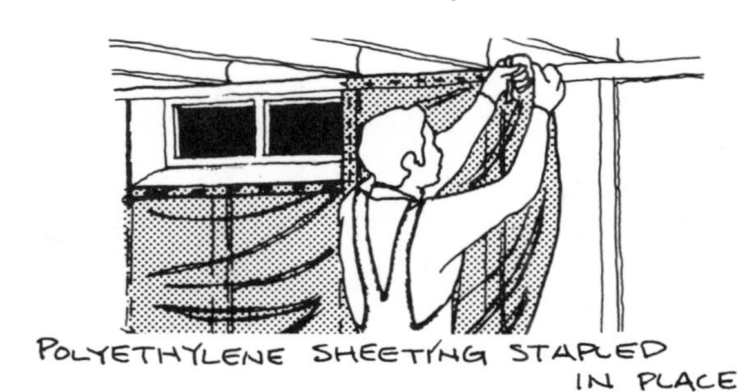

POLYETHYLENE SHEETING STAPLED IN PLACE

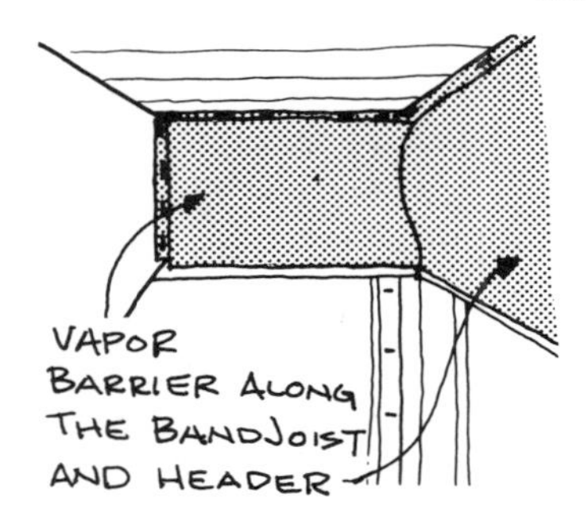

7 Nail gypsum board to the wood framework. Gypsum board can be glued to the framing if a vapor barrier sheet has not been installed covering the framing. The gypsum board should extend the entire height of the wall including the bandjoist and header. You can "finish" the basement by painting the gypsum board or by covering it with wood paneling.

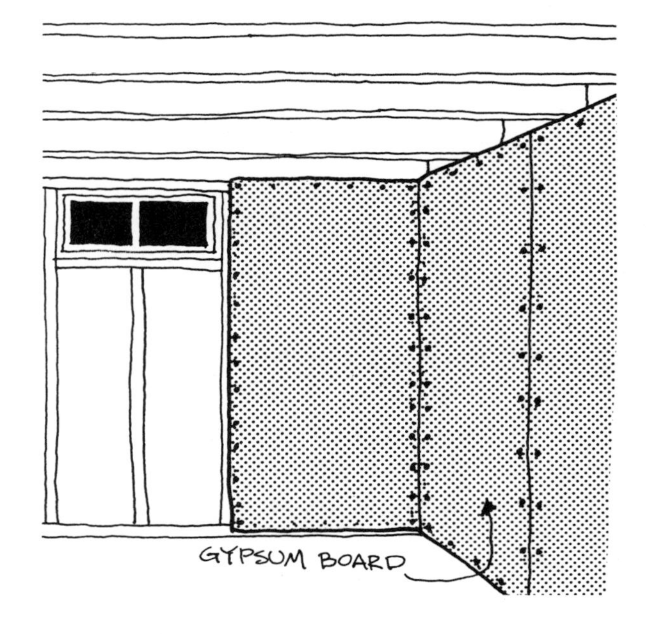

NOTE:

An alternative to a polyethylene sheet is foil faced gypsum board or a vapor barrier paint. Apply the paint, either an aluminized, oil based enamel, or latex vapor barrier paint, over the exterior of the gypsum board after it's installed.

RIGID INSULATION

1 Rigid insulation should be installed between furring strips (if the insulation is thicker than 1-1/2", use 2"x 4" lumber or metal studs as described for Roll/batt insulation, step 1, in this chapter). Attach the furring strips to the wall with an adhesive and/or masonry nails. Use shims for uneven wall surfaces. Attach new furring over existing furring to achieve the required depth for the insulation.

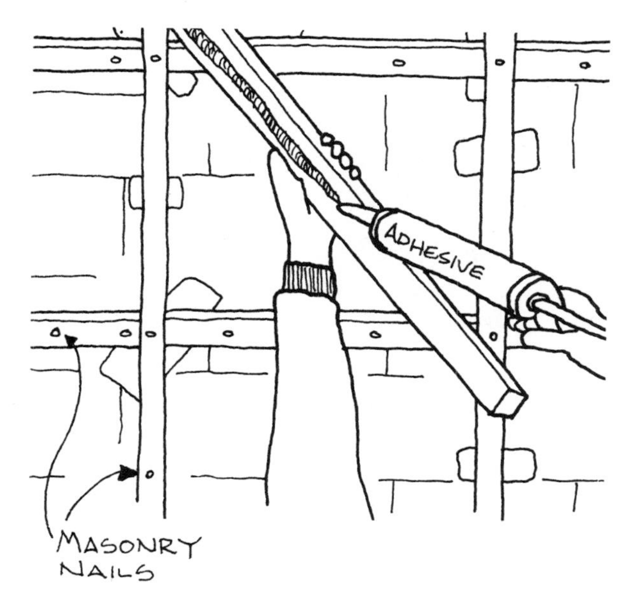

2 Attach the vertical strips to the wall with a maximum spacing of 4' on center (the long edges of 4' x 8' gypsum board panels will have wood backing).

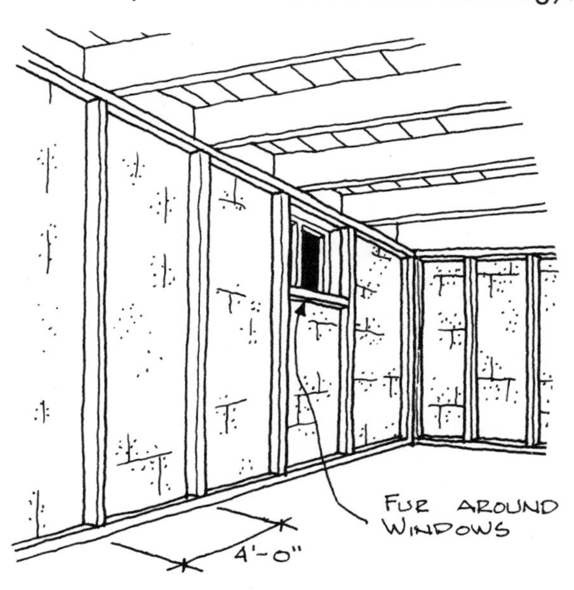

Rigid insulation is usually sized (16" x 48", 16" x 24", 24" x 48") to fit between furring spaced 24" o.c. (or some multiple of that). Check the dimensions of the insulation that you'll be using prior to installing the furring to minimize cutting the insulation. The important thing to remember is to provide vertical support for the gypsum board. Be sure to fur out around windows, pipes, and other openings or obstructions to provide support for the gypsum board.

An alternative to furring strips are metal "Z" channels. The channels are available in varying depths to accommodate different insulation thicknesses.

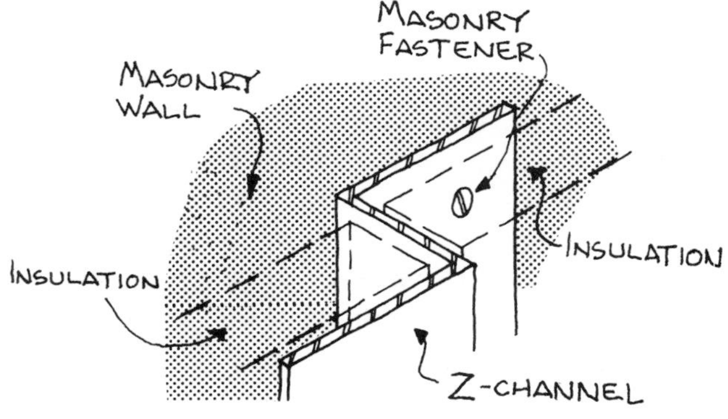

3 Insulate the bandjoist/header with roll/batt insulation as previously described (see Roll/batt insulation, step 2). Rigid insulation can be used here, but roll/batt insulation may be easier to install.

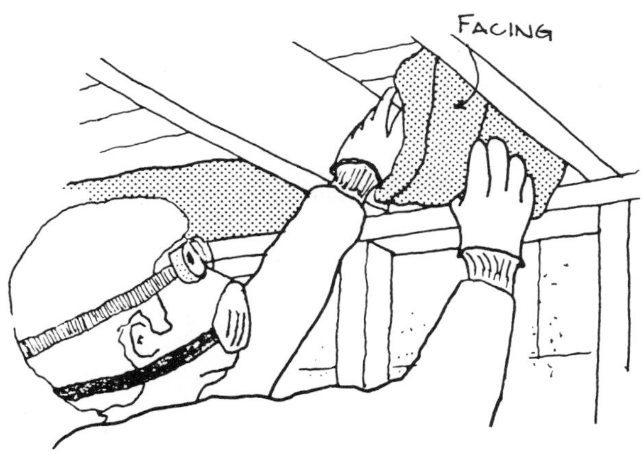

4 Attach rigid insulation directly to the wall with an adhesive. Be sure the adhesive is compatible with the insulation! Apply the adhesive along the perimeter of the board with spot "dabs" on the interior. Firmly press the insulation in place so that it fully contacts the wall surface. The boards should butt tightly together.

ATTACH RIGID INSULATION TO THE WALL BETWEEN FRAMING LUMBER

5 Some foam board is available with an integral foil face that serves as a vapor barrier (the foil is to face out towards you). Otherwise, add a vapor barrier to the rigid insulation as described for Roll/batt insulation, step 6, in this chapter.

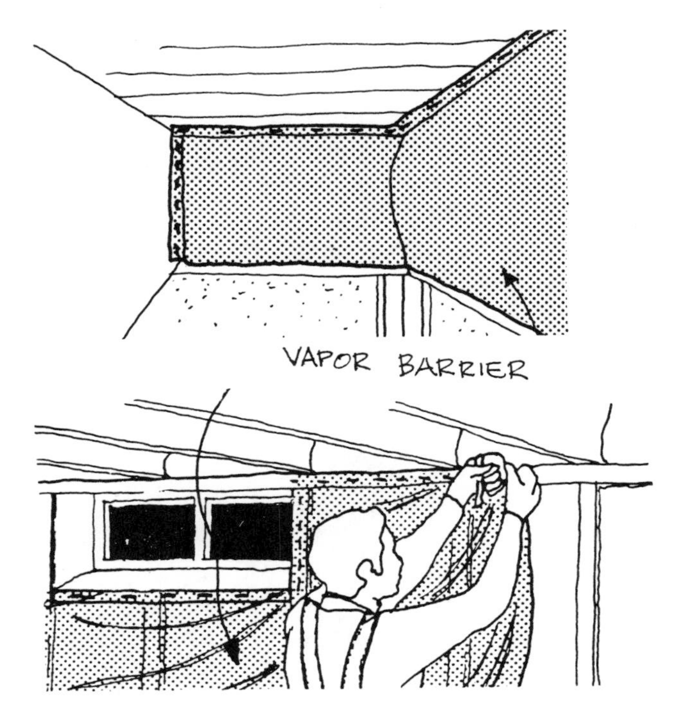

6 Nail or glue gypsum board to the furring strips. The gypsum board should extend the entire height of the wall. You can

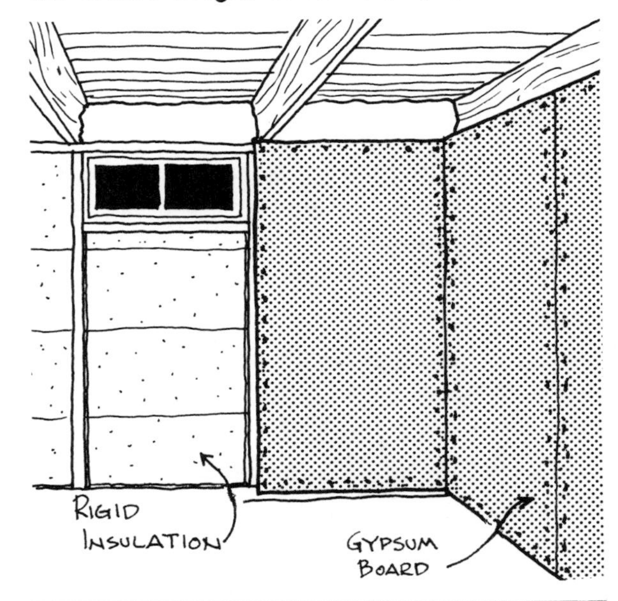

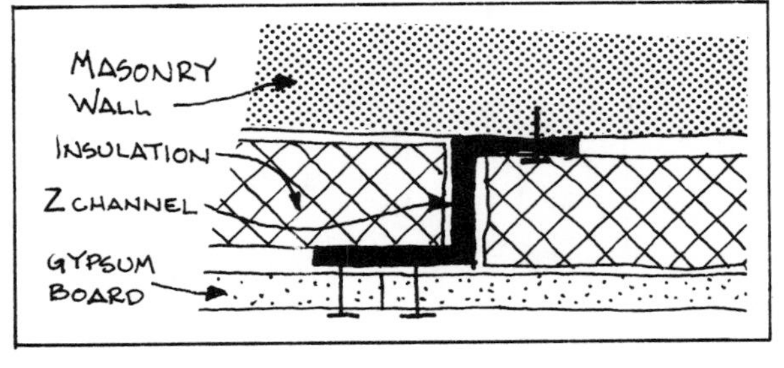

"finish" the basement by painting the gypsum board or by covering it with wood paneling.

BLOWN IN PLACE INSULATION

1 If the basement walls are finished with an air space behind the finish, insulation can be blown into the air space. Special equipment is needed for this. You may have to hire a trained installer, although you can rent the equipment and do it yourself. All cracks and holes in the finish should be sealed prior to installing the insulation.

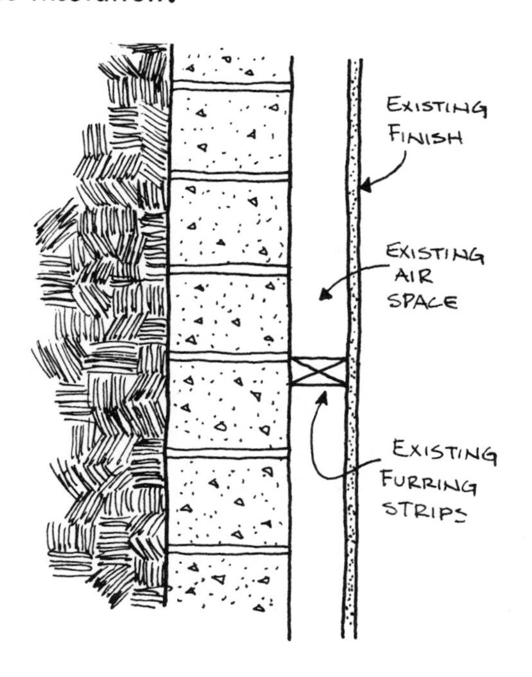

NOTE:

If your home is subject to a high water table where leakage could re-occur, cellulose should not be blown into the wall.

2 Drill two holes (1-1/2" or 2" diameter) through the finish per stud space. Locate one hole near the ceiling with the other hole 4' above the floor or

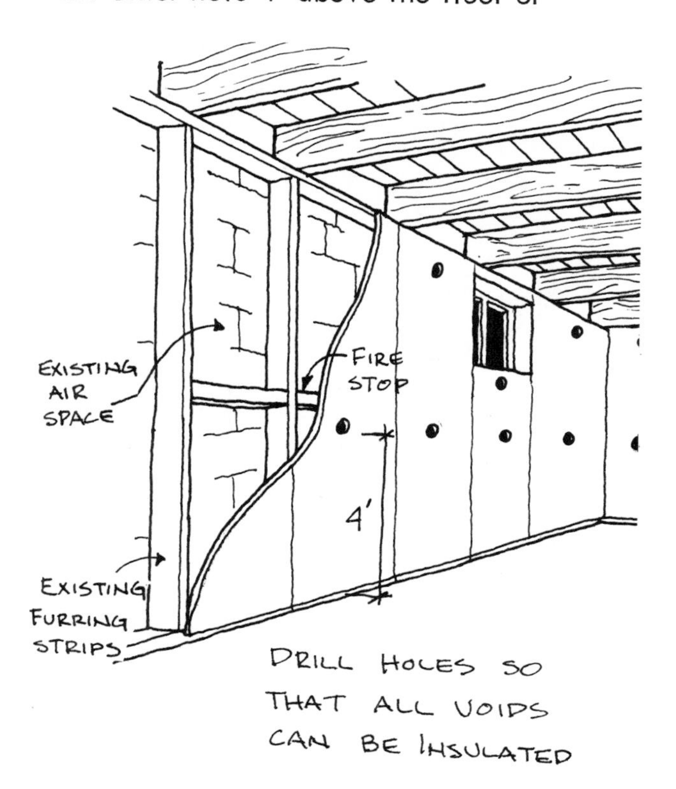

just below a fire stop (it is recommended that insulation not be blown more than 4' down or 1' up). To insure that there are no obstructions in the wall (such as fire stops), lower a weight through the top access hole until it touches the bottom of the cavity. Walls that contain wiring not armored in conduit should not be insulated.

3 Begin blowing insulation into the wall through the hole closest to the floor. Finish the job by blowing insulation through the hole near the ceiling. If there are more than 2 holes per stud space, work from the bottom to the top. Fill all irregularly framed spaces around windows, pipes, and other obstructions on the wall. If necessary, pack the area occupied by the nozzle with insulation.

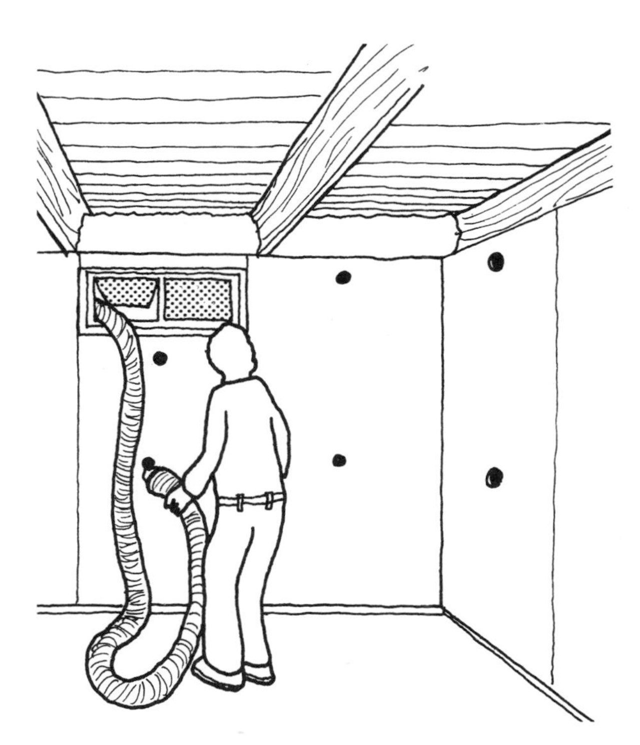

4 Walls that have a narrow fur space, conduit, protruding mortar, or nails within the fur space may be difficult to insulate. Examine the insulation as it is installed to determine the extent of fill. If there are obstructions within the fur space, you may have to drill additional access holes. Rapping the wall with a cushioned tool may cause the insulation to drop through the fur space.

5 After blowing through the finish, patch the holes with gypsum plaster or plastic inserts which are made specifically for plugging access holes (pieces of gypsum board and "joint compound" can also be used).

6 To add a vapor barrier over blown in place insulation, use an aluminized,

oil based enamel, or latex vapor barrier paint. Apply one or more coats of paint over the finish.

22

insulate basement walls on the exterior

description

Insulating basement walls on the exterior is an alternative to the previous retrofit. You may want to consider this retrofit if you're insulating the exterior of masonry walls of upper floors, where the interior basement walls are finished, or in very cold climates.

It's much easier to insulate the exterior of basement walls when the home is being built. Installing insulation after (as for this retrofit) involves excavating around the basement. This may be difficult to do because of plantings, sidewalks, or driveways that have been located around the home. In addition, you have to find a place to keep the dirt while the work is being done.

Excavation is necessary down to the frost line.* The walls should be cleaned of all dirt and loose material. Seal all cracks and joints in the wall and apply waterproofing. Attach rigid insulation board to the wall with an adhesive (some liquid vapor barriers also double as an adhesive). Insulation above grade should be covered with an exterior finish.

In severely cold climates such as Northern Minnesota and Maine, basement walls which are insulated on the inside may freeze against the surrounding earth. If the earth "heaves" when it freezes, it may lift the upper part of the basement causing cracks in the basement walls, leaks, and structural damage. This is one reason for insulating on the exterior. Check with your building department or local HUD/FHA office.

materials

rigid insulation

- extruded polystyrene
- polyurethane

waterproofing-adhesives

- asphalt emulsion
- fibrated mastic
- liquid epoxy
- hot melt asphalt
- tar modified urethane
- panel adhesive

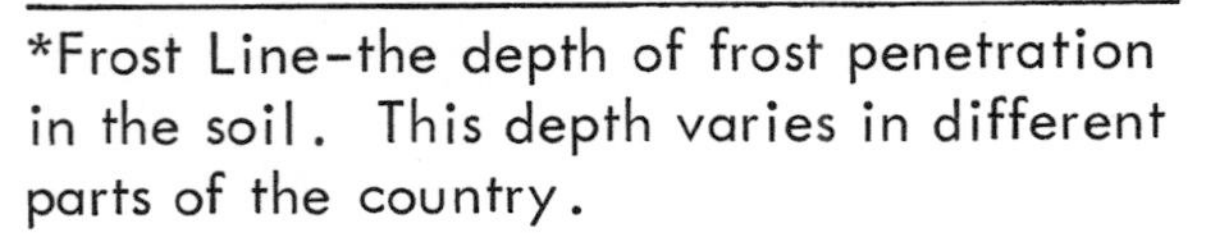

*Frost Line-the depth of frost penetration in the soil. This depth varies in different parts of the country.

masonry cleaners

- acidic type

other materials

- packing

- caulk

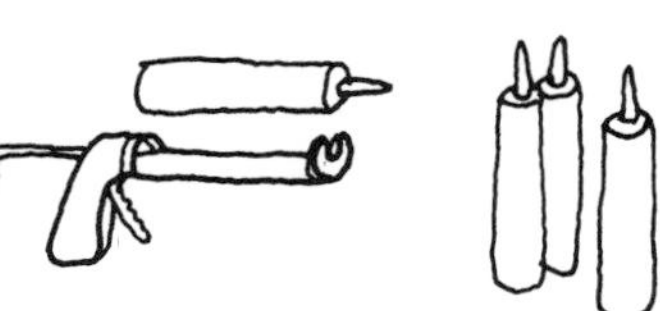

- patching cement

- flashing

- furring strips

exterior finish

- lap siding
- cement asbestos sheets
- latex based stucco
- polymer finish
- vinyl cement
- sheet vinyl

preparation

The necessary excavation will disrupt any planting (grass, flowers, shrubs, etc.) that you have done around the home. For the duration of the work there will be a trench around most of the home. Precautions should be taken to keep people from stumbling into it. Put stakes and cloth "flags" around the edges of the trench whenever work is completed for the day. If a contractor is doing the work, you should understand who is responsible for restoring the yard to its "original" appearance (final grading, re-planting, sidewalk or driveway repair, etc.). Will it be done by you or a contractor?

Before excavating, note where the various services (gas, water, sewer, electric, telephone) enter the home to avoid rupturing pipes and/or snapping wires. Also, some homes have drainage tile that are fed directly from downspouts. These tile will have to be reinstalled in the exact level and location (to the outside of the new insulation) they were in to avoid drainage problems around the home. Refer to Chapter 6, "A Word about Safety", for additional details on excavation.

1 Excavate around the home to the depth of the frost line (check with the local building department for this depth). Place dirt in a spot to make backfilling easy, preferably, right next to the trench. Avoid placing dirt beneath trees as this can damage branches. The excavation can be done with a "back-hoe", power trencher, or ambitious workers equipped with picks and shovels. The trench should be wide enough for a person to waterproof and install insulation to the wall (18" - 24").

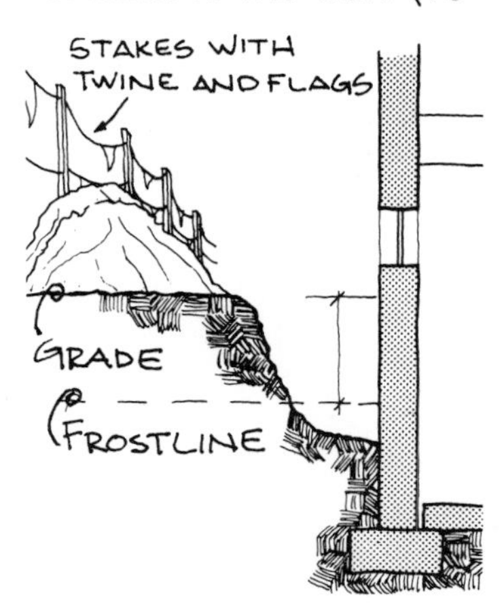

2 Clean all dirt and loose material from the wall with a stiff brush and water. A mild acidic masonry cleaner can also be used, but avoid skin contact with the cleaner by wearing rubber gloves and a long sleeve shirt.

CLEAN THE WALL OF ALL DIRT AND LOOSE MATERIAL

3 Allow the wall surface to dry before sealing the joints and cracks with caulk and, if necessary, packing. Before caulking, cracks should be routed 1/4" or more in width and depth. Working cracks (use an elastomeric sealant here) should be routed 3/4" to 1" in width and depth. Patching cement can be used for large dormant cracks and uneven wall surfaces. Refer to Chapter 5, "A Word about Caulks", for details regarding "dormant" and "working" cracks.

installation procedures

1 Begin by applying a liquid vapor barrier to the wall if one does not exist or if the existing one allows water to penetrate into the basement. If the liquid vapor barrier doesn't double as an adhesive, you'll need to select an adhesive for bonding the insulation to the wall that is suitable for exterior application. The vapor barrier and/or adhesive must also be compatible with the insulation. Apply the vapor barrier/adhesive to the wall in such a fashion so as to fill any uneveness in the wall surface.

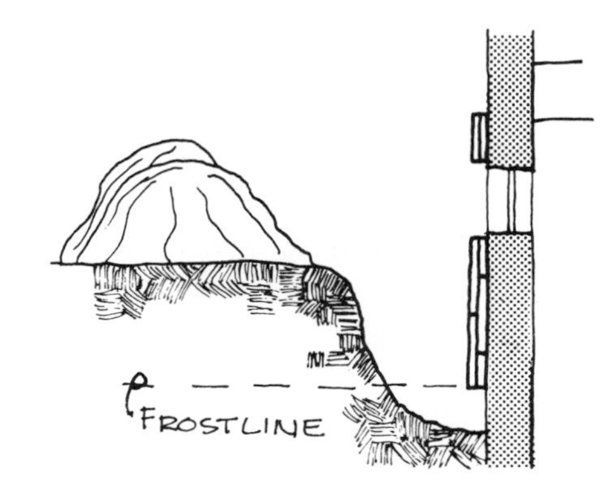

2 Apply the insulation to the wall down to the frost line. In order to achieve a certain "R" value, you may have to apply two or more layers of insulation. If this is the case, stagger all vertical and horizontal joints. Apply the adhesive, approximately 2" wide by 1/4" or 3/8" thick, to the perimeter of each board. Then apply dabs of the adhesive to the interior of each board approximately 8" on center.

STAGGER THE JOINTS IF TWO OR MORE LAYERS OF INSULATION IS USED

3 Insulation above grade should be covered with a suitable exterior finish (see step 6 in this chapter). Consequently, you may need to place furring strips between vertical joints of the insulation

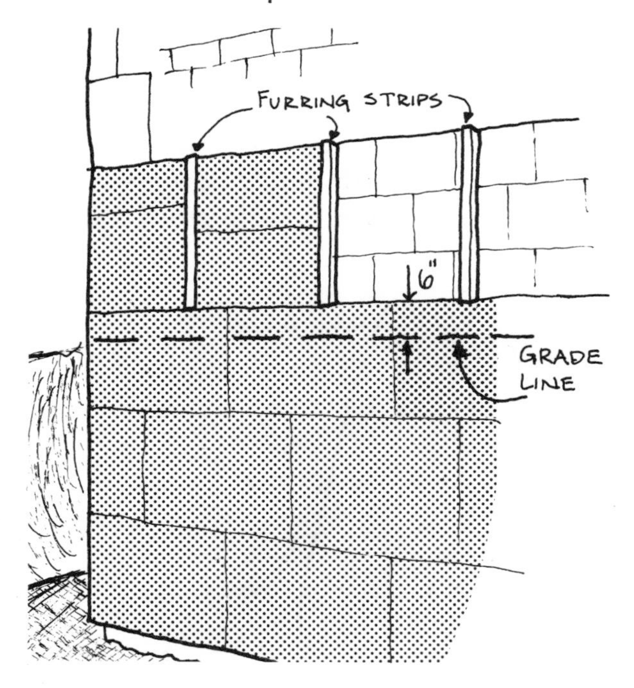

boards. Attach the furring strips to the wall with masonry nails. The furring strips should be flush to the face of the insulation board and must be installed so that they remain a minimum of 6" above grade when the trench is backfilled.

4 Be sure to fur out around basement windows, doors, vents, and other openings through the wall. Where the joint between the strips, wall, and insulation is exposed to the weather, seal it with caulk.

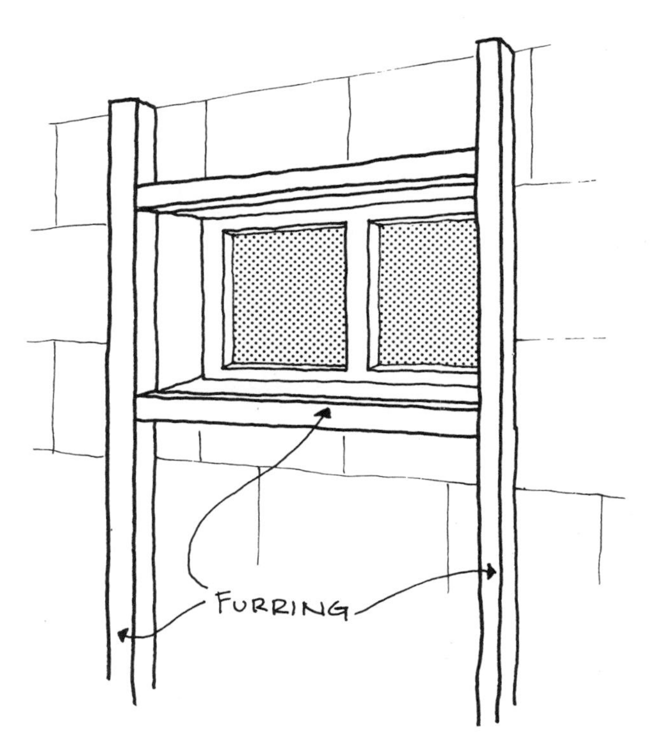

5 Fasten the exterior finish to the furring strips. Start at the grade level and work your way up. Wood or metal must be stopped so that it does not touch the ground (check local building codes). Most codes do not allow siding to come closer than 6" above grade. Building felt or asphalt can be used to cover the insulation where there is no finish. Install finish in strict accordance with manufacturer's installation procedures.

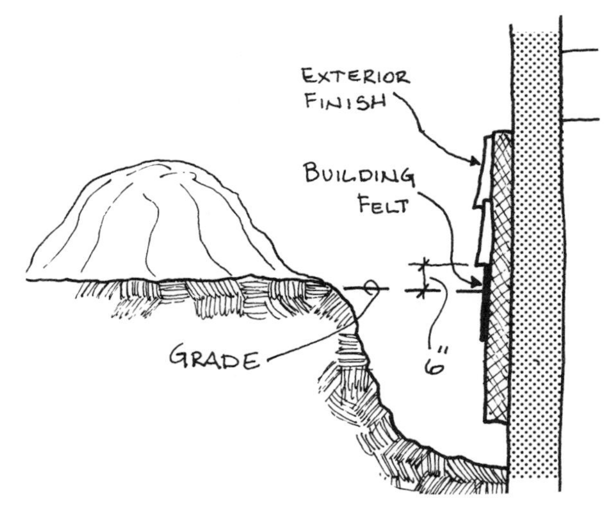

NOTE:

New products may be available to you for weatherproofing exposed rigid insulation. Also, insulation with a weatherproof covering on one side may be available.

6 A gravel backfill should be used against the insulation board (this will let water drain away from the insulation). Lay in gravel as you are backfilling the trench. Place a little extra dirt over the trench to allow for settling. Taper the fill away from the home so that water drains away from it.

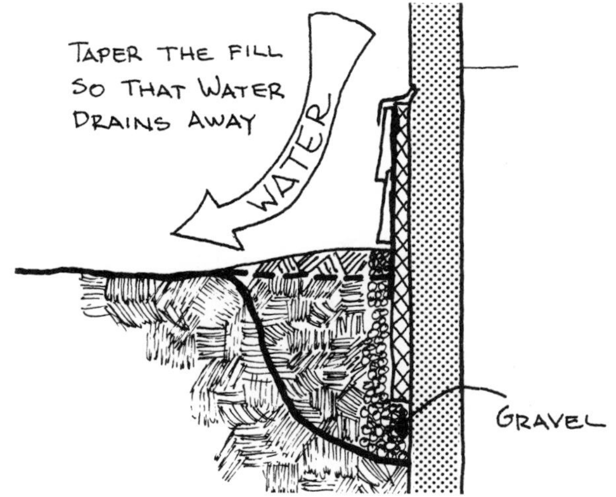

7 Attach flashing over the top edge of the insulation. For frame walls (except stucco), tuck the flashing under the siding a minimum of 1-1/2" or the width of the siding, whichever is less.

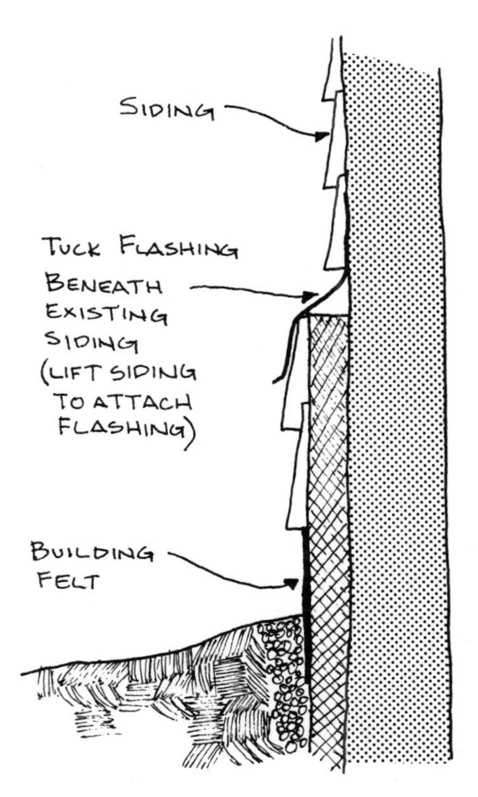

With stucco walls, bed the top of the flashing on the stucco with a high performance caulk (see Chapter 5, "A Word about Caulks", for properties of caulks).

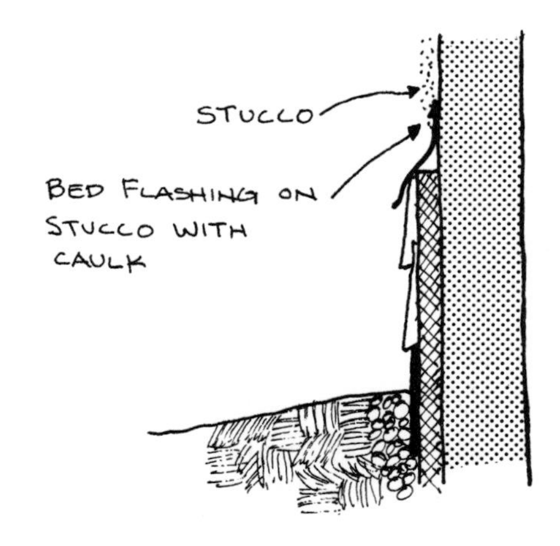

With masonry walls, cut a 3/4" reglet* in the wall just above the top edge of the insulation. After the flashing has been installed in the reglet, seal it with a continuous bead of caulk.

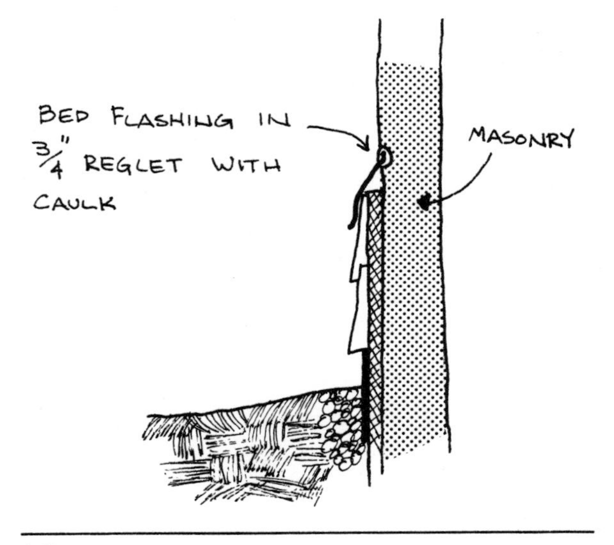

*Groove

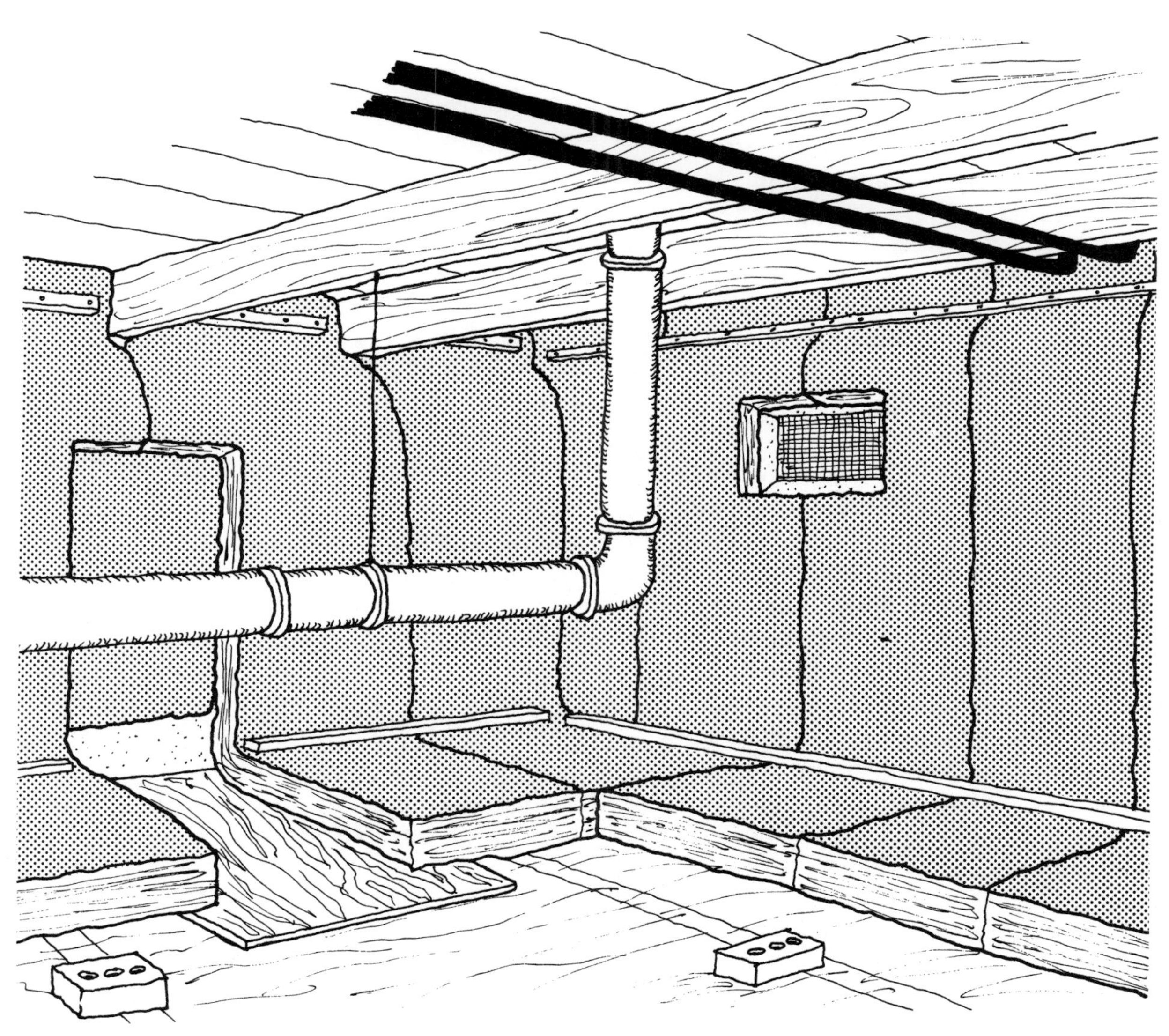

23

insulate crawl space walls on the interior

description

The air in an uninsulated crawl space can get quite cold during the winter, making the floor above it uncomfortably cool. The temperature of the crawl space can be modified somewhat by installing insulation and closing the vents during the winter. There are a number of ways to insulate a crawl space.

If there are pipes or heating ducts in the crawl space, insulation can be added to the interior of the walls. In this manner, the crawl space stays relatively warm, preventing pipes from freezing. In extreme cold climates (Northern Minnesota and Maine), however, insulating the interior crawl space walls can cause the walls to "heave" because of deep frost penetration. If this is the case, insulating the exterior of crawl space walls (see Chapter 24) may be recommended. Check with your building department or local HUD/FHA office.

The floor above the crawl space can be insulated directly by installing the insulation between the floor joists (see Chapter 25). Doing this will significantly reduce the crawl space temperature in the winter. Therefore, if there are pipes or ducts in the crawl space, the insulation should be placed on the cold side (beneath) them, or they should be insulated separately with pipe and duct insulation (see Chapter 43). Insulating crawl space walls on the interior solves the problem of adding an acceptable exterior protective finish as would be necessary with insulating on the outside. However, work space may be restricted and access may have to be provided (see Chapter 26).

Moisture buildup in a crawl space can be a problem. To prevent moisture migration from the ground, a polyethylene sheet or other vapor barrier should be laid over the ground. In addition, a crawl space should be properly vented to allow moisture that does get into crawl space to escape to the outside. This is usually a problem in the summer when the humidity is high. Consequently, if you closed the vents for the winter, you should open them in the summer. If moisture is not vented from the crawl space, the wood may begin to rot. Additionally, it is wise to periodically check the crawl space for signs of moisture during the winter.

materials

roll/batt insulation

- glass fiber
- rock wool
  (foil or kraft paper faced)

rigid insulation

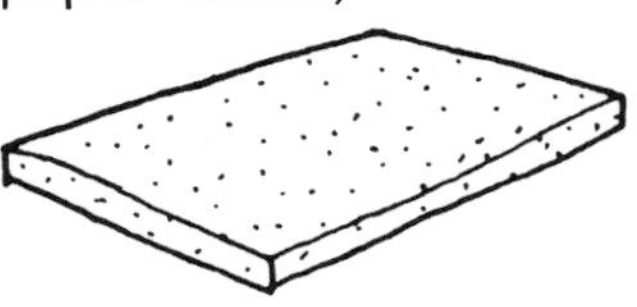

- polystyrene
- polyurethane

vapor barrier

- polyethylene film
  (6 mil minimum is recommended)

waterproofing - adhesives

- asphalt emulsion
- fibrated mastic
- liquid epoxy
- panel adhesive

other materials

- masonry cleaner

- caulk (or patching cement)

- packing 
- fasteners 
- vents and vent covers 

preparation

Before adding insulation to the crawl space wall, any water seepage through the wall must be stopped by water-proofing the exterior of it. Crawl space walls which show evidence of leakage should not be insulated before corrections or repairs are made.

If the water seepage is below grade, excavating the perimeter of the crawl space becomes necessary (see Chapter 24, "Insulate Crawl Space Walls on the Exterior", for details). Waterproofing is necessary down to the interior grade level rather than the frost line as discussed for that retrofit.

1 Clean all dirt and loose material from the wall with a stiff brush and water. A mild acidic masonry cleaner can also be used. After the wall has dried, seal all cracks and joints with caulk and, if necessary, packing. Cracks should be routed 1/4" or more in width and depth. Working cracks (use an elastomeric sealant here) should be routed 3/4" to 1" in width and depth. Large dormant cracks can be sealed with patching cement. Refer to Chapter 5, "A Word about Caulks", for details regarding "dormant" and "working" cracks.

NOTE: 

Avoid skin contact with the masonry cleaner by wearing rubber gloves and long sleeve shirts.

---

2 Apply waterproofing to the wall from the exterior grade line down to the interior grade line. For backfilling procedures, see Chapter 24.

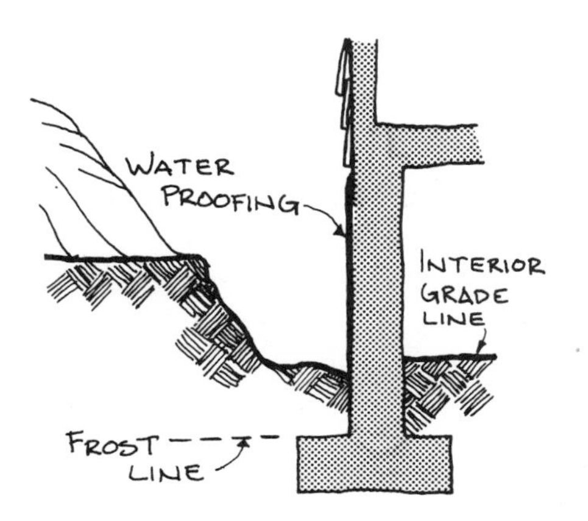

installation procedures

WITH ROLL/BATT INSULATION

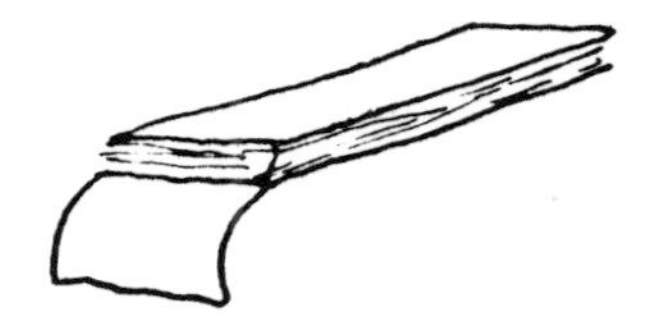

1 Begin with the walls that are perpendicular to the joists. Lay a sheet of polyethylene along the base of this wall (it should extend out from the foundation wall by at least 30" and up the wall by at least 6"). While you're working, fold the vapor barrier back to avoid puncturing or tearing it. Use polyethylene or duct tape to attach it to the wall or anchor it to the wall with dirt.

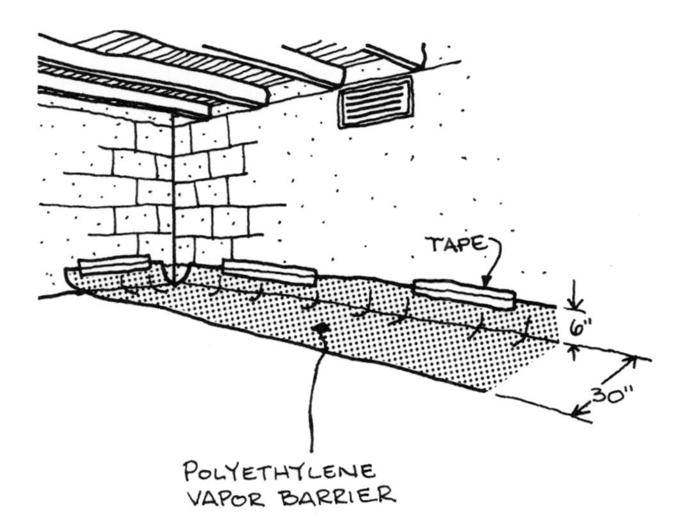

2 Cut roll/batt insulation to extend from the underside of the floor, against the headers, down the foundation wall, and out from the wall at least 2 feet. Place dry sand between the insulation and the vapor barrier.

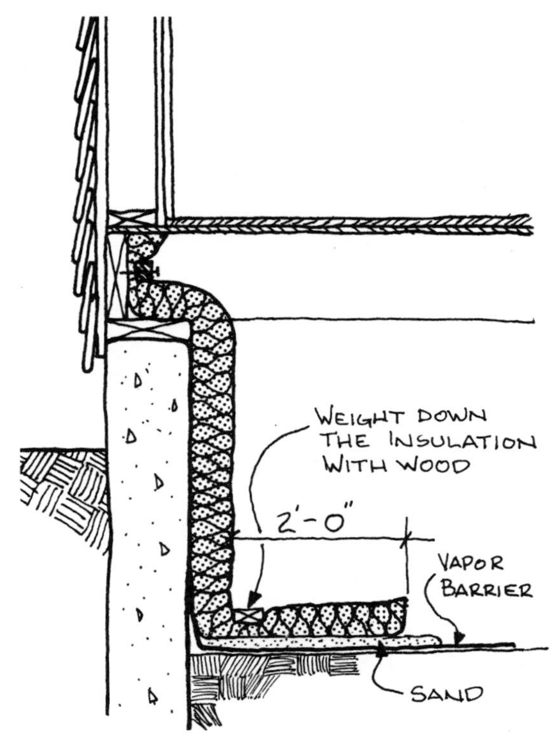

CUT INSULATION SO THAT AFTER IT'S INSTALLED, IT WILL EXTEND FROM THE BOTTOM OF THE FLOOR, DOWN THE WALL, AND OUT INTO THE CRAWL SPACE.

3 Begin installing the insulation in one corner with 1" x 2" furring strips. Nail through the strips and insulation into the header. The top edge of the insulation should press firmly against the floor deck. Foil or kraft paper backing on the insulation is to face towards you.

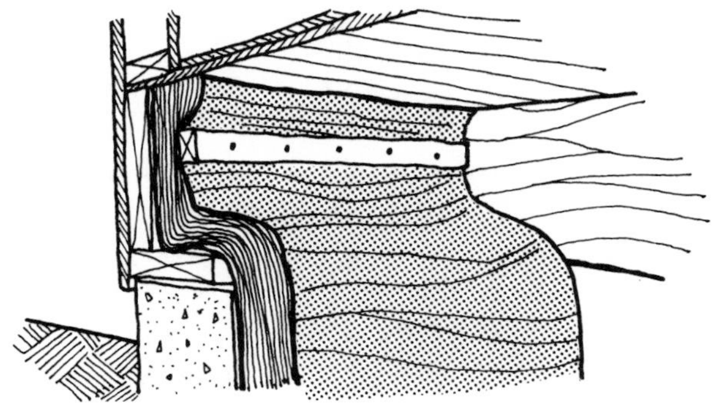

NOTE:

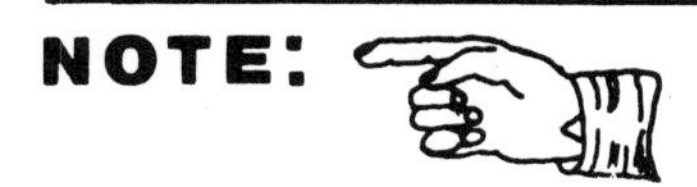

Where termite shields exist, use the method pictured below.

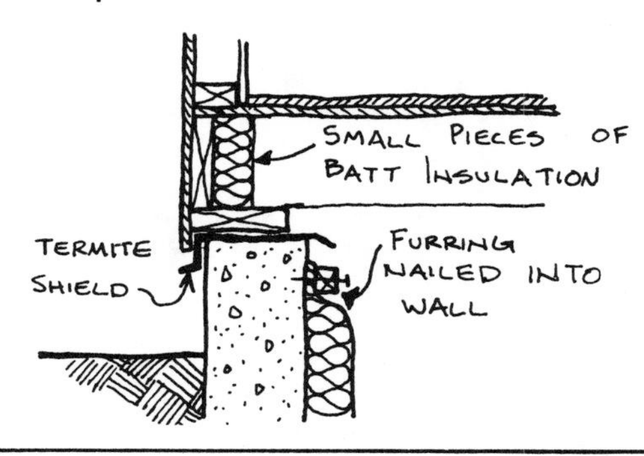

4 Alternatively, you can cut small pieces of insulation to fit snugly against the header and nail the longer batts to the sill plate.

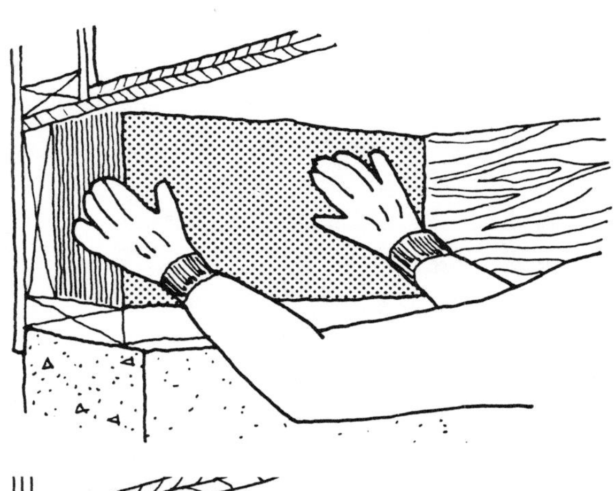

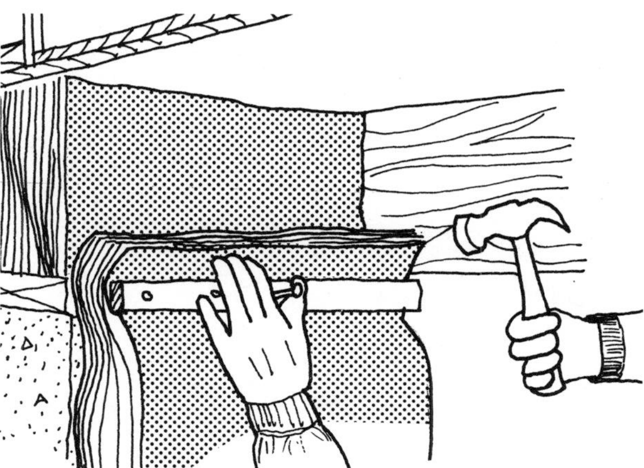

5 Cover the entire wall in this fashion. Make sure each batt butts firmly against the adjacent piece of insulation. Repeat steps 1 through 5 for the other wall (or walls) that is (are) perpendicular to the floor joists.

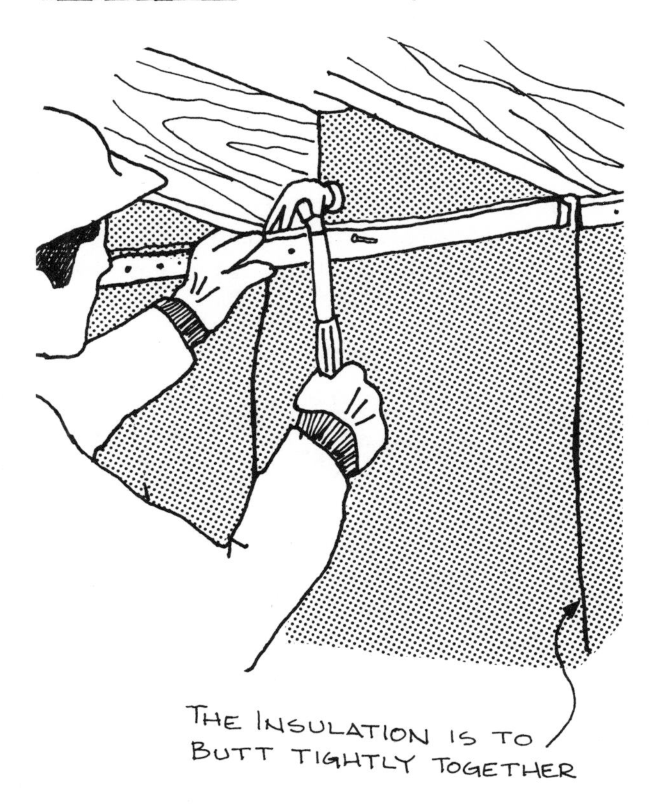

6 Finish laying a vapor barrier across the ground. Lap the sheet edges at

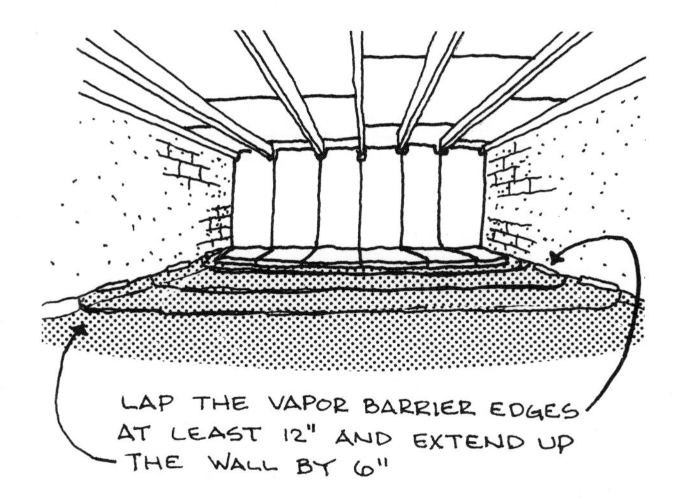

least 12". Use rocks or bricks to hold down the sheet. Use duct or polyethylene tape to attach the ends of the vapor barrier to the foundation wall. Bring the vapor barrier up the side of the wall at least 6".

7 To insulate the bandjoist where the walls are parallel to the joists, cut long pieces of insulation and nail them directly to the bandjoist or staple the flanges on the insulation to the floor to avoid compressing the insulation. Then insulate the walls as described in the previous steps.

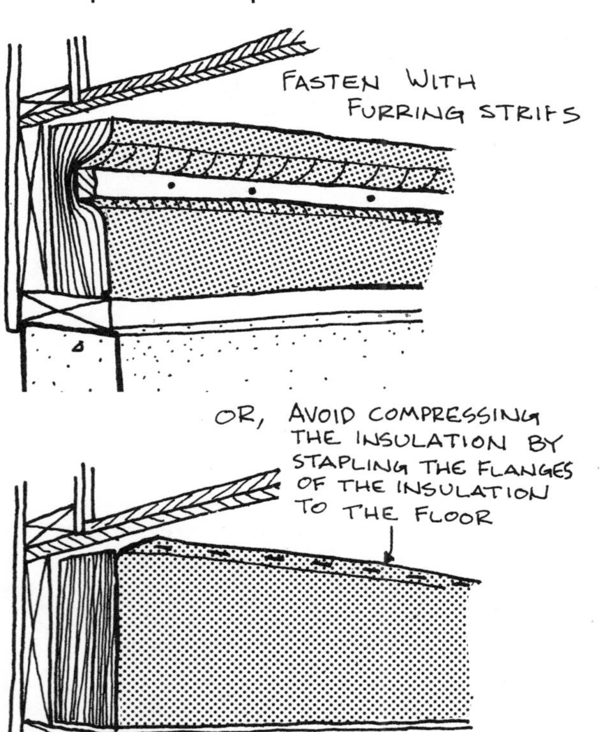

8 Fibrous insulation can be draped over vents. If existing, however, strip off the facing of the insulation that covers the vent. Covering the vent in this manner reduces the "free" area of the vent. Consequently, the crawl space should be checked for moisture during the summer. If moisture is present, it

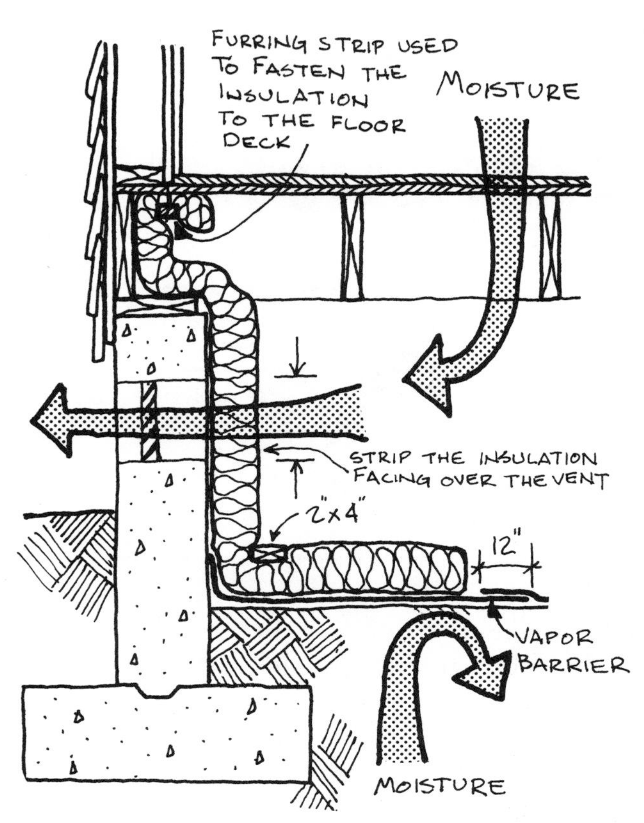

will be necessary to either remove the insulation from the face of the vent or install additional vents.

NOTE:

Avoid puncturing holes in the vapor barrier as you're working. All holes in the vapor barrier should be patched with duct or polyethylene tape.

WITH RIGID INSULATION

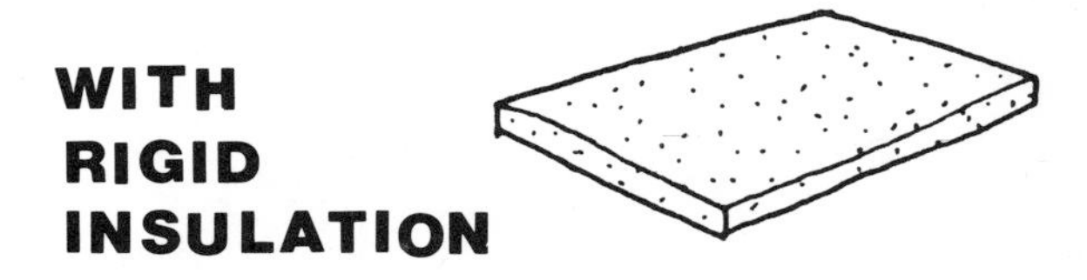

1 Lay a vapor barrier across the ground in the same fashion as previously described in steps 1 and 6.

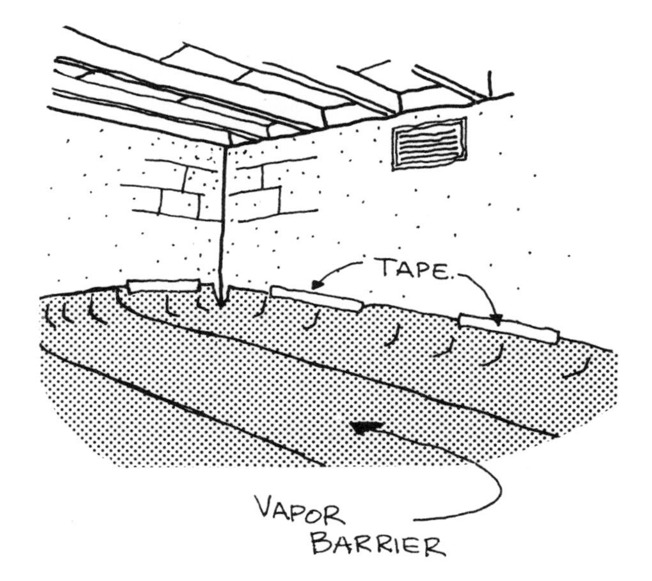

2 Attach rigid insulation directly to the wall down to the ground with a compatible adhesive. Apply the adhesive along the perimeter of the board with spot "dabs" on the interior of it. Press the insulation in place so it firmly contacts the wall surface. Butt the boards together tightly.

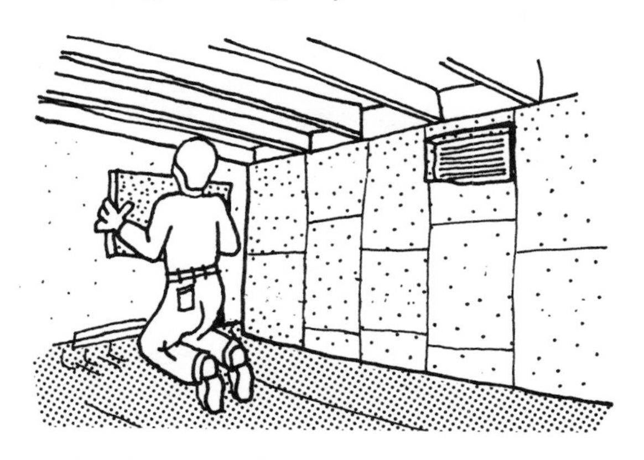

AN IMPORTANT NOTE :

Check with your building department about covering rigid insulation with a fire resistant material when installed in the crawl space. If it must be protected, add furring strips to the wall. (see Chapter 21, "Insulate Basement Walls on the Interior", for details).

3 Insulate the header and bandjoist with pieces of batt or rigid insulation. Cut small pieces of insulation to fit against the header between the floor joists. Use longer lengths of insulation and fasten them directly to the bandjoist.

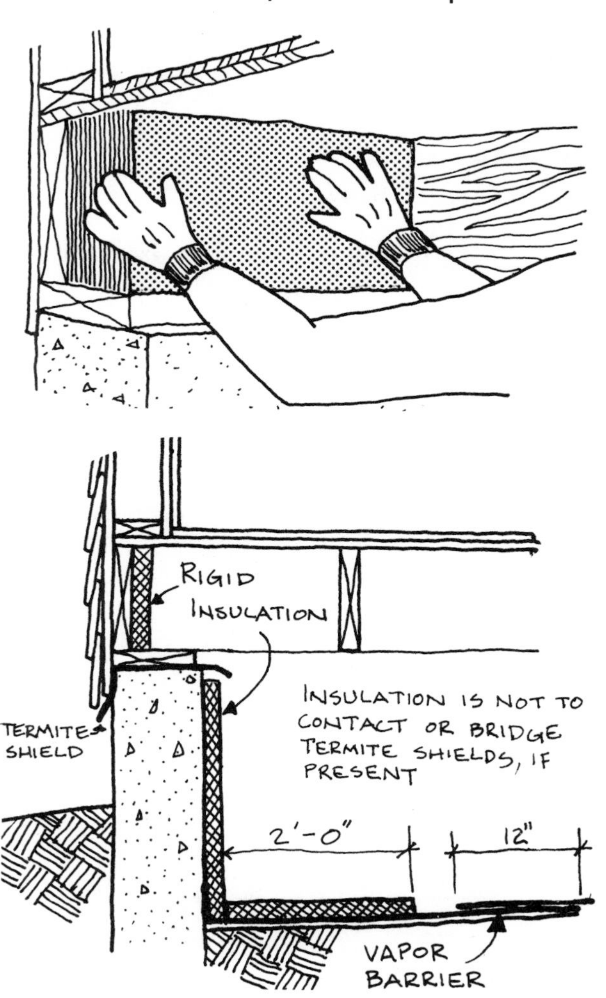

4 Lay additional pieces of insulation on the ground so they extend out 2 feet from the wall as described for Roll/batt insulation, step 2.

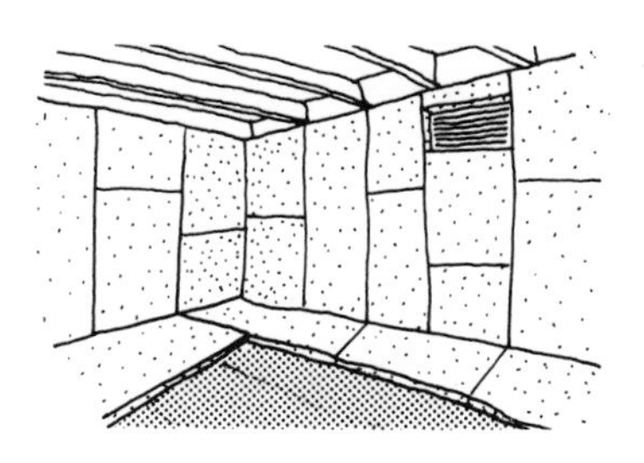

NOTE: 

Again, patch all holes in the vapor barrier with duct or polyethylene tape.

CRAWL SPACE VENTILATION

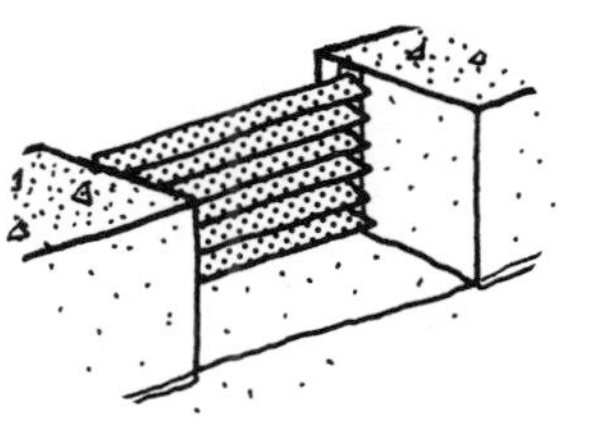

The crawl space should be properly vented to avoid moisture build-up which can be a problem in summer. Crawl space vents are usually opened during the summer (when the relative humidity is high) and closed* (and insulated - see Chapter 41) during the winter. Vents are available that correspond to the length and height dimensions of concrete block. Use a type which can be closed in winter - otherwise insulating the walls doesn't make sense. Check occasionally for moisture or leaks into the crawl space. If this becomes a problem, the vents can be opened in winter to get rid of such moisture.

*The vents should be opened year round if the heating plant is located in the crawl space. Or, provide a sheet metal duct from the vent to the point at which the heating plant takes in combustion air (this is known as "outdoor air combustion").

Crawl space	Ratio of total net ventilating area to floor area[1]	Minimum number of ventilators[2]
w/o vapor barrier	1/150	4
w/ vapor barrier	1/1500	2

[1] The actual area of the ventilators depends on the type of louvers & size of screen used.

[2] Foundation ventilators should be distributed around foundation to provide best air movement. When (2) are used, place one toward the winded side & opposite side.

Obstructions in ventilators - louvers and screen[1]	To determine total area of ventilators, multiply required net area (sqft) by
1/4-inch-mesh hardware cloth	1
1/8-inch-mesh screen	1-1/4
No. 16-mesh insect screen (with or w/o plain metal louvers)	2
Wood louvers & 1/4-inch-mesh hardware cloth	2
Wood louvers & 1/8-inch-mesh screen	2-1/4
Wood louvers & No. 16-mesh insect screen	3

[1] In crawl-space ventilators, screen mesh should not be larger than 1/4 inch

Source: "Condensation Problems In Your House: Prevention and Solution", U.S. Department of Agriculture, Forest Service, 1974

To install this type of vent, remove a block and set the vent in place (caulk

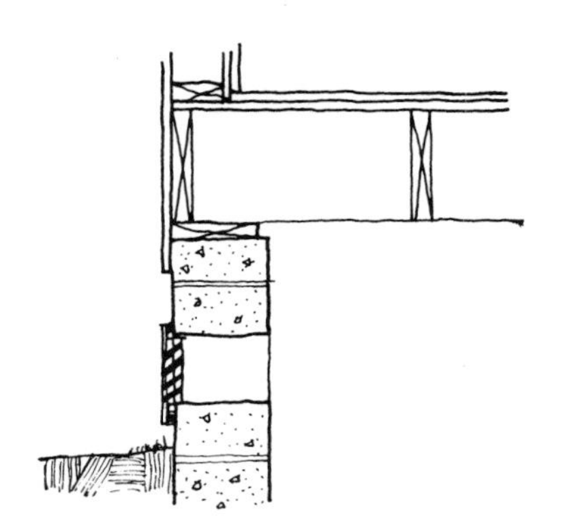

around the vent). Use a hammer to knock out the block (wear safety goggles while doing this). Avoid massive blows with the hammer as you may crack other blocks in the wall.

To determine the number and size of vents needed, refer to the tables on the previous page.

For additional information on ventilation, see Chapter 36, "But What about Attic Ventilation?".

24

insulate crawl space walls on the exterior

description

Similarly to basement walls, the alternative to insulating crawl space walls in the interior is insulating the walls on the exterior. You may want to consider this retrofit if you're insulating the exterior of masonry walls (Chapter 29) and the building has a crawl space, or if the building is located in a very cold climate. In severely cold climates such as northern Minnesota and Maine, crawl space walls that are insulated on the interior below grade may permit the ground to freeze against the walls. If the earth "heaves" when it freezes, it may lift the crawl space walls, causing structural damage.

It's much easier to insulate the exterior of crawl space walls when the home is being built. Installing insulation after (as for this retrofit) involves excavating around the crawl space.

This may be difficult to do because of plantings, sidewalks, or driveways that have been located around the home. In addition, you have to find a place to keep the dirt while the work is being done.

Excavation is necessary down to the frost line* or footing, whichever is shallower. The walls should be cleaned of all dirt and loose material. Seal all cracks and joints in the wall and apply waterproofing. Attach rigid insulation board to the wall with a compatible adhesive (some liquid vapor barriers also double as an adhesive). Insulation above grade should be covered with a suitable exterior finish.

materials

rigid insulation

- extruded polystyrene
- polyurethane

waterproofing - adhesives

- asphalt materials
- fibrated mastic
- liquid epoxy
- hot melt asphalt
- tar modified urethane
- panel adhesive

---

*Frost line – the depth of frost penetration in the soil. This depth varies in different parts of the country. Refer to maps of climate conditions to get an estimate of frost penetration depth for your area. Local building officials or the county agricultural extension service will know how deep the ground can freeze in your area.

masonry cleaners

- acidic type

other materials

- packing

- caulk

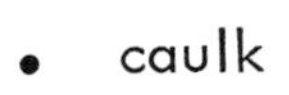

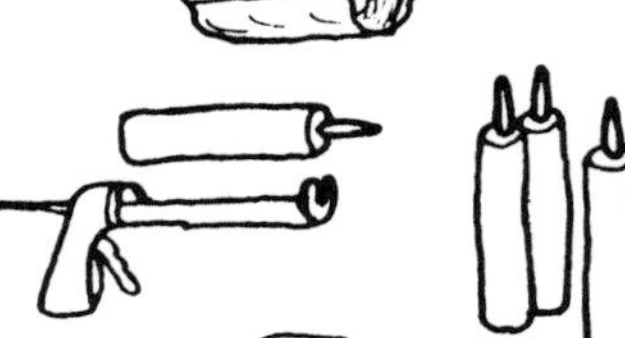

- patching cement

- flashing

- furring strips

exterior finish

- lap siding
- cement asbestos sheets
- latex based stucco
- polymer finish
- vinyl cement
- sheet vinyl

preparation

The necessary excavation will disrupt any planting (grass, flowers, shrubs, etc.) that you have done around the home. For the duration of the work, there will be a trench around most of the home. Precautions should be taken to keep people from stumbling into it. Put stakes and cloth "flags" around the edges of the trench whenever work is completed for the day. If a contractor is doing the work, you should understand who is responsible for restoring the yard to its "original" condition (final grading, re-planting, sidewalk or driveway repair, etc.). Will it be done by you or a contractor?

Before excavating, note where the various services (gas, water, sewer, electric, telephone) enter the home to avoid rupturing pipes and/or snapping wires. Also, some homes have drainage tile that are fed directly from downspouts. These tile will have to be reinstalled in the exact level and location (to the outside of the new insulation) they were in to avoid drainage problems around the home. Refer to Chapter 6, "A Word about Safety", for additional details on excavation.

1 Excavate around the crawl space to the depth of the frost line or top of the footing, whichever is shallower. Place the dirt in a spot to make back-filling easy, preferably right next to the trench. Avoid placing the dirt underneath the branches of a tree because the tree roots extend as far as the branches and will be damaged. The excavation can be done with a "back-hoe", power trencher, or ambitious workers equipped with picks and shovels. The trench should be wide enough for a person to water-proof and install insulation to the wall (18" - 24").

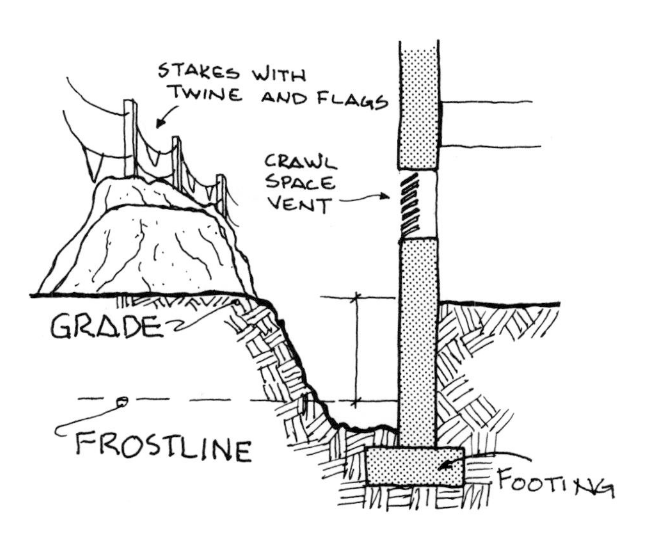

2 Clean all dirt and loose material from the wall with a stiff brush and water. A mild acidic masonry cleaner can also be used, but avoid skin contact with the cleaner by wearing rubber gloves and a long sleeve shirt.

3 Allow the wall surface to dry before sealing the joints and cracks with caulk and, if necessary, packing.

Before caulking, cracks should be routed 1/4" or more in width and depth. Working cracks (use an elastomeric sealant here) should be routed 3/4" to 1" in width and depth. Patching cement can be used for large dormant cracks and to smooth uneven wall surfaces. Refer to Chapter 5, "A Word about Caulks", for details regarding "dormant" and "working" cracks.

installation procedures

1 Begin by applying a liquid vapor barrier to the wall if one does not exist or if the existing one allows water to penetrate into the crawl space. If the liquid vapor barrier doesn't double as an adhesive, you'll need to select an adhesive for bonding the insulation to the wall that is suitable for exterior application. The vapor barrier and/or adhesive must be compatible with the insulation.

2 If an adhesive is needed in addition to the vapor barrier, apply the adhesive to the insulation board. Apply the adhesive, approximately 2" wide by 1/4" or 3/8" thick, to the entire perimeter of each board. In addition, apply dabs of the adhesive of the same thickness to the interior area of the board approximately 8" on center. Stagger all vertical joints (if two or more layers are needed to obtain a certain "R" value, stagger all vertical and horizontal joints). Apply pressure over the entire surface of the board to insure uniform and tight contact with the wall. Be sure that all joints butt tightly together. Install the insulation down to the frost line.

3 Insulation above grade should be covered with a suitable exterior finish (step 6 in this chapter). Consequently, you should place furring strips between the vertical joints of the insulation boards. Attach the furring strips to the wall with masonry nails. The furring strips should be flush to the insulation board and must be installed so that they remain a minimum of 6" above grade when the trench is backfilled.

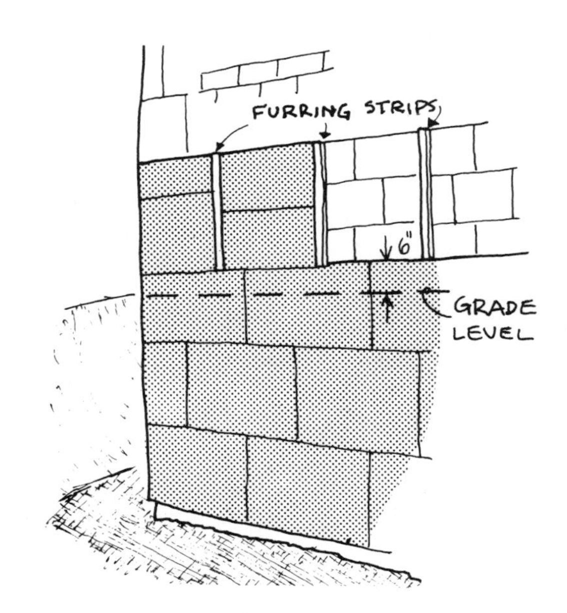

4 Be sure to fur out around crawl space vents, access panel, and other openings in the wall. Where the joint between the strips, wall, and insulation is exposed to the weather, seal it with caulk.

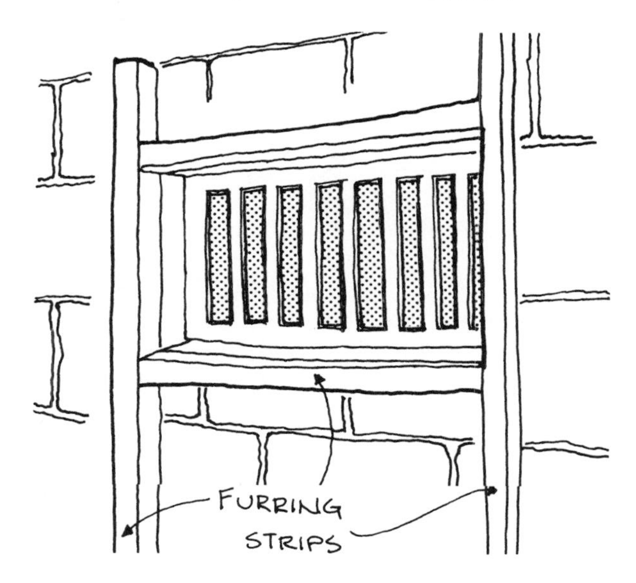

5 Fasten the exterior finish to the furring strips. Start at grade level and work your way up. Wood or metal must be stopped, however, so that it does not touch the ground (check local building codes). Most codes do not allow siding to come closer than 6" above grade. Building felt or asphalt can be used to cover the insulation where there is no finish. Install finish in strict accordance with manufacturer's installation procedures.

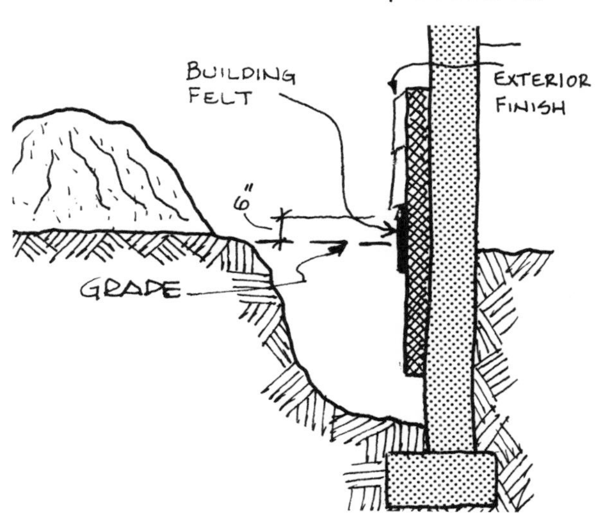

NOTE:

New products may be available to you for weatherproofing exposed rigid insulation. Also, insulation with a weatherproof covering on one side may be available.

6 A gravel backfill should be used against the insulation board (this will let water drain away from the insulation). Lay in gravel as you are backfilling the trench. Place a little extra dirt over the trench to allow for settling. Taper the fill away from the home so that water drains away from it.

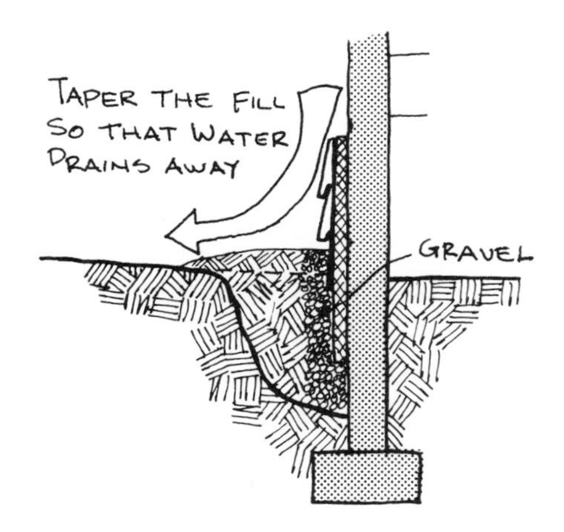

7 Attach flashing over the top edge of the insulation. For frame walls (except

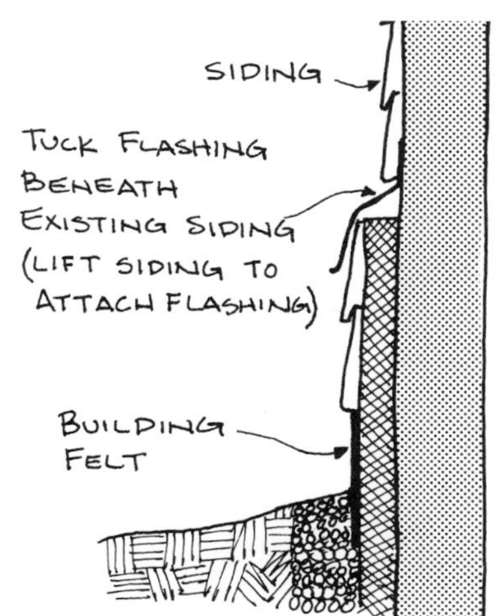

stucco), tuck the flashing under the siding a minimum of 1-1/2" or the width of the siding, whichever is less.

With stucco walls, bed the top of the flashing on the stucco with a high performance caulk (see Chapter 5, "A Word about Caulks", for properties of caulks).

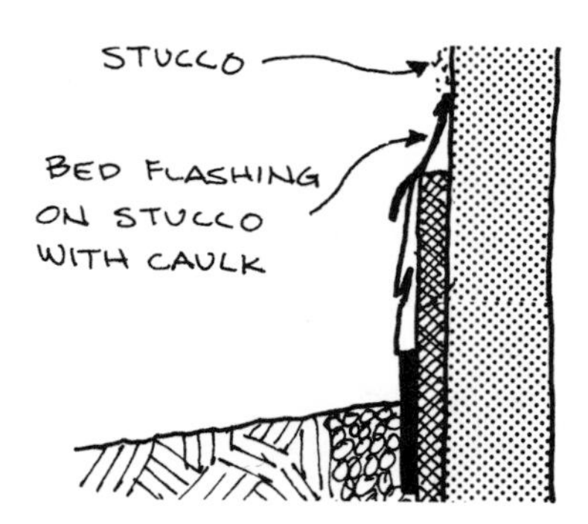

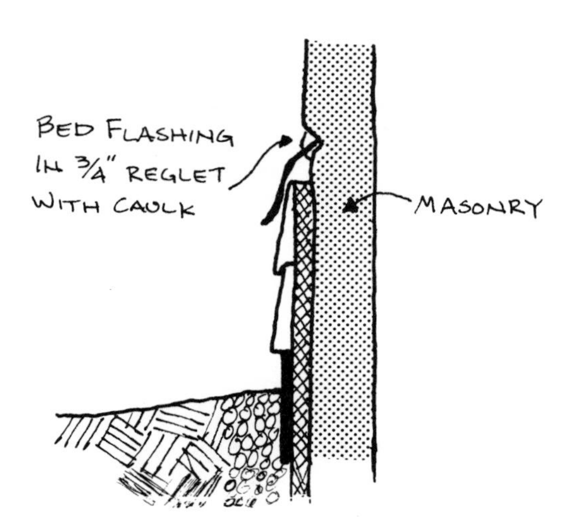

With masonry walls, cut a 3/4" reglet (groove) in the wall just above the top edge of the insulation. After the flashing has been installed, seal it with a continuous bead of caulk.

25

insulate floor above crawl space

description

An alternative to insulating the walls of the crawl space is insulating the floor above the crawl space. This is done by attaching the insulation to the floor joists (batt or rigid insulation can be used - check local building codes for use of rigid insulation in the crawl space). A variety of methods can be used to hold the insulation in place. If a vapor barrier is used here, install so that it faces up towards the heated space. In other words, once you have finished installing the insulation and the vapor barrier, you will not be able to see the vapor barrier. If a ground cover is required, see Chapter 23 for installation procedures.

NOTE: 

Insulating the crawl space floor reduces the temperature of the crawl space during the winter, subjecting exposed water pipes to freezing. If you intend to insulate the crawl space in this fashion, be sure to place the insulation on the cold side of pipes and ducts. Alternatively, you can insulate the pipes and ducts with pipe and/or duct insulation (see Chapter 43). Also, any cooling effect from the crawl space to the home during the summer is lost by adding insulation to the floor.

materials

for rigid insulation

Rigid insulation

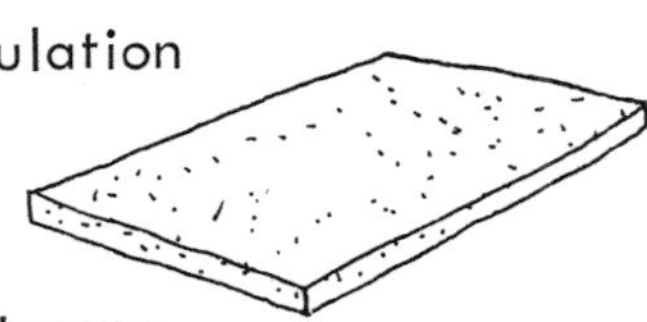

- polystyrene
- polyurethane
- polyisocyanurate
- fibrous sheathing board

- nails, washers

- caulk

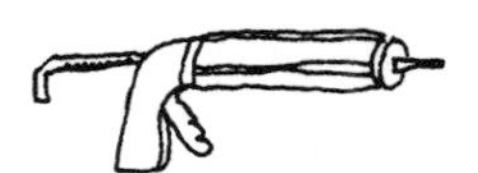

- scraps of batt insulation

for roll / batt insulation

Roll or Batt Insulation

- glass fiber
- rock wool

  (foil faced, kraft paper faced, unfaced, reverse flange)

- vapor barrier (polyethylene)
- tape (polyethylene)

- staples

- wire mesh

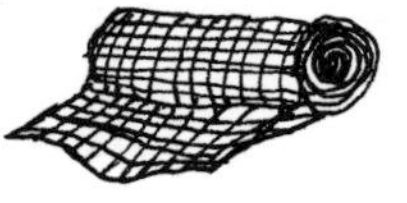

- nails

- wire

- wood bridging

installation procedures

ROLL / BATT INSULATION WITH FACING

Faced insulation is installed with the facing placed up towards the floor boards (heated space). A special type of batt insulation (reverse flange) with stapling flanges is available and is installed by stapling the flanges to the joists. If you use insulation without the stapling flanges, be sure the insulation fits snugly between the joists. The insulation can be held in place by one of the following methods:

- wire mesh
- wire laced onto nails
- wood bridging strips
- mastic adhesive

Remember, if you're not insulating pipes in the crawl space, wrap the insulation on the cold side of them.

wire mesh

1 Begin by cutting the mesh into widths that correspond to the center-line joist dimension.

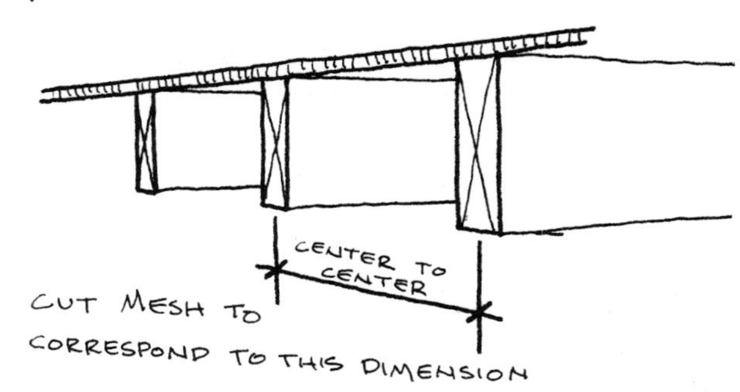

2 At the end of the joist run, span two joists by stapling the mesh that you just cut to the underside of the joists. Then, stuff the end of a batt onto the sill plate up against the header and rest it on the mesh.

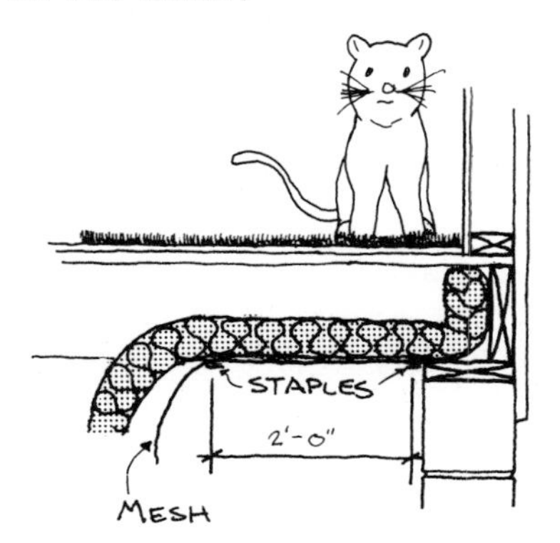

3 Now, move in behind the mesh and insulation and, with one hand, push the mesh and insulation up, stapling the mesh to the joists with the other hand.

4 When you've reached the end of one batt, staple an additional 9" of mesh to the joists to serve as a support for your next batt.

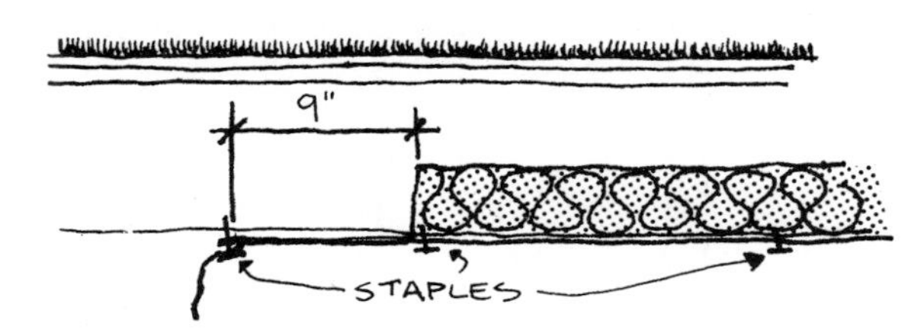

5 You may be able to do more than one joist run at a time (just cut the mesh to match 2, 3, or 4 joist widths). After practice, you could probably handle up to four joist runs. The procedure is the same; it's just a bit more cumbersome.

wire laced onto nails

1 Get yourself started by pounding nails into the bottom of the floor joists at

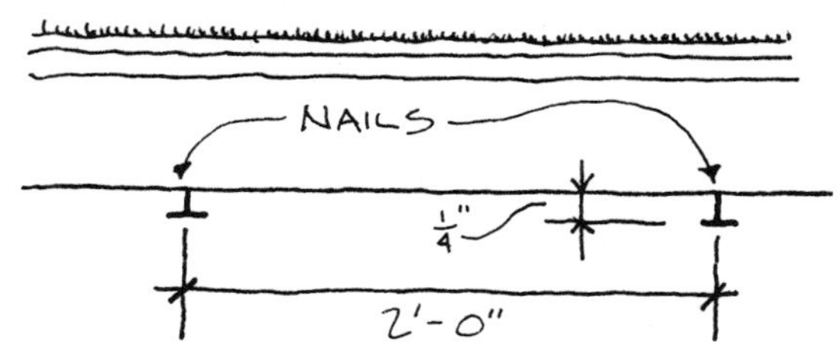

approximately 2 ft. on center leaving 1/4" of the nail head exposed.

NOTE:

Reverse flange insulation can simply be stapled to the joists.

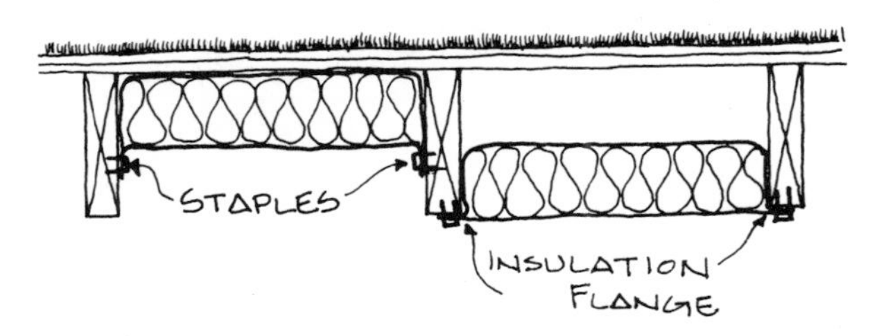

2 Stagger the nails on alternate joists.

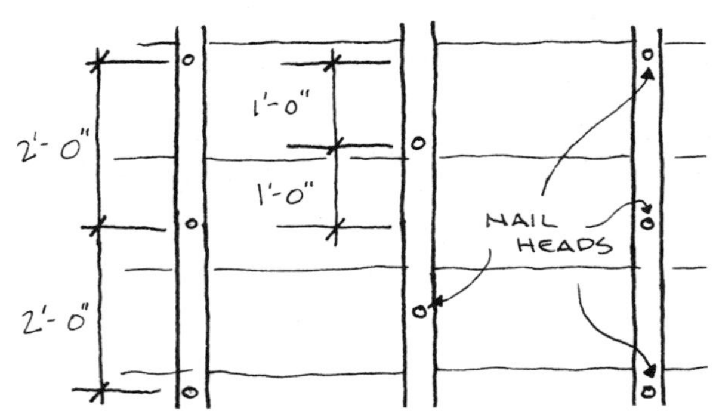

View of the floor joists looking up (you're lying on your back)

3 Once again, stuff the end of the batt back onto the sill plate up against the header. Move in behind the insulation and, with one hand, hold the insulation up while lacing the wire from joist to joist with the other hand.

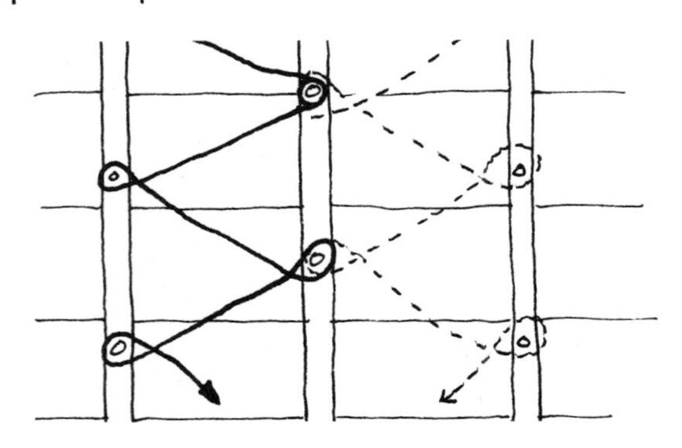

(Note that practically all nails will have two wires wrapped around them.)

Unlike wire mesh, you'll probably be able to do only one joist run at a time.

wood bridging strips

1 Essentially, wood bridging strips are "snapped" between two floor joists with the insulation resting on top of the strips. The bridging can be made from furring strips (1" x 2" or smaller) and are cut to a dimension slightly wider than the joist space.

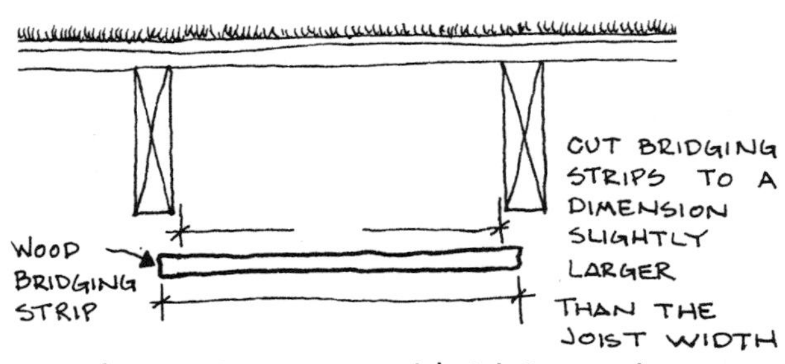

An alternative to wood bridging strips is "tiger teeth" (pieces of rather rigid wire that are pointed on both ends). These are made specifically for this purpose.

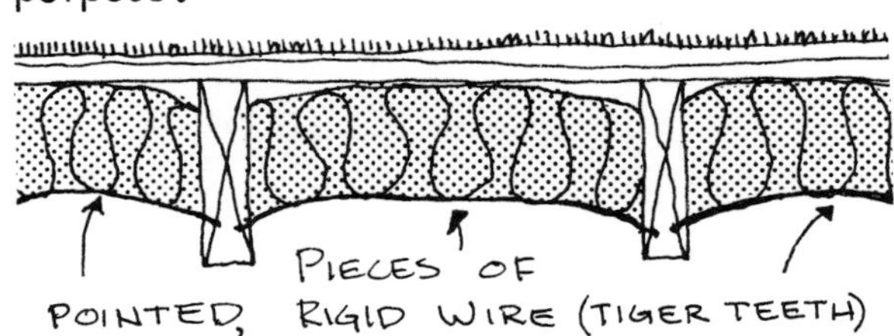

2 When installing the insulation in this manner, you may have to temporarily use your head to hold the insulation in place while you use both hands to "snap" the bridging strips or "tiger teeth" in place. Install the strips approximately 2' apart.

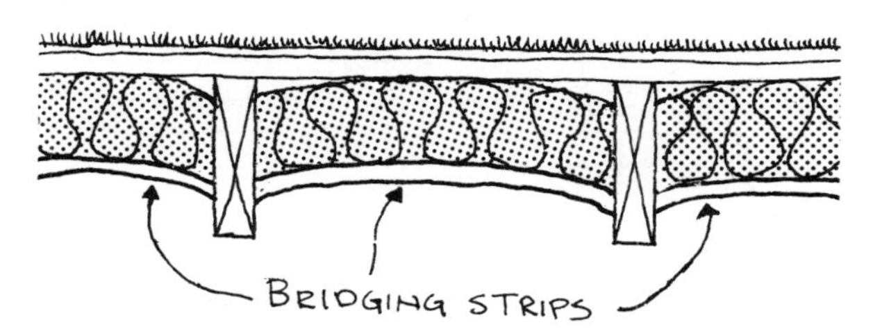

adhesive

An adhesive may also be used to hold the insulation in place, but, the insulation may not hold to the floor. It is suggested that if an adhesive is used, one of the previous techniques of holding the insulation in place also be used. Apply dabs of the adhesive approximately 6" apart along the length of the batt then press the insulation firmly against the floor boards.

NOTE:

The adhesive should be applied to the facing of faced insulation only. Adhesives will not work if applied to unfaced insulation.

ROLL/BATT INSULATION WITHOUT FACING

Unfaced insulation is installed in the same manner as faced insulation. However, in some areas of the country, it is recommended practice to install a vapor barrier (usually a polyethylene sheet) along with the unfaced insulation. The vapor barrier should not lap over the underside of the joists; instead, it should be cut short and stapled to the side of the joists. In this manner, moisture can escape into the crawl space and be vented to the outside.

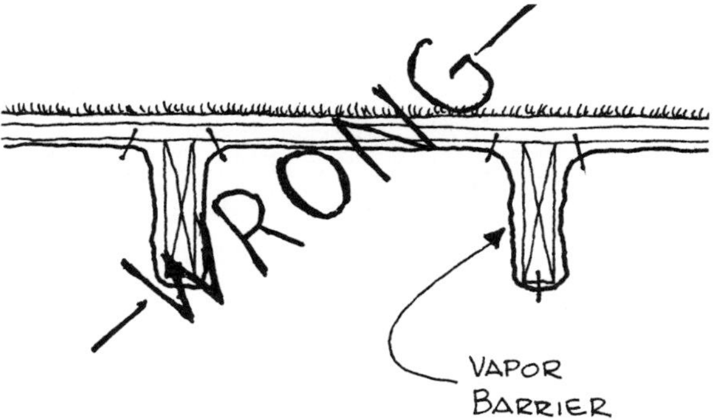

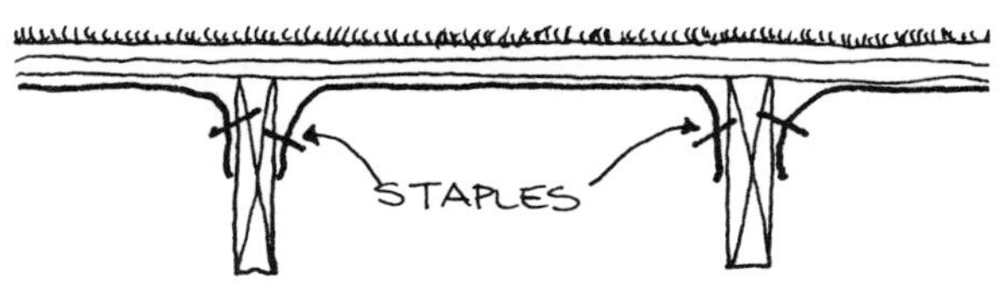

RIGID INSULATION

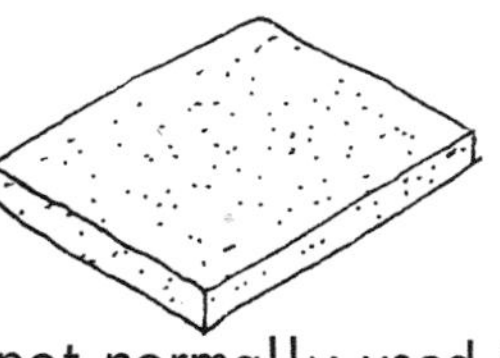

1 Rigid insulation is not normally used to insulate floors above crawl spaces. Check local building codes for use of rigid insulation in crawl spaces. However, if you do use it, attach it directly to the underside of the floor joists with large head nails, washers, and adhesives. The joint formed between the sill plate and the rigid insulation should be sealed with caulk or with an asphalt emulsion. Vapor barriers do not have to be installed with rigid insulation in this situation, since the crawl space can be vented to keep it dry (a vapor impermeable ground cover may be recommended, however).

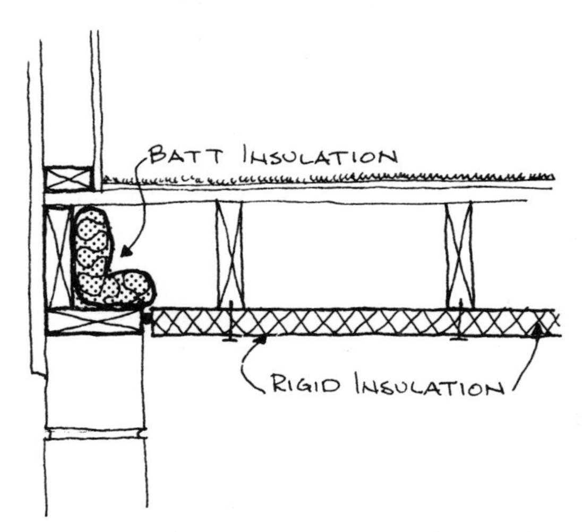

2 Prior to installing the rigid insulation, you should be certain to insulate the bandjoists and headers with batt insulation (foil or kraft paper backing should face the interior of the crawl space). If this is not done, heat will continue to escape through the bandjoists and headers, thus reducing the effectiveness of the newly added insulation.

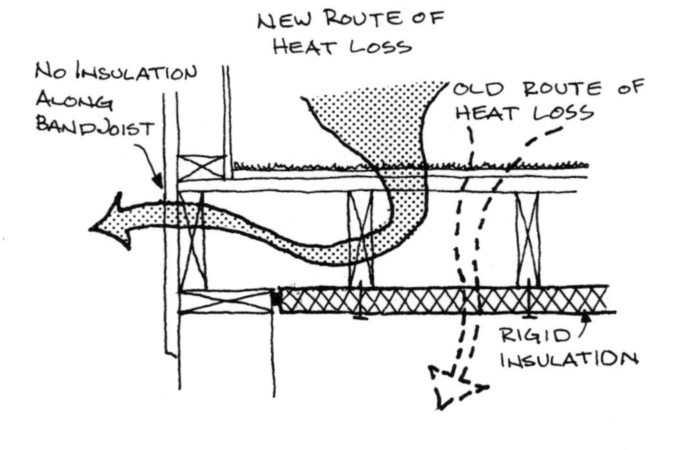

A FINAL NOTE:

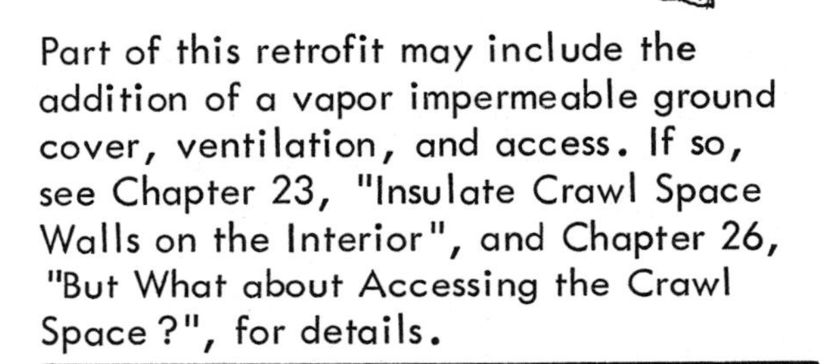

Part of this retrofit may include the addition of a vapor impermeable ground cover, ventilation, and access. If so, see Chapter 23, "Insulate Crawl Space Walls on the Interior", and Chapter 26, "But What about Accessing the Crawl Space?", for details.

26

but WHAT about accessing the crawl space?

description

To install insulation in a crawl space, there must be some means of access, i.e., a way of getting into it. Even if insulation is installed on the exterior of the foundation walls (see Chapter 24), there may not be a ground cover vapor barrier inside the crawl space and you must access the crawl space to lay this vapor barrier.

To access the crawl space, first examine the foundation wall to see if there is an access door or hatch. Perhaps there's a vent that, if temporarily removed, would provide a large enough opening for access.

If you have to open the foundation wall for access, check to see if vent installation is required. One oversized vent opening can be used for access. An oversized vent can be installed or the opening can be made smaller after insulation work is completed. Or, perhaps there is a framed-in portion of foundation wall under steps or porches, which can be more easily opened for access.

INSTALLING A WALL HATCH

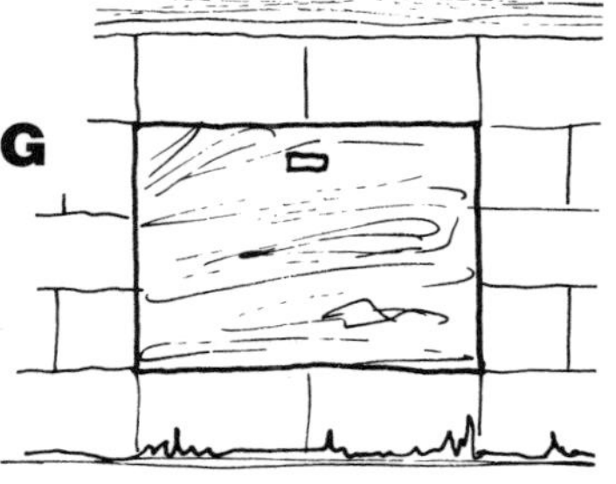

Crawl spaces are usually enclosed by concrete block foundation walls (skip to "Installing a floor hatch" in this chapter if there is a poured concrete wall).

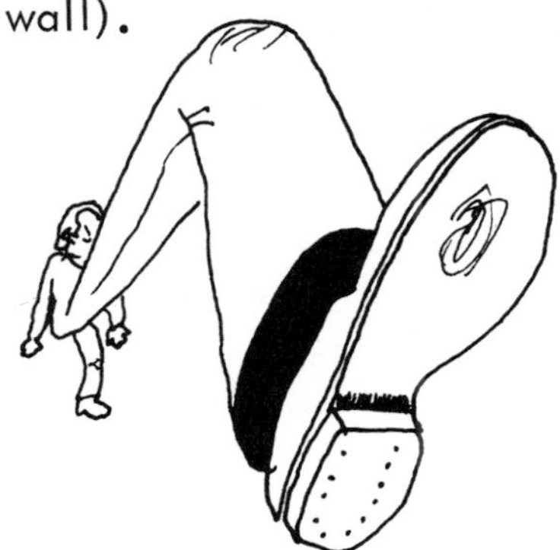

1 On top of the foundation wall is a "sill plate" which may be separated

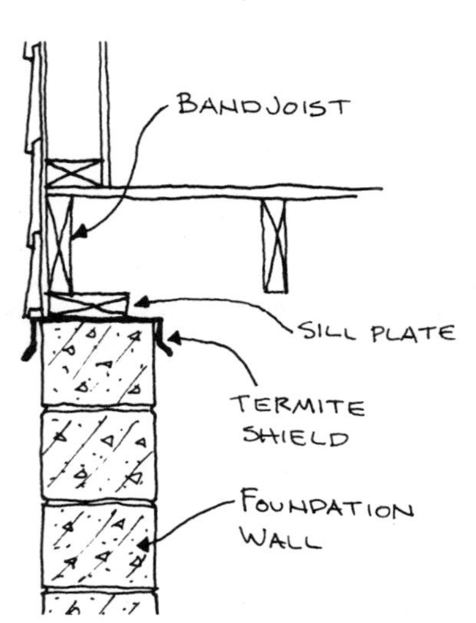

from the top of the foundation wall by a termite shield. It is preferable to open the foundation wall on a side of the home where there are "bandjoists". That is, the joist which rests on top of the sill plate and is parallel to the floor joists is the bandjoist.

The edge joists into which the floor joists frame (perpendicular to the floor joists) is called the "header", although it too is sometimes referred to as the bandjoist.

In order to build an access hatch in the foundation wall, you must remove structural support from beneath the bandjoist or header. Since the bandjoist carries just the wall load while the header carries both wall and floor loads, it is better to cut the opening in the wall beneath the bandjoist.

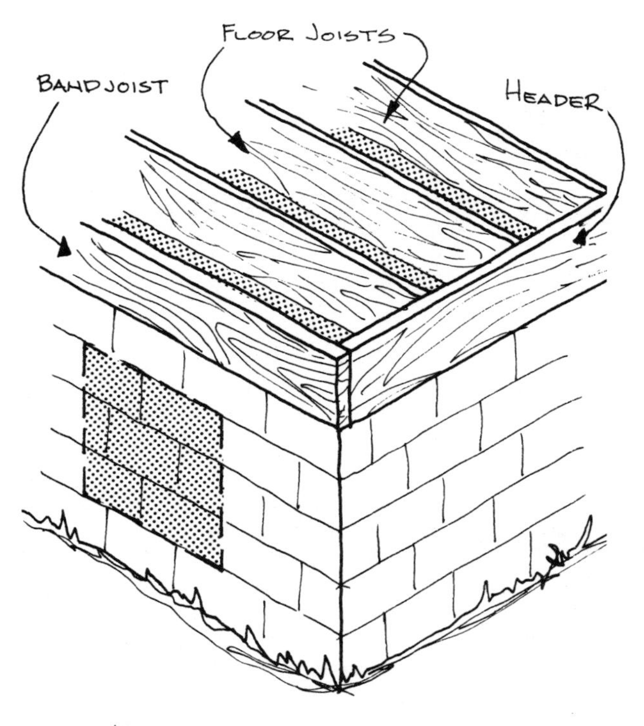

IT IS BETTER TO OPEN A CRAWL SPACE HATCH IN THE FOUNDATION WALL WITH THE BANDJOIST

2 Measure the size of the opening to be cut and mark off this dimension on the foundation wall block. It is better to open the wall along lines corresponding to the joints of the blocks. Block is ordinarily 8" high and 12" to 16" in length. An opening of 16" x 24" should prove adequate. A masonry "hardened" chisel and awl are the proper tools to use.

AN IMPORTANT NOTE:

Shatterproof protective eyewear must be worn at all times by all persons who do this work or are near to the work site. Workers who are cutting into the foundation wall must be warned to discontinue if persons not wearing protective eyewear get close to the work area.

3 If there is adequate wall area, cut the opening leaving one course* of block below the sill plate. Provide intermediate support to keep this row of block in place if necessary. For instance, remove block D, F, and G and put in a 2" x 4" to support blocks A and B before cutting half of blocks C and E away.

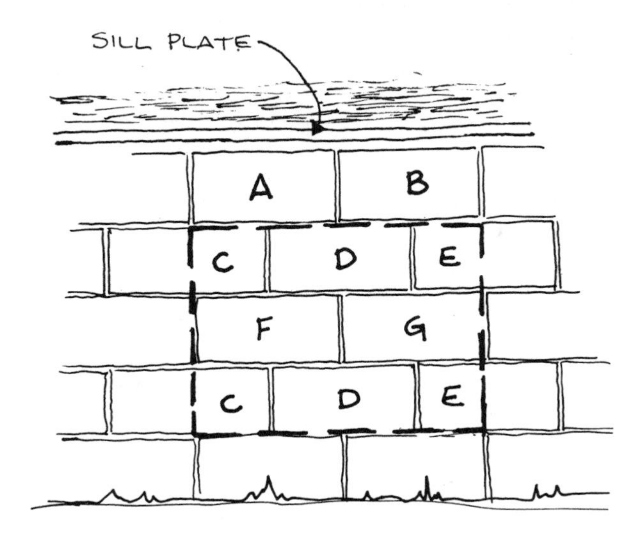

- REMOVE BLOCKS D, F, & G FIRST
- IF AN INTERMEDIATE SUPPORT IS NEEDED TO HOLD A & B IN PLACE, ADD IT BEFORE REMOVING C & E

*A row of masonry units, i.e., bricks, concrete block, etc.

4 Frame the opening by inserting a "lintel" and then putting in "posts" to hold up the lintel at its ends.

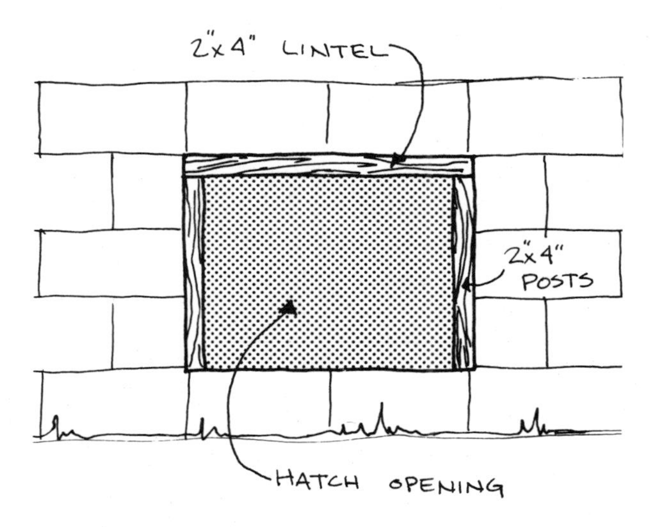

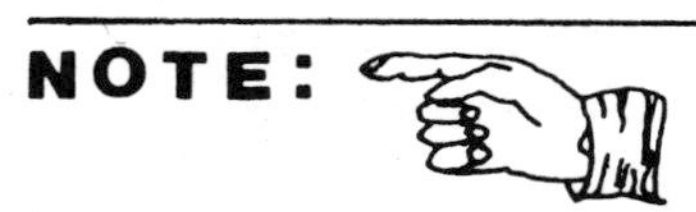

NOTE:

To install a vent the size of the access hole when you're through with the insulation work, see Chapter 23, "Insulate Crawl Space Walls on the Interior".

5 If you want to keep an exterior access, then install a door cut to fit the opening. Hinge the door at the top (along the 2" x 4" lintel) or on the side (hinge it to one of the posts).

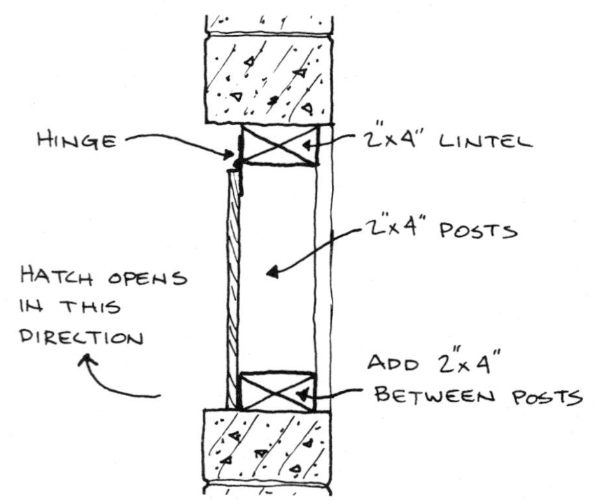

Fasten another 2" x 4" between the bottom ends of the "posts". Plywood (exterior grade - outerface 'A' grade) is suitable for a door. Install a latch to hold the door closed.

6 If the crawl space walls are insulated, insulate the inside of the access door and weatherstrip the edges. Polystyrene board of 1" or greater thickness is a convenient material for insulating the access door and adhesive backed foam tape makes a good weatherseal. See Chapter 41, "Seal Foundation Crawl Space Vents", for additional details.

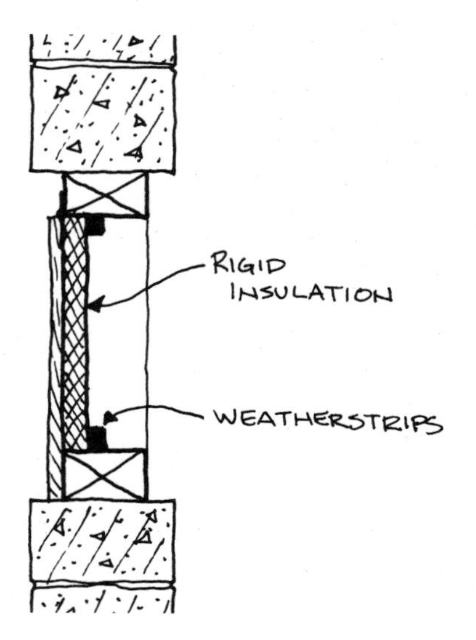

If you don't want to keep an exterior hatch to the crawl space, nail the plywood to the 2" x 4" blocking.

INSTALLING A FLOOR HATCH

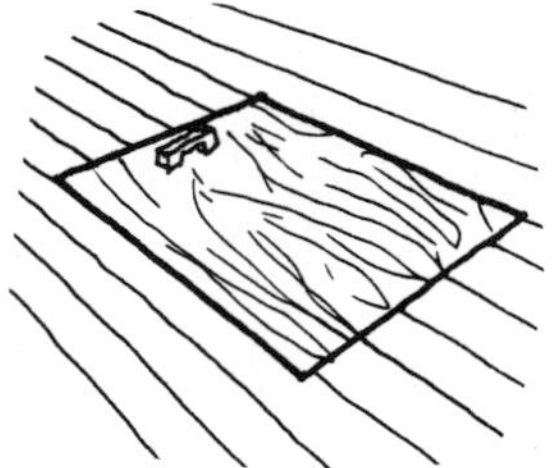

You may prefer to access the crawl space from above, or, you may encounter a poured concrete foundation wall, badly deteriorated foundation wall or sill plate, or other problems that would prevent cutting an access hatch from the outside. In this case, an access hatch can be cut through the floor over the crawl space. Find an area of floor out of the path of travel. Closet floors are a good place to locate an access hatch.

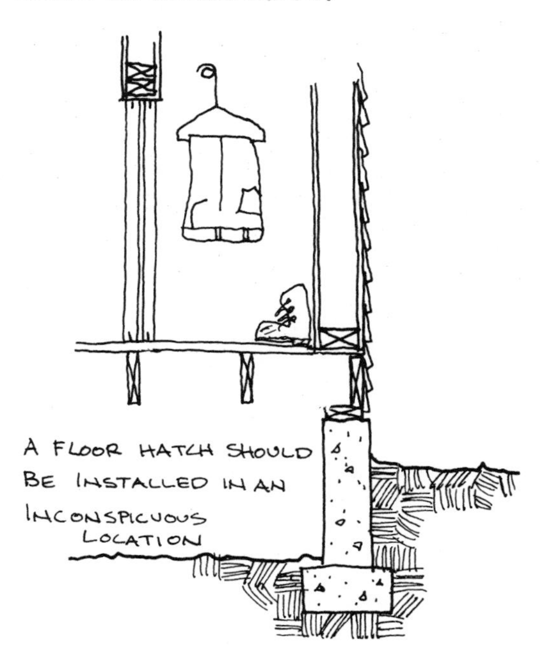

1 Locate existing floor joists and cut between them. Ordinarily, this will provide an access hole of minimum 14"-15" width which should be adequate.

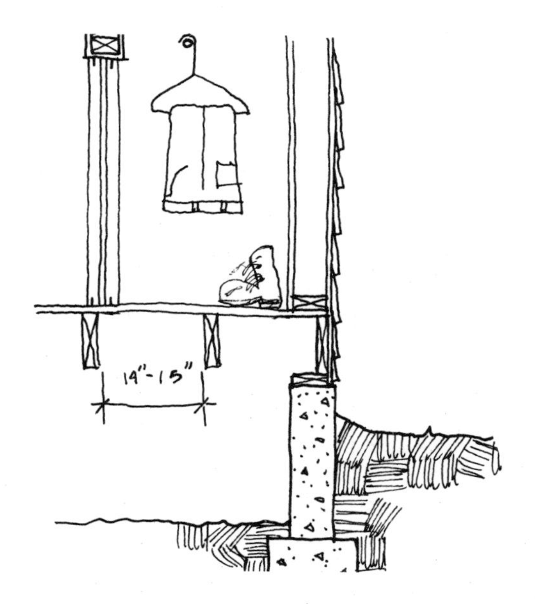

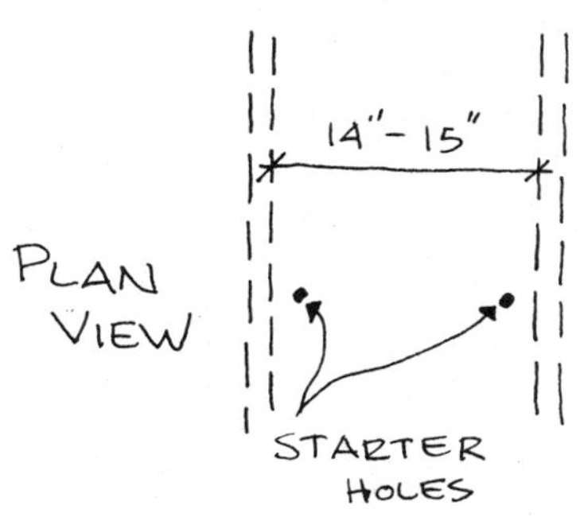

2 Drill two holes adjacent to a floor joist at a distance equal to the "long" dimension of the access opening.

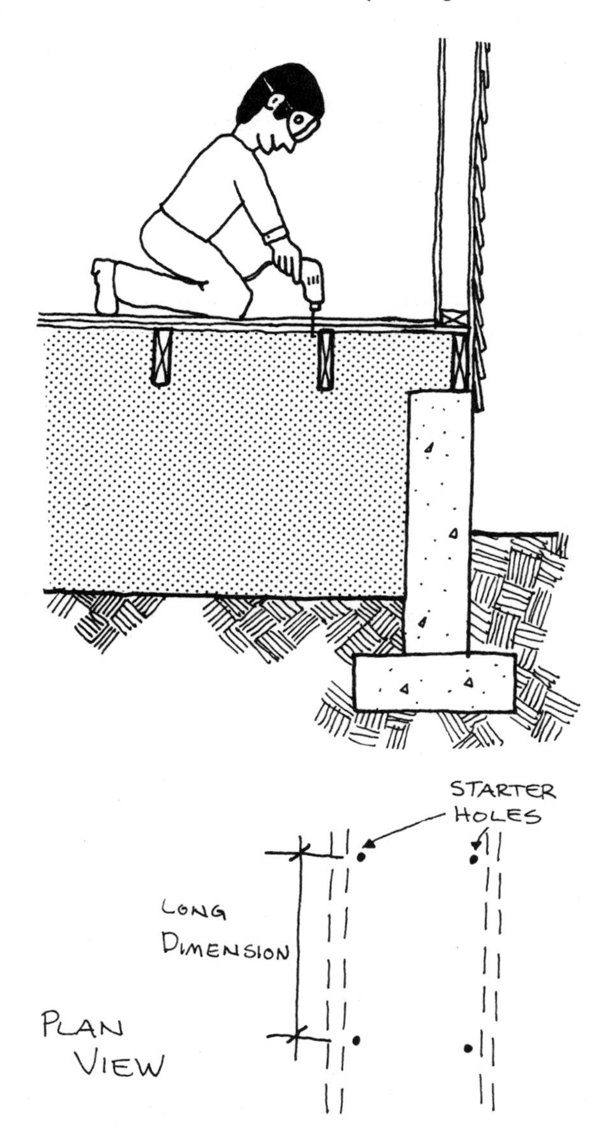

3 No matter what sort of saw is used, mount a board to serve as a guide while cutting between holes.

NOTE:

It is advisable to not cut floor joists.

4 Be sure to make straight cuts. A sabre saw is easiest for this work. Otherwise, a combination of "coping" or "keyhole" saw and a cross cut saw can be used. With the sabre saw, set the blade to cut through any sub-flooring so that one pass of the saw completes the cut.

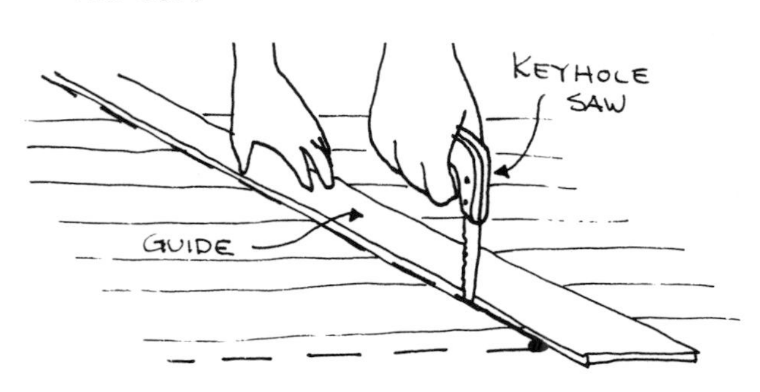

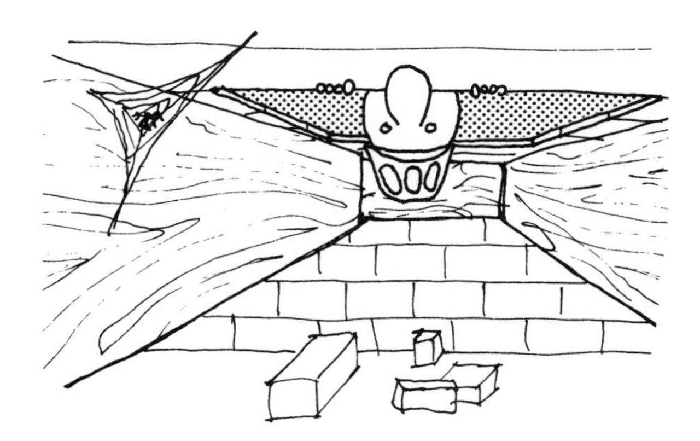

5 Before beginning insulation work, install framing around the access hole as shown. When job is complete, cut an access door to lay flush with the existing floor and bearing on all four framing pieces. If insulation was added to the floor, be sure to add insulation to the hatch.

VIEW LOOKING UP FROM CRAWL SPACE

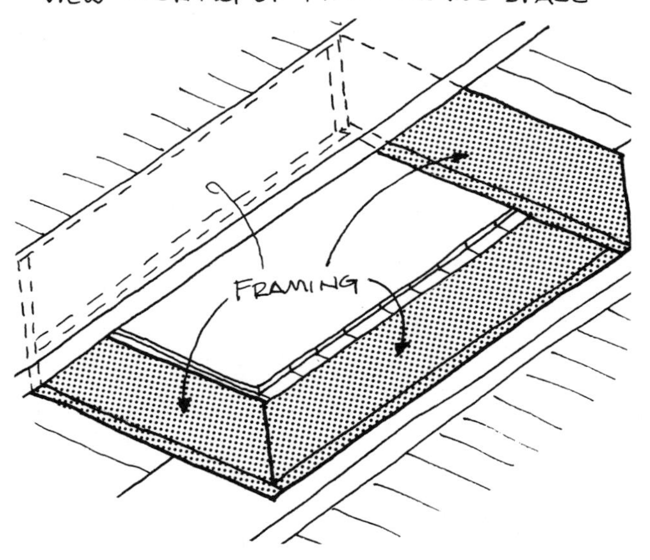

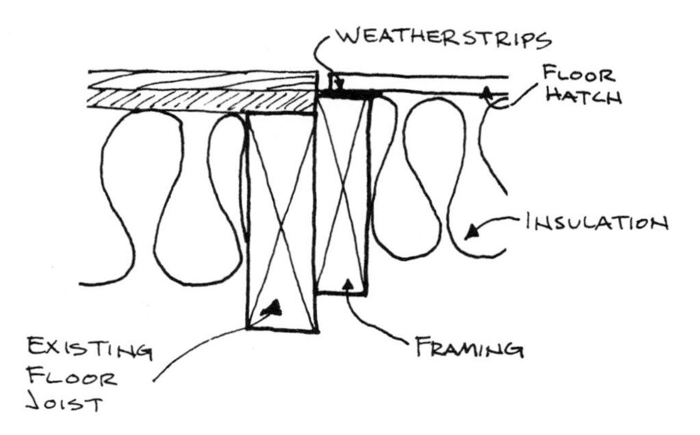

27

insulate slab on grade

description

People who live in homes with concrete floors built directly on the ground (slab on grade) may complain of "cold floors". If the slab is not insulated, its edge is exposed to the outside air. This exposed edge acts like a "drain", pouring heat to the outside, thus making the slab feel cold. In addition, the slab edge and foundation is exposed to the ground.

At grade level, the temperature of the ground approaches that of the air (this represents quite a temperature "swing" between summer and winter). As you move down in the ground, this temperature "swing" is not as drastic as the ground temperature becomes more stable. Therefore, it's not only important to insulate the slab edge with rigid insulation, but also the foundation.

As with the previous two exterior, below grade, wall insulation retrofits (Chapters 22 and 24), excavation is necessary down to the frost line or footing, whichever is shallower. The slab edge and foundation should be cleaned of all dirt and loose material prior to bonding rigid insulation to the slab and foundation with a compatible adhesive. Any insulation that will be exposed to the weather after backfilling must be protected by a weatherproof finish and/or flashing.

uninsulated

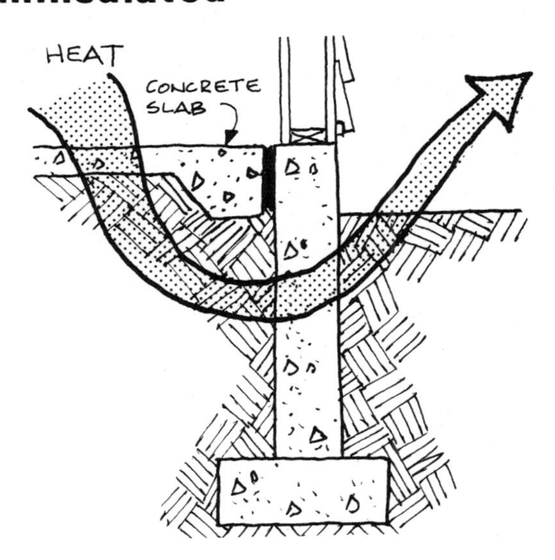

insulated

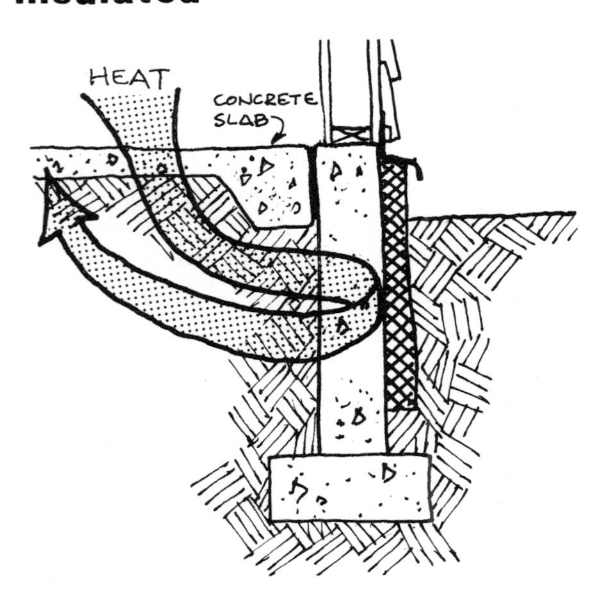

materials

rigid insulation

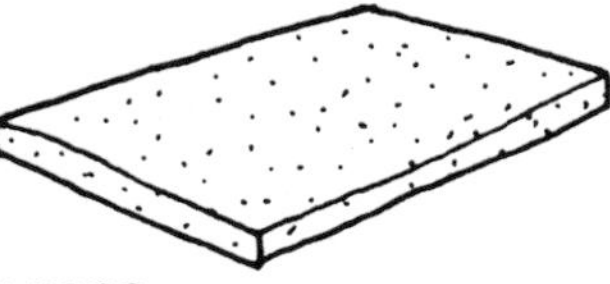

- extruded polystyrene
- polyurethane

waterproofing-adhesives

- asphalt emulsion
- fibrated mastic
- liquid epoxy
- hot melt asphalt
- tar modified urethane
- panel adhesive

masonry cleaners

- acidic type

exterior finish

- siding
- cement asbestos sheet
- latex based stucco
- polymer finish
- sheet vinyl
- vinyl cement

other materials

- caulk

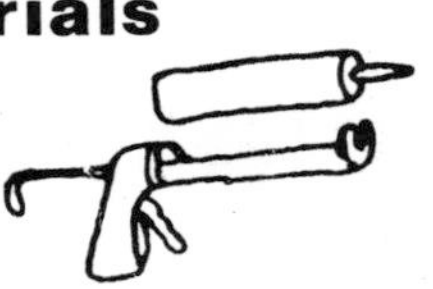

- packing

- flashing

preparation

The necessary excavation will disrupt any planting that you have done around the home. For the duration of the work, there will be a trench around most of the home. Precautions should be taken to keep people from stumbling into it. Put stakes and cloth "flags" around the edges of the trench whenever work is completed for the day. If a contractor is doing the work, you should understand who is responsible for restoring the yard to its "original" condition (final grading, re-planting, sidewalk or driveway repair, etc.). Will it be done by you or a contractor?

Before excavating, note where the various services (gas, water, sewer, electric, telephone) enter the home to avoid rupturing pipes and/or snapping wires. Also, some homes have drainage tile that are fed directly from downspouts. These tile will have to be reinstalled in the exact level and location (to the outside of the new insulation) they were in to avoid drainage problems around the home.

Refer to Chapter 6, "A Word about Safety", for additional details on excavation.

1 Excavate around the slab to the depth of the frost line or top of the footing, whichever is shallower (check with the local building department for frost line depth). Place the dirt in a spot to make backfilling easy, preferably right next to the trench. Never pile dirt underneath the branches of a tree because the tree roots extend as far out as the branches and will be damaged. The excavation can be done with a "back-hoe", power trencher, or ambitious workers equipped with picks and shovels. The trench should be wide enough for a person to install insulation to the slab edge and footing (18" - 24").

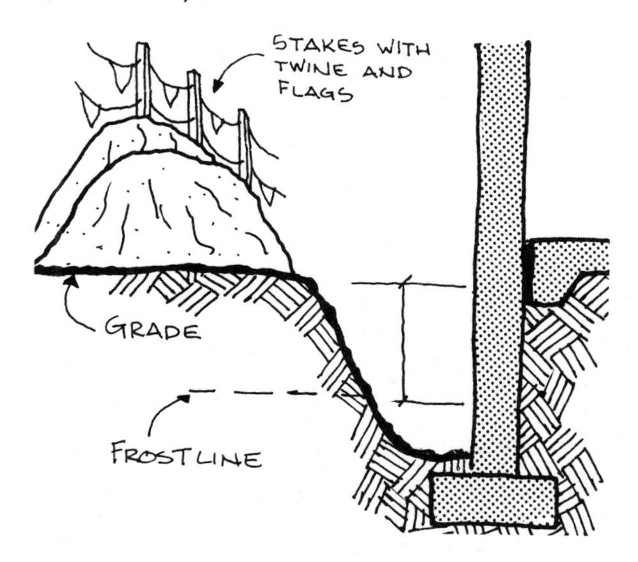

2 Clean all dirt and loose material from the slab and foundation with a stiff brush and water. A mild acidic masonry cleaner can also be used, but avoid skin contact with the cleaner by wearing rubber gloves and a long sleeve shirt.

installation procedures

1 Apply the adhesive to the insulation board, approximately 2" wide by 1/4" or 3/8" thick, to the entire perimeter of each board. Be sure the adhesive is compatible with the insulation. In addition, apply dabs of the adhesive of the same thickness to the interior area of the board approximately 8" on center. Stagger all vertical joints (if two or more layers are needed to obtain a certain "R" value, stagger all vertical and horizontal joints). Apply pressure over the entire surface of the board to insure uniform and tight contact with the foundation. Be sure that all joints butt tightly together.

2 Any insulation that will be exposed to the weather after backfilling should be weatherproofed. Use a suitable exterior finish and install in strict accordance with manufacturer's procedures. A wood finish, however, must be stopped a minimum of 6" above grade. If necessary, re-install drainage tile after insulation has been placed and weatherproofed.

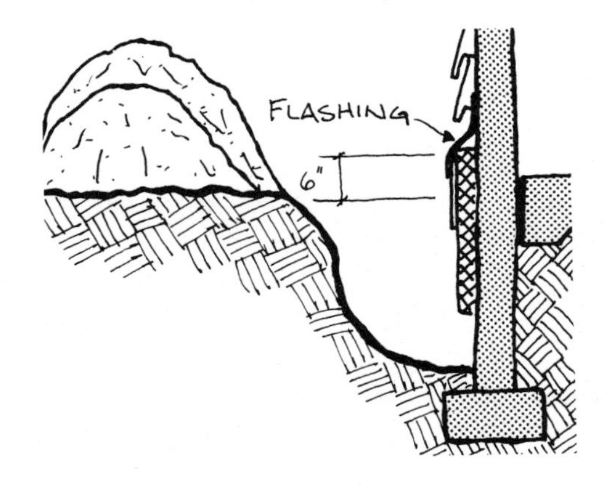

NOTE:

New products may be available to you for weatherproofing exposed rigid insulation. Also, insulation with a weatherproof covering on one side may be available for use.

3 A gravel backfill should be used against the insulation board (this will

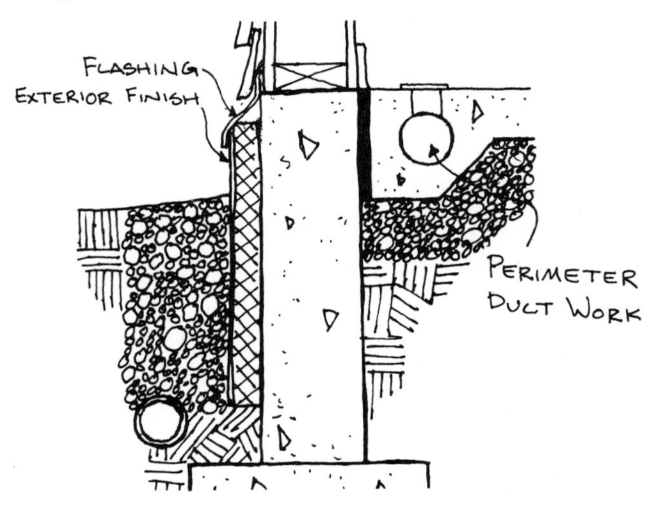

let water drain away from the insulation). Lay in gravel as you are backfilling the trench. Place a little extra dirt over the trench to allow for settling. Taper the fill away from the home so that water drains away from it.

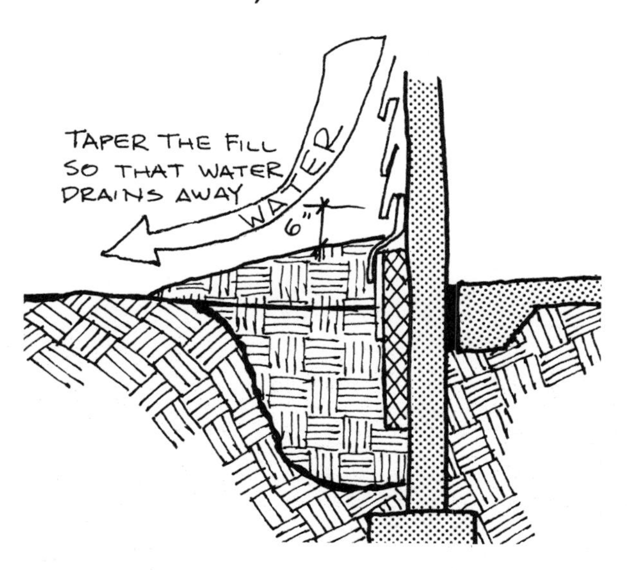

4 Attach flashing over the top edge of the insulation. For frame walls (except stucco), tuck the flashing under the siding a minimum of 1-1/2" or the width of the siding, whichever is less.

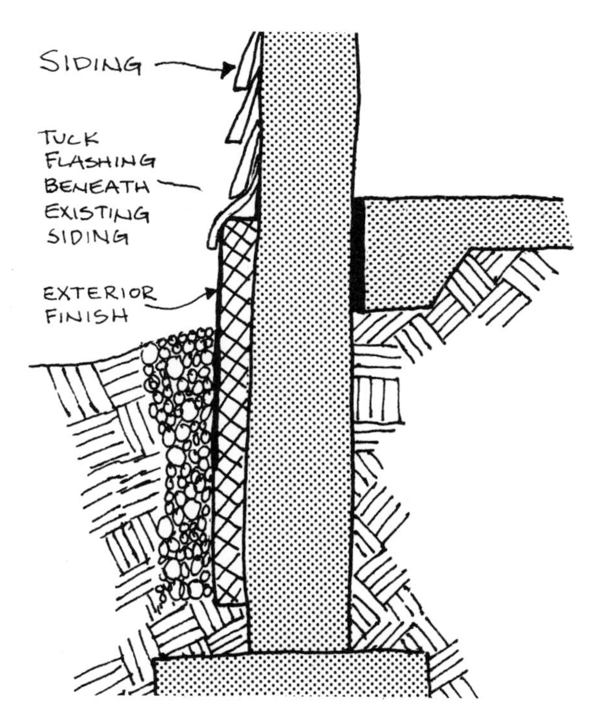

With stucco walls, bed the top of the flashing on the stucco with a high performance caulk (see Chapter 5, "A Word about Caulks", for properties of caulks).

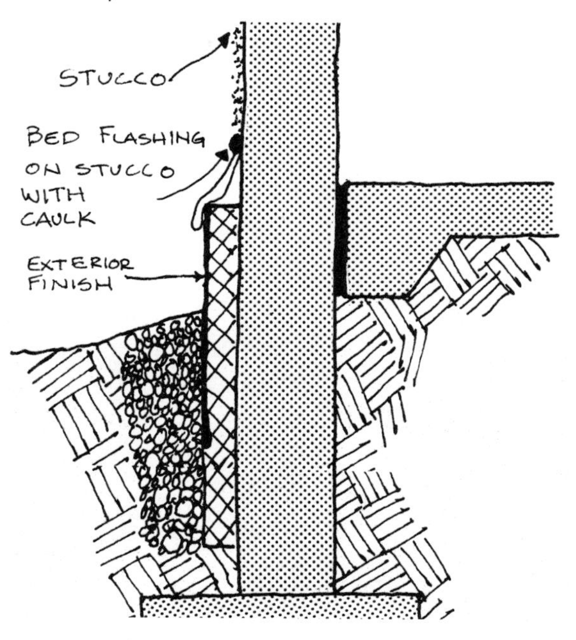

With masonry walls, cut a 3/4" reglet (groove) in the wall just above the top edge of the insulation. After the flashing has been installed, seal it with a continuous bead of caulk.

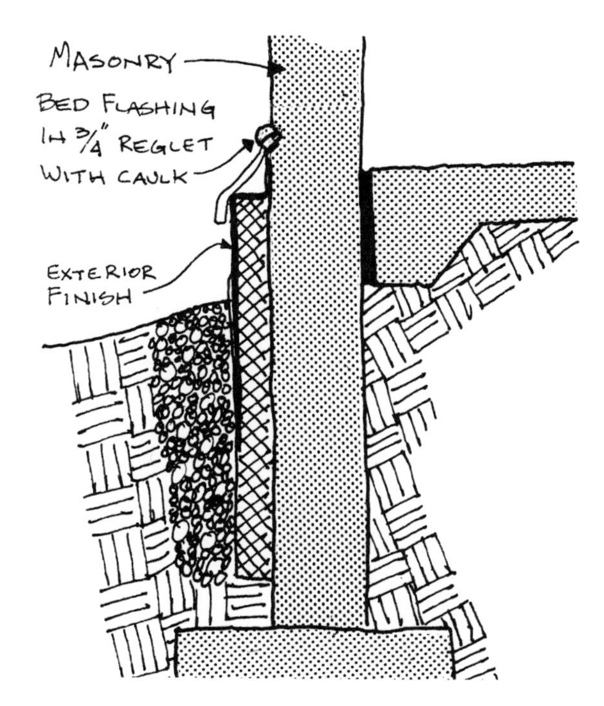

28

insulate masonry wall on the interior

description

Yes, existing masonry walls can be insulated. The approach is different from frame walls, but insulation can be added to the interior or exterior of masonry walls. The insulation is then covered with a new interior or exterior finish (see Chapter 29, "Insulate Masonry Wall on the Exterior").

The main difference between insulating a frame wall and a masonry wall is that masonry walls do not have open spaces between studs to blow or fit insulation. Brick surfaces are uneven, so you'll ordinarily find narrow (3/4" - 1-1/2" thick) furring strips and plaster or gypsum board finish. The walls can be stripped to the masonry and insulation furred in from the masonry, or the insulation can be furred in over the existing finish. In this case, the rooms will be noticeably smaller. Special detailing is needed around windows, doors, switches, and receptacles to bring them even with the new wall surface when insulation is installed on the interior. This will require skilled carpentry for trim and finish work.

Insulation can sometimes be blown into the existing fur space if it is at least 1-1/2" deep. Also, insulation can sometimes be poured in place successfully in fur spaces even less than 1-1/2" deep. In this situation, a new interior finish is not required unless needed to boost the fire rating.

A complete job involves insulating the portion of wall at the joist space (between floor and ceiling of space below). Several techniques are possible and their use in encouraged.

This retrofit involves high labor skills. It should only be done by a person skilled in carpentry or by a qualified contractor.

NOTE:

Some building codes prohibit any "combustible" insulation material (even though wood furring itself is flammable). In other cases, combustible insulation such as polystyrene can be used but it must be covered with a 15 or 30 minute fire rated finish. Check with your building department.

---

materials

roll/batt insulation

- glass fiber
- rock wool
  (both available as foil faced, kraft faced, unfaced, or reverse flange)

rigid insulation

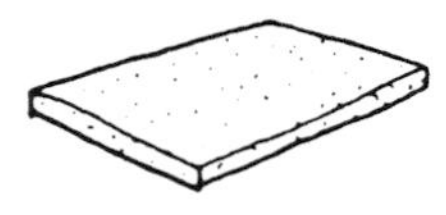

- polystyrene
- polyurethane
- polyisocyanurate
- mineral or wood fiber
  (sheathing board)

poured in place insulation

- cellulose
- perlite
- vermiculite

blown in place insulation

- cellulose
- rock wool
- glass fiber
- urea-formaldehyde
- perlite
- vermiculite

vapor barrier

- aluminized or oil based enamel
- latex vapor barrier paint
- polyethylene film
- foil faced gypsum board

furring / framing

- 2" x framing lumber
- furring strips
- metal channels or studs
- gypsum board or other fire retardant
  material - check local codes

waterproofing-adhesives

- fibrated mastic
- asphalt emulsion
- liquid epoxy
- panel adhesive

other materials

- masonry cleaner
- caulk (or patching cement)
- packing
- fasteners
- hole plugs

preparation

The perimeter walls can be stripped back to the masonry with the insulation furred out from that point, or the insulation can be furred out over the existing interior finish. If either of these two techniques is used (rather than blowing or pouring insulation into the existing furred air space), you'll need to do some preparatory work to the walls.

Increase fur space depth by removing existing finish and furring and installing new furring or by attaching new furring on top of existing finish and furring. Depth of fur space must be at least as great as thickness of insulation to be installed (accounting for masonry wall unevenness). However, fur depth at least 90% of thickness of fibrous insulation to be installed does not cause serious reduction of the thermal resistance of the insulation.

Always check for opening of fur space to attic (air bypass) at level of top floor ceiling. Pack opening with pieces of insulation or seal with foam (see Chapter 39, "Seal Air Bypasses") before installing insulation.

To insulate back to the inside surface of the masonry, mark interior trim which rests over finish and remove in an orderly fashion for later reinstallation. This includes ceiling molding, baseboards, and window and door trim. For example, casing around a window should be marked R (right), L (left), and T (top) and should be bound and set aside. Remove fixture and outlet covers. Remove interior finish and fasteners from existing furring and place on the outside of the home to be hauled away.

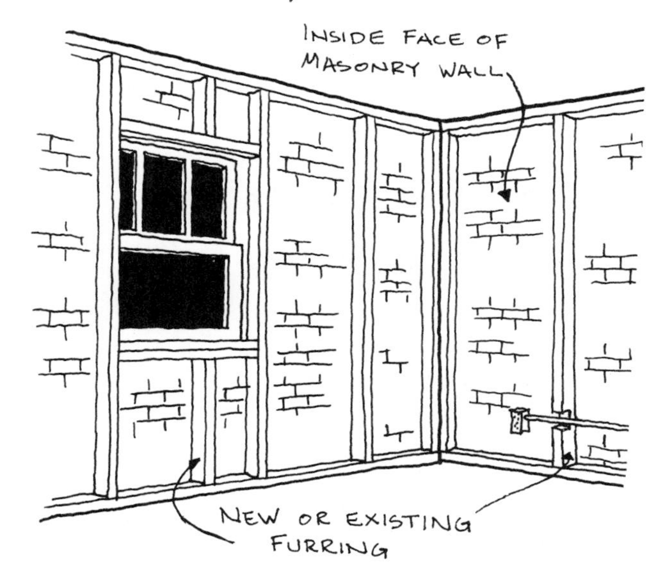

If the insulation will be added over the existing finish, remove the ceiling molding, baseboards, and window and door trim and reinstall with new finish. In addition, electrical boxes will have

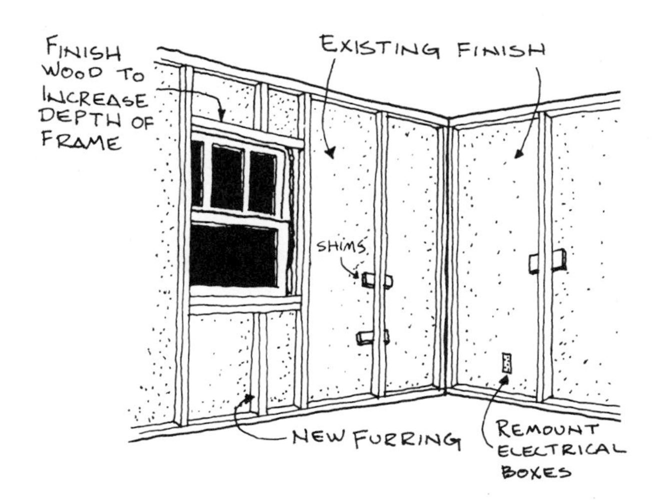

to be remounted so that they will be flush to the new interior finish. Shimming may be required when furring is added to uneven wall surfaces. Where possible, center furring strips over existing furring strips.

NOTE: 

If a vapor barrier is integral to the existing wall finish, it must be stripped off before insulating over this existing finish.

installation procedures

ROLL/BATT INSULATION

Roll/batt insulation, whether it's faced or unfaced, is installed between framing members. Essentially, this means building a frame wall over the masonry with 2" x 4" lumber (thinner "R-7" insulation blankets can be placed between 2" x 2" furring) or metal channels. The insulation, as well as the finish material, is then attached to this framing.

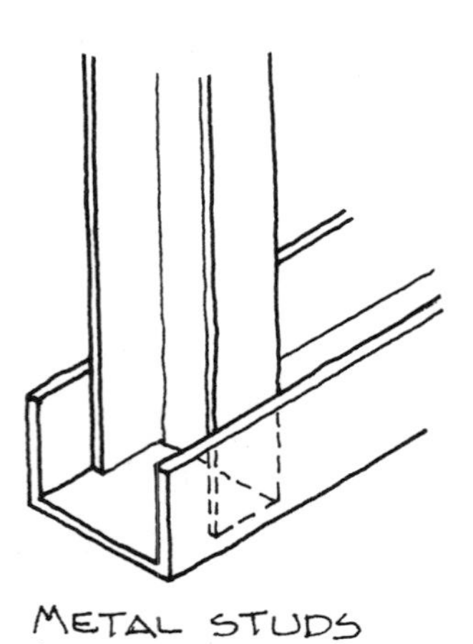

METAL STUDS

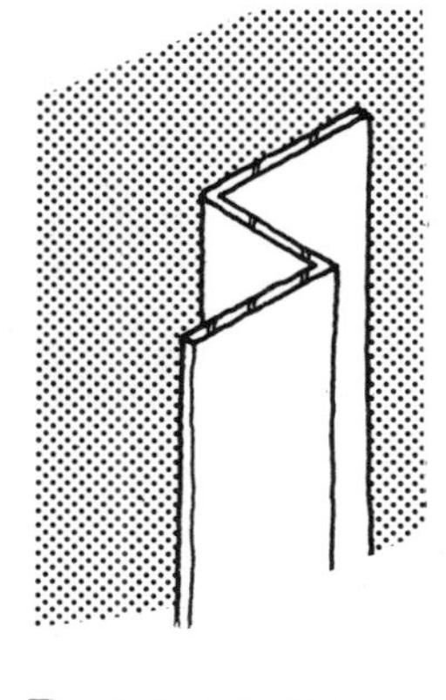

Z-CHANNEL

1 Begin by building the framework. Space the framing to correspond to the width of the insulation. Typically, this is 16" or 22" center to center. Nail the bottom plate to the floor and the top plate to the ceiling (nail into the joists). Then toenail the studs into place. Framing can be "tilted up" after being built on the floor just as a housing contractor does. If this is done, be sure to find the tightest dimension between the floor and ceiling and space the plates 1/4" or so less than this dimension.

If you stripped the wall finish back to the masonry, there may be an opening between the floor and the wall and between the ceiling and the wall. These openings should be sealed and insulated with packing and/or insulation. Do this work before adding the framework.

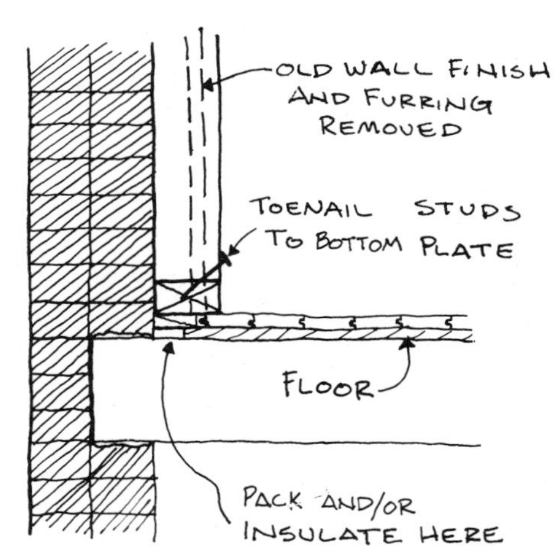

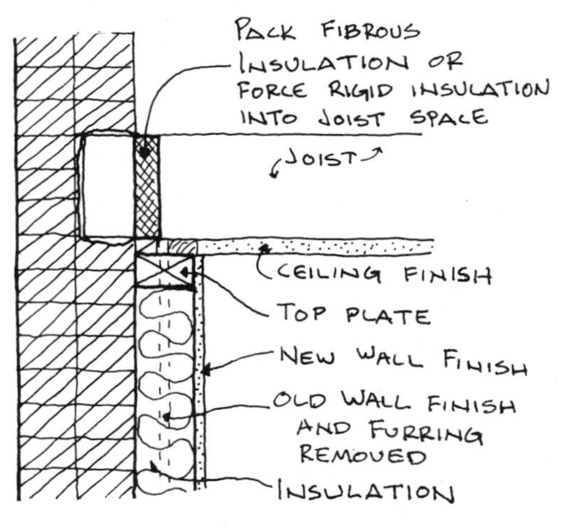

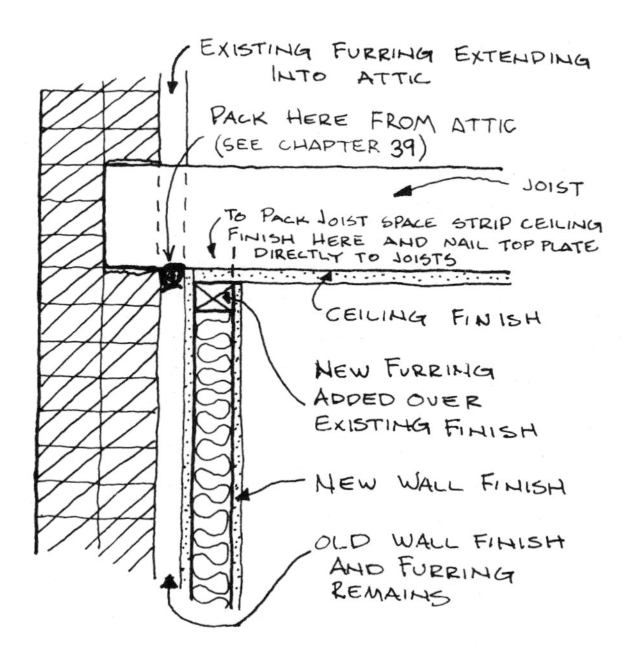

2 Install the insulation between the studs. The insulation is to extend the full height of the wall. If the insulation has facing, install it so that it faces inside (towards you). Cut the insulation with a utility knife or utility shears and a straight edge (it's easier to cut with the facing placed up). Pack narrow spaces, especially around windows, doors, and electrical boxes with scraps of insulation.

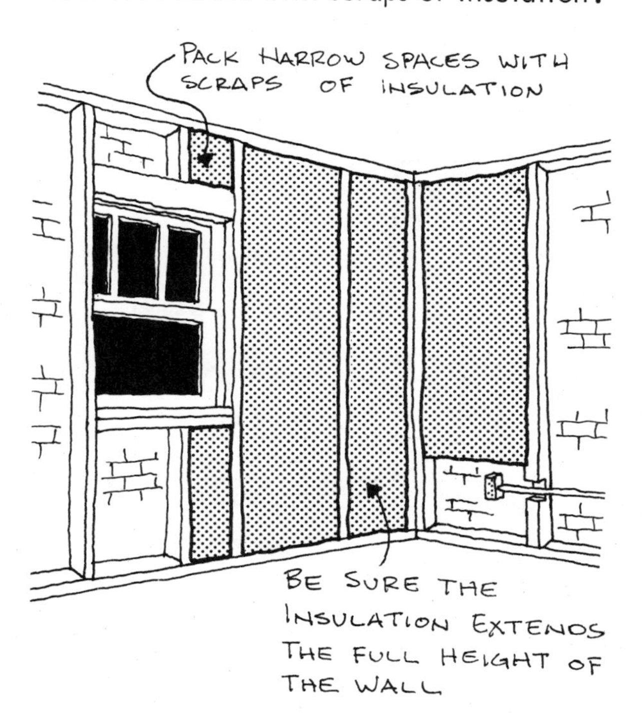

3 Place the insulation behind wiring and pipes. Do not place insulation in contact with exposed wiring. Secure insulation a minimum of three inches away from exposed wiring.

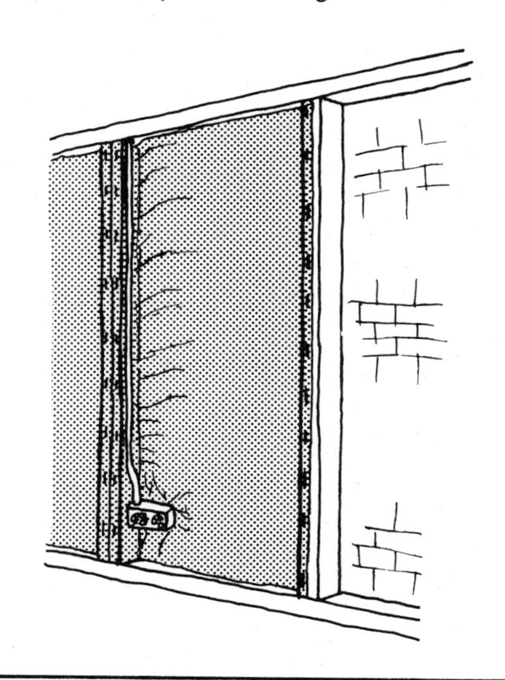

NOTE:

Install insulation no closer than 3" from recessed electical fixtures such as lights and fans. Check local codes.

4 Staple, or tape, the insulation to the studs. If staples are used, space them

PULL AND STAPLE ALTERNATE LENGTHS OF THE FLANGE TO ACHIEVE A TIGHT NEAT ATTACHMENT

6" apart and cover them with polyethylene tape if vapor protection is required. Butt insulation edges to prevent voids between pieces of insulation and framing members. Avoid creases, folds, and "fishmouths" in the flanges of the insulation.

NOTE: 

Foil faced insulation will increase the insulation value (resistance) if a 3/4" air space is provided between the foil and the interior finish.

5 Wedge unfaced insulation between the studs. Some batts are semi-rigid and made to fit tightly between framing members by friction. A vapor barrier (usually a polyethylene sheet) can then be installed over the unfaced insulation. Staple the vapor barrier to the studs. Then go back and tape over the staples with polyethylene tape to keep moisture out. The foil or asphalt impregnated kraft paper on faced insulation serves as a vapor barrier.

NOTE:

An alternative to a polyethylene sheet is foil faced gypsum board or vapor barrier paint. Apply the paint, either an aluminized, oil based enamel, or latex vapor barrier paint, over the exterior of the gypsum board (see step 6). Vapor barrier faced insulation can be used to protect against condensation at the floor joist space, but it will be very difficult to obtain a continuous barrier even at stringer joists.

Tape the sheeting to the ceiling finish and to the flooring beneath the position

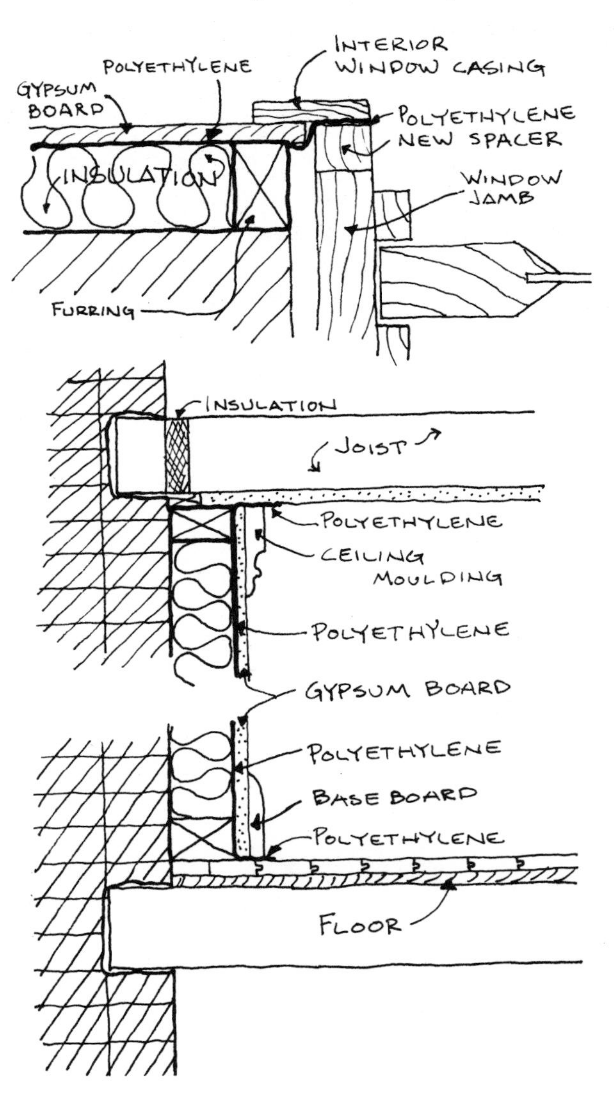

of the floor molding to be replaced to provide continuous coverage. Likewise, extend vapor barrier over frames of openings to have casing and finish reinstalled.

6 Nail gypsum board to the wood framework. You can glue gypsum board to the framing if a vapor barrier sheet has not been installed covering the framing. The gypsum board should extend the entire height of the wall. Follow instructions in manufacturer's installation manual for interior finishing details for gypsum board.

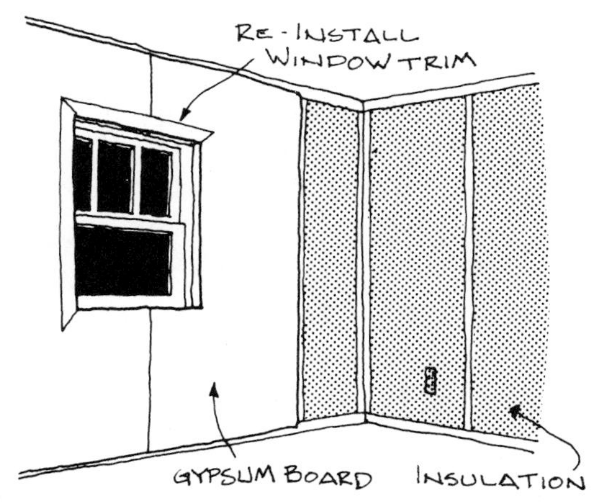

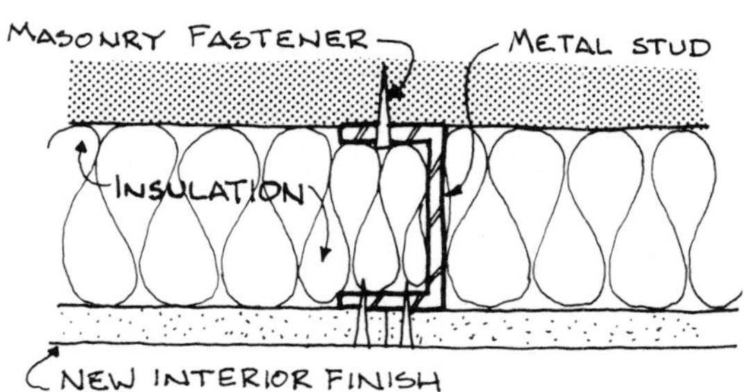

RIGID INSULATION

1 Rigid insulation should be installed between furring strips (if the insulation is thicker than 1-1/2", use framing as described for Roll/batt insulation, step 1. Attach the furring strips to the wall with an adhesive and/or masonry nails. Use shims for uneven wall surfaces. When adding furring strips over the existing finish, try to center them over the existing furring strips for nailing.

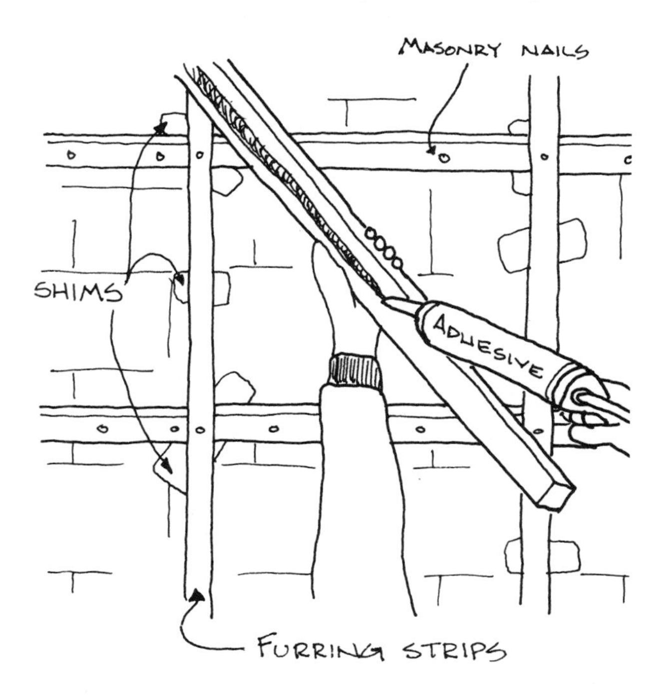

2 Attach vertical furring strips to the wall with a maximum spacing of 4' on center (the long edges of 4' x 8' gypsum board panels will have wood backing). Rigid insulation is usually sized to fit between furring spaced 24" on center, or some multiple of that (16" x 48", 16" x 24", 24" x 48"). Check the dimensions of the

insulation that you'll be using, however, prior to installing the furring to minimize cutting the insulation. The important thing to remember is to provide vertical support for the gypsum board. Be sure to fur out around windows, doors, and other obstructions you may find on the wall.

An alternative to furring strips are metal "Z" channels. The channels are available in varying depths to accommodate different insulation thicknesses.

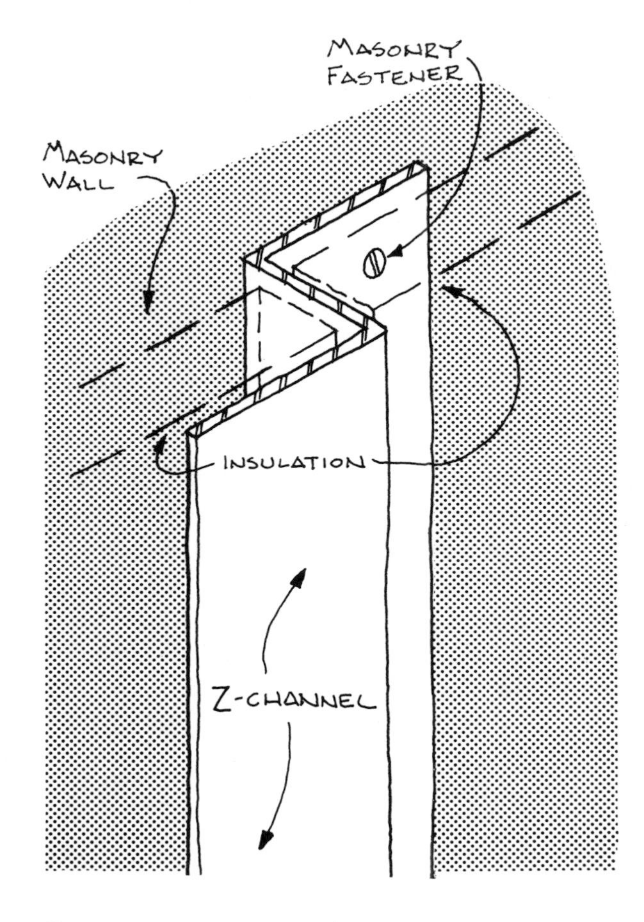

3 Attach rigid insulation directly to the wall with a compatible panel board adhesive. Apply the adhesive along the perimeter of the board with "spot dabs" on the interior. Firmly press the insulatior in place so that it fully contacts the wall surface. The boards should butt tightly together.

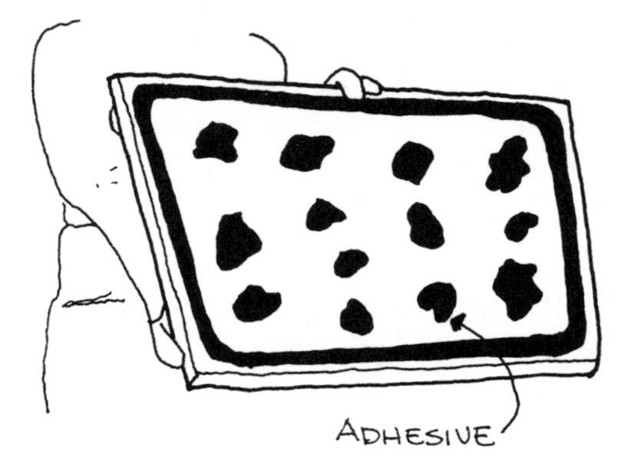

4 Some rigid insulation board is available with an integral foil face that serves as a vapor barrier (the foil is to face towards you). Otherwise, add a vapor barrier to the rigid insulation in the same manner as described for Roll/batt insulation, step 5, in this chapter.

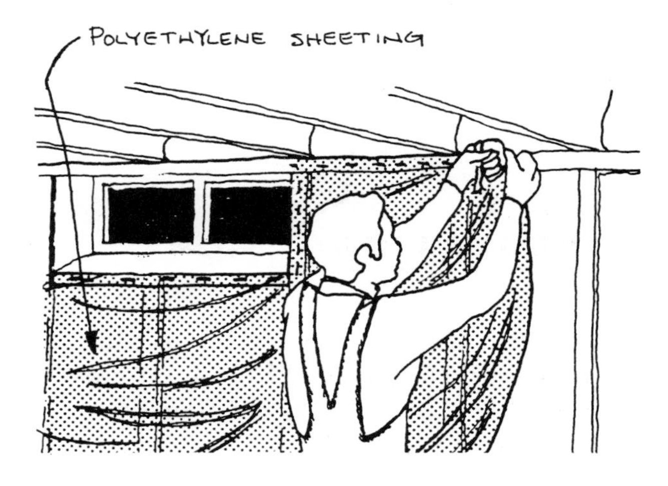

5 See step 6 for Roll/batt insulation in this chapter for details on installing a gypsum board wall finish.

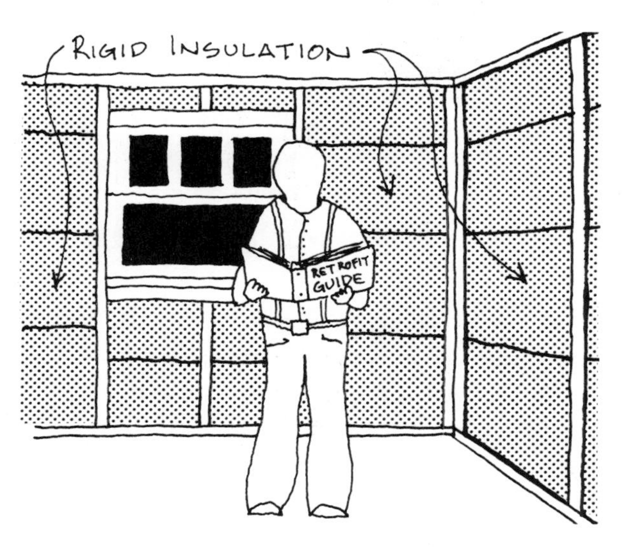

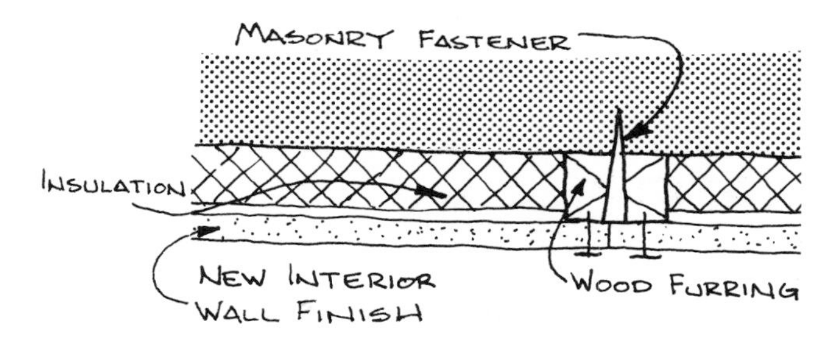

BLOWN IN PLACE INSULATION

1 If the walls are finished with an air space behind the finish, insulation can be blown into the air space. Special equipment is needed for this. You may have to hire a trained installer, although you can rent the equipment and do it yourself (urea-formaldehyde will have to be installed by a trained installer). All cracks and holes in the exterior masonry wall and the interior finish should be sealed prior to installing the insulation.

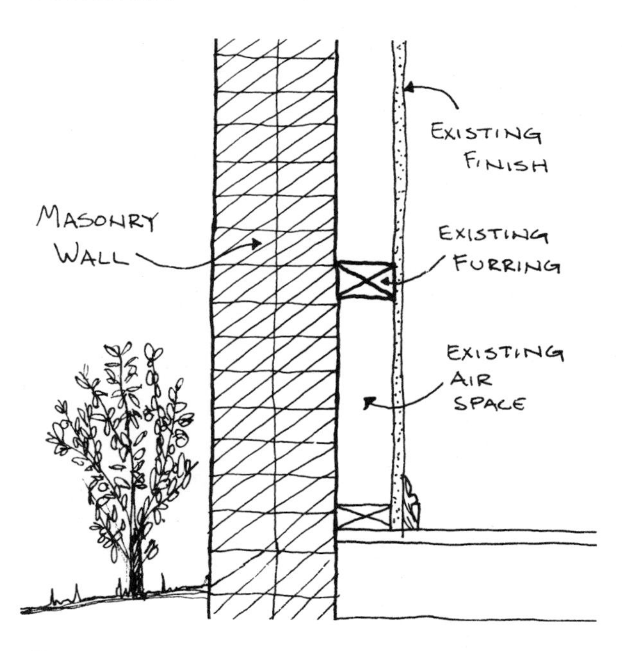

2 Drill two holes (1-1/2" or 2" diameter) through the interior finish per stud space. Locate one hole near the ceiling with the other 4' above the floor or just below the fire stop. It is recommended that insulation not be blown more than 4' down or 1' up. With urea-formaldehyde foam, locate the first hole 2' above the

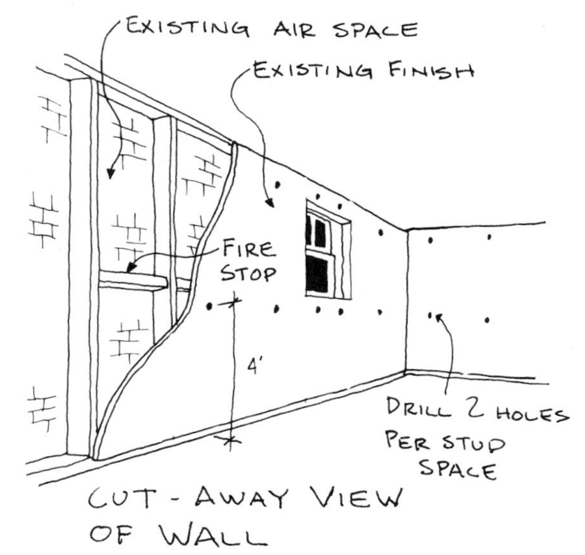

floor with the other holes located upwards at 4' intervals. To insure that there are no obstructions in the wall (such as fire stops), lower a weight through the top access hole to examine for obstructions. Walls that contain wiring not armored or in conduit should not be insulated.

3 Begin blowing insulation into the wall through the hole closest to the floor. Finish the job by blowing insulation through the hole near the ceiling. Fill all irregularly framed spaces around windows, doors, and other obstructions on the wall. If necessary, pack the area occupied by the nozzle with insulation.

NOTE:

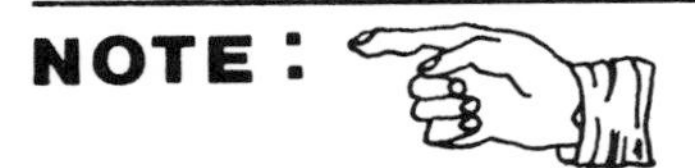

Urea-formaldehyde must be installed into relatively airtight spaces.

4 Urea-formaldehyde can also be installed over open furring strips or studs with a special applicator – only minor hand troweling is required. Treat the foam as a combustible material and cover it with gypsum board or another fire retardant finish.

5 After blowing through the finish, patch the holes with gypsum plaster or plastic inserts which are made especially for plugging access holes (pieces of gypsum board and "joint compound" can also be used).

NOTE:

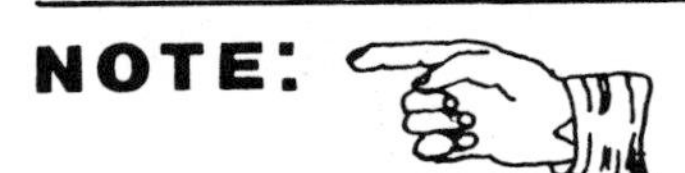

Walls with a narrow fur space, conduit, protruding mortar, or nails within the fur space may be difficult to insulate. Examine the insulation as it is installed to determine the extent of the fill. If there are obstructions within the fur space, you may have to drill additional holes to install the insulation. Rapping the wall with a cushioned tool may cause the insulation to drop through the fur space.

6 To add a vapor barrier over blown in place insulation, use an aluminized paint, oil based enamel paint, or latex vapor barrier paint. Apply one or more coats of paint over the finish. Urea-formaldehyde foam releases moisture as it "cures," therefore, if a vapor retardant paint is used, allow the foam to dry for at least one week before painting.

29

insulate masonry wall on the exterior

description

The alternative to insulating masonry and brick veneer walls on the interior is insulating these walls on the exterior with rigid insulation. Siding, either wood, aluminum, or vinyl, can be used to protect the insulation. A stucco-like material, with wire mesh used as reinforcement, is also available to provide the necessary weather and moisture protection ("Dryvit®" is one such system). As with interior masonry wall insulation, exterior masonry wall insulation will increase the wall thickness. If there's a basement or crawl space, consider this retrofit along with the retrofits described in Chapter 22 ("Insulate Basement Walls on the Exterior") or Chapter 24 ("Insulate Crawl Space Walls on the Exterior").

You especially may want to consider this retrofit if the masonry is in extremely poor condition and in need of extensive repair. These repairs may include sealing cracks and joints, tuck pointing (replacing mortar), and patching. These are time consuming repairs and chances are you won't be successful in sealing the entire wall. Therefore, it may make more sense to cover the wall with insulation and a new exterior finish.

This retrofit is also especially helpful if interior space is at a premium. That is, adding insulation to the interior of walls will make the rooms noticeably smaller but adding exterior insulation will not. If the rooms are small to begin with, you may not want to make them smaller by adding interior insulation.

Carefully consider use of exterior insulation as regards to the exterior building appearance. Several available insulation systems which utilize a cementitious coating will change the masonry surface to a smooth or stucco-like surface finish. Window and door openings will be more deeply set.

NOTE:

This retrofit involves high labor skills. It should be done only by a skilled contractor.

---

materials

rigid insulation

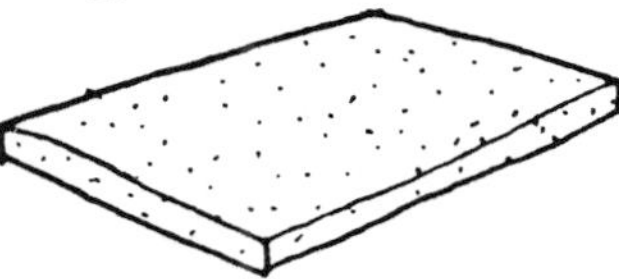

- polystyrene
- polyurethane
- polyisocyanurate foam board
- cellulose wood fiber board

dampproofing — adhesive liquids

- asphalt emulsion
- fibrated mastic
- liquid epoxy
- hot melt asphalt
- tar modified urethane
- panel adhesive

masonry cleaners

- acidic type

furring framing

- wood strips
- metal channels

other materials

- packing
- caulk
- patching cement
- flashing (corrosion resistant metal, glass fiber, ceramic, etc.)

exterior finish

- wood, metal, or vinyl siding
- stucco or exterior type plaster

dryvit® system materials

- expanded polystyrene
- reinforcing fabric (glass fiber mesh)

primus/adhesive (plaster material
mixed with portland cement)
"quarzputz®" (stucco-like material;
various colors available)

preparation

1 Clean all dirt, oil, residue, and loose material from the wall with a stiff brush and water. A mild acidic masonry cleaner can also be used. However, avoid skin contact with the cleaner by wearing rubber gloves and a long sleeve shirt.

CLEAN THE WALL WITH AN ACIDIC CLEANER

2 Since you're "tightening up" the building with a new exterior finish, it is usually not necessary to seal cracks and joints in the wall. However, the wall surface should have no surface irregularities. Use patching cement to level any surface depressions.

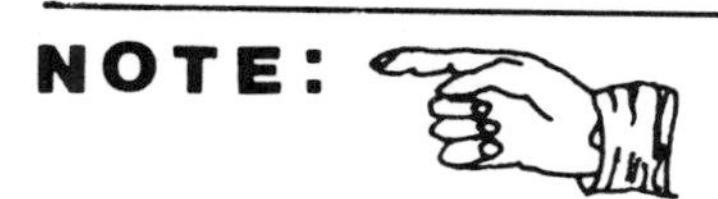

NOTE:

Refer to siding and insulation manufacturer's installation specifications, or contact manufacturer, for review of unique installation procedures which may be required as you do this work.

installation procedures

1 Begin by applying a liquid vapor barrier to the wall. If the liquid vapor barrier doesn't double as an adhesive, you'll need to select an adhesive for bonding

the insulation to the wall that is suitable for exterior application. Apply the vapor barrier/adhesive to the wall in such a fashion so as to fill any unevenness in the wall surface. The vapor barrier and/or adhesive must be compatible with the insulation!

2 Attach furring strips to the wall with masonry nails. Space the furring strips so that they will fall between the vertical joints of the insulation when it's applied. Be sure to fur out around windows, doors, and other openings in the wall. The furring strips should be as thick as the insulation to obtain a flush surface for the installation of the exterior finish. Some exterior finishes do not require furring. Check with the manufacturer's installation procedures for the particular exterior finish that you're using.

3 If an adhesive is needed in addition to the vapor barrier, apply the adhesive to the insulation board. Be sure that the adhesive is compatible with the insulation.

Apply the adhesive, approximately 2" wide by 1/4" or 3/8" thick, to the entire perimeter of each board. In addition, apply dabs of the adhesive of the same thickness to the interior area approximately 8" on center. Stagger all vertical joints (if two or more layers are needed to obtain a certain "R" value, stagger all vertical and horizontal joints). Apply pressure over the entire surface of the board to insure uniform and tight contact with the wall. Be sure that all joints butt tightly together.

NOTE: 

Flashing is required to cover all exposed top and side edges of the insulation, over the coping, and beneath window sills to keep water from getting down behind it. See "Flashing details" in this chapter.

4 Fasten siding to the furring strips. Start just above grade and work your way up. Wood or metal must be stopped so that it does not touch the ground (check local buiding codes). Most codes do not permit siding to be installed closer than 6" above grade. If any insulation extends below this 6", cover it with building felt, asphalt, or vinyl cement and, if necessary, fireproof with a material such as asbestos cement board.

THE DRYVIT® SYSTEM

1 If Dryvit® is being used, there is no need for applying a liquid vapor barrier or furring strips (as discussed previously in steps 1 and 2). Dryvit® has its own stucco-like finish that bonds directly to the insulation.

Dryvit® Primus/adhesive is mixed with Portland cement to obtain the adhesive that is used to bond the insulation to the wall. This same mixture is used to embed the reinforcing fabric (step 3).

2 Bond the insulation to the wall in the same fashion as described previously (step 3).

3 Use the same mixture for applying the reinforcing fabric (glass fiber mesh) to the insulation that you used for bonding the insulation to the wall. This mixture should be trowelled on to a thickness of 1/16", firmly embedding the fabric in it.

EMBED THE REINFORCING FABRIC IN THE PRIMUS/ADHESIVE

4 Finally, trowel the Quarzputz® finish over the insulation. This finish is available in various colors. Prior to application, however, the finish needs to be mixed with a high speed mixer.

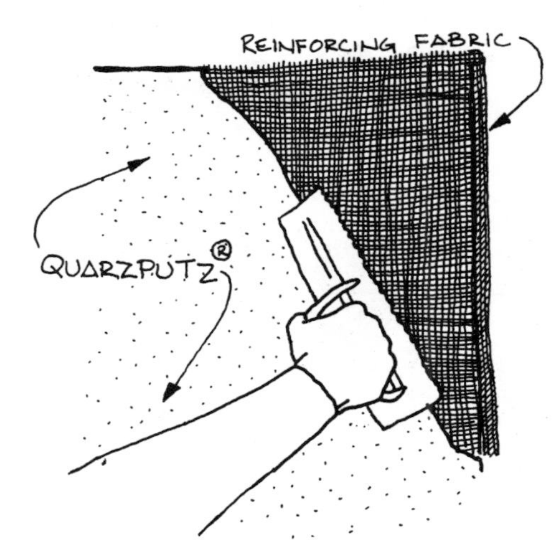

TROWEL THE QUARTZPUTZ OVER THE REINFORCING FABRIC/INSULATION

FLASHING DETAILS

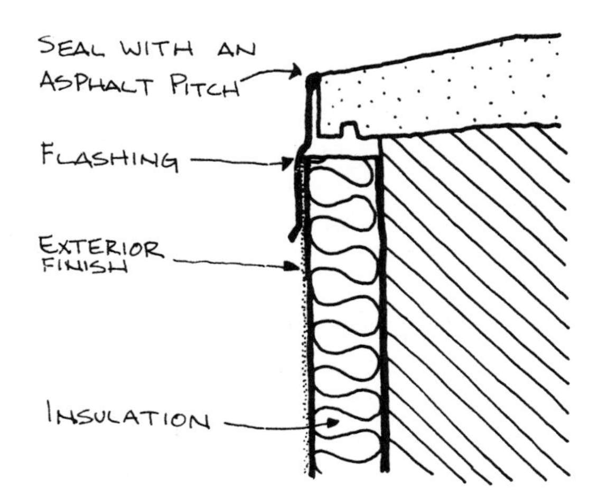

WINDOW SILL

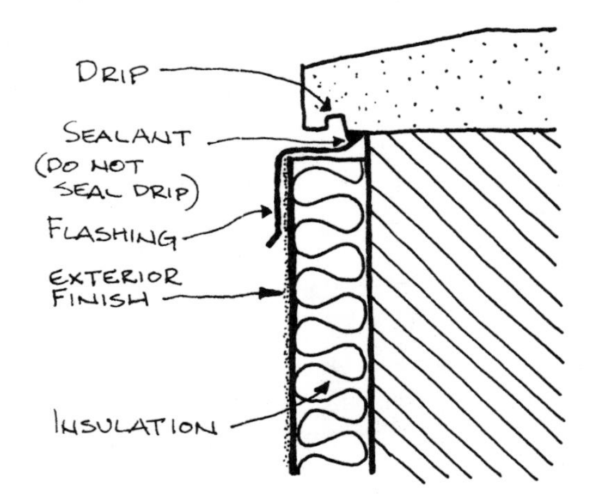

WINDOW SILL (ALTERNATE)

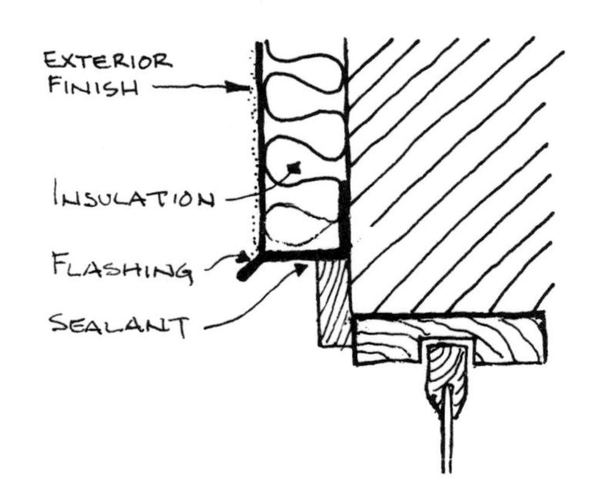

JAMB

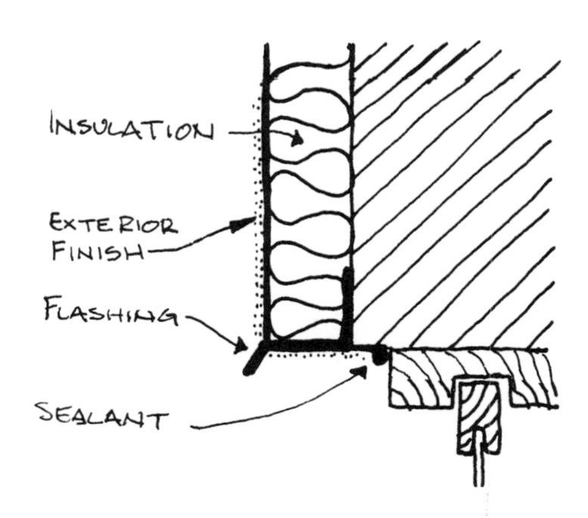

HEAD JAMB

FOUNDATION CONDITIONS

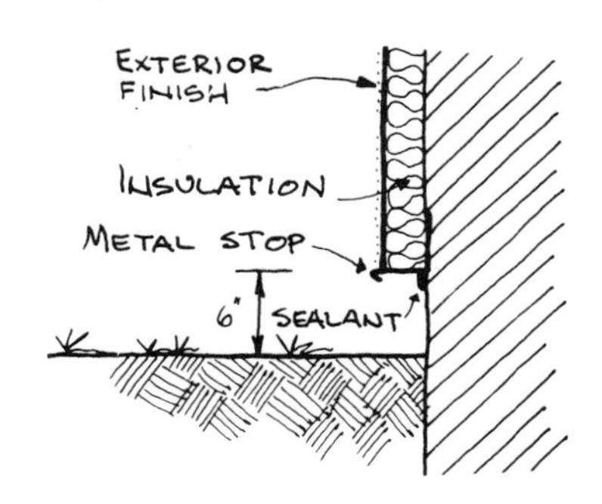

Dryvit® exterior finish can be continuous and extend down to and below grade.

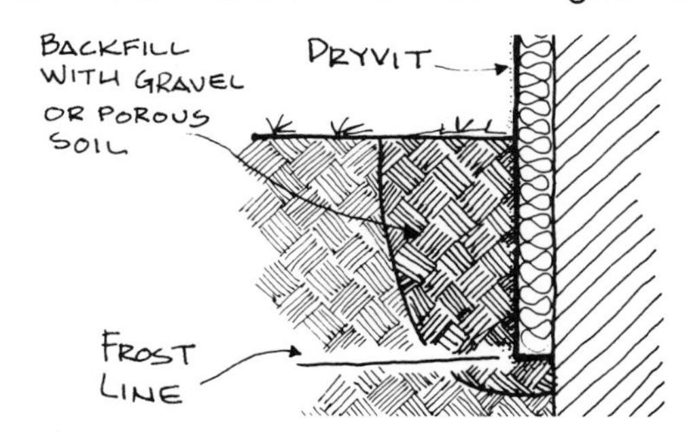

PARAPET

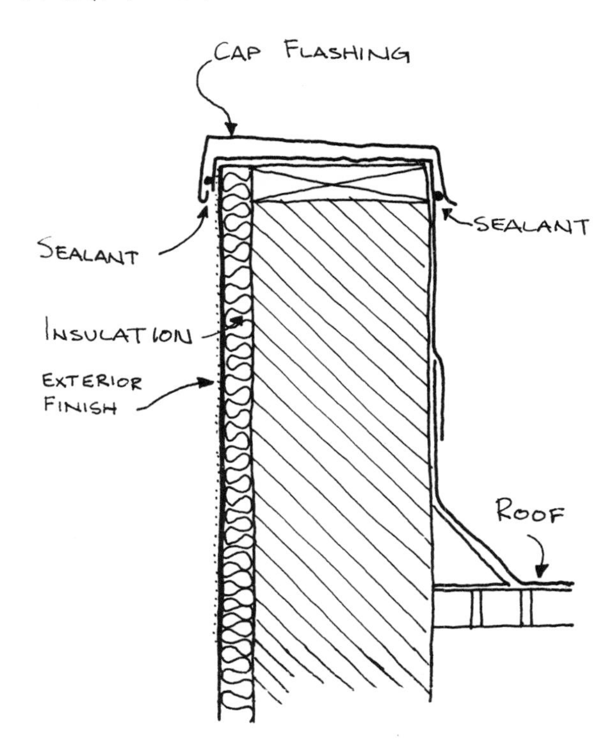

NOTE:

The cap flashing must be continuous and bedded in pitch (Primus/adhesive for Dryvit®) to assure that no moisture can penetrate behind the insulation.

NOTE:

Standard exterior finish pieces, such as stops, corner beads, and flashing strips, are available for wood, metal, or vinyl siding.

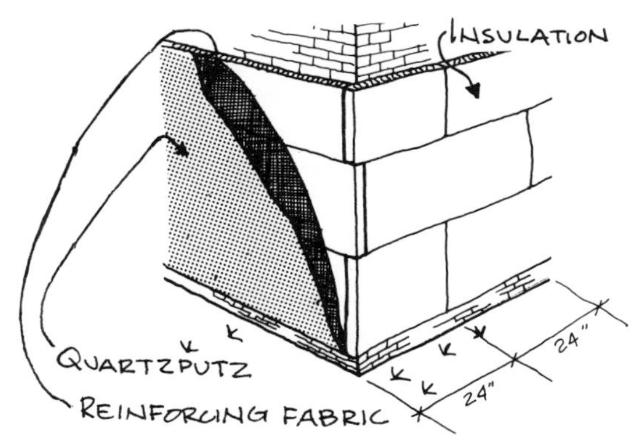

For additional information about the Dryvit® System, write:

Dryvit® System, Inc.
420 Lincoln Ave.
Warwick, Rhode Island 02888

30

insulate open frame wall

description

Next to an unfinished attic, insulating an open, or unfinished, frame wall is probably the easiest insulation retrofit because of easy access. Open frame walls to be insulated are those walls that separate a heated space from the outside or from an unheated space. The only time you would insulate an open frame wall (or closed frame wall) separating two heated spaces would be for sound deadening, which is not considered in this guide. The open frame wall may be part of an unfinished section of the home facing outdoors, a wall separating a garage or utility room from a living space, or knee walls in a finished attic.

Roll/batt, rigid, or foam insulation can be used to insulate an open frame wall. In all cases, you'll need to cover the insulation with a "15 minute or 30 minute" fire-retardant wall surface (15 minute flame spread rating, minimum). Check with your building department.

As you will see, your job is essentially finishing the wall.

materials

roll/batt insulation

- glass fiber
- rock wool

  (both available as foil faced, kraft paper faced, or unfaced)

rigid board insulation

- polystyrene
- polyurethane
- polyisocyanurate

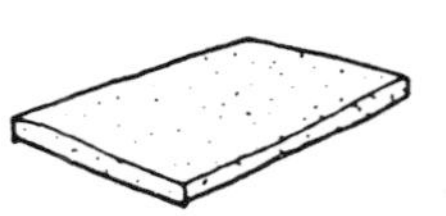

foamed in place insulation

- urea-formaldehyde

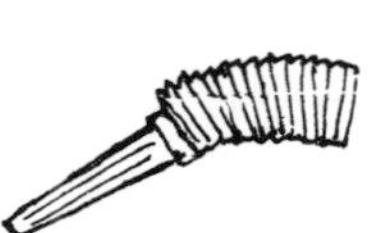

vapor barrier

- polyethylene film
- aluminized or oil based enamel
- latex vapor barrier paint
- foil faced gypsum board

furring/framing

- furring strips
- gypsum board or other fire retardant material -- check local codes

other material

- caulk

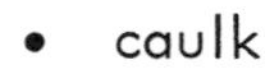

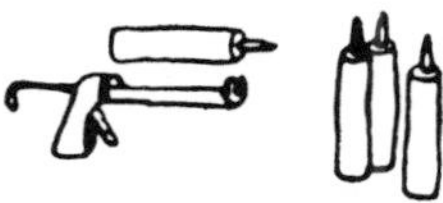

- packing

- fasteners

preparation

1 Before adding insulation to the wall, any water seepage through the wall must be stopped by sealing or making other necessary repairs to the exterior wall surface. Most wall moisture problems encountered are cases of water leaking or wind-driven into the wall around gutters or loose siding.

CRACKS CAN BE SEALED WITH CAULK

2 Existing studs may have to be furred out to accommodate the thickness of the insulation that you're going to use. Nail furring strips over the studs to achieve the desired wall depth. A wall depth less than 90% of fibrous insulation thickness is not adequate to develop the thermal resistance of the insulation. If rigid insulation is going to be used, the stud depth must be at least as thick as the insulation. These pictures are looking down at the wall cut horizontally.

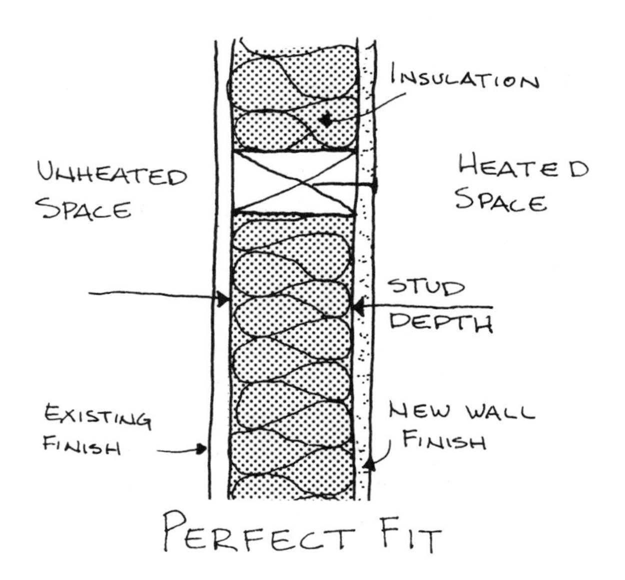

PERFECT FIT

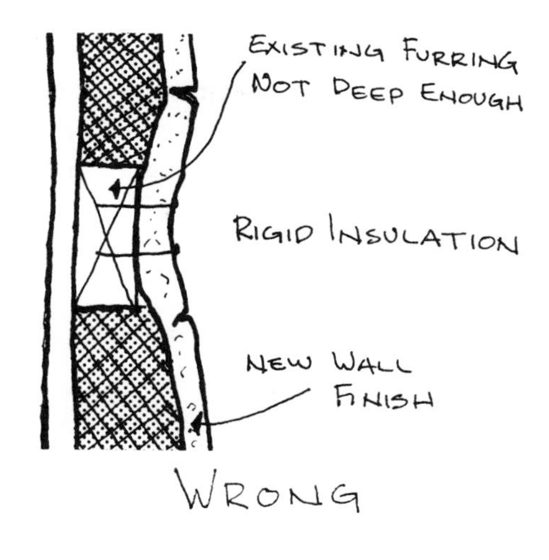

WRONG

CORRECT

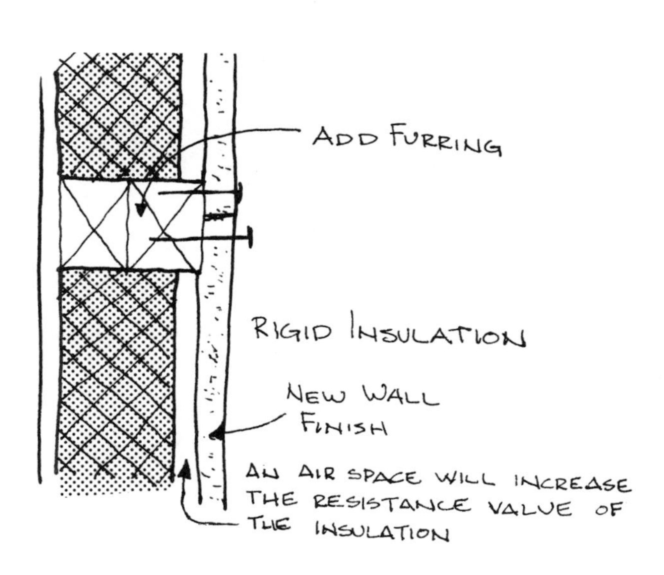

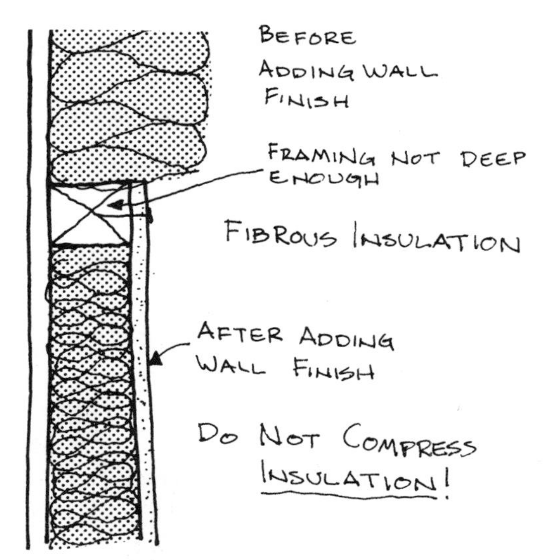

WRONG

CORRECT

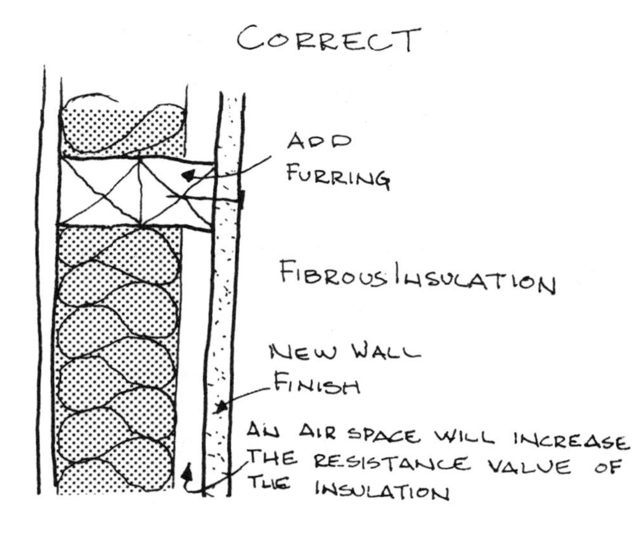

installation procedures

ROLL/BATT INSULATION

1 Install the insulation between the studs. The insulation is to extend the full height of the wall. If the insulation has a vapor barrier facing, install it so that it FACES TOWARDS THE HEATED SPACE (see step 4). Vapor barrier facing or installation of vapor barrier film is recommended whenever insulation is placed into an open wall. Cut the insulation with a utility knife or shears and a straight edge. Pack narrow spaces, especially around windows and doors, with scraps of insulation.

2 Place the insulation behind wiring and pipes. Do not place insulation in contact with wiring where the copper or aluminum is exposed.

NOTE: 

Install insulation no closer than 3" from recessed electrical fixtures such as lights and fans. Check local codes.

3 Staple, or tape, the insulation to the studs. If staples are used, space them approximately 6" apart and cover them with polyethylene tape if vapor protection is required. Be sure that there are no voids between pieces of insulation that butt together. Be sure to butt the insulation tightly to the sole plate and top plate. Avoid creases, folds, and "fishmouths" in the flanges of the insulation. Pull and staple alternate lengths of the flange to achieve a tight, neat attachment.

(SOLE PLATE IS ON THE FLOOR)

NOTE: 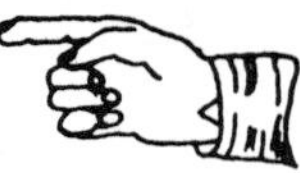

Foil faced insulation will increase the insulation value if a 3/4" air space is provided between the foil and the interior finish.

4 If you are in an unheated space installing the insulation, the vapor barrier will be facing away from you. In other words, you won't be able to see the vapor barrier after you install the insulation. Additionally, you won't be able to use staples or tape to hold the insulation in place unless the insulation is enclosed in paper. Use wire (laced from stud to stud), wire mesh, or "tiger teeth" to hold the insulation in place. See Chapter 25, "Insulate Floor Above Crawl Space", for installation techniques with these materials.

INSULATION CAN BE HELD IN PLACE WITH WIRE MESH

5 Wedge unfaced insulation between the studs. Some batts are semi-rigid and made to fit tightly between frame members by friction. If necessary, use one of the installation techniques discussed in Chapter 25 to hold the insulation in place. Install a vapor barrier (usually a polyethylene sheet) over the unfaced insulation. Staple the vapor barrier to the studs, then go back and tape over the staples with polyethylene tape to keep moisture out.

Again, if you're installing the insulation from the unheated space, install the vapor barrier prior to wedging the unfaced insulation between the studs.

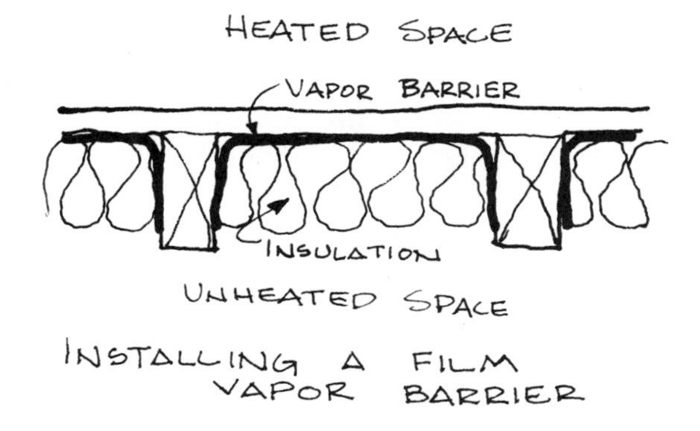

INSTALLING A FILM VAPOR BARRIER

6 Nail gypsum board to the studs. You can glue gypsum board to the studs if a vapor barrier sheet has not been installed covering the studs. In this case, flanges of insulation are stapled to the sides of the studs. Studs are not covered by the vapor barrier so this is not a good practice in very cold climates. The gypsum board is to extend from the flooring to the ceiling finish. Any plastic, foil backed, or kraft impregnated paper vapor barrier must be covered with a "15 minute" fire resistant interior finish (minimum). Foil backed gypsum board can be used to provide a vapor barrier in this case.

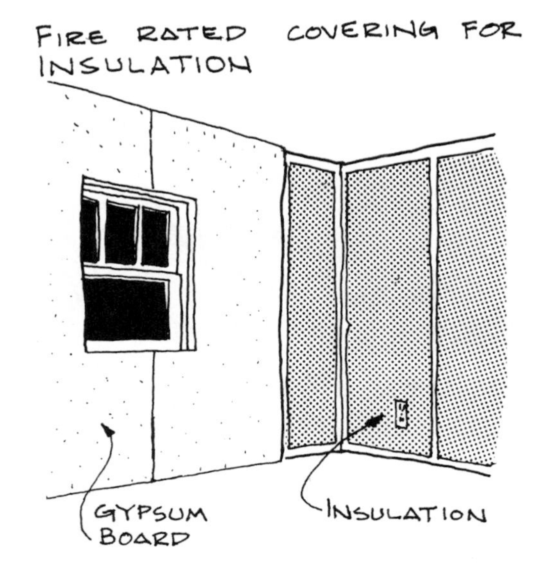

RIGID INSULATION

Attach rigid insulation directly to the wall with a panel board adhesive.* Apply the adhesive along the perimeter of the board with "spot dabs" on the interior. Firmly press the insulation in place so that it fully contacts the wall surface. The boards are to butt tightly together. Install gypsum board as described previously in step 6.

*Several "construction adhesives" can be used to glue foam insulation. Check with your supplier for an adhesive which will not damage the insulation. Test the adhesive with the insulation before using it on the job.

UREA – FORMALDEHYDE

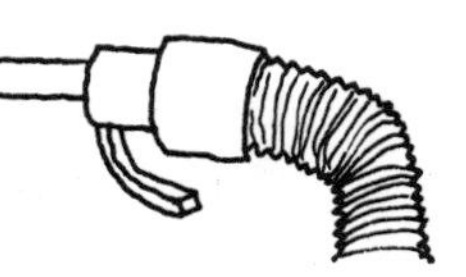

1 Urea-formaldehyde foam can be installed between open studs with a special applicator -- only minor hand troweling is required. Treat the foam as a combustible material and cover it with gypsum board. You will probably need the services of someone skilled in this installation technique for this work.

2 A vapor barrier should also be used over urea-formaldehyde. Since urea-formaldehyde releases moisture as it "cures", the vapor barrier should be between the foam and the finish. This can be achieved with foil backed gypsum board. Finish must be applied immediately after foaming to enclose the foam in an airtight cavity to properly cure. Allow the foam to dry for at least one week before painting the finish.

COVERING AND FINISHING THE WALL

31
insulate closed frame wall

description

Chances are that if you're retrofitting an older home, the exterior walls have either an insufficient amount of insulation or no insulation at all. Insulation can be blown into the wall by drilling holes either through the interior or exterior finish or sheathing. In any event, these walls are difficult to insulate, requiring special skills, techniques, and equipment.

If the walls are clad with aluminum, vinyl, or steel siding, the siding pieces have to be pried away from the wall and holes drilled through the sheathing. Asbestos and slate shingles are very fragile and must be removed from the wall where the holes are to be drilled. Holes can be drilled directly through wood siding (and patched afterwards), but to avoid problems with appearance, siding pieces can be removed with the holes drilled through the sheathing.

Brick and stone veneer walls may be accessed by drilling through the interior finish. Individual bricks can be chiseled out or corners of mortar joints can be drilled, but again there may be problems with appearance of the completed job. Drilling holes through walls in homes built with balloon framing may not be necessary (the insulation can often be added to the wall cavity from the attic).

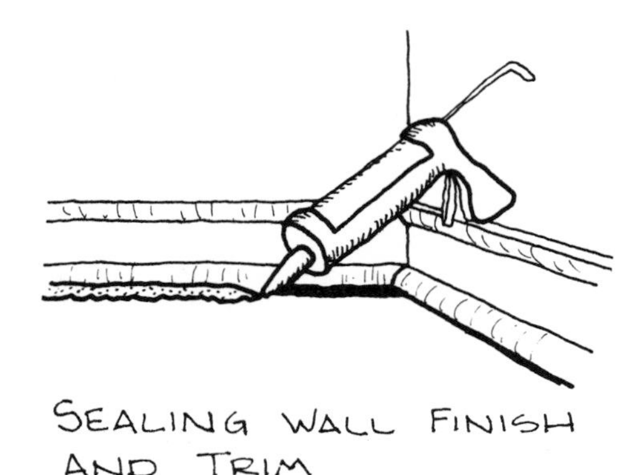

materials

blown in place insulation

- cellulose
- glass fiber
- rock wool
- vermiculite
- perlite

foamed in place insulation

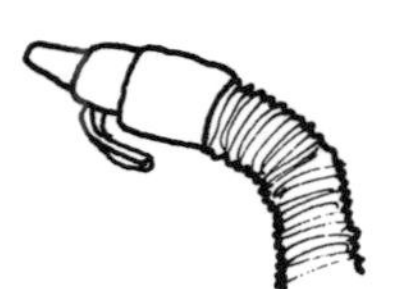

- urea-formaldehyde

other materials

- vapor barrier paint
- caulk

- hole plugs

preparation

1 Cracks where the interior wall surface meets the floor or ceiling are to be sealed before insulation is added to the wall. Use a clear or matching color caulk for this purpose. For a neater appearance, remove the baseboard prior to sealing the crack.

2 Holes and other cracks in the interior wall finish should also be sealed before installing the insulation (see "On patching holes" in this chapter for details on repairing holes in plaster or gypsum board walls). These patches should be painted with a vapor retardant paint whether or not the remaining wall surface is painted.

NOTE:

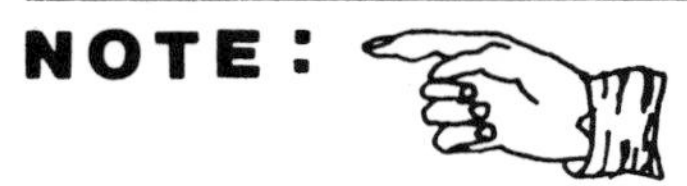

Urea-formaldehyde will deteriorate if exposed to air. Therefore, do not install UF foam if the walls are in such poor condition that an airtight cavity cannot be achieved. In this regard, walls with plank sheathing which have cracks between the planks and no building paper are poor candidates for an effective foam job.

CHECK FOR BALLOON FRAMING

Balloon framing is sometimes found in older homes. Unlike newer construction, the wall studs in balloon framed homes are continuous from the foundation wall to the attic. There may not be a plate along the top of the wall studs. If this is the case, the insulation can be blown into the wall cavity from the attic so drilling holes through the exterior finish is not necessary. Balloon framing may extend down to the foundation wall without intermediate plates. Unless the stud cavity at the first floor line is sealed or the framed portion of the basement wall is finished, this will have to be sealed, or you'll have a basement full of insulation. If UF foam is used, both the bottom and top of the wall cavity must be sealed.

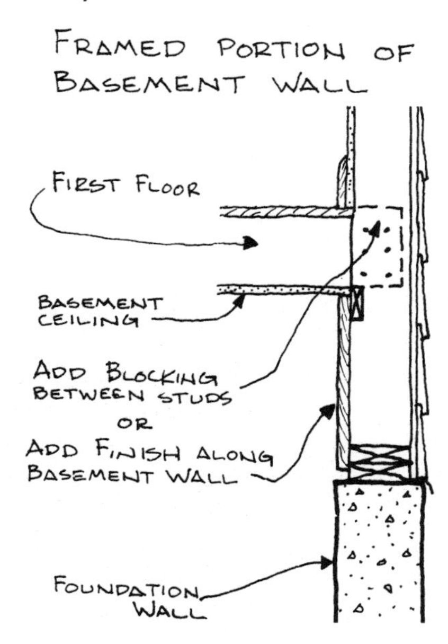

SEAL WITH BLOCKING IF UREA-FORMALDEHYDE IS USED
NO TOP PLATE
BRACING
STUDS
JOISTS
1x4 RIBBON STRIP
BRACING 1x4'S LET INTO FACES OF STUDS
FOUNDATION WALL
SEAL WITH BLOCKING PRIOR TO ADDING INSULATION

TYPICAL BALLOON FRAMING

ON DRILLING HOLES

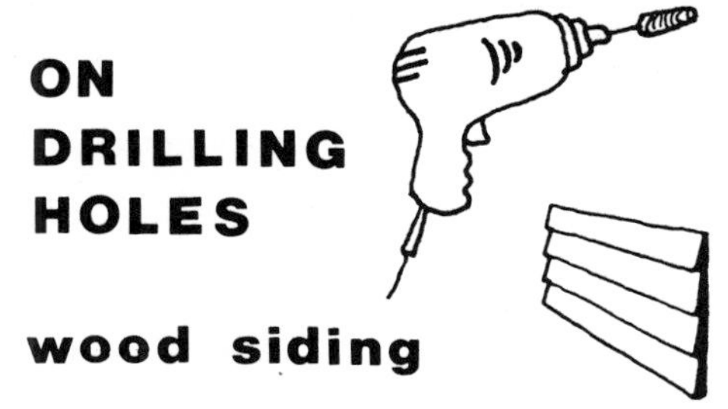

wood siding

An entire section of wood siding can be removed exposing the building paper and/or sheathing.

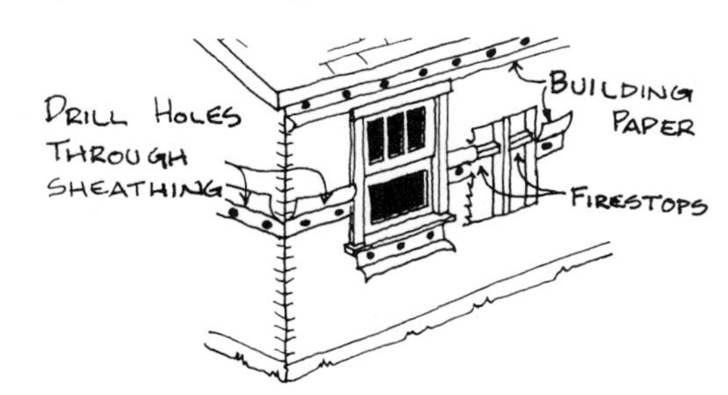

To remove the siding, insert a piece of sheet metal under the piece of siding to be removed. Wedge a pry bar under the siding and over the sheet metal (the sheet metal will protect the siding below).

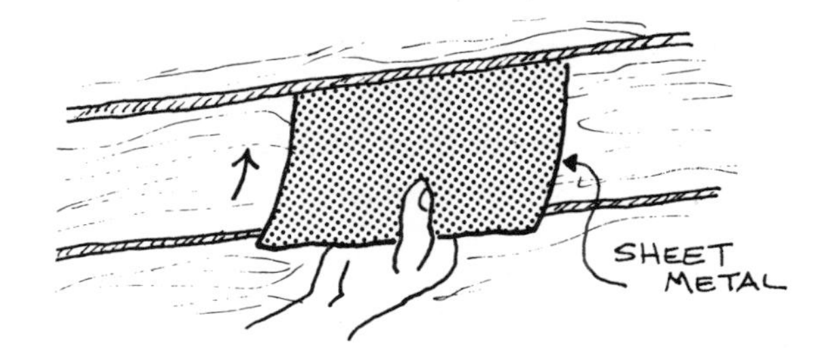

NOTE:

Be sure to read "On patching holes" at the end of this chapter.

Gently pry up the siding about 1/2". Remove the pry bar and push the siding back in place (tap down with a mallet if necessary). Remove the nails that are sticking out. Repeat the prying every 16" to remove all the nails. If the siding does not come off after this, look for a missed nail or paint seal still holding the siding.

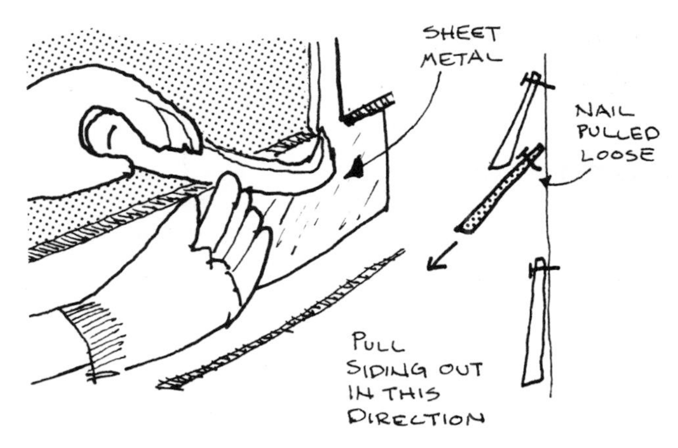

If there is building paper, slit it along the sides and bottom exposing a location for the hole in the sheathing. Fold the building paper up.

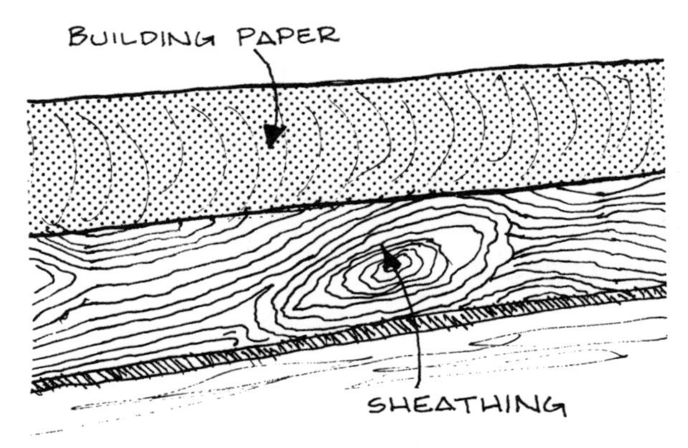

aluminum, steel, vinyl siding

Holes should not be drilled through this type of siding, as the siding could crack and it cannot be easily patched. Entire pieces of siding should be removed, as per wood siding, with the holes drilled through the sheathing. If building paper is present, fold it up.

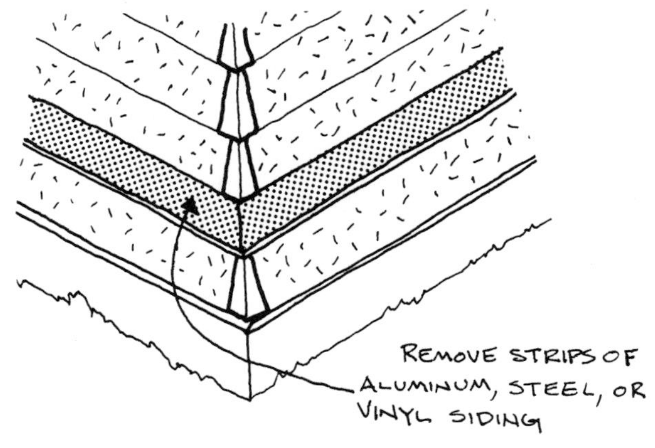

stucco

For stucco, the holes can be drilled directly through the finish. When the job is finished, the holes are patched with cement to match the adjacent surface.

wood shingles

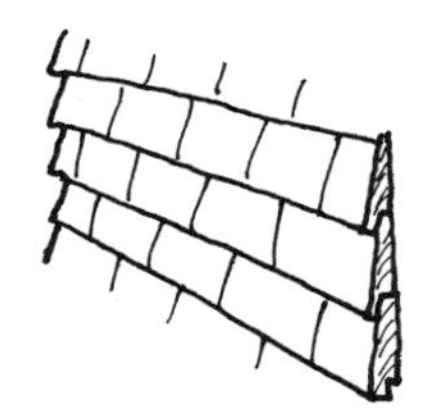

Remove an entire shingle so that the hole can be drilled through the sheathing. Two holes can sometimes be drilled into two adjacent stud cavities if shingles are removed at stud "center lines". Lift the bottom of the shingle with a nail puller to a point where the nail can be removed. Remove the shingle and wedge between shingles on the wall near the spot where it belongs so that it can be replaced after insulating the wall.

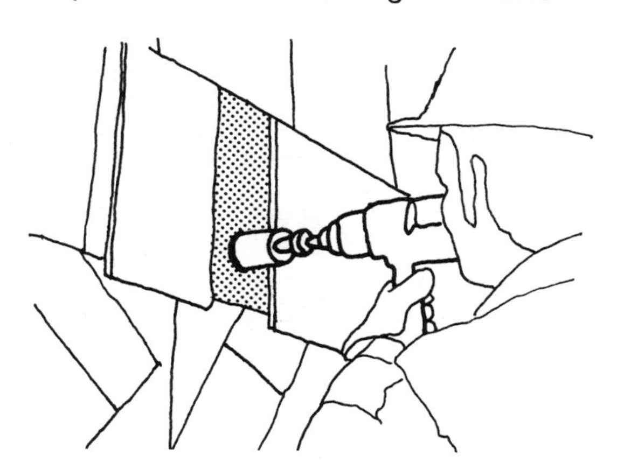

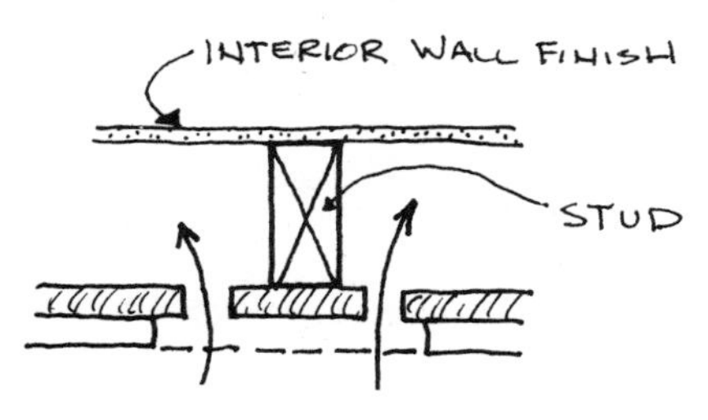

SHINGLE REMOVED AT STUD "CENTER LIN TO ACCESS TWO STUD CAVITIES

asbestos or slate shingles

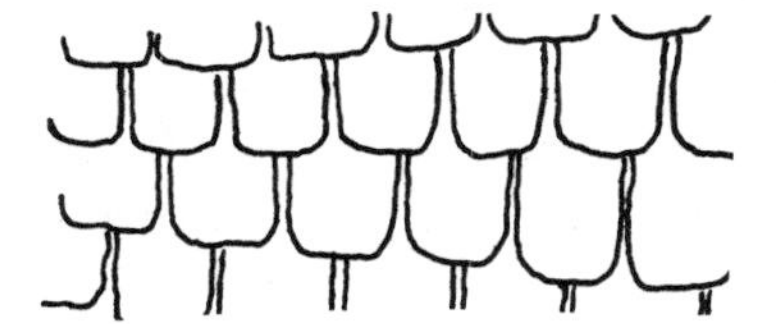

These shingles are quite fragile and may break easily if pried loose as recommended for other sidings. Therefore, prying the shingles out is not recommended. Try clipping the nail heads with flat-head snippers or cutting the nails on the underside of the shingle with a hacksaw blade. Once the shingle is out, remove the remainder of the nail.

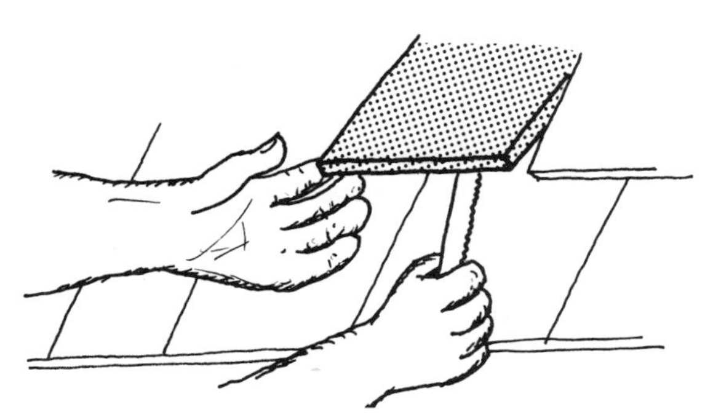

CUT THE NAILS WITH A HACKSAW BLADE

brick or stone (veneer)

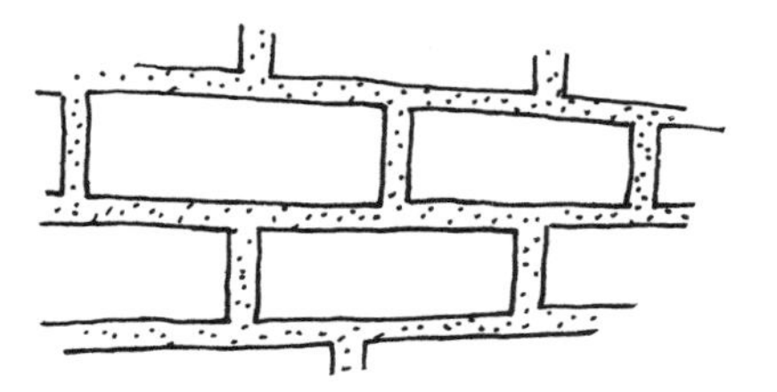

Entire pieces of brick or stone can be removed (two stud spaces can be accessed if holes are drilled on both sides of an area behind a brick or stone which covers a stud as for wood, asbestos, and slate shingles). Additionally, mortar joints at corners of bricks or stones can be drilled, using a masonry drill bit. Slower drill speeds are often necessary for safety in using power tools.

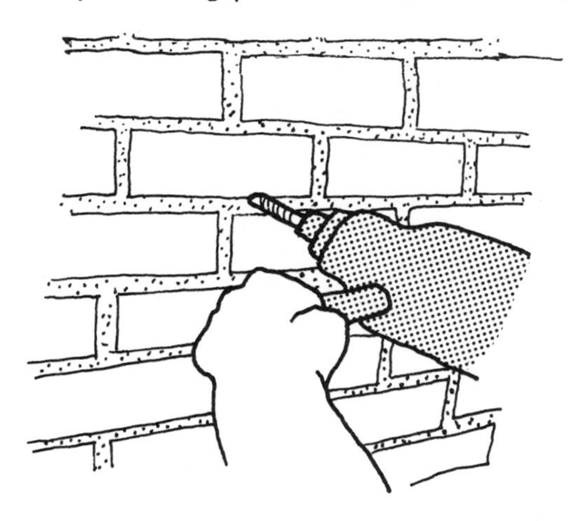

NOTE :

Accessing a veneer, metal, asbestos, and slate shingle wall on the exterior can be quite difficult. An alternative is accessing the wall from the interior by drilling through the interior finish. To conceal holes, you may be able to detach ceiling molding and window skirting and replace over holes after installation. Procedures for installing the insulation remain the same.

installation procedures

1 Now the holes can be drilled. The holes should be slightly larger than the end of the blower hose, generally between 1-1/2" to 3" in diameter.

The stud spacing can be determined from the outside by observing horizontal nail spacing intervals where sheathing or siding is nailed into studs. Use a stiff wire to "feel" for narrow stud cavities on either side of windows and doors or at corners of walls. Drill and fill these narrow and irregular cavities seperately.

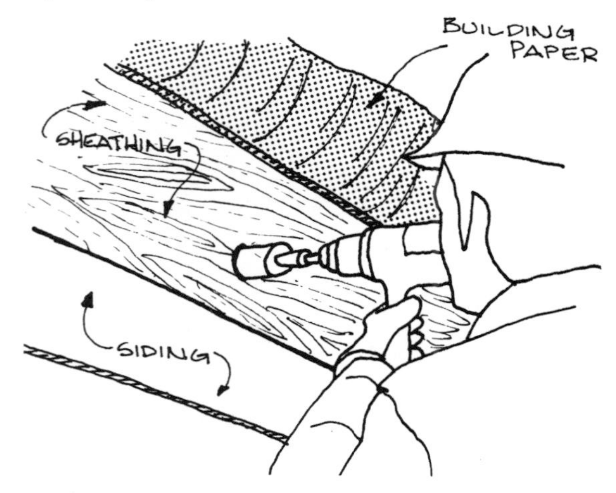

2 Remember, the idea is to access every cavity. This can usually be done with two holes per stud space, one located near the ceiling with the other approximately 4' off the floor (with UF foam, drill a hole 2' above the floor and then upwards at 4' intervals). It is recommended that insulation not be blown more than 4' down or 1' up.

If a building has fire stops, holes are required above and below the stops. Holes must also be drilled above and below windows. To insure that there are no obstructions (such as fire stops) in the wall, lower a weight through the access hole until it touches bottom.

3 Before preparing a wall for insulation, remove several outlet covers and/or fixtures in the wall. If the electrical wiring is not in conduit or metal armored, it is recommended that insulation not be installed. It is also recommended that sidewall insulation not be done if the home has problems with "short circuits". This is very difficult to determine, since wiring is concealed and you cannot usually "see" a short circuit. If there has been a fire of unknown cause, unreliable operation of appliances, flickering lights, frequent blown fuses, or if there are oversized fuses or metal plugs in fuse sockets, then it is likely there may be a short circuit or other wiring problems.

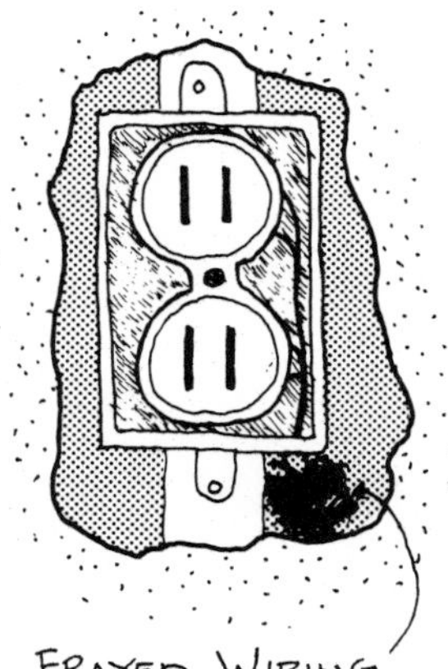

4 Look for recessed lights, fans, and other heat generating devices mounted into the wall (check both interior and exterior of the home). Blown and foam insulation must not fill in around these fixtures.

You may frequently find kitchen exhaust fans and flush wall lights. Since these will usually be mounted high on the wall, you may be able to insulate the lower stud cavity below them.

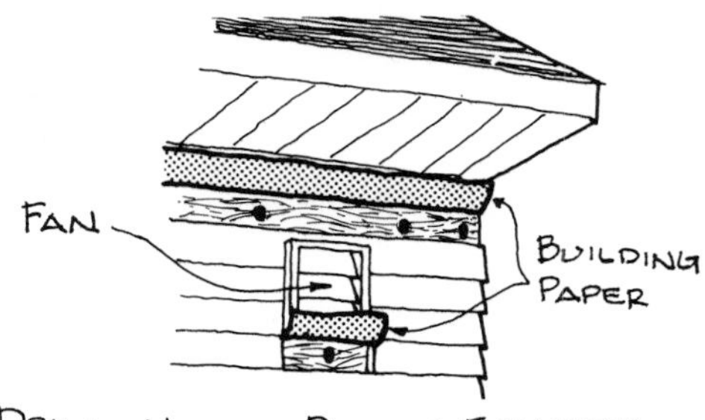

DRILL HOLES BELOW EXHAUST FANS AND FLUSH WALL LIGHTS

5 Apply a vapor barrier wall finish over the interior wall. Use an aluminized paint, oil based enamel paint, or a latex based vapor barrier paint. Apply one or more coats of paint over the entire interior wall finish. Vinyl "wall paper" also acts as a vapor barrier. If there is an existing vinyl wall covering or if you know that a vapor barrier type paint was already applied to the wall then additional vapor protection is not needed over that existing vapor barrier. Urea-formaldehyde foam releases moisture as it "cures", therefore, if a vapor retardant paint is to be applied, allow the foam to dry for at least one week before painting. Do not omit areas in closets and behind cabinets. Kitchens and bathrooms are critical in this regard.

AN IMPORTANT NOTE :

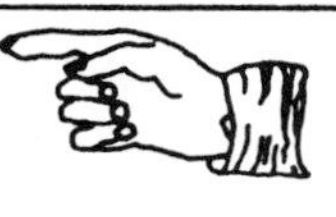

If urea-formaldehyde foam is used, there is a possibility of objectionable odors and/or burning of eyes for up to two weeks after installation of foam. The installer should be required to submit a written warranty that no odor or noxious fumes will persist after two weeks. For additional information regarding UF foam, see the section entitled "Urea-formaldehyde" in Chapter 2, "A Word about Insulation".

ON PATCHING HOLES

exterior

Plug holes in sheathing with airtight plugs approximately 72 hours after installing urea-formaldehyde insulation. Screen or fiberboard plugs are suitable for plugging holes in sheathing after insulating with fibrous materials. Reinstall building paper over the sheathing by taping or stapling a "patch" under the existing paper and folding existing paper down over patch. Carefully reinstall the exterior finish so as to present the same appearance as it originally had.

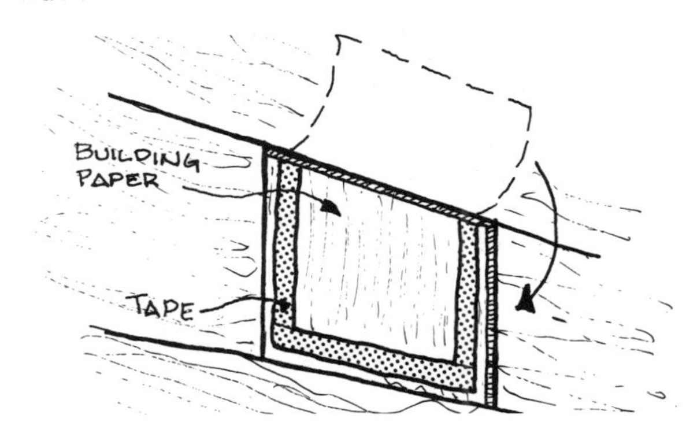

NOTE:

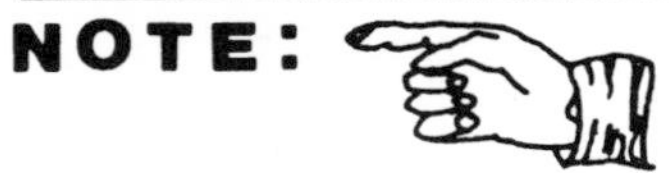

For additional patching details, see Chapter 40, "Repair Structural Hole".

interior

The following procedure can be used to patch holes that were drilled through plaster as well as for existing holes in plaster.

First, clean all loose plaster and dirt around the hole. Tie a piece of string or wire (12" long) to the center of a piece of plaster lath or hardware cloth that is slightly larger than the hole to act as backing.

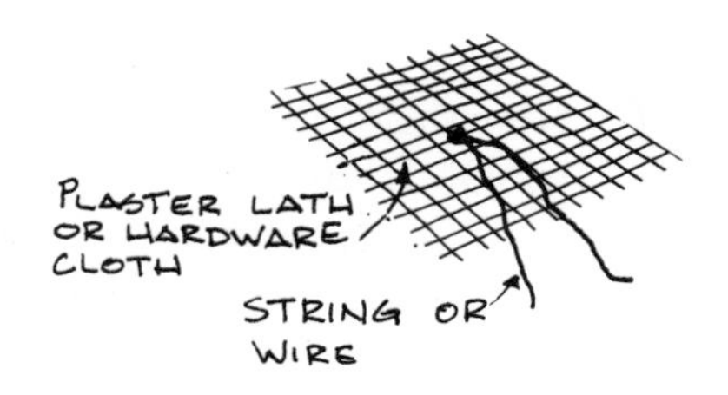

Hold onto the string or wire and place the backing through the hole. Pull so that the backing is flush to the back of the hole. Place a small piece of wood across the hole and tie the string or wire to it so that it holds the backing firmly in place.

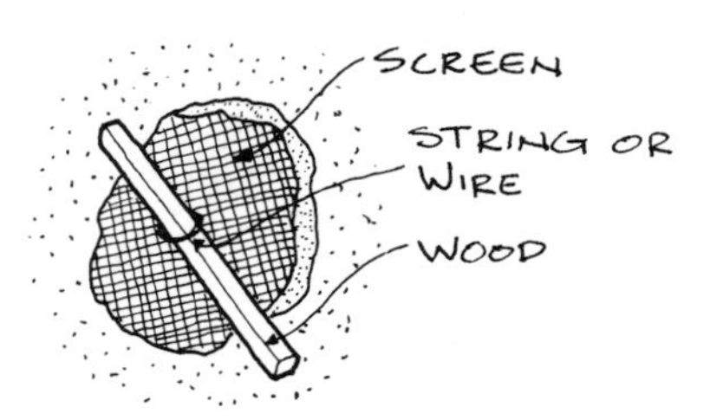

Prepare the patching material according to the manufacturer's instructions and fill the hole to within 1/8" of the surface. Wet the edges of the hole prior to patching (a spray mist bottle is good for this) to keep the wall from drawing water out of the patching material.

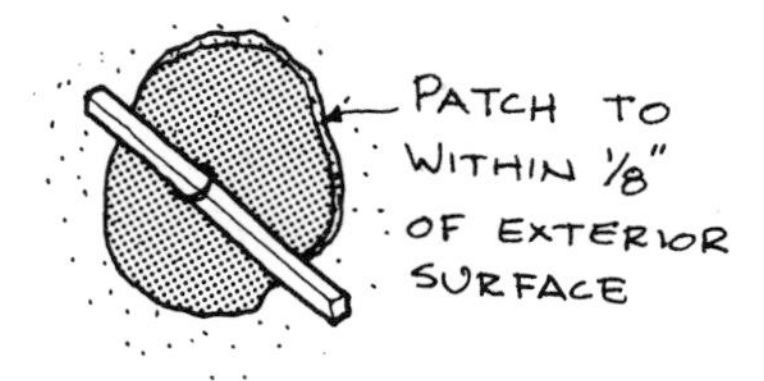

After the hole has dried, remove the wood by cutting the string or wire. Apply a second coat of patching material to the hole so it is flush to the wall finish. After it has dried, sand and paint with a vapor retardant paint.

NOTE:

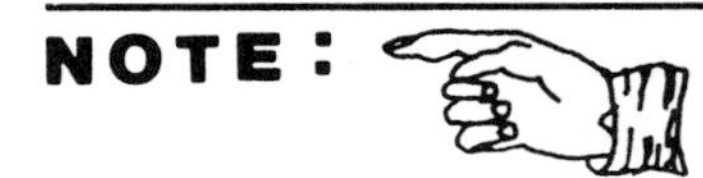

Wood and other wall finishes can be plugged with plastic inserts made especially for this purpose.

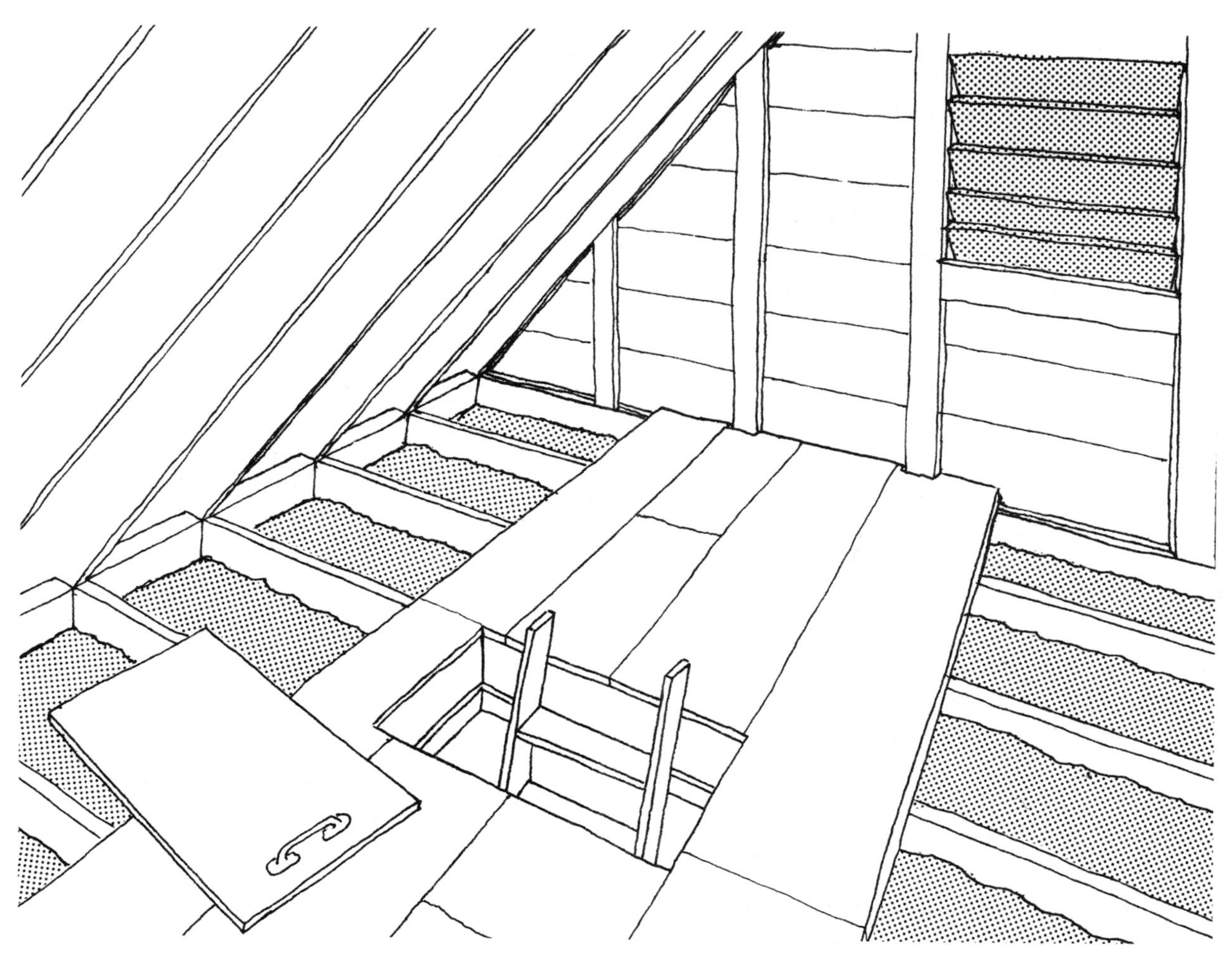

32

insulate unfinished attic

description

The next time there's a good snowfall, take a look at some of the roofs around your home. You may find some roofs where the snow has melted and others where it hasn't. Your first impression might be that the homes where the snow hasn't melted are colder than the homes where it has melted. In fact heat escaping through uninsulated attics is causing the snow to melt while insulated attics keep the heat in the home (and the snow on the roof). The idea of attic insulation is to keep the attic cool by keeping heat in the home.

A home loses a significant amount of heat through an uninsulated attic. This can be reduced with the addition of attic insulation. Attic insulation also helps keep the home cooler during the summer by keeping heat out. Before adding attic insulation, be sure to read Chapter 39, "Seal Air Bypasses".

In attics without flooring, roll/batt or loose fill insulation can be placed between ceiling joists. In attics with flooring, sections of the flooring can be removed and loose fill insulation blown between the joists. There may be situations where roll/batt insulation is placed between the rafters (see "Insulate Finished Attic Outer Sections", Chapter 34, for details).

An important aspect of energy conservation is attic ventilation. With proper attic ventilation, condensation and summer heat build-up in the attic can be prevented. Vents should have an open area of not less than 1/600 of the attic floor area and not less than 1/300 of the attic floor area for flat roofs. See Chapter 36, "But What about Attic Ventilation?", for details.

In some parts of the country, it is standard practice to install a vapor barrier with attic insulation (check with your local building department). However, if you're placing insulation with a vapor barrier facing over existing insulation, slash the facing (6" at 2' intervals) before placing the insulation over existing insulation. If moisture problems are apparent that require the installation of a vapor barrier, move the existing insulation and install the barrier (or vapor barrier facing), not slashed, directly against the ceiling finish. See Chapter 3, "A Word about Vapor Barriers" and Chapter 36, "But What about Attic Ventilation?", for clues on moisture problems.

WHICH HOME HAS ATTIC INSULATION?

materials

roll/batt insulation

- glass fiber
- rock wool

  (both available as foil faced, kraft paper faced, or unfaced)

loose fill insulation

- cellulose
- glass fiber
- rock wool
- vermiculite
- perlite

vapor barrier

- polyethylene film
- foil faced insulation
- asphalt impregnated kraft paper
- aluminized or oil based enamel
- latex vapor barrier paint

other materials

- lumber

- fasteners

- vents

- adhesives

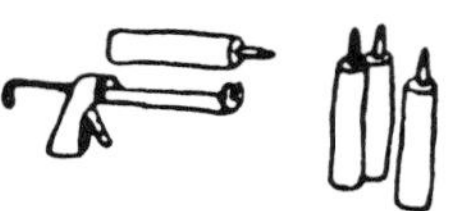

SOME WORDS OF CAUTION

1 Watch where you step.

2 Watch for protruding nails. In a case like this, wearing a hard hat is recommended.

3 Exercise care in handling electrical wiring.

preparation

1 Here are some suggestions you should consider prior to installing attic insulation. If the attic has no hatch, see Chapter 35, "But What about Accessing the Attic?".

Provide adequate lighting.

2 Install temporary flooring. The ceiling boards will not support your weight and it may be awkward to continually balance yourself on the ceiling joists.

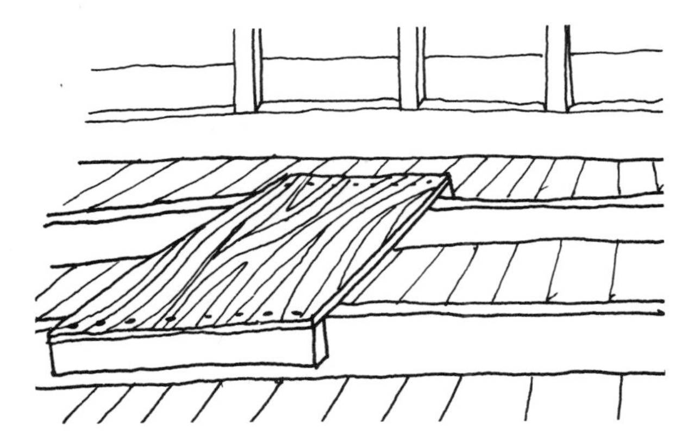

3 Seal cracks and openings that allow warm air to bypass the insulation (see "Seal Air Bypasses", Chapter 39, for details).

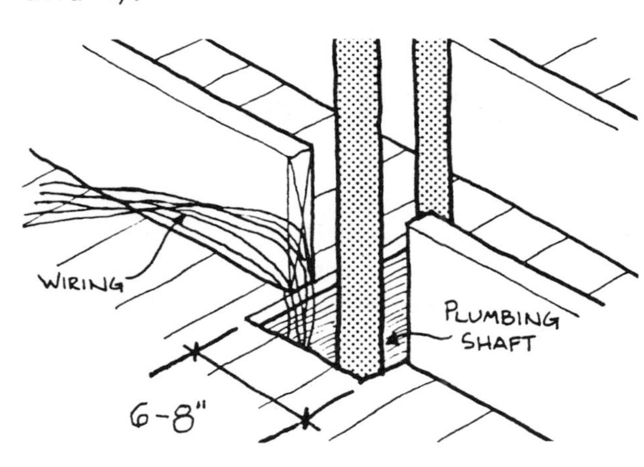

4 Repair leaks in the roof.

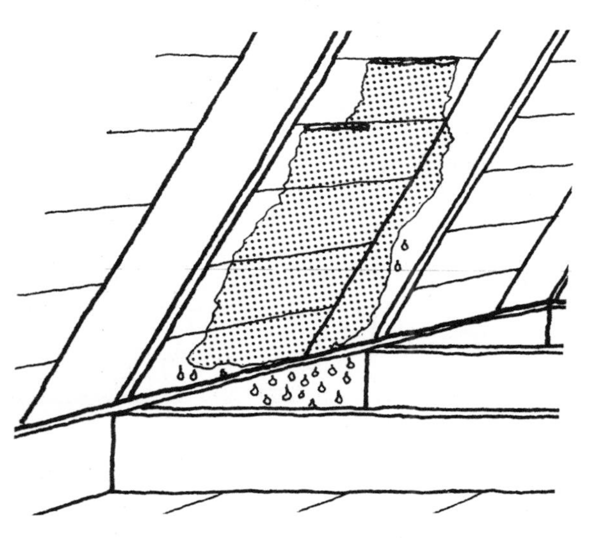

installation procedures

ROLL/BATT INSULATION WITH FACING (preferable in attics without insulation and without flooring)

ROLL/BATT INSULATION WITHOUT FACING (preferable in attics with insulation and without flooring)

LOOSE FILL INSULATION (can be used in attics with or without insulation and with or without flooring)

ROLL / BATT INSULATION WITH FACING

1 Lay the insulation between the ceiling joists with the foil or kraft paper facing towards the living space (facing downward). Fluff the insulation before laying it down (remember, it's the air spaces trapped between the fibers that give insulation its "R" value).

VAPOR BARRIER SIDE DOWN

2 Fit the ends of roll/batt insulation tightly together. Also, cut rolls/batts to fit around bridging strips.

NOTE: 

You'll notice that the rolls/batts are slightly wider than the joist spacing-- this insures a snug fit against the joists.

3 Cut the insulation with a utility knife on a hard surface. It's easier to cut with the facing placed up. Compress the insulation as you cut.

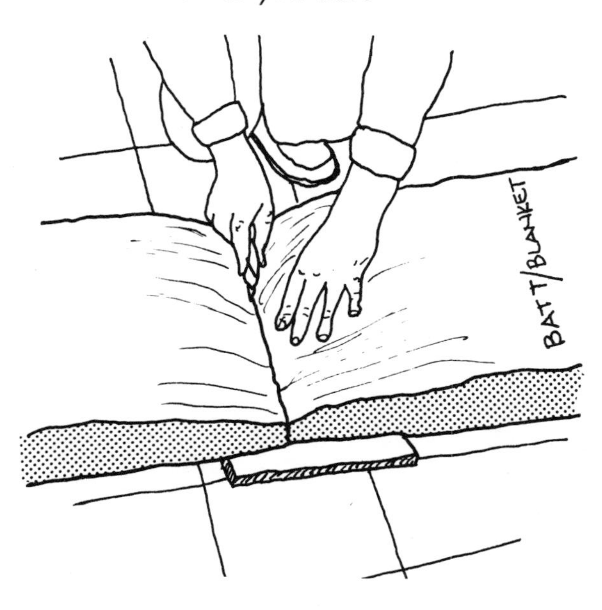

NOTE: 

Don't compress the insulation after it's installed.

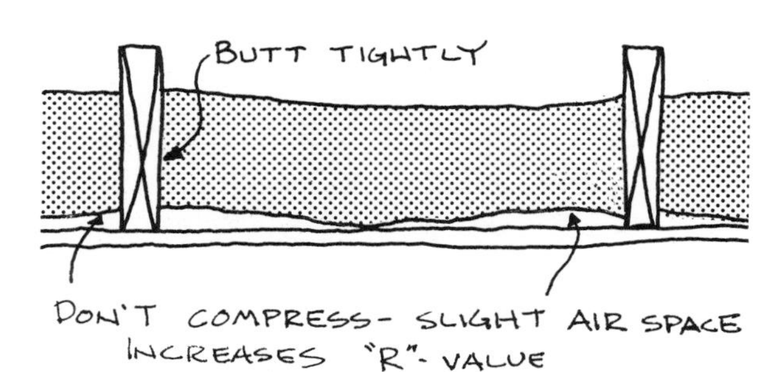

4 At the end of the joist run, lay the roll/batt so as to cover the top plate, but do not extend the insulation out in such a manner so as to restrict air flow paths from eave vents.

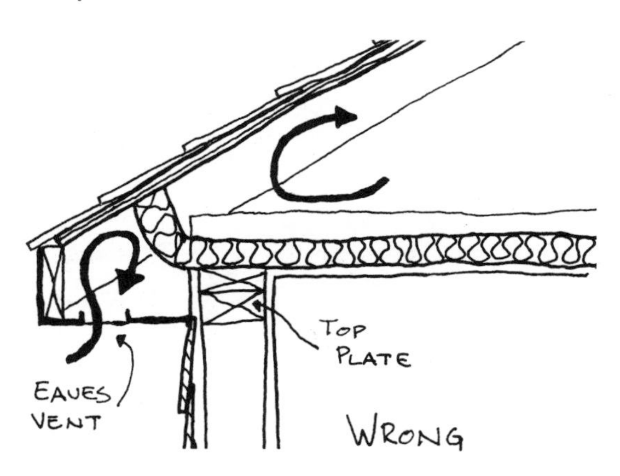

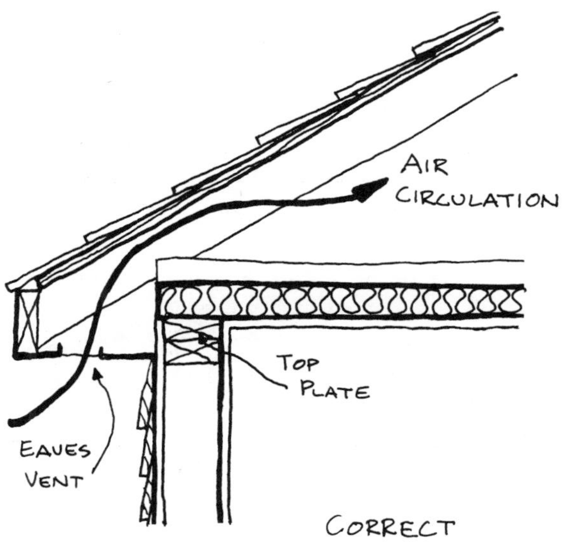

5 Keep insulation 3" away from electrical fixtures (lights, fans, or other heat generating fixtures).

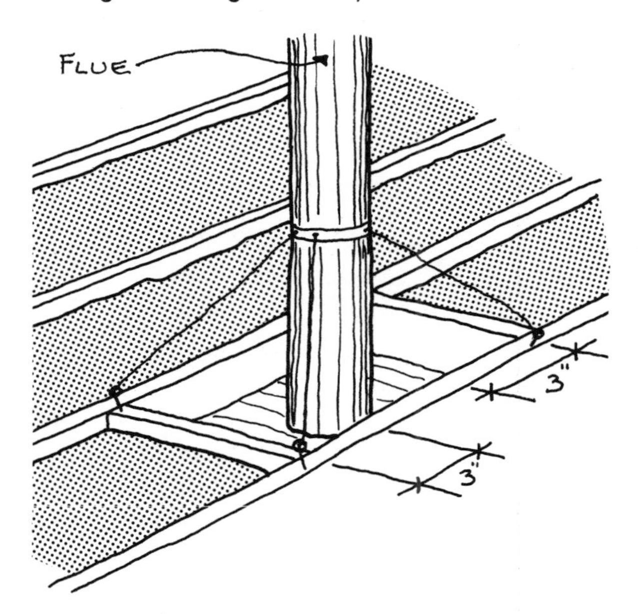

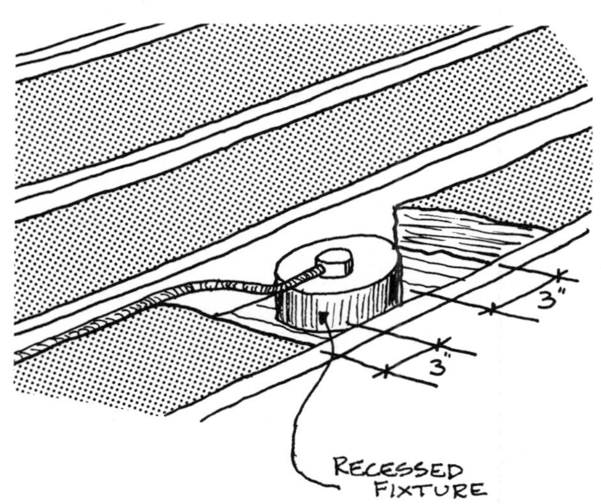

6 If you cover the fixture with a vent cone or sheet metal, the insulation can be butted to the covering.

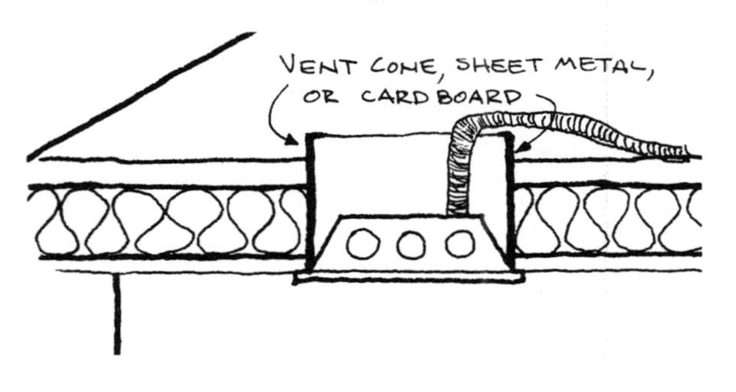

AN IMPORTANT NOTE:

If you are laying faced insulation over existing insulation, slash the facing with a knife. This will prevent moisture from condensing within the insulation. If the existing insulation has no vapor barrier and one is required, re-install this insulation over the vapor barrier or vapor barrier faced insulation.

7 Don't throw the scraps away. Use them to pack narrow spaces that you may find in the attic (or how about the bandjoist and headers in the basement or crawl space?).

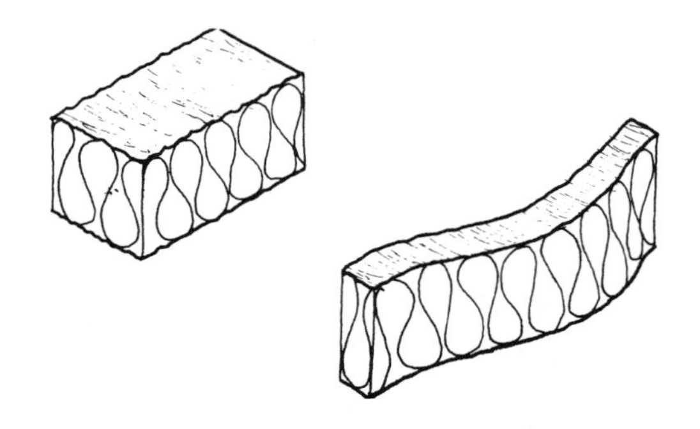

NOTE:

Narrow spaces between wood framing and a chimney are to be filled with non-combustible materials ONLY (glass fiber, rock wool, vermiculite). DO NOT USE CELLULOSE HERE.

8 Don't forget to insulate the attic hatch or door.

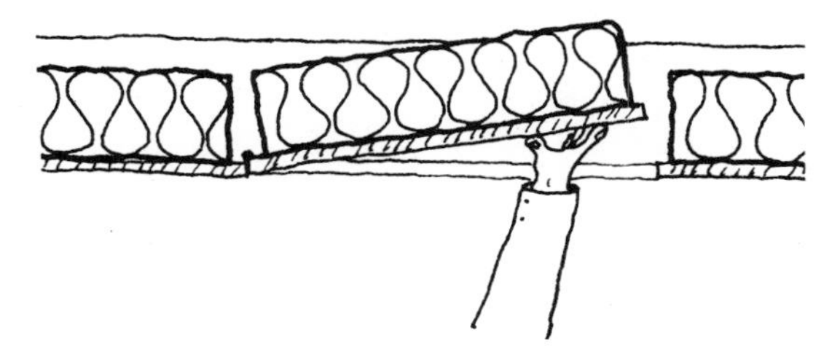

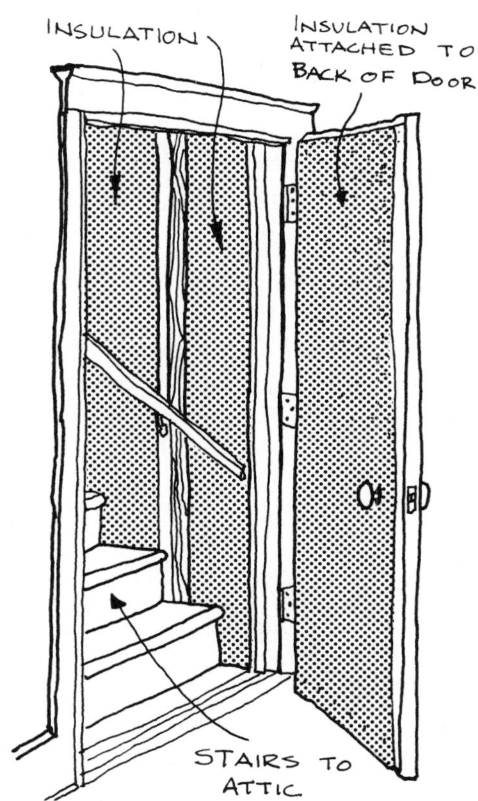

ROLL/BATT INSULATION WITHOUT FACING

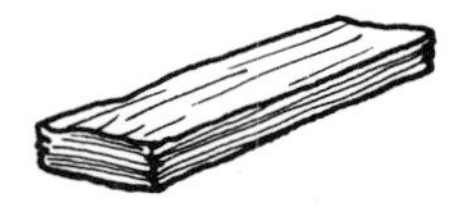

Unfaced insulation is installed in the same manner as faced insulation. However, in some areas of the country, it is standard practice to install a vapor barrier (usually a polyethylene sheet) with the unfaced insulation. The vapor barrier should not lap over the top of the joists; instead, it should be cut and laid between or stapled to the side of the joists. In this manner, moisture can pass alongside the joists and escape into the attic and be vented to the outside.

An alternative to the polyethylene sheet is a vapor barrier paint. The paint is applied to the interior ceiling surface (it even comes in colors).

LOOSE FILL

Loose fill insulation can be used in attics with or without existing insulation.

If there is existing insulation, apply the loose fill directly over it. If there is no existing insulation and if it is standard practice to install vapor barriers, install the vapor barrier as per "Roll/batt insulation without facing".

In attics with flooring, loose fill insulation is generally used because of its ease of installation (that is, it can be blown under the floor boards).

In attics without flooring, loose fill insulation can be blown in place, or poured in place.

To keep insulation fibers from drifting into the house, turn on a window fan so that air is blowing into the house and close windows and doors. Leave attic access open so that air is blown through and out of attic.

Always vacuum insulation; never wash down sinks or drains.

1 Pour or blow the insulation between the ceiling joists to a depth that gives you the required " R " value. Spread the insulation evenly for efficiency and to prevent condensation on the ceiling surface below areas where the insulation might be spread too thin. Use a rake, a piece of wood, or a template (made out of wood or card-board) to spread the insulation.

SPREAD THE INSULATION EVENLY WITH A TEMPLATE OR RAKE

2 If wood flooring exists over the ceiling joists, remove the boards at 4' intervals prior to blowing the insulation. Replace the boards when you've finished. Use your hand as a baffle to control the flow of insulation. For hard to reach spots, attach a pole to the hose.

ATTACH A POLE TO THE HOSE FOR HARD TO REACH AREAS

3 At the end of the joist runs, install baffles (wood, cardboard, or scraps of roll/batt insulation) to keep the insulation from obstructing air flow paths.

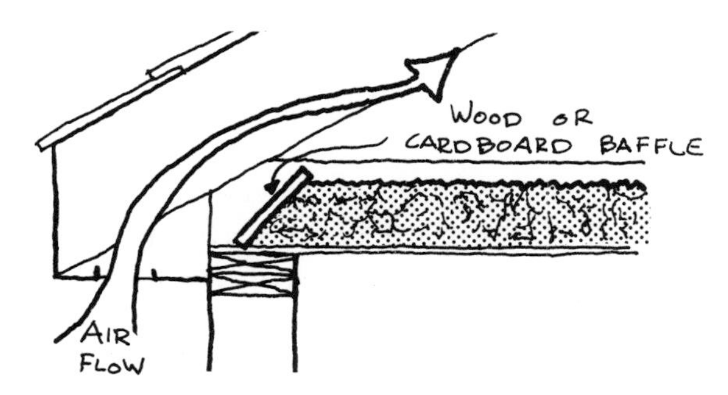

4 It is important that you keep the insulation away from heat generating fixtures (lights, fans, chimneys, etc.). Use vent cones or sheet metal for this purpose or staple roll/batt insulation to secure it away from the heat source at least 3" on all sides.

5 Be sure to insulate the attic hatch. Build a simple box frame on the hatch, fill it with loose fill and cover it with plywood or cardboard.

6 Installing loose fill insulation can generate a lot dust. Wear "surgical" masks to avoid inhaling the dust. Also, try to avoid getting dust in living areas.

AVOID THIS

SOME TIPS ON BLOWING MACHINES

Place the machine on a piece of plywood or tarp. Stack the bags of insulation near the machine.

Clean the trap beneath the hopper of insulation.

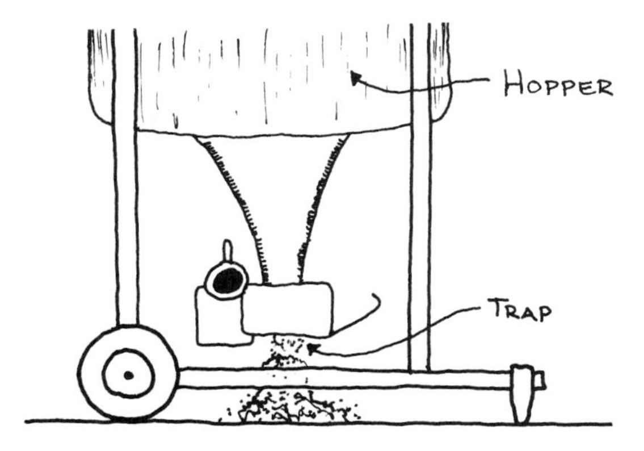

Before attaching the hose, plug in the machine and turn it on. Hold your hand over the vent to make sure it works. Then attach the hose.

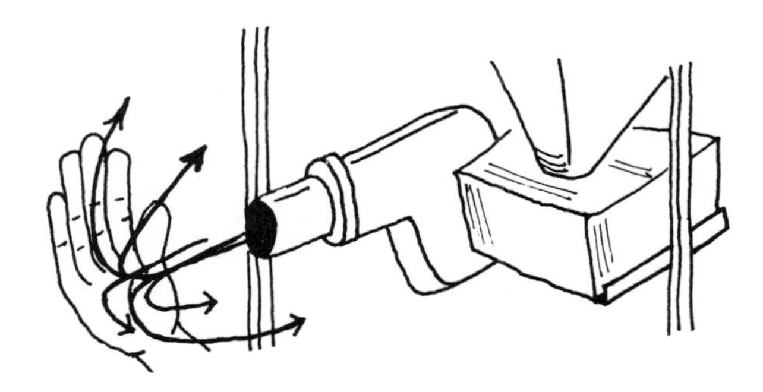

If possible, rest the insulation on the handle to feed the machine.

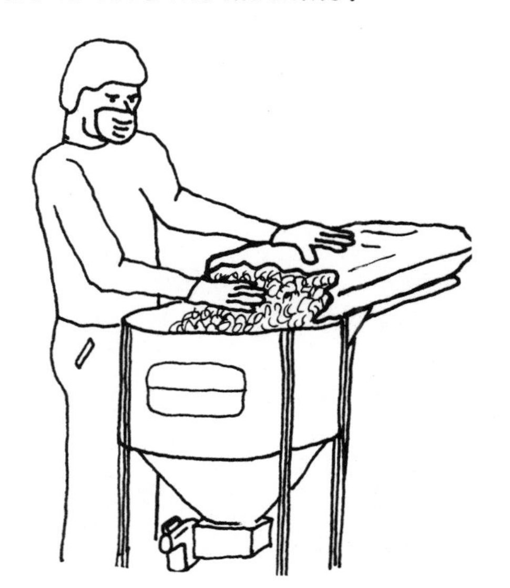

The machine, by virtue of the rotating choppers, will fluff the insulation. However, it helps if the person feeding the machine breaks the large clumps of insulation into smaller pieces.

NOTE:

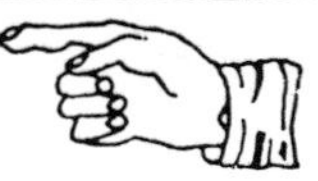

Never leave insulation blower unattended if there are small children, pets, or pranksters who might tamper with the machine.

33

insulate finished attic

description

Your biggest problem in insulating a finished attic will probably be gaining access where insulation is needed (unless access panels already exist). To do a complete job, you'll need to insulate the outer attic ceiling joists, the knee walls, the ceiling finished against the roof, the collar beams, and the end walls. This could mean as many as three access panels.

Remember, the idea is to insulate between heated and unheated spaces. Oftentimes, you'll find dormers in finished attics that should be insulated. If, however, the space above the outer ceiling joists is being used, you'll want to insulate the rafters instead (see Chapter 34). If there are no knee walls or no collar beams, also refer to Chapter 34.

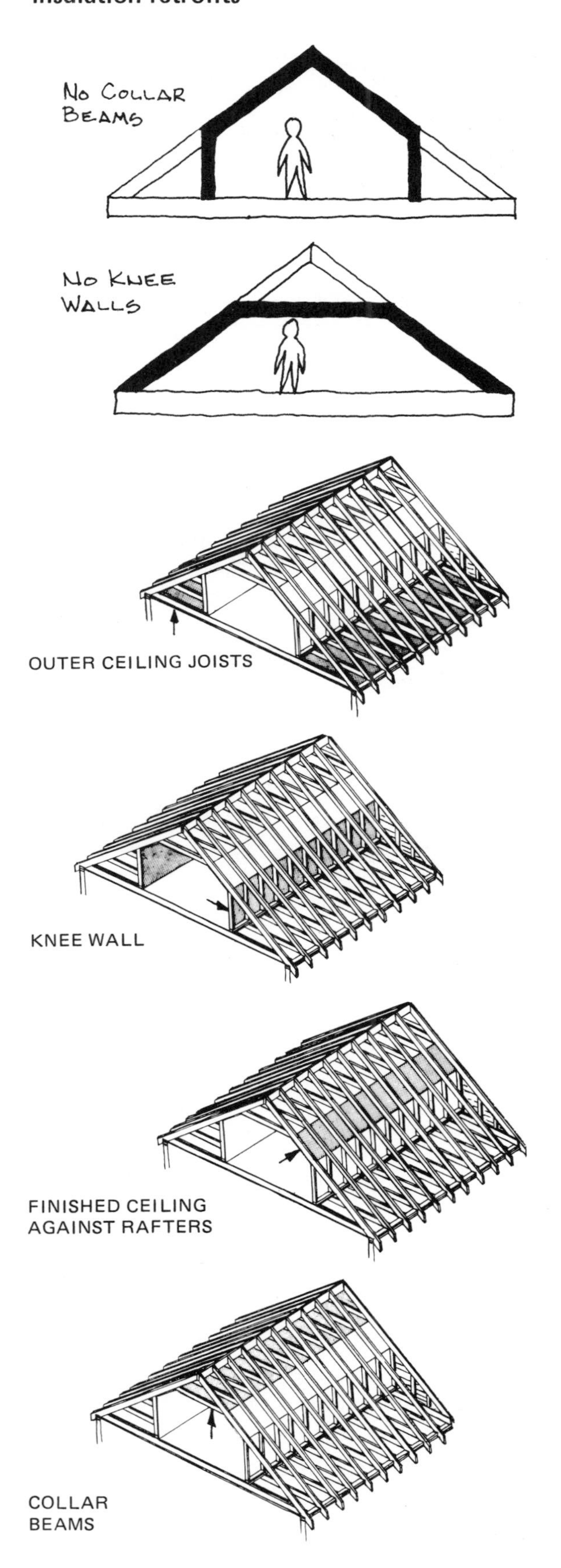

materials

roll/batt insulation

- glass fiber
- rock wool
  (both available as foil faced, kraft paper faced, or unfaced)

blown in place insulation

- cellulose
- glass fiber
- rock wool
- vermiculite
- perlite

vapor barrier

- polyethylene film
- aluminized or oil based enamel
- latex vapor barrier paint
- foil faced insulation
- asphalt impregnated kraft paper
- foil faced gypsum board

wood

- furring strips
- framing lumber

other materials

- fasteners

- adhesives

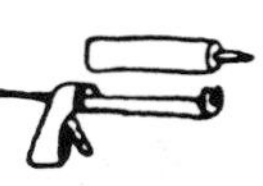

- vents

preparation

1 Before doing any insulation work, all leaks in the roof must be patched. Insulation that becomes soaked with water presents a hazard of ceiling collapse and moisture damage to wood structural members.

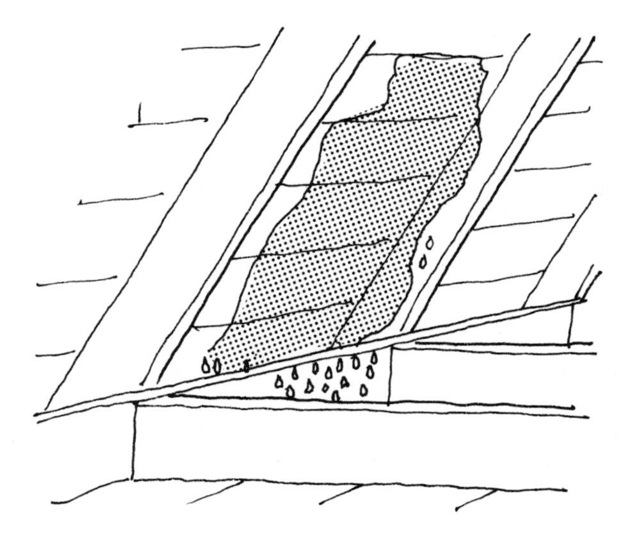

NOTE:

See Chapter 32, "Insulate Unfinished Attic", for some important words of caution about working in attics.

2 Seal all cracks and openings within the attic floor that allow warm air to bypass the insulation. These bypass routes are simply paths by which warm air

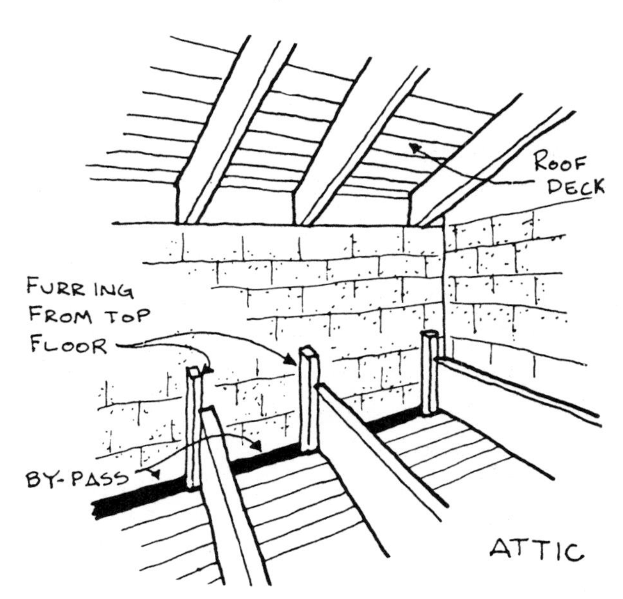

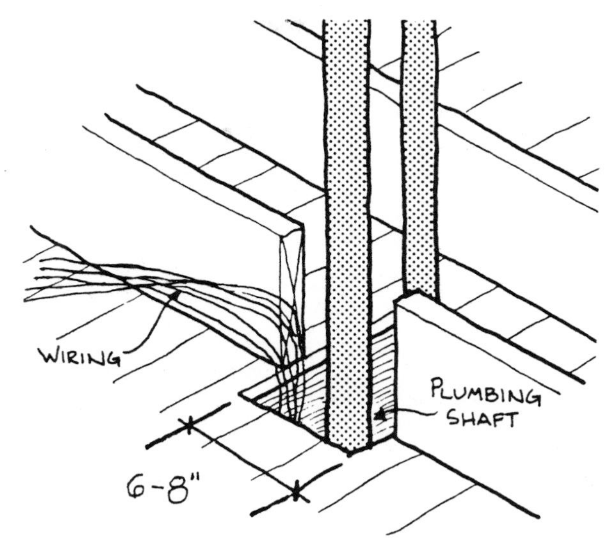

escapes from the living space to the attic, bypassing attic insulation. Bypass routes are usually a function of building construction. See Chapter 39, "Seal Air Bypasses", for a description of bypasses and where to look for them.

3 Provide adequate lighting.

4 It is a good idea to lay planks over outer ceiling joists and over collar beams when installing insulation in these areas. They provide better footing and make it less likely that you'll step through the ceiling finish. Also, planks help to spread your weight over several joists in case the joists are not strong.

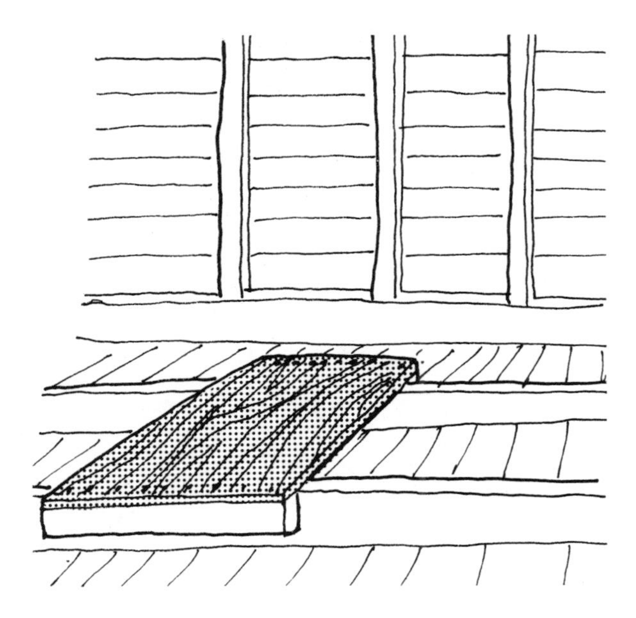

5 Existing studs may have to be furred out to accommodate the thickness of the insulation that you're going to use. Nail furring strips over the studs to achieve the desired thickness. This may be necessary, for example, when insulating between studs of a knee wall if the insulation is thicker than the stud depth.

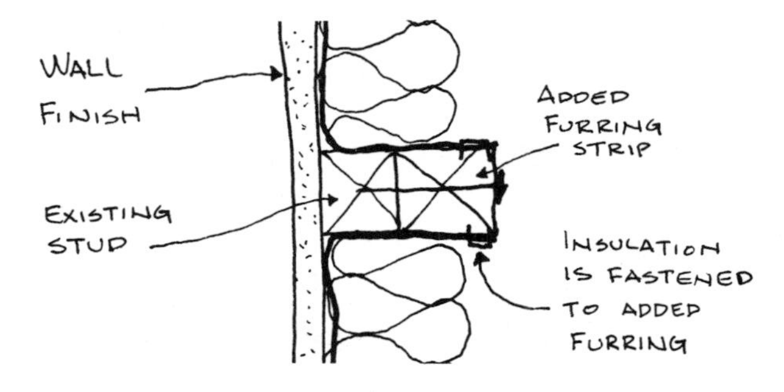

installation procedures

- **outer ceiling joists**
- **ceiling finished against the roof (batt insulation)**
- **knee wall**

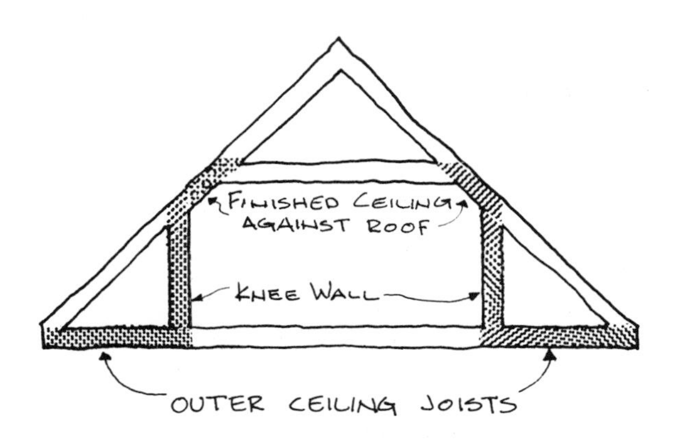

1 To insulate these sections, you need to gain access through both knee walls. If access panels do not exist in the knee walls, they can be added (for additional information on hatches, refer to Chapter 35, "But What about Accessing the Attic?"). For each knee wall, cut a section of wallboard between two studs with a saber saw (save the wallboard). The resulting opening should be about 15" wide -- large enough for an average size person to fit through.

2 First insulate the outer ceiling joists with roll/batt or loose fill insulation. For details, see Chapter 32, "Insulate Unfinished Attic".

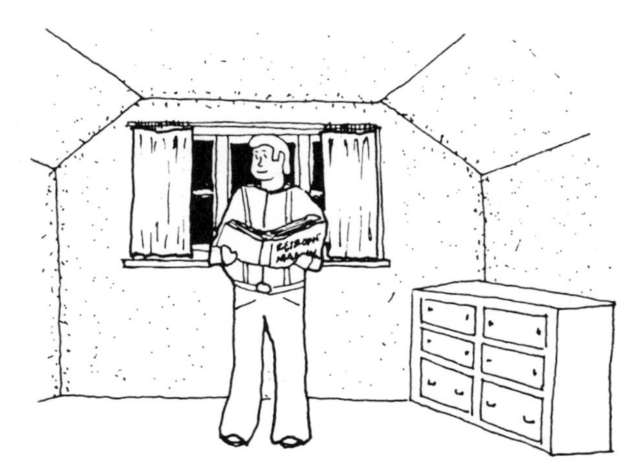

3 Now the tricky section -- the ceiling finished against the roof. Essentially you're insulating the length of the attic rafters between the collar beams and the knee walls. Cut pieces of batt insulation that correspond to this length and shove the insulation in place. If this is too long or tight, attach two furring strips or other light wood strips to one end of the insulation (along the edge). "Walk" the insulation in place by successively forcing each side with the strips.

NOTE:

The above methods for insulating the ceiling finished against the roof will also work if accessed from the collar beams (assuming there's enough room for a person to work). Alternatively, loose fill insulation can be poured or blown into this cavity - see step 8 for details.

4 If you used wood to "walk" the insulation in place, twist the wood strips 90° to slightly compress the insulation once the insulation is in place. This will provide two air circulation routes between the roof deck and the insulation between the roof eaves and ridge. You need about 1" between the deck and insulation for adequate air circulation. If there is at least 1" of air space between the insulation and the roof deck for air circulation, the furring strips will not be needed.

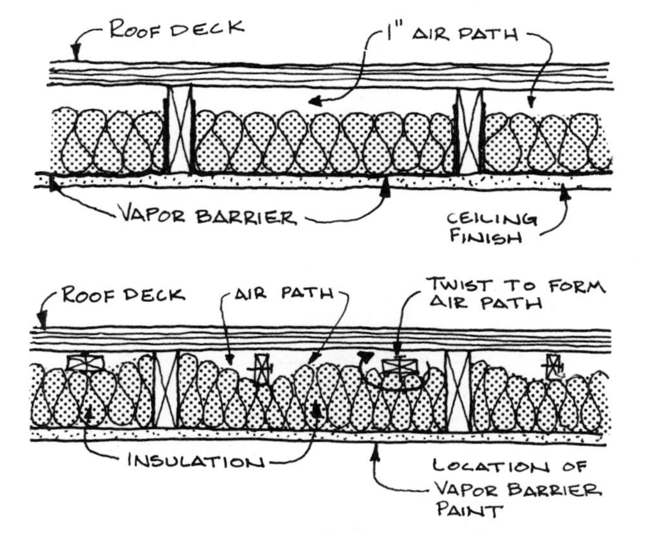

If there's difficulty in placing the insulation due to sheathing nails or plaster, try placing heavy polyethylene between the insulation and the rough area. Once the insulation is in place, the polyethylene can be pulled out.

5 The knee walls can now be insulated. Remember, if you're using a vapor barrier, it is to face towards the living space (on the warm-in-winter side of the insulation). The insulation is to extend the entire height of the knee wall to a point approximately 1" below the roof deck (to allow for air circulation). If you can't staple the insulation in place, use wire (laced from stud to stud), wire mesh, or "tiger teeth" to hold the insulation. For additional information on these installation techniques, refer to Chapter 25, "Insulate Floor Above Crawl Space". If existing, be sure to seal the opening beneath the knee wall between the ceiling joists.

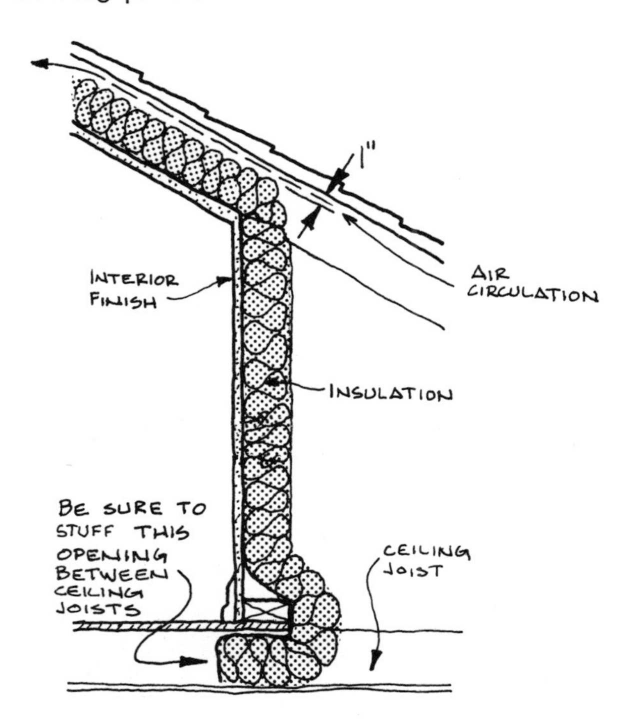

collar beams

ceiling finished against the roof (loose fill)

end walls

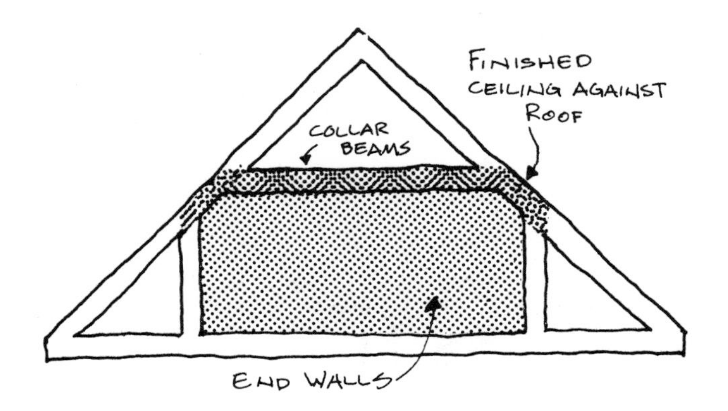

6 You'll need to gain access above the collar beams in order to insulate them. In some finished attics, the collar beams are long and there is space above them for a person to work. In other attics, the collar beams are short (2'-3' long). You can usually estimate the amount of room up there by looking at the end walls and pitch (angle) of the roof.

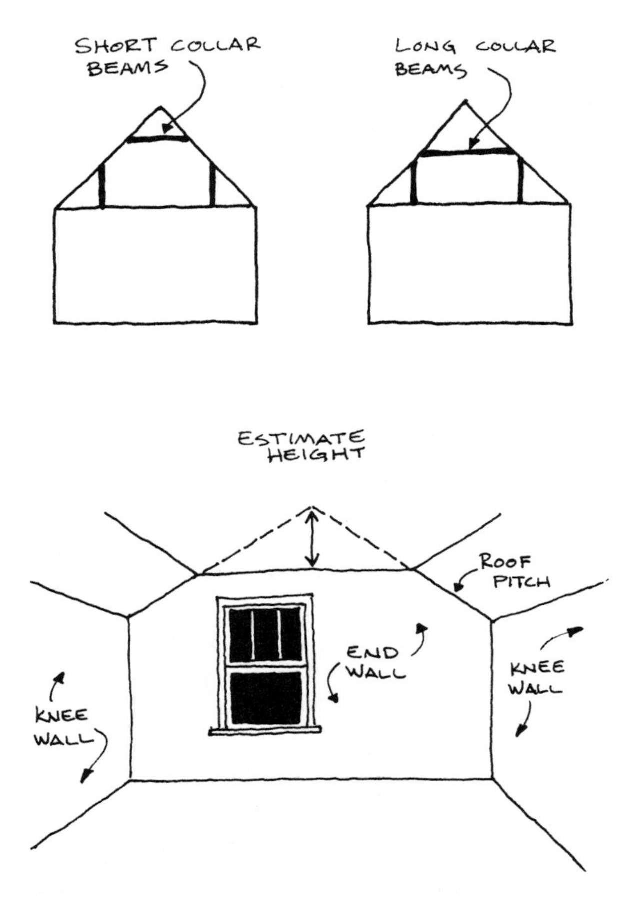

7 If there is room to work above the collar beams (long collar beams), consider installing an attic hatch in the ceiling (see Chapter 35, "But What about Accessing the Attic?") if there is no access panel.

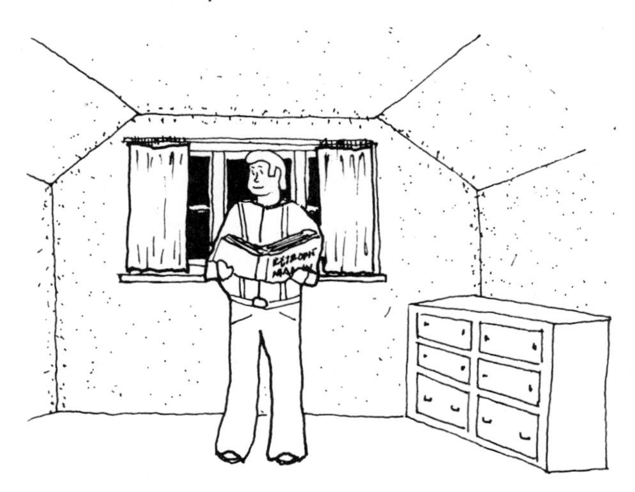

8 Now that you've gained access to the collar beams, you can insulate this section in the same fashion as the outer ceiling joists with roll/batt or loose fill insulation (see Chapter 32, "Insulate Unfinished Attic", for details). If you're using batts, it may be easier to insulate the ceiling finished against the roof from up here by shoving the batts down between the roof rafters.

If you're using loose fill, you can pour or blow the insulation down between the rafters. With loose fill, you'll need something at the other end to hold the insulation in place. Batt insulation used to insulate the knee walls can serve this purpose. It is important, however, to maintain an air space between the insulation and the roof deck. Thus, be careful that the insulation does not "run down" or settle at the top of knee walls preventing air circulation.

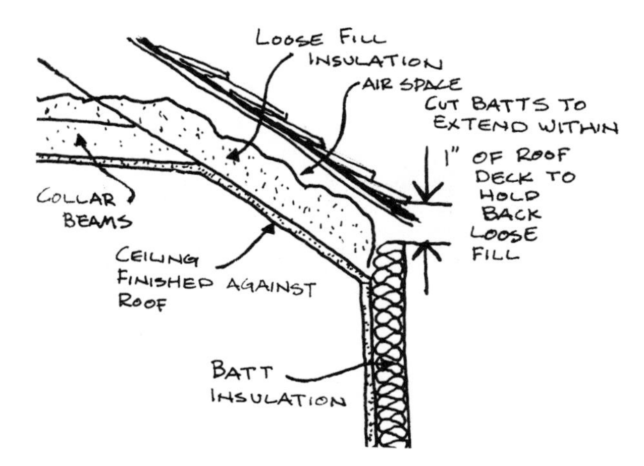

9 If the finished attic has short collar beams, cutting an access hatch in the ceiling may not work. Access holes (just large enough for an insulation hose) can be drilled at the top end of each rafter cavity. From these holes, the collar beams and the ceiling finished against the roof can be insulated

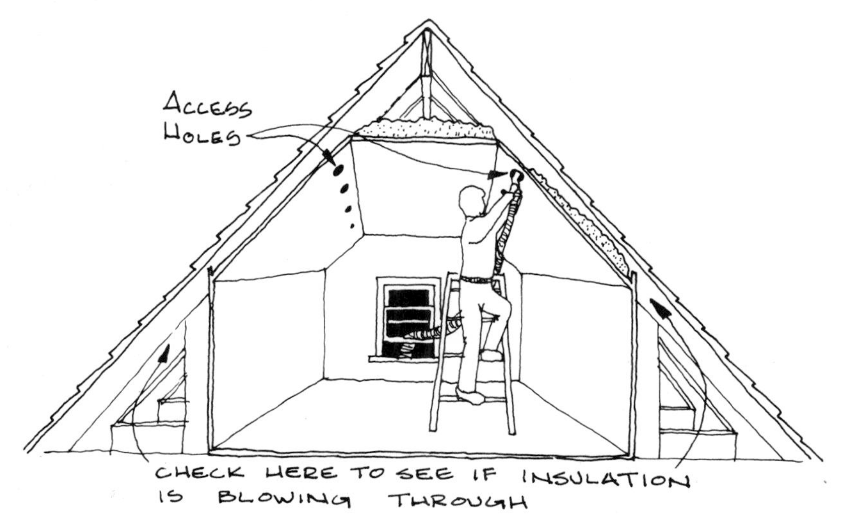

with loose fill. Check occasionally behind the knee walls to be sure that the insulation isn't blowing out the other end of the rafters.

10 To do a complete insulating job in the finished attic, both end walls should be insulated. If the studs are exposed, see Chapter 30, "Insulate Open Frame Wall". If not, see Chapter 31, "Insulate Closed Frame Wall".

PATCHING ACCESS HOLES

1 If you would like to access or use the outer attic (storage, for example) at some time in the future, install a removable hatch. For details, see "But What about Accessing the Attic?", Chapter 35.

2 If you don't want to use the outer attic, patch the access opening(s) to match the existing interior finish (if possible, use the wallboard that you saved). Before beginning, make sure you've removed planking and all the tools from the outer attic. Nail 2" x 2"'s around the perimeter of the opening between the existing studs. Recess the 2" x 2"'s the thickness of the gypsum board so that when you install the gypsum board, it will be flush to the existing finish. Staple a piece of batt insulation across the opening and then nail the original piece of gypsum board (or cut a piece to match) to the 2" x 2"'s.

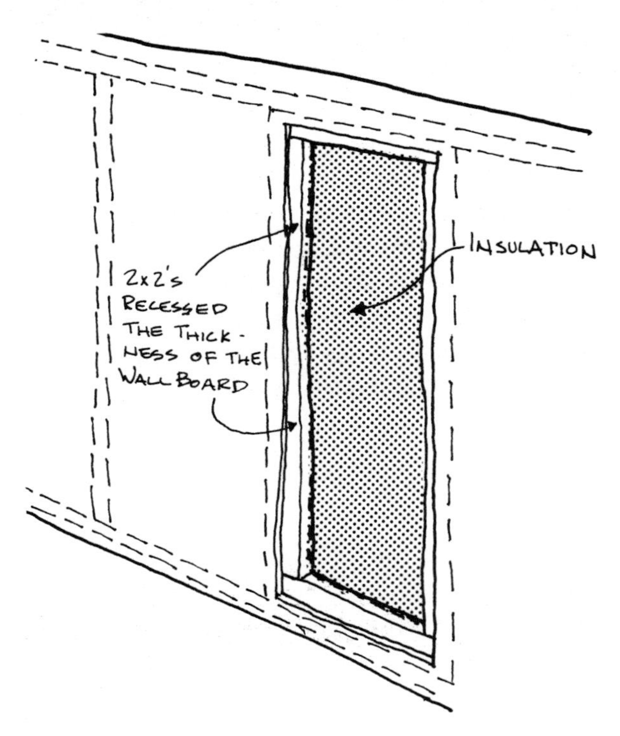

To seal the joints and cover the nail "dimples", apply a smooth "bedding" coat of wallboard taping compound over all joints and nail "dimples" with a 6" drywall knife. Before the compound dries, bed drywall tape in the compound. When this dries, apply a second and third coat of compound (sanding between coats) and then paint to match the interior finish. Refer to a home repair manual for detailed procedures for installing wallboard.

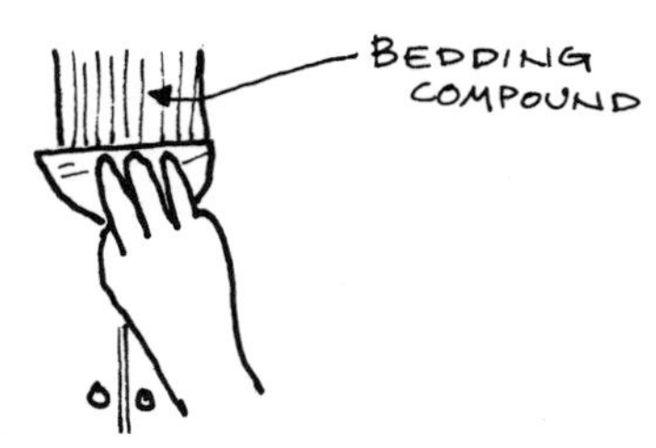

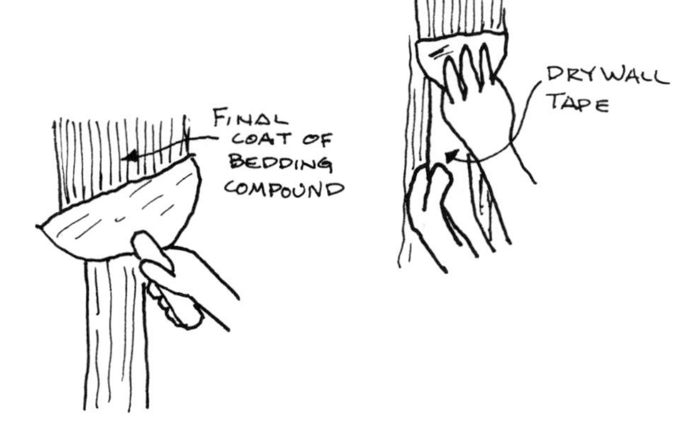

For additional details, see Chapter 40, "Repair Structural Hole".

34

insulate finished attic outer sections

description

If the outer attic, or the space behind the knee walls and above the collar beams, is used, or if there are uninsulated ducts or warm water piping running in these spaces, placement of insulation in the finished attic will be different from that discussed in the previous chapter. Depending on what part of the outer attic is being used, the outer end walls and rafters are insulated in combination with some of the finished attic sections discussed in the previous chapter.

The outer attic may be used for storage purposes, in which case you probably don't mind keeping the space relatively warm. Or, there may be heating ducts (or even a furnace) in the outer attic

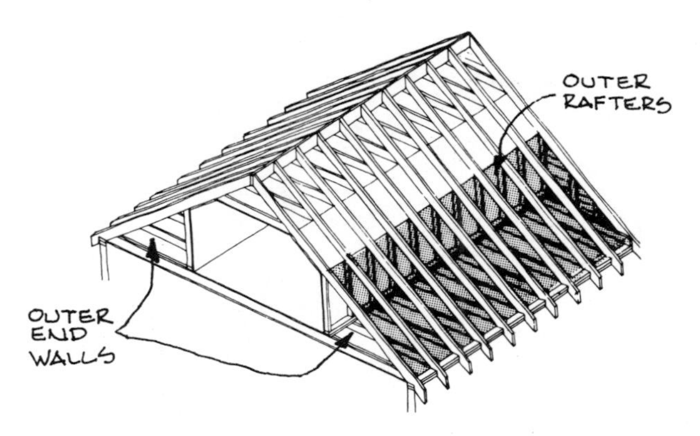

already keeping the outer attic warm. When considering what sections to insulate, keep in mind that you want to add insulation between warm (used) and cold (not used), or the outdoor, spaces.

EXAMPLES

A.

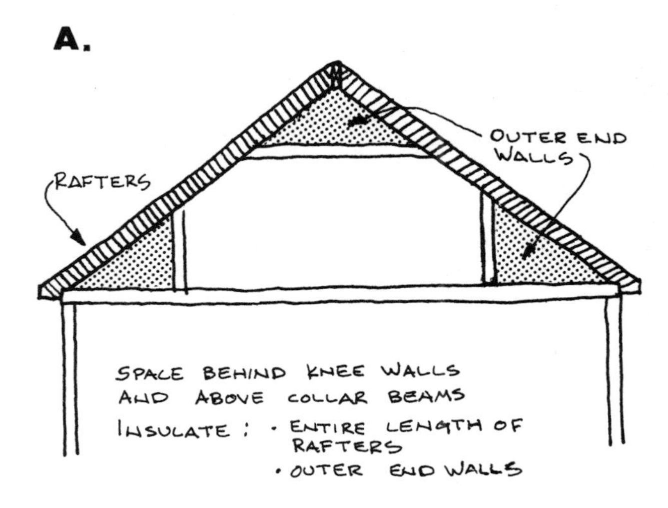

B.

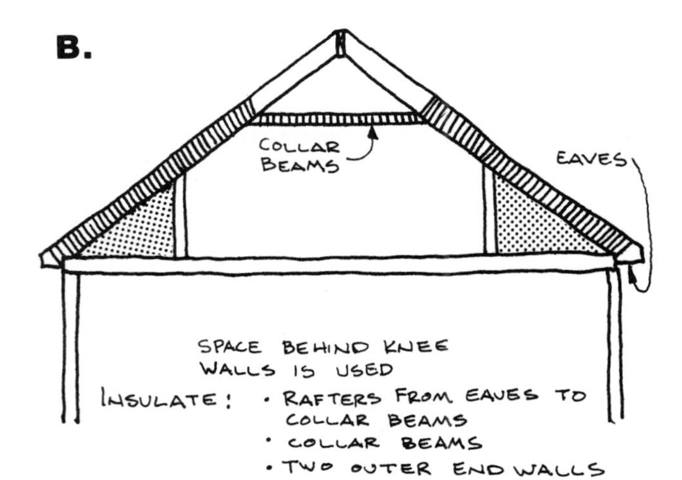

C.

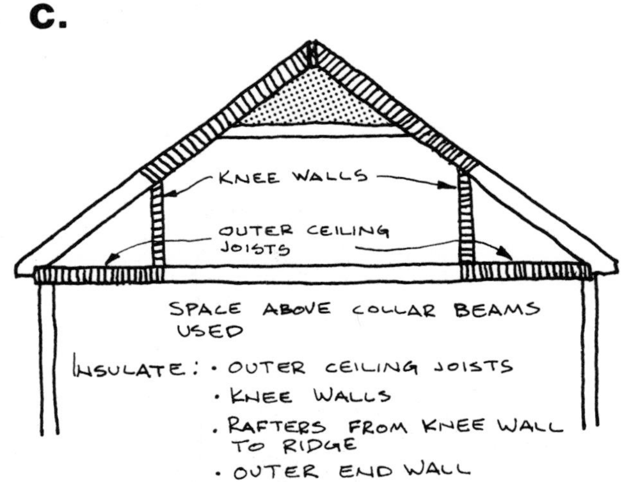

D.

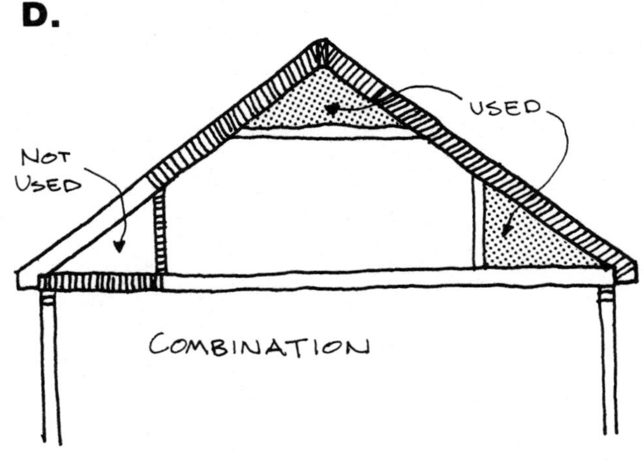

NOTE: 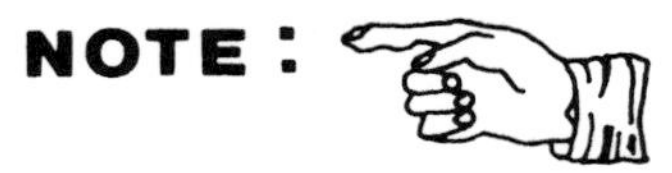

In all of these examples, a complete insulating job would also include the end walls on the interior of the living space.

materials

roll/batt insulation

- glass fiber
- rock wool

  (both available as foil faced, kraft faced, unfaced, or reverse flange)

blown in place

- cellulose
- glass fiber
- rock wool
- vermiculite
- perlite

vapor barrier

- polyethylene film
- aluminized or oil based enamel
- latex vapor barrier paint
- foil faced insulation
- asphalt impregnated kraft paper
- foil faced gypsum board

wood

- furring strips
- framing lumber

other materials

- fasteners
- vents

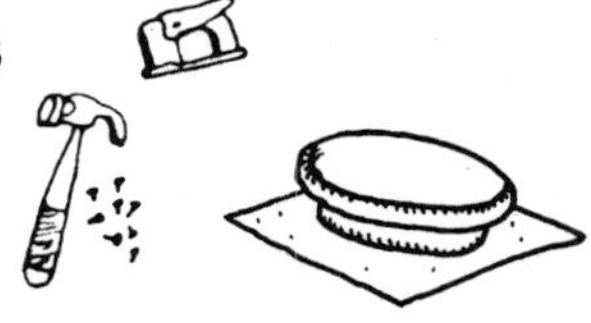

preparation

1 Before adding any insulation, all leaks in the roof must be patched.

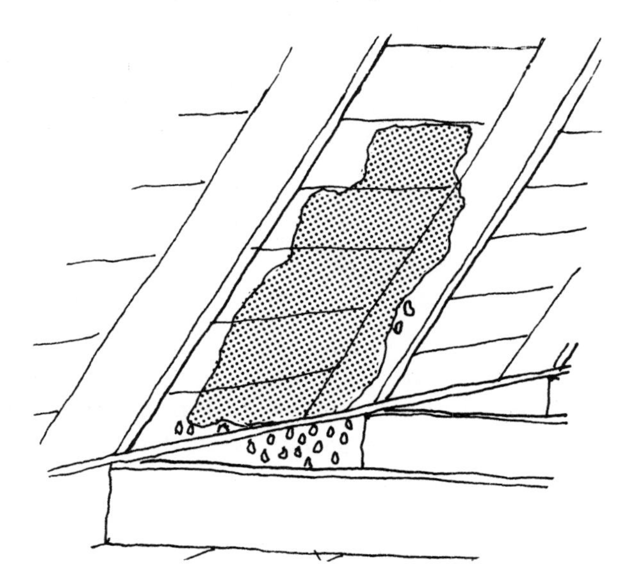

NOTE:

See Chapter 32, "Insulate Unfinished Attic", for some important words of caution about working in attics.

2 Provide adequate lighting.

3 It is a good idea to lay planks over outer ceiling joists and over collar beams when working in these areas. They provide better footing and make it less likely that you'll step through the ceiling finish. Also, planks help to spread your weight over several joists in case the joists are not strong.

4 Existing studs and/or rafters may have to be furred out to accommodate the thickness of the insulation that you're going to use. Nail furring strips over the studs/rafters to achieve the desired thickness.

Remember, it is desirable to provide a route for air venting on the cold side of insulation. Therefore, if the depth of rafters would provide less than 1" of space between roof deck and insulation, rafters should be furred out or a thinner insulation material should be used.

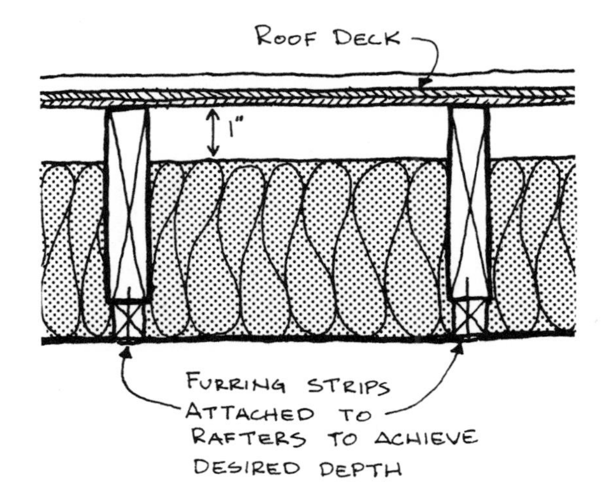

installation procedures

- **rafters**
- **outer end walls**

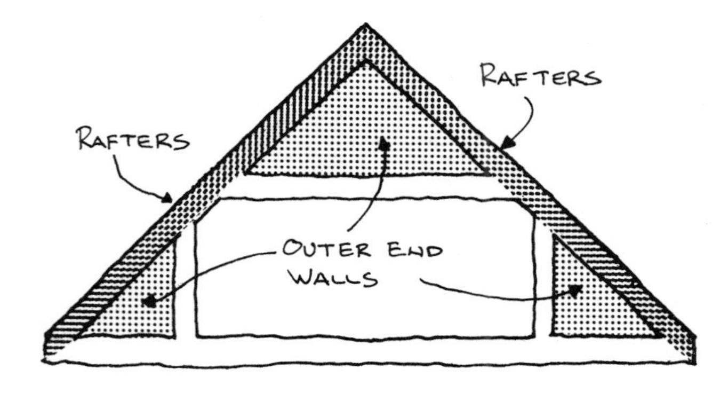

If finished attic sections other than rafters and outer end walls need to be insulated, see Chapter 33, "Insulate Finished Attic".

1 You'll need to gain access through the knee walls and/or the collar beams to insulate the rafters and outer end walls. If access panels do not exist, they can be added. (For more information on hatches, refer to Chapter 35, "But What about Accessing the Attic?"). Cut your access panel between two studs or joists with a sabre saw (save the wallboard). The resulting opening should be about 15" wide -- large enough for an average size person to fit through.

2 If you're using roll/batt insulation with an integral vapor barrier (foil or asphalt impregnated kraft paper), install it so that it faces in towards you. This applies to the rafters as well as the outer end walls. When insulating the rafters, be sure to leave a 1" air space between the insulation and the roof deck.

Vapor barrier faced insulation may be used, but it is poor practice to rely on a vapor barrier to control moisture migration into wood members because of complicated framing connections between rafters, plates, joists, and knee wall studs. Therefore, it is important to maintain air circulation between the insulation and the roof deck.

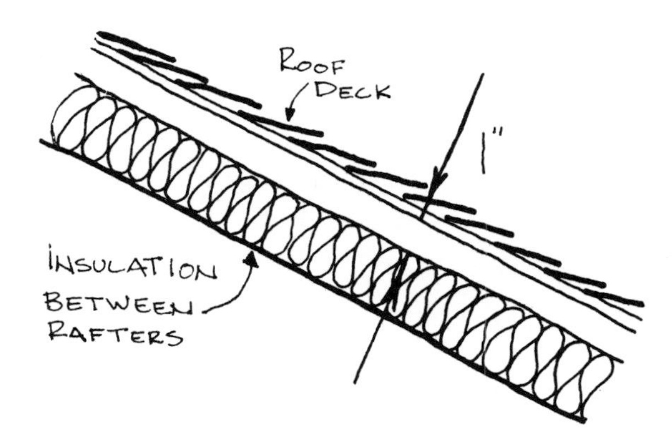

3 Use tape or staples to fasten the insulation to the rafters. If staples are used to install insulation with an integral vapor barrier, use polyethylene tape to cover the staples to maintain the integrity of the vapor barrier.

4 If unfaced insulation is used, apply polyethylene sheets over the insulation for vapor protection. Use staples to attach the sheeting to the rafters. Again, to protect the integrity of the vapor barrier, go back and cover the staples with polyethylene tape.

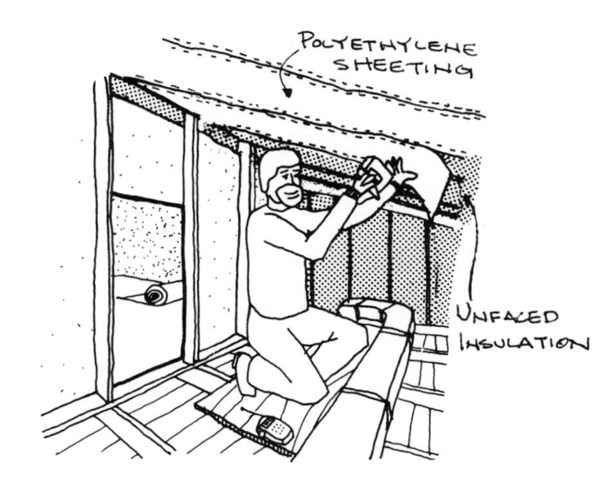

5 For examples A, C, and D shown on the second page of this chapter, a ridge vent or gable vents are needed above the collar beams. If a ridge vent is used, butt the insulation against the ridge beam. However, if gable vents are used, an air path to the gables must be continuous along the length of the ridge beam.

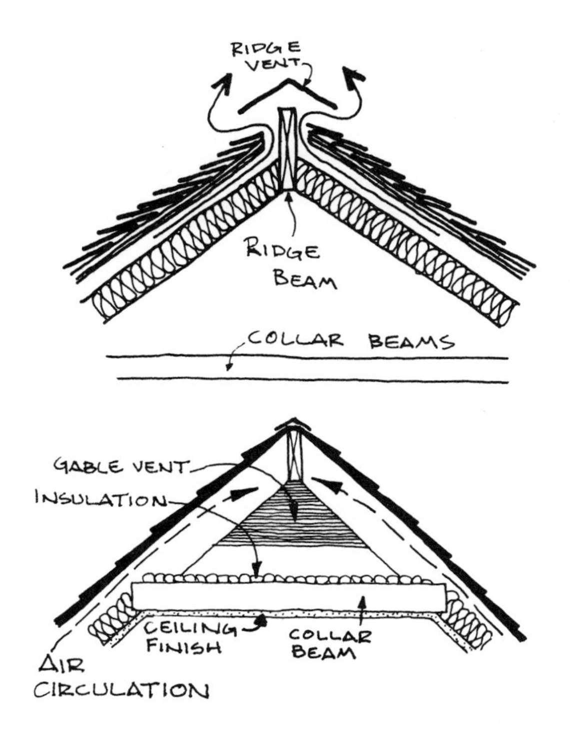

6 For examples A, B, and D shown on the second page of this chapter, it is important to allow for outside air circulation between the insulation and the roof deck from eave vents up to the ridge or gable vents. Stop insulating at the end of the joist runs and fold the insulation down to the "plate".

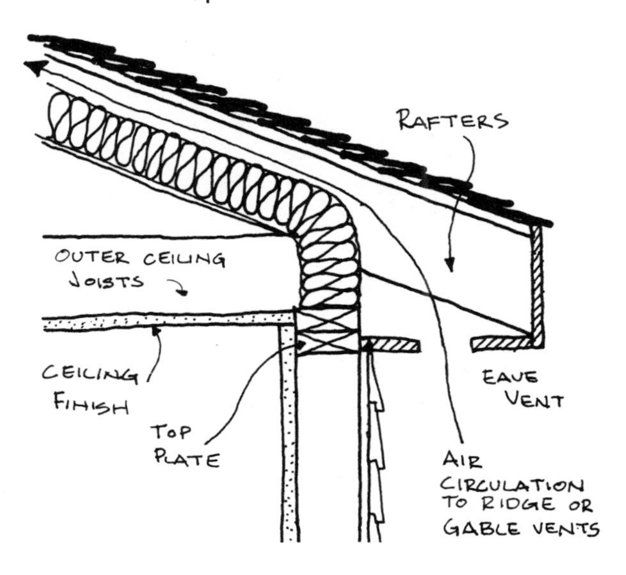

NOTE: 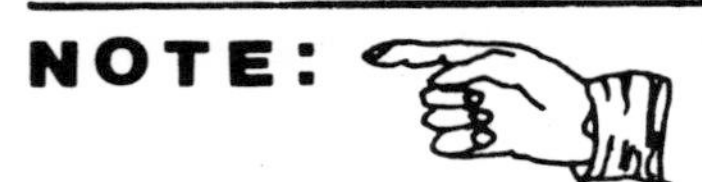

See "But What about Attic Ventilation?", Chapter 36, for details on installing attic vents.

7 Refer to Chapter 30, "Insulate Open Frame Wall", for details on insulating the outer end walls.

8 See Chapter 33, "Insulate Finished Attic", for details on patching access openings in knee walls and ceilings.

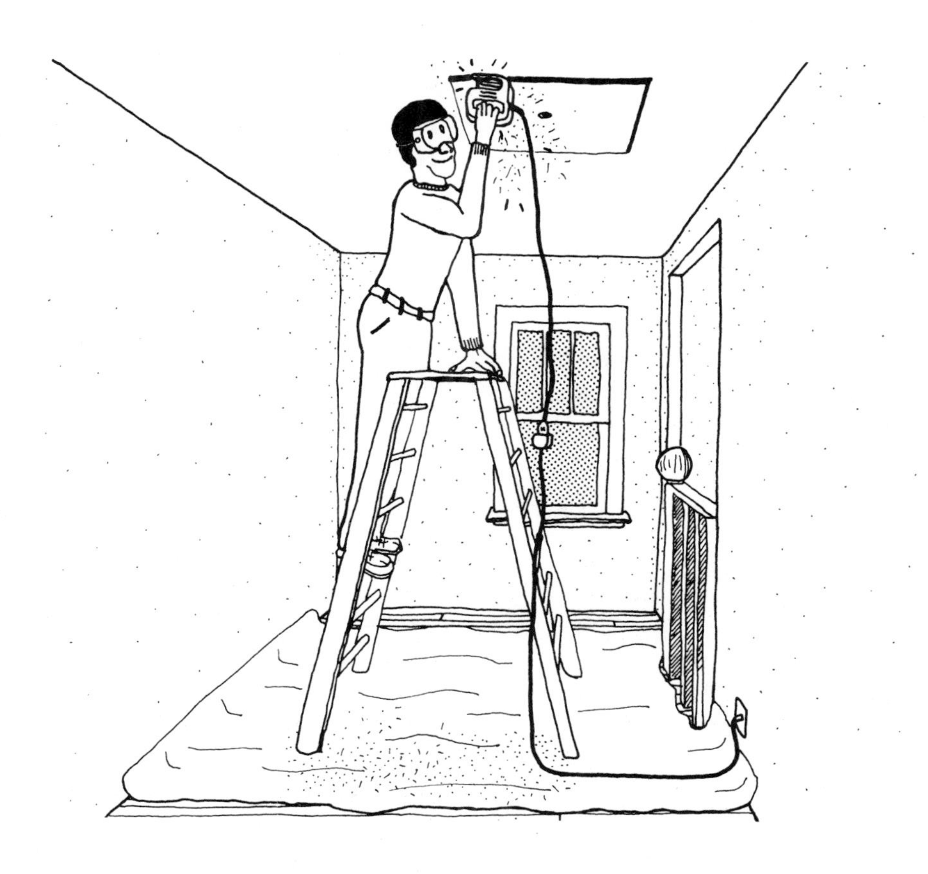

35

but WHAT about accessing the attic?

description

Insulation must often be installed in an unfinished attic (or roof crawl space) or above collar beams in a finished attic. These spaces may not be accessible and openings must be made to provide access.

An opening may be made through the ceiling of a room below the attic so that an installer and insulation may pass through to the attic. The procedures for installing an attic hatch should be followed with care so that unnecessary damage and unnecessary work can be avoided. An opening required for installation of vents may also be used as access (see Chapter 36, "But What about Attic Ventilation?").

Another advantage of access to attic spaces is to check for the presence of moisture if the roofing develops leaks. It's relatively easy to find the source of roofing problems by examining the bottom side of roof decking.

1 To find a good location for the hatch, take care to find an area of the ceiling which has enough attic height above it to allow for moving around. Attics which have a shallow roof should be accessed close to the line of the ridge of the roof. If there is a flat roof, it is usually pitched slightly to drain. Flat roofs usually pitch upward toward the front of the building. Access these roofs toward the front of the house. For appearance, you may desire to place the hatch in a closet ceiling.

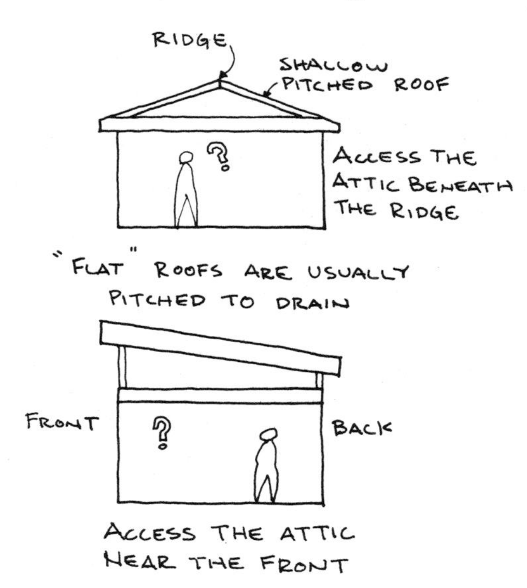

2 Make a preliminary check on attic height by drilling a small hole and probing for the bottom of roof joists or roof deck using a straightened coat hanger.

AN IMPORTANT NOTE :

If you determine by probing that the roof deck rests on the ceiling joists, then DO NOT INSTALL A HATCH. This type of roof must have insulation installed between each pair of joists separately (see Chapter 37, "Insulate Finished Ceiling Against Roof").

---

3 After finding a location for the attic hatch, the location of structural members must be determined. Structural members will usually run parallel to or perpendicular to the walls. Tapping the ceiling lightly with a hammer, screwdriver handle, or your knuckles, you can usually "trace" the position of ceiling joists. Joists are usually 16" apart (16" on center), so after locating one joist, adjacent joists can be readily found.

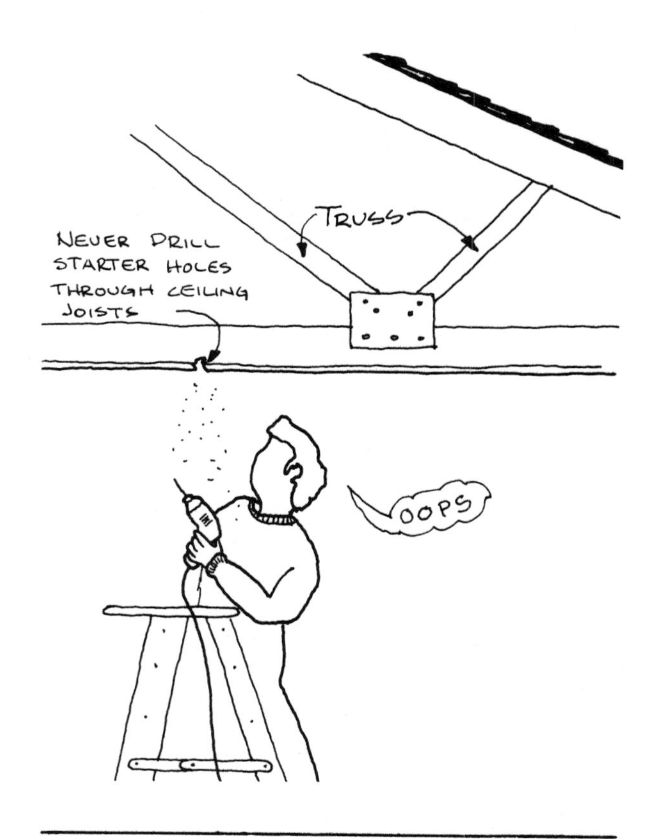

NOTE:

Since ceiling joists usually do not have any weight to support other than the ceiling finish, there is little danger of weakening the structure by cutting the joists. However, if the roof is supported on trusses or if there is bearing through ceiling joists to rafters, cutting a joist may cause serious weakening and/or structural damage. Since you don't know whether the joists which you located help to support the roof until after the hole is cut, NEVER CUT THROUGH A CEILING JOIST TO MAKE AN INITIAL HOLE.

4 Make only straight, clean cuts through the ceiling finish. Since the ceiling finish may be plaster on wood lath, drill a "starter hole" through the finish. This hole should be at the edge of a ceiling joist. Make another hole at a distance equal to the desired edge length of the opening plus an additional 3" (for framing the opening). Connect the two holes with a straight "guide line." Then use a "keyhole" saw or a sabre saw to make a cut following the guide line (if available, a larger power "reciprocating saw" works best).

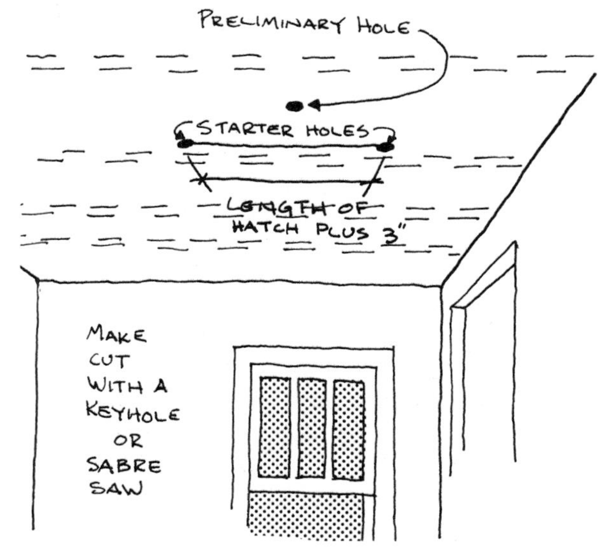

5 Strike lines at a right angle from this cut from the "starter holes." Cut toward the edge of the adjacent joist stopping as the edge of the joist is reached (see step 9).

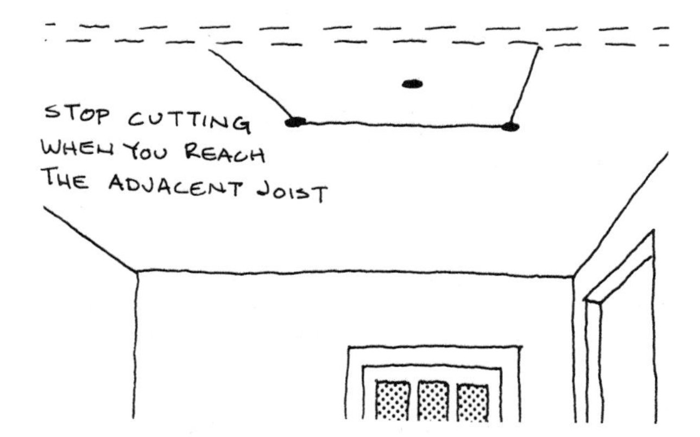

6 Finish cutting the hatch by drilling starter holes at the ends of the two previous cuts, marking a guide line, and sawing along the line. A piece of heavy tape over the first cut will keep the cut-out ceiling finish from dropping on the floor.

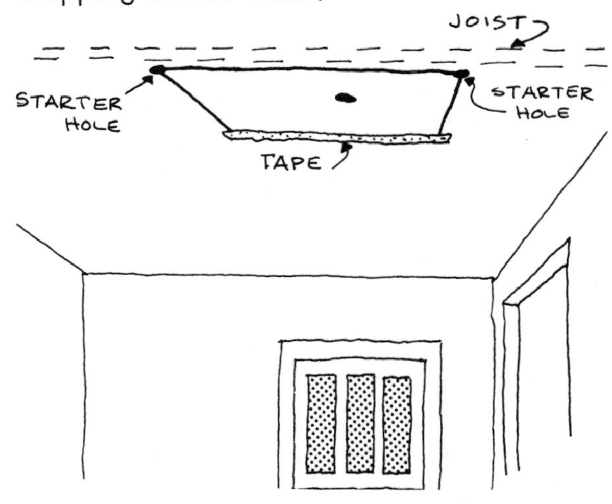

NOTE:

Both gypsum board and lath and plaster will create a lot of dust when drilled or sawn. Wear a dust mask and protective goggles when drilling and sawing. Electrical wiring may be resting on the ceiling finish above. Wear insulating gloves and use double insulated grounding tools. The main switch at the circuit panel should be thrown to cut off power if the previous precautions cannot be taken.

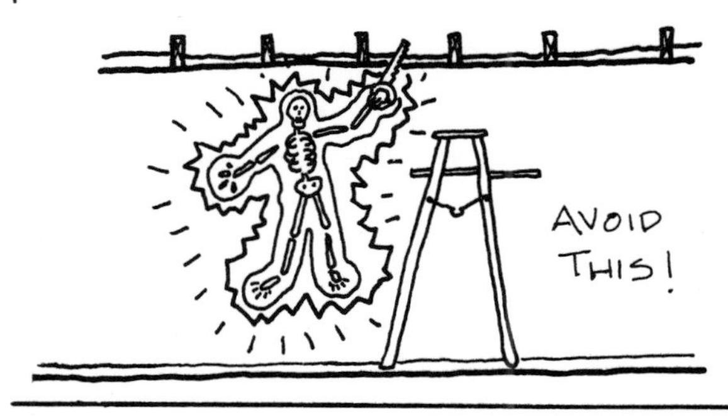

7 If the access hole needed is no wider than the distance between two joists, skip to step 12 for fabricating and mounting the hatch.

8 Now you can examine the roof support to determine if a hatch wider than the joists spacing may be installed. If the roof is supported by trusses, then the ceiling joists are also the "bottom chords" of the trusses and are part of the structural support of the roof. DO NOT CUT ANY PART OF A TRUSS. If there are any connections between the joists bordering the hatch and rafters or roof joists above, it's safer to leave them alone.

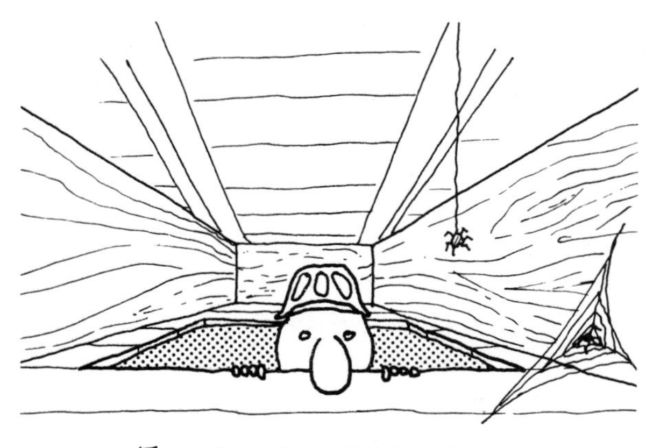

EXAMINE THE ROOF STRUCTURAL SUPPORTS

9 If you must cut a joist, first attach framing lumber to the tops of three joists with the joist to be cut in the middle. If the ceiling finish is not secure, then drill and screw this lumber in place to avoid hammering. These

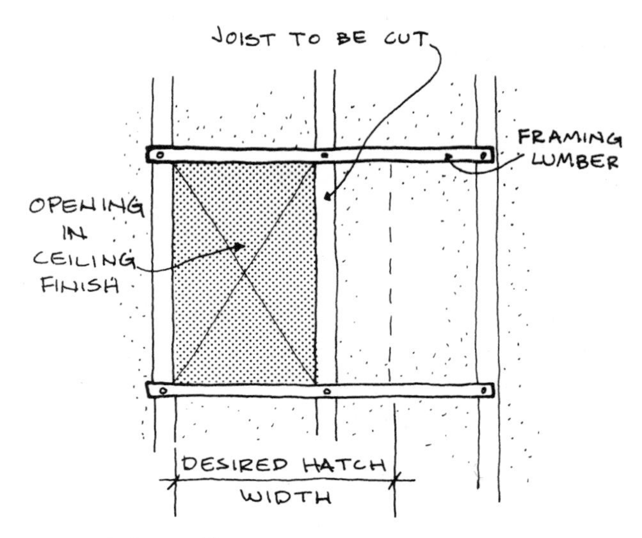

ATTIC PLAN - LOOKING DOWN

pieces of lumber should line up with the previous cuts which were made that are perpendicular to the joists.

AN IMPORTANT NOTE:

See "Some precautions to take before cutting ceiling joists" in this chapter before cutting any joists.

---

10 Now cut through the joist using the top framing and the opening in the ceiling finish to guide the saw.

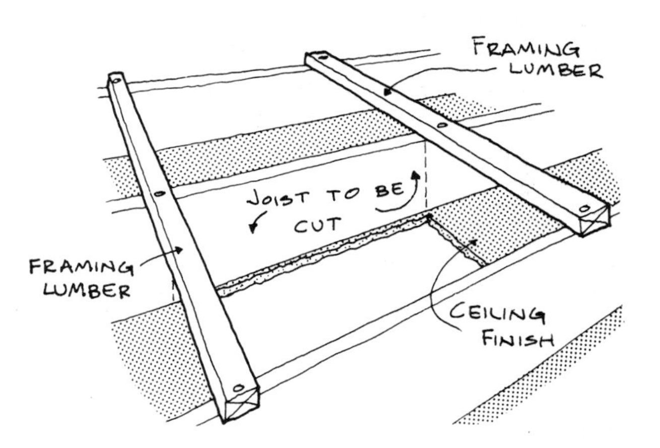

11 Cut some lumber, preferably of the same dimensions as the joists, to span between the uncut joists for use as "headers". Nail the headers in place.

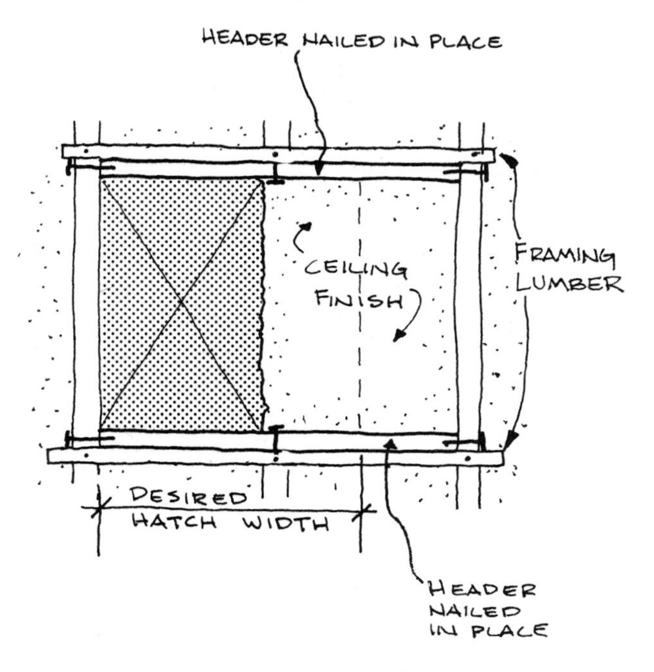

ATTIC PLAN - LOOKING DOWN

12 A piece (or pieces) of blocking can then be inserted to line up with the desired edge for the hatch. Cut the ceiling finish to this blocking as well as along the inside edge of the blocking.

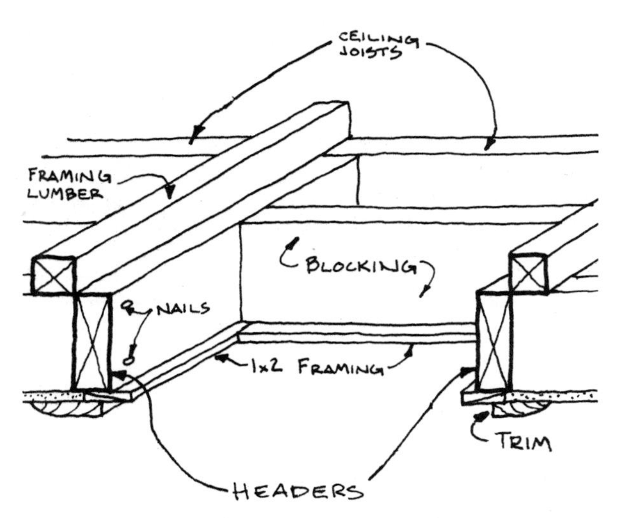

1" x 2" framing can be used to form an edge support for the hatch and to form a base for finish trim to cover the cuts through the ceiling finish.

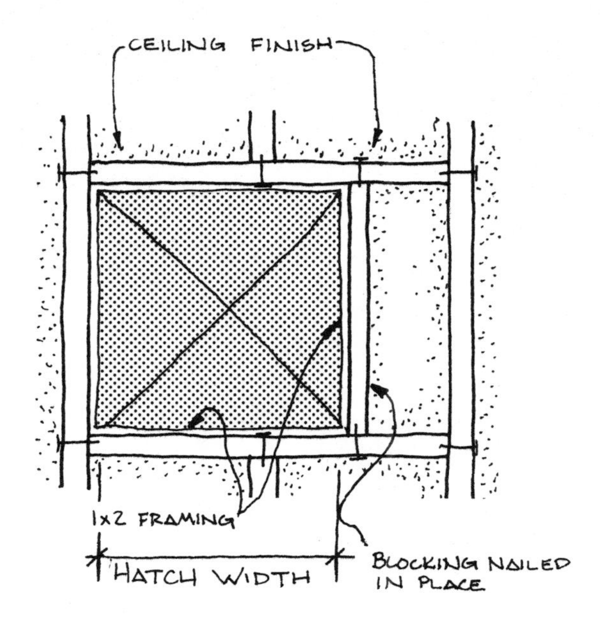

ATTIC PLAN - LOOKING DOWN

13 The hatch can be cut to fit inside the framed opening and rest on the 1" x 2" framing. 1/2" thick plywood should be used so as to support a person's weight should someone step on the hatch. Mount rigid insulation or staple blanket insulation on the cold side of the hatch to reduce heat loss through it.

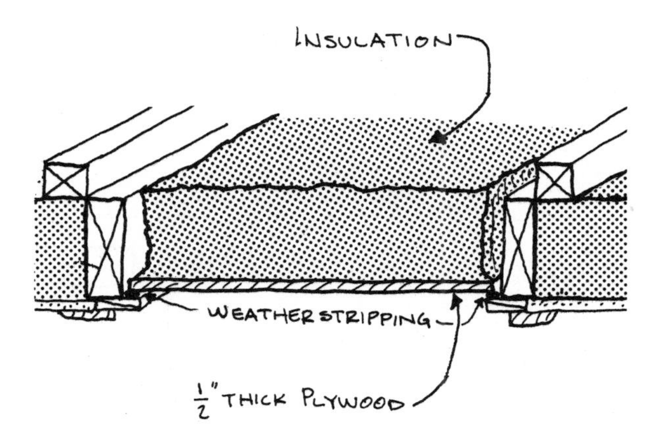

To prevent infiltration, caulk the joints of the framing pieces from the attic. Place adhesive backed foam around the hatch edges so that when the hatch is in place the foam presses against the 1" x 2" hatch supports.

NOTE:

Window latches may be installed to compress the hatch against the support frame to seal it shut and to provide security.

Test and trim the hatch for a tight fit and convenient operation. Leave the hatch tightly closed when the insulation job has been completed.

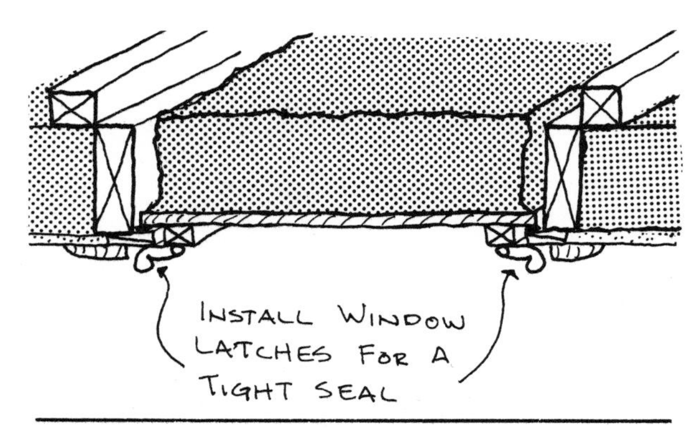

some precautions to take before cutting ceiling joists

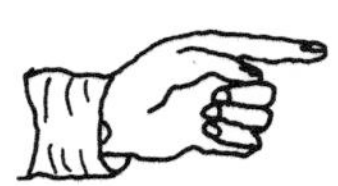

Examine the condition of the attic framing and the type and stability of roof supports before cutting the joist.

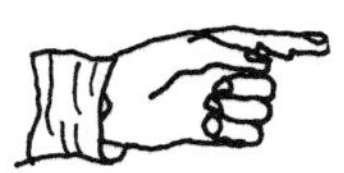

If the ceiling joists are decayed, cracked, or have slumped (deflected), do not cut any joists.

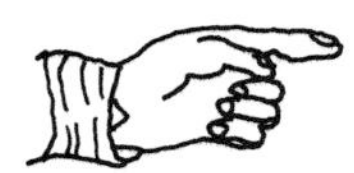

If the side walls have been pushed outward, do not cut any joists.

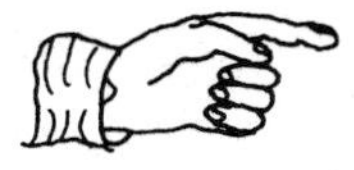

If lumber blocking connects the roof joists (flat roof) or rafters (pitched roof) to the joists, do not cut any joists.

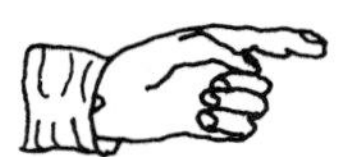

If there are trusses, do not cut a joist because it is acting as the bottom chord of the truss.

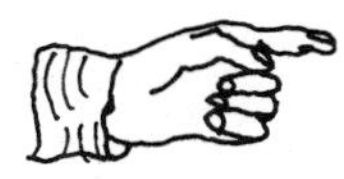

If you encounter wiring, find another hatch location or have a qualified person reposition the wiring.

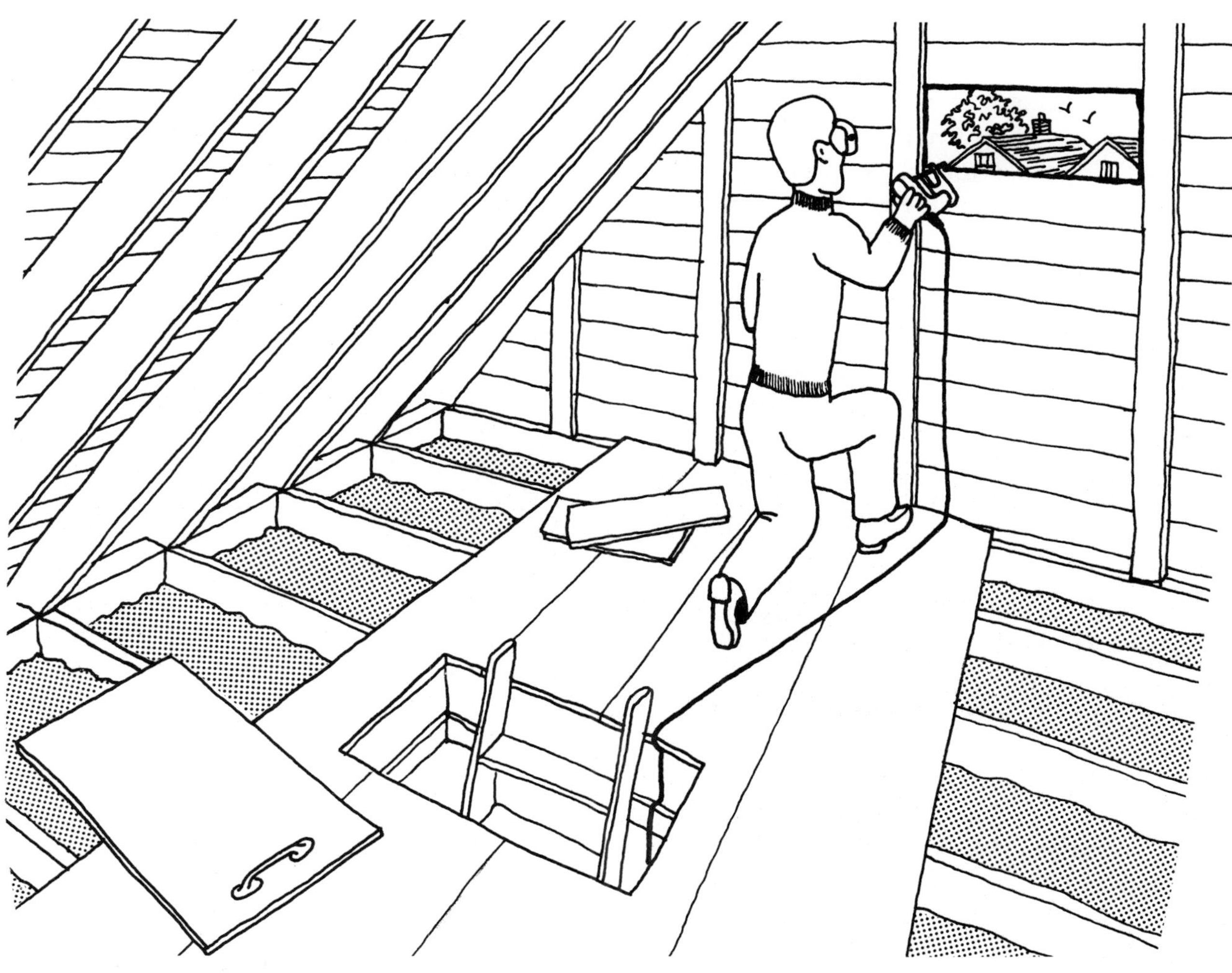

36

but WHAT about attic ventilation?

description

If the attic is not adequately vented, then additional vents should be installed. The required vent areas depend on the geographic location of the home. This is due to the tendency for more moisture to accumulate in buildings which are located in areas with severely cold weather (see Chapter 3, "A Word about Vapor Barriers").

There are some quick ways to tell whether an attic has accumulated moisture in the past. If we know the path moisture travels, we can find signs of its presence. As water vapor in the house moves up through the attic, it can condense on any surface which is at or below the "dew point" (see Chapter 4, "A Word about Condensation"). The ceiling is usually warm enough that vapor passes through. However, vapor in the attic will usually condense on the rafters and roof deck, and perhaps on the attic end walls when temperatures are in the 35° F and below range. How much moisture builds up depends on many things, but where temperatures stay below 35° F for a long time, moisture has a chance to continually accumulate.

WHAT TO LOOK FOR

The coldest surfaces of the attic tend to accumulate the most moisture. One of the coldest surfaces will be roofing nails. Points of roofing nails can usually be seen sticking through the roof deck. If these nails are rusted or if the roof deck is stained in a circle around the nails, then enough moisture has accumulated on these nails to indicate that the attic is not adequately vented.

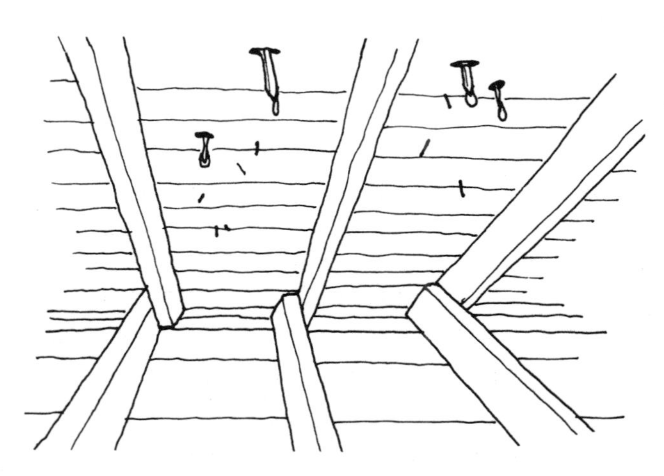

LOOK FOR MOISTURE AROUND ROOFING NAILS

If you examine the attic in cold weather, you may actually see frost or icicles on nails, rafters, and decking. Naturally, when the attic warms up, the ice melts. The water may drip onto the ceiling or be absorbed by the wood.

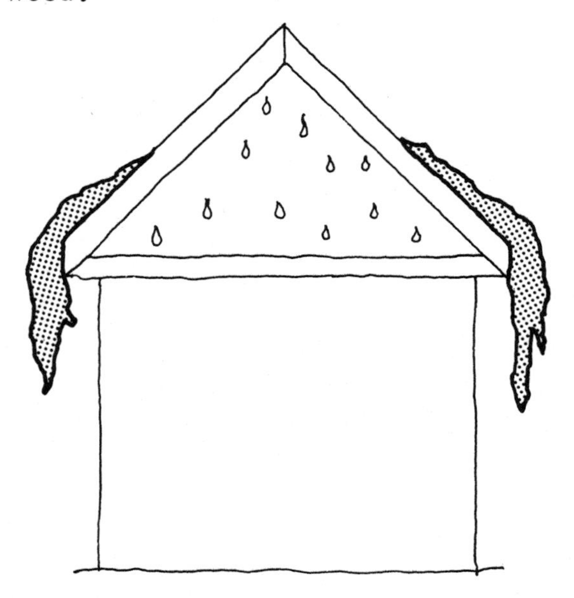

FROST MAY APPEAR ON NAILS, RAFTERS, AND DECKING

Look for staining or peeling paint in spots of the ceiling below or wood dry rot on rafters or decking. Dry rot appears like spider webs clinging to the wood when it is active. After rotting has occurred, the wood is soft and can be rubbed loose or easily pierced with a screwdriver.

LOOK FOR PEELING PAINT

NOTE:

Ceiling damage and rotting are signs of serious damage. If it appears that ceiling finish may fall (letting warm air past the insulation that you are about ot install) or if the roof could develop a leak which would wet the insulation and cause ceiling damage, then ceiling and roof repair should be done before installing insulation.

MOISTURE PROBLEMS

If the existing vents provide the recommended vent area (see table in this chapter) and moisture accumulation is still a problem in the attic, then "surplus" water vapor may be the problem. Try to solve this by:

1 If you use a vaporizer or humidifier during the winter, it is advisable to cut back their use. Humidity need not be higher than 30% relative humidity for comfort. Humidifiers are available with a humidity control.

2 High water vapor areas are kitchens and bathrooms. You may have to ventilate these rooms when the humidity goes up. This may be accomplished by opening a window slightly for 15 minutes after cooking or bathing.

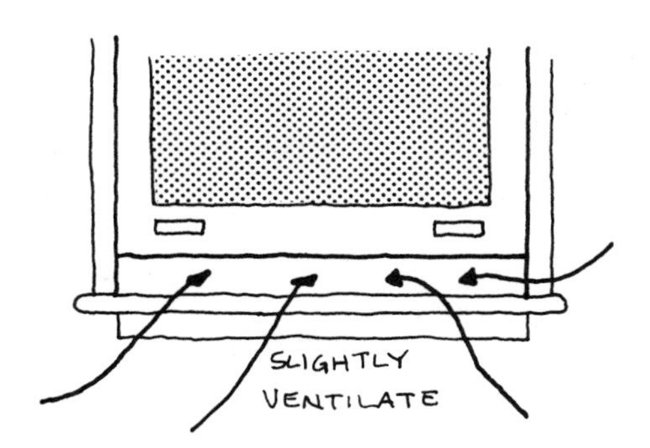

NOTE:

For very tight houses one can obtain an "air-to-air" heat exchanger. This device, which can be fitted into a window, exhausts warm overly humid or stale air for fresh air, but the heat of the exhausted air is transferred to the fresh incoming air.

3 Check for moisture accumulation in the basement or crawl space. If the crawl space (or floor slab) does not have a vapor barrier, the moisture could be coming up from the ground. Install a vapor barrier over the earth floor of the crawl space if none exists (see Chapter 23 for details).

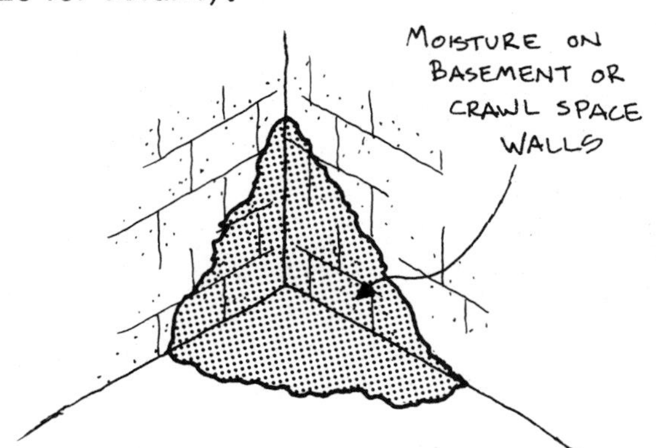

4 You may have difficulty tracking down other possible sources of excess moisture. When you don't think that other measures would really solve the problem, then install a vapor barrier in the ceiling when you insulate.

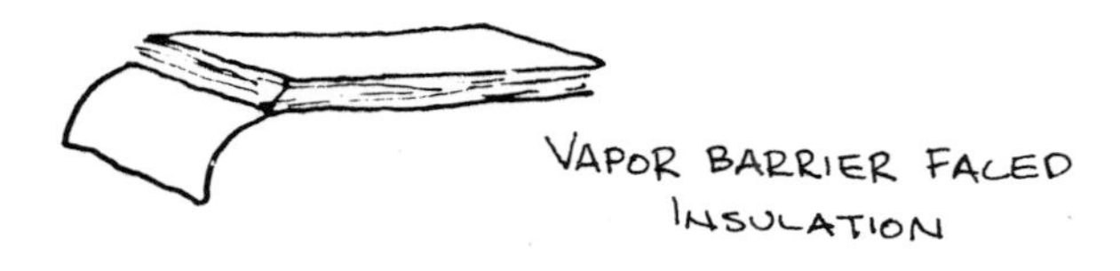

AREA OF ATTIC VENTS

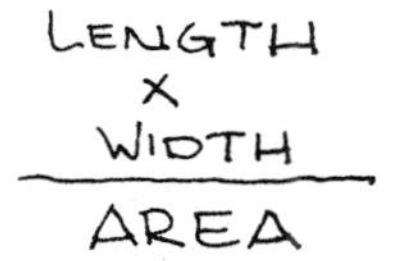

In addition to venting moisture during the winter, proper attic ventilation also reduces heat build up during the summer. This, in turn, helps keep the temperature of the home cooler. Proper attic ventilation is necessary to keep moisture from condensing during the winter and preventing heat build-up during the summer.

Two concepts are important for properly venting an attic;

1) AREA of attic vents

2) LOCATION of attic vents

Total unobstructed open vent area is to be 1/600 of attic floor area (1/300 of attic floor area for flat roofs).

For example, if your attic measures 50' x 30', you would need 2.5 square feet of vents.

50' × 30' = 1500 ft²
1500 ft²/600 ft² = 2.5 ft²

If you're using No. 16 mesh insect screen, you will need 5.0 square feet of vent area.

2.5 ft² × 2 (FROM TABLE) = 5.0 ft²

Ventilating area increase required if louvers and screening are used

Obstructions in Ventilators– louvers and screens	To determine total area of ventilators, multiply required net area in ft² by –
1/4" mesh hardware cloth	1
1/8" mesh screen	1-1/4
No. 16 mesh insect screen (with or w/o plain metal louvers)	2
Wood louvers & 1/4" mesh cloth	2
Wood louvers & 1/8" mesh screen	2-1/4
Wood louvers & #16 mesh ins. screen	3

- if metal louvers have drip edges that reduce the opening, use same ratio as shown for wood louvers

Source: "Problems In Your House: Prevention and Solution", Agriculture Information Bulletin No. 373, September 1974, pp. 12-13, Washington, D.C.: U.S. Department of Agriculture Forest Service.

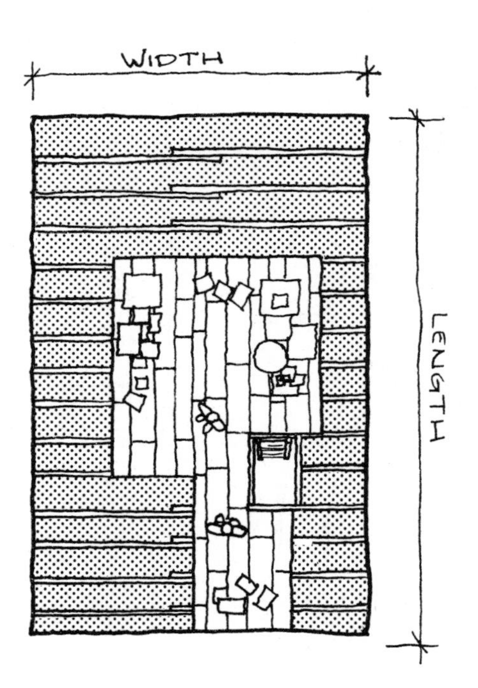

LOCATION OF ATTIC VENTS

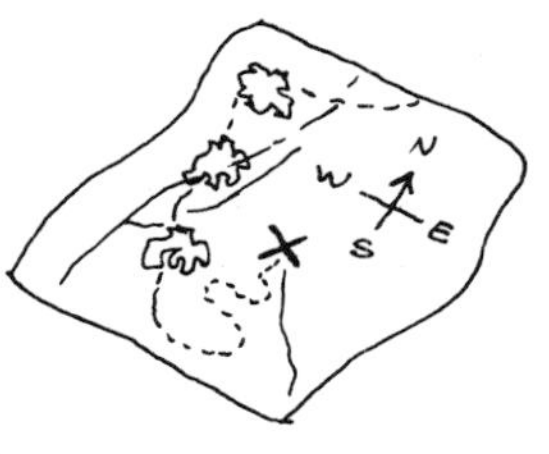

It is recommended that both inlet and outlet ventilation be used. Inlet vents are those that are placed low on the roof, generally in the overhangs. Outlet vents are those that are placed high on the roof, generally along the ridge of the roof or at the top of the gables. A combination of inlet and outlet vents assure attic ventilation by wind pressure and/or the "stack" or "chimney" effect if the vents are located properly.

wind pressure

Wind causes a positive (high) pressure on the windward side and a negative (low) pressure on the leeward side of the home. If vents are located on these sides of the home, the attic will vent.

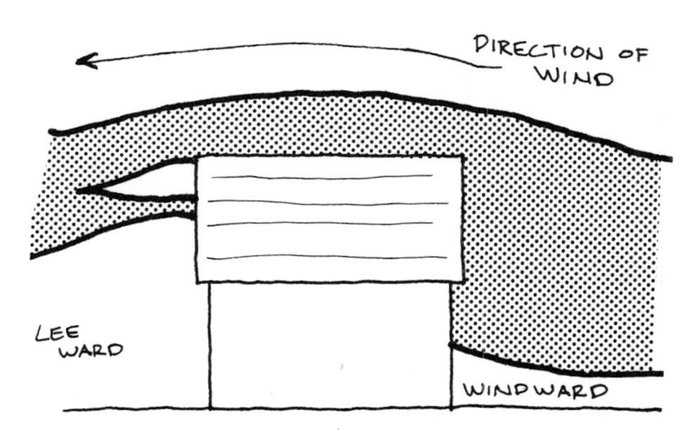

stack effect

This is airflow caused by the difference in elevation between the lowest and the highest vent openings. As the warm air rises through the outlet vents, relatively cooler air is drawn into the attic through the inlet vents.

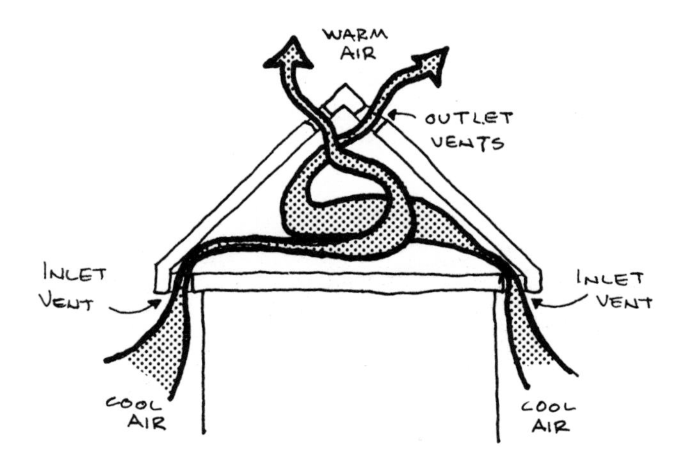

NOTE:

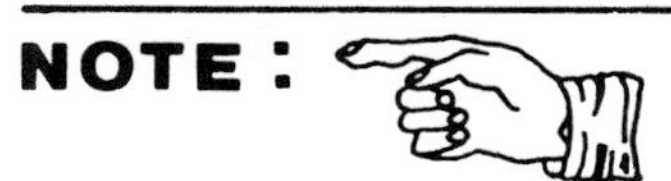

Vents which are installed may also be used for accessing the attic if access is not otherwise available. This may mean that vent opening area could be greater on one side of the attic than another or high openings may have more area than low openings.

Roofs which have excess vent area in one location must have an adequate number and area of vents in other locations as though there were only minimum vent area where the excess of vent area exists.

As a rule, regardless of the area and location of existing vents or the area and placement of vents which will be used for access, there should be an "even " distribution of vents.

TYPES OF ATTIC VENTS

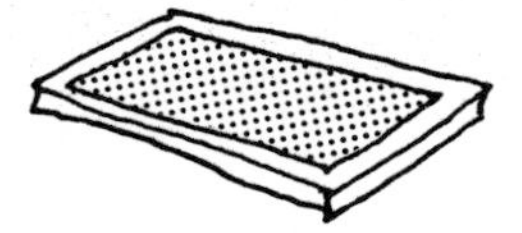

eave vents

There are a variety of eave vents of which a "framed" vent is recommended for retrofitting. Space eave vents one every twenty feet or so and at least one for each face of the building. Eave vents (frame and continuous) are considered as "low" or inlet vents.

FRAME VENT

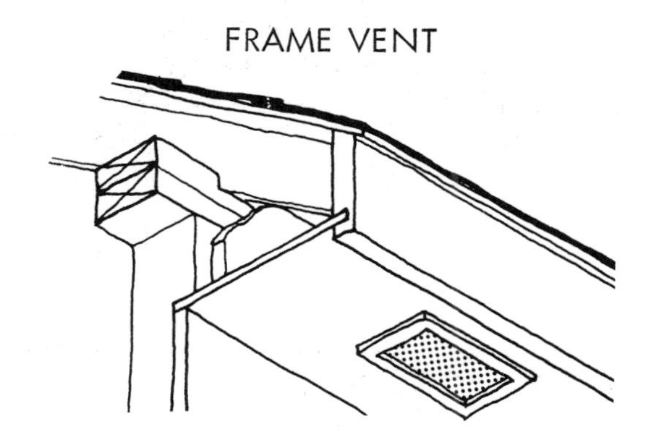

CONTINUOUS VENT

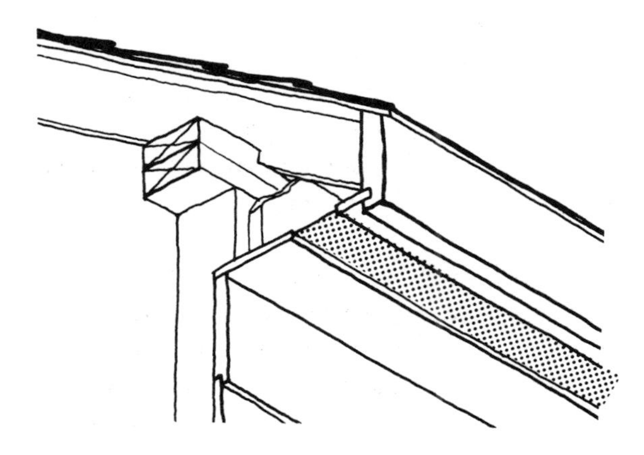

SPECIAL CONDITIONS

The eave line may have different conditions which call for different approaches to vent installation. Most eave lines will have a soffit into which vent holes and screen can be set. After making an opening in the soffit, make sure that an air passage into the attic exists from the soffit/eave enclosure.

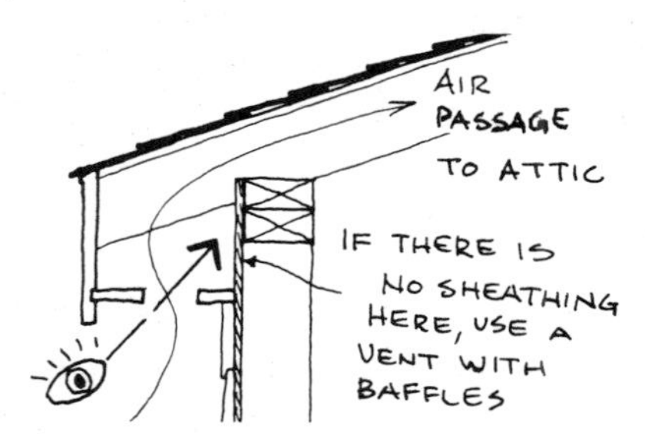

Some eaves don't have overhanging rafters. Drill a test opening as before. If an air passage can be obtained, install vents with deep louvers or baffles.

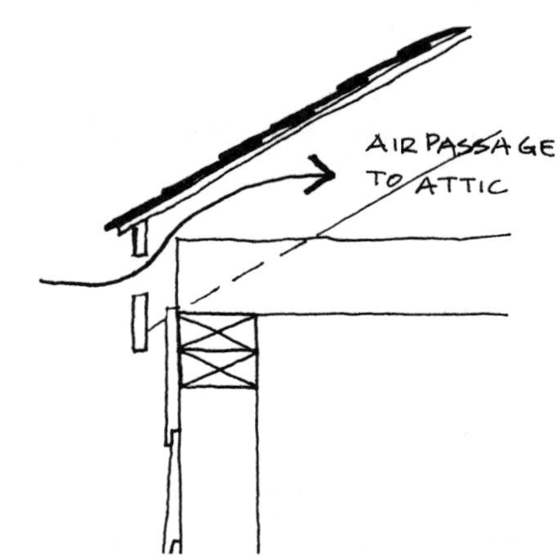

AN EAVE WITHOUT OVERHANGING RAFTERS

Some eaves have a soffit attached to the rafters. Open this soffit the same way as for previous conditions.

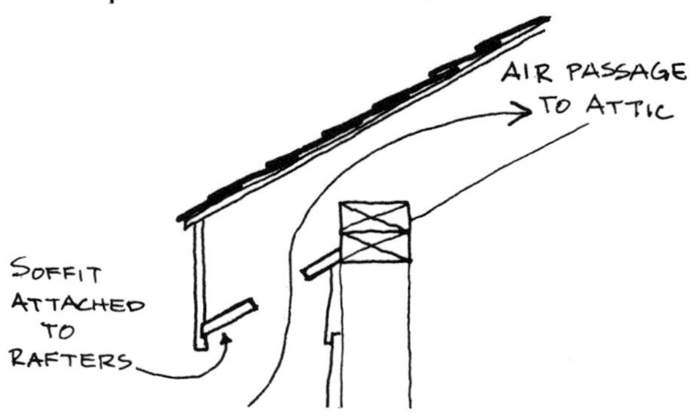

Round vents can be installed in the soffit.

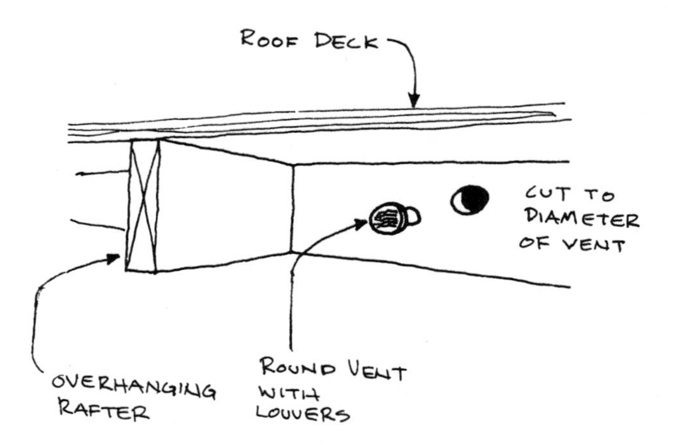

Some eaves have siding which runs up between the protruding rafters. Round vents can be installed in this siding.

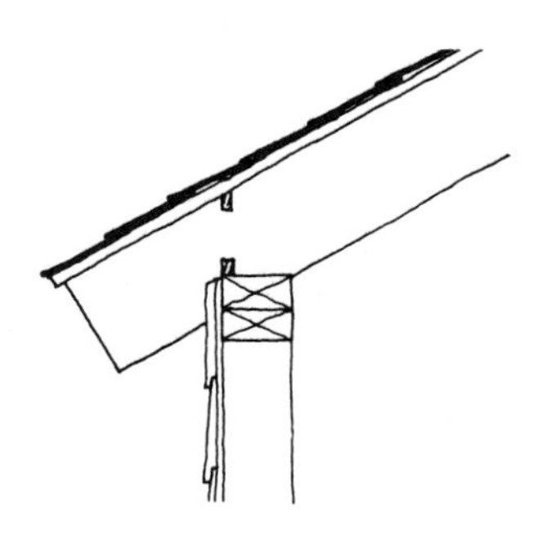

roof vents

Roof vents are considered as "high" or outlet vents. The roof vent should be weatherproofed to keep rain and snow out of the attic. The higher the roof vent is placed on the roof, the better the ventilation due to the "stack"* effect.

STANDING VENT

Space the same as for eave vents. If a standing vent is located <u>less than half</u> way up the roof, it can be considered a low vent.

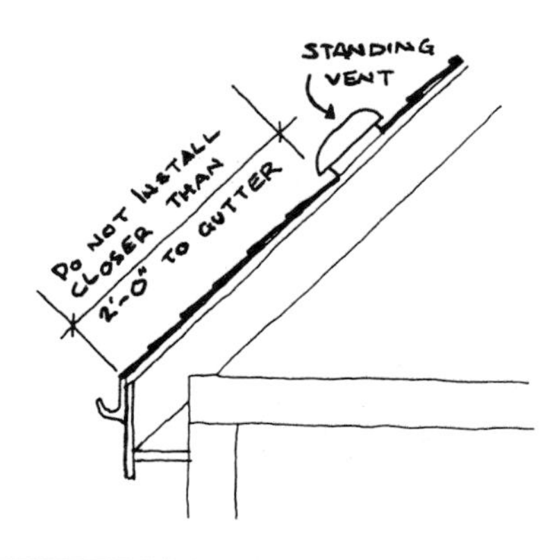

* Stack effect is the movement of lighter air upwards with its volume replaced by cooler heavier air. Stack effect can be used to maintain an air circulation pattern.

TURBINE VENT

A slight breeze will cause the blades in this vent to spin, drawing air out of the attic. On a warm day, you may notice the blades spinning without an apparent breeze. This is caused by the "stack" effect within the attic. This vent will usually not get covered by snow assuring attic ventilation during the winter.

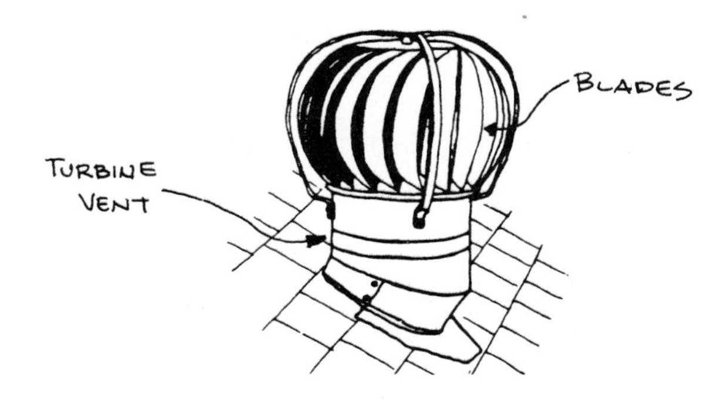

RIDGE VENT

This vent may be difficult to install for retrofit work. However, where there is a need to vent between each rafter, a ridge vent may be the best solution.

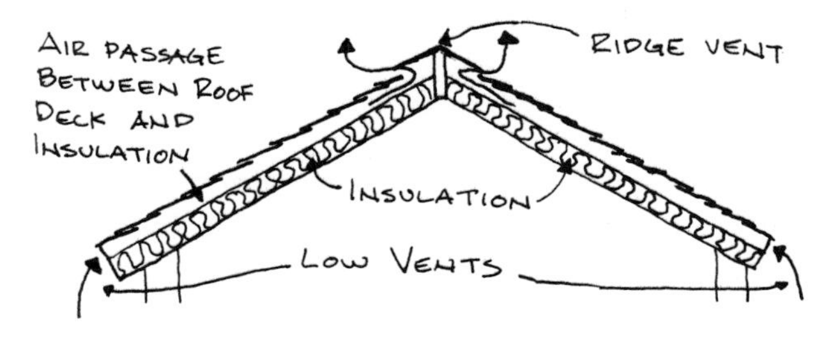

Building with "cathedral" ceiling or "post and beam" construction.

gable vents

Gable vents are also considered as "high" vents and will vent the attic by "stack" effect if eave vents are also used. A gable vent should be placed in each gable. If this happens to be on the windward and leeward sides of the home, the attic will also vent by wind driven air.

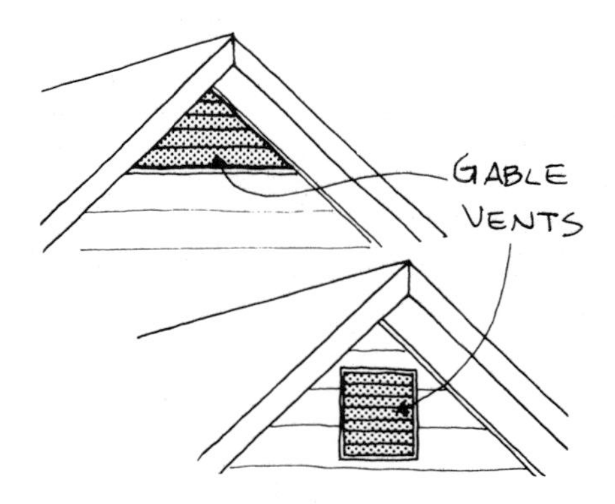

For hipped roofs, install standing vents in place of gable vents.

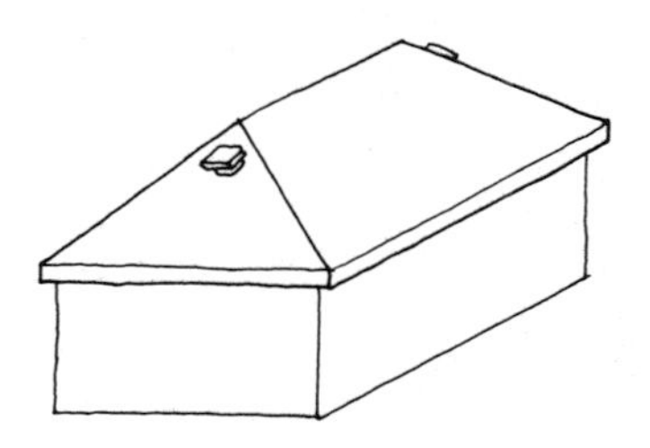

INSTALLING VENTS

eave vent

Locate eave vents between rafters. Drill starter holes and cut the opening with a keyhole or saber saw. Apply a bead of caulk to the frame of the vent before screwing it in place.

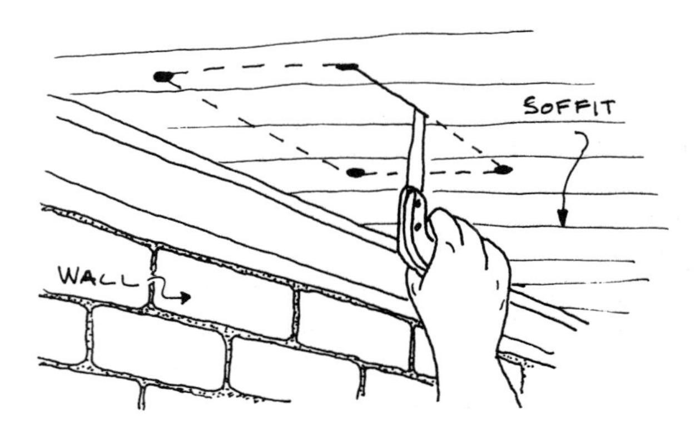

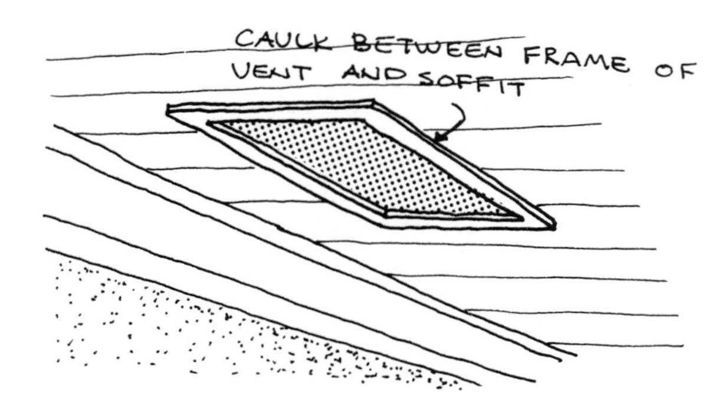

Winds may whip and gust at roof eaves. Be sure to check to see if snow or rain has been forced into the attic, soffit space, or perimeter wall. Try smaller mesh screen and/or baffles to control this problem.

roof vent

Roof vents are installed by cutting a hole through the roof and nailing the vent and flashing in place. Caulk the nails and flashing to prevent water leakage.

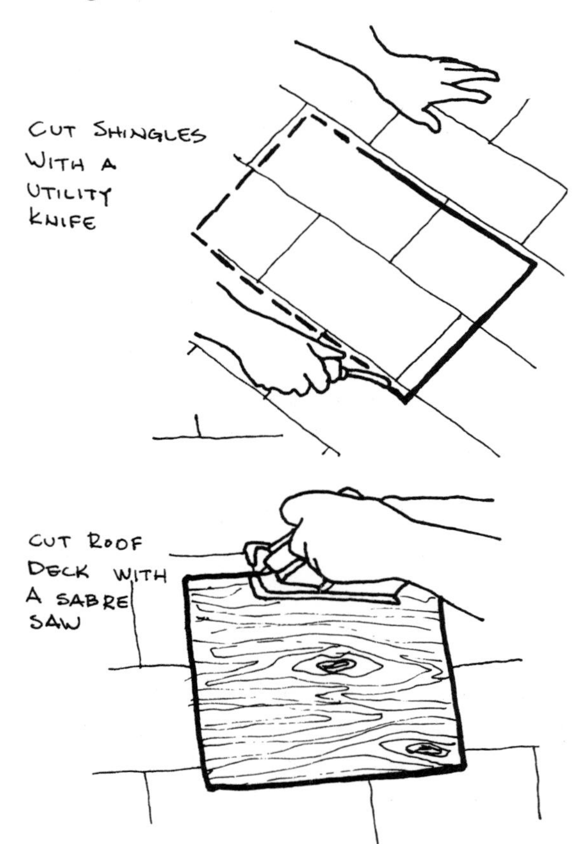

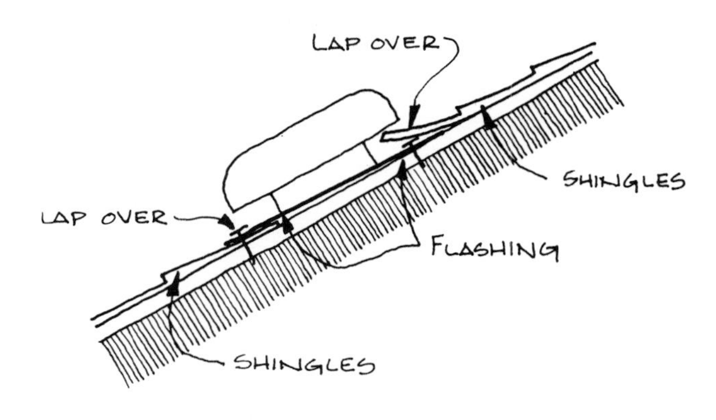

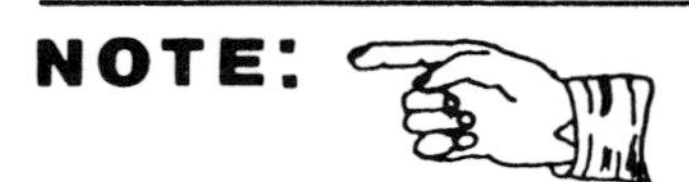

NOTE:

If you're installing a turbine vent, you'll need to level the ventilator. Do this by adjusting the base.

gable

Cut through the siding and sheathing as needed to install gable vents.

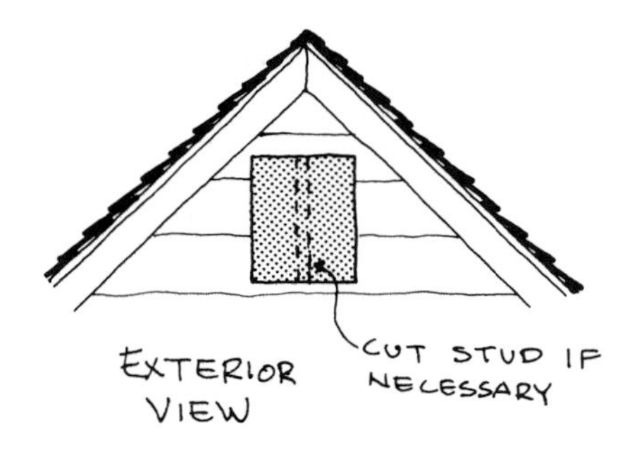

Install a "header" and sill piece spanning between the end wall studs.

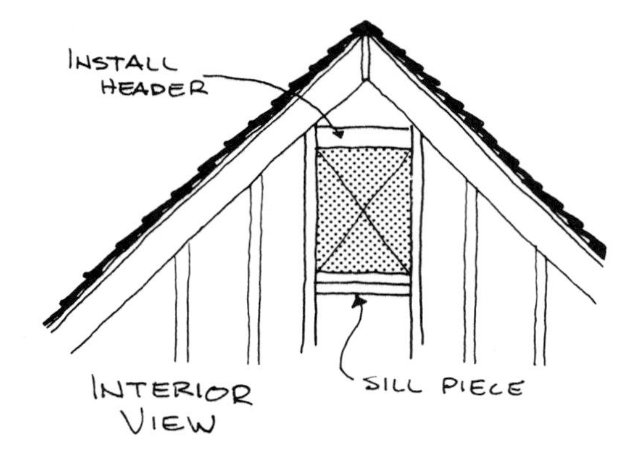

37

insulate finished ceiling against roof

description

What about homes without attics? With no attic, the ceiling finish is attached directly to the roof structural system. This type of ceiling condition is typical in mobile homes. Other examples are homes with cathedral ceilings (as pictured here) or flat roofs.

There are three options for insulating this type of ceiling. The insulation can be added over the roof deck, in the cavity between the ceiling finish and the roof deck (if it exists), or onto the existing ceiling finish.

Insulating over the existing roof deck essentially means adding a new roof.

This involves a high degree of labor skill and is not covered in this guide.

materials

rigid insulation

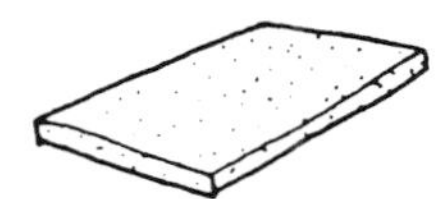

- polystyrene
- polyurethane
- glass fiber

roll/batt insulation

- glass fiber
- rock wool
  (both available as foil faced, kraft faced, unfaced, or reverse flange)

blown in place insulation

- cellulose
- glass fiber
- rock wool
- vermiculite
- perlite

vapor barrier

- polyethylene film
- aluminized or oil based enamel or latex vapor barrier paint
- foil faced insulation
- asphalt impregnated kraft paper
- foil faced gypsum board

other materials

- quarter round or other appropriate molding
- panel adhesive
- vents

preparation

1 Cathedral ceilings are ceilings that are finished against the roof deck and the structural system of the roof is open to the living area (you can see the beams). Since there is no cavity between the ceiling finish and the roof deck, the insulation must be added over the existing roofing or to the underside of the roof deck.

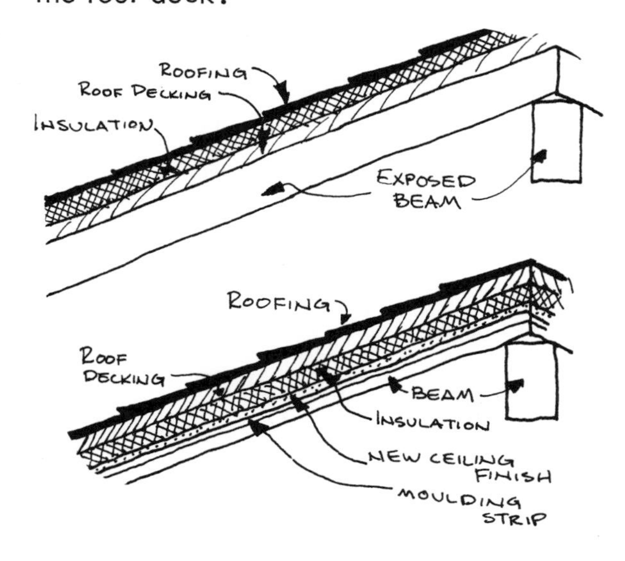

NOTE:

Roofing must be regarded as a vapor barrier. If insulation is installed against the bottom of the roof deck, moisture may accumulate in the decking which is next to the roofing and on the "cold" side of the insulation. Since the structural beams can allow moisture to pass through, they must be painted with a vapor retardant paint before installation.

2 Ceilings finished against the roof with a cavity between the ceiling finish and the roof deck can be insulated by blowing insulation into this rafter cavity. This can be done by removing the facia board and/or drilling holes through the interior ceiling finish.

These types of ceilings can also be insulated in the manner described previously for cathedral ceilings.

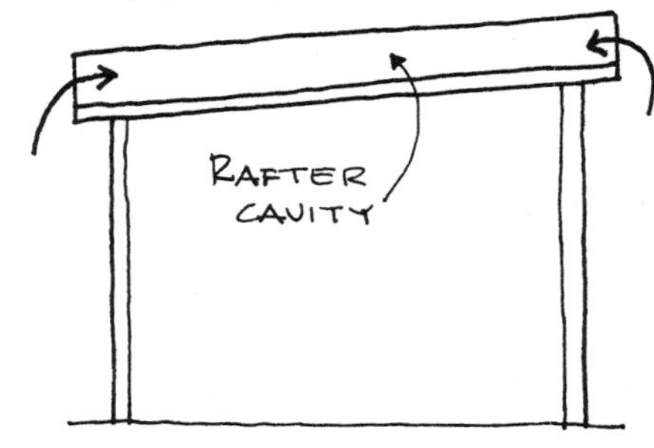

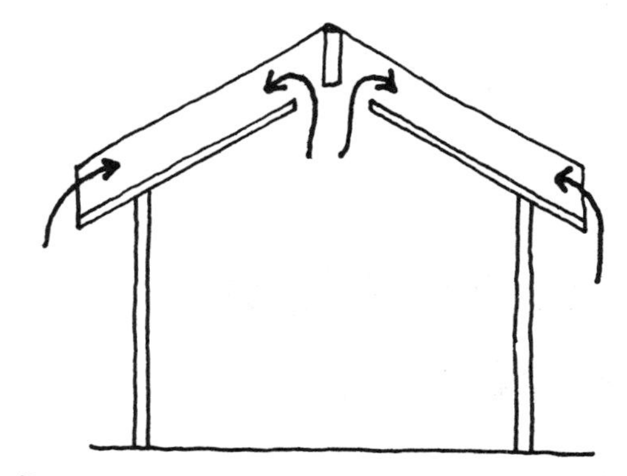

DRILL HOLES THROUGH INTERIOR FINISH AND REMOVE FACIA BOARDS TO ACCESS RAFTER CAVITY

3 It is preferable to provide venting of moisture in this cavity from an air layer above the insulation and below the roof deck. A ridge (for pitched roofs) or "strip" vent (for flat roofs) should be installed to provide venting.*

*If your area is less than 6000 degree days and you know that unvented roofs of this type show no evidence of moisture accumulation/damage if the ceiling is coated with vapor retardant paint, then vents are not necessary--although they are still preferred if only because the roof stays much cooler in summer.

This vent must open to all rafter runs. Therefore, install at right angles to the rafters. If the vent opening is cut before insulating, you may find that insulation can be installed from the roof through the opening facilitating access and making for a faster and better quality installation.

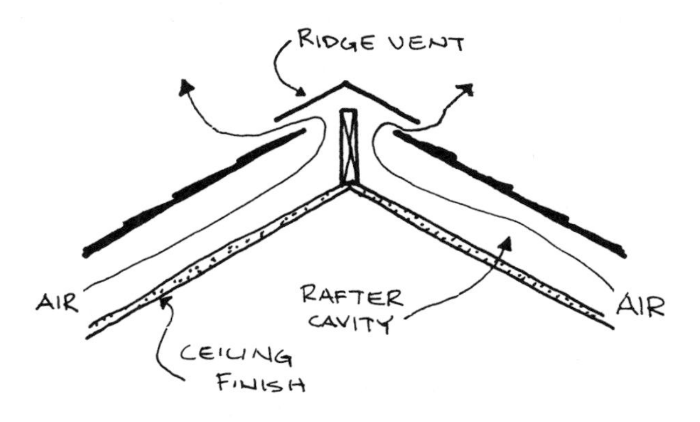

SHADED AREA INDICATES 6000 OR MORE DEGREE DAYS

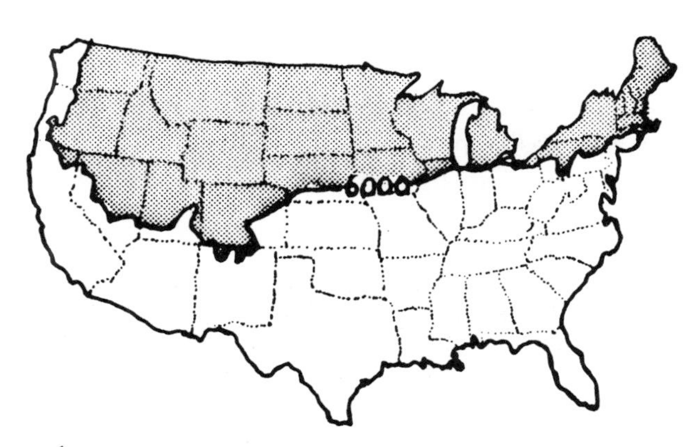

SOURCE: U.S. DEPT. OF COMMERCE CLIMATIC ATLAS OF THE UNITED STATES

4 Sometimes, a home with a ceiling finished against rafters has some space above the ceiling near the center of the home that is enclosed by interior walls and can be accessed through a grille. This space may contain a return air plenum to the furnace which should also be insulated. If you're insulating the rafter cavity, this space might be used to gain access to the rafters near the ridge.

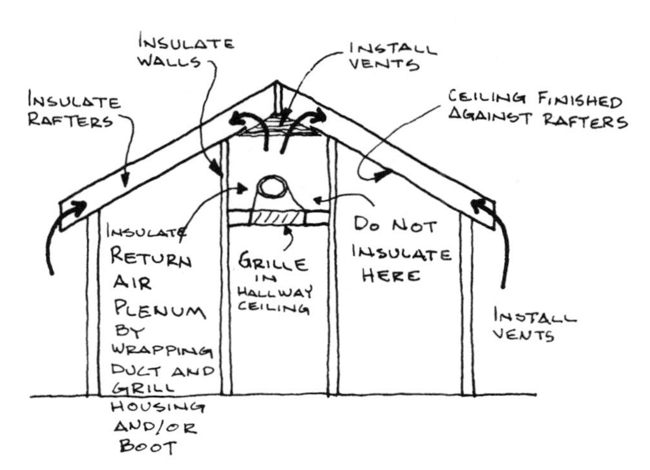

5 Rigid or batt insulation can be used to insulate cathedral ceilings by installing the insulation directly to the ceiling finish. If these types of insulation are used on ceilings where the roof structural system (rafters) are not exposed, you'll first have to add furring strips or framing lumber to the ceiling. In either case, a new ceiling finish is added over the insulation. This finish should be fire resistant -- check with your local building department.

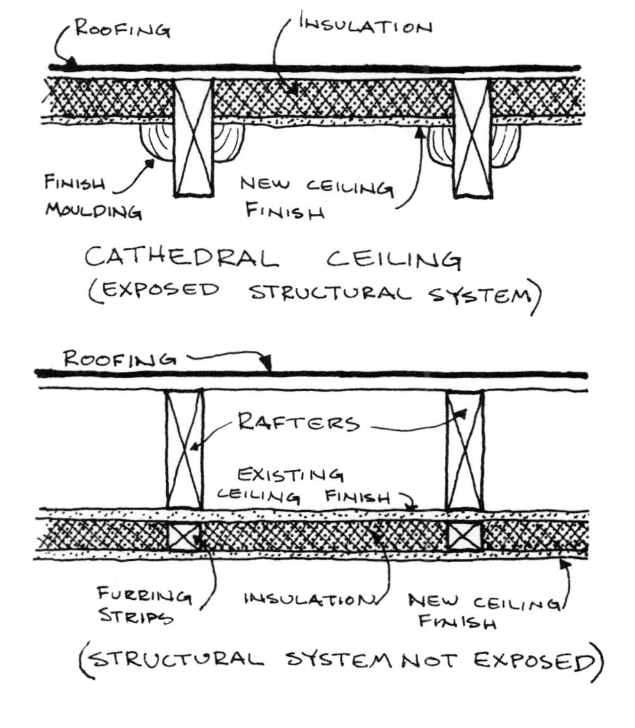

6 Blown in place insulation can be used to insulate ceilings where the rafters are not exposed. However, keep in mind that you should provide a route for air ventilation on the "cold" side of the insulation. This may be difficult to do if the insulation "runs down" or settles at the bottom of the rafter cavity.

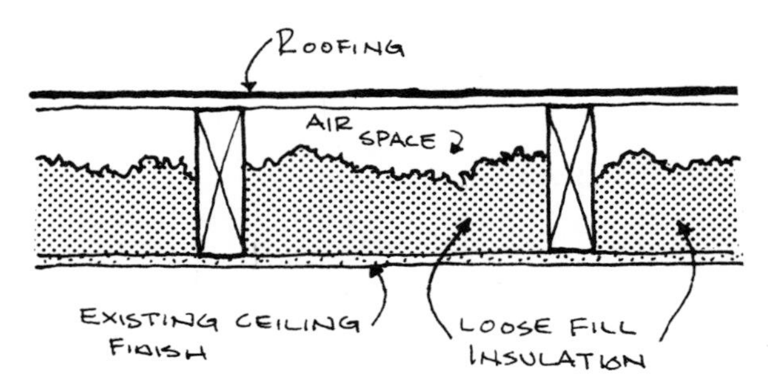

installation procedures

RIGID INSULATION

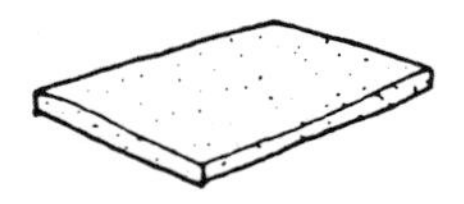

1 Cut rigid insulation to fit between the rafters, roof beams, or furring strips. Apply dabs of panel adhesive to the board and firmly press it in place against the roof deck.

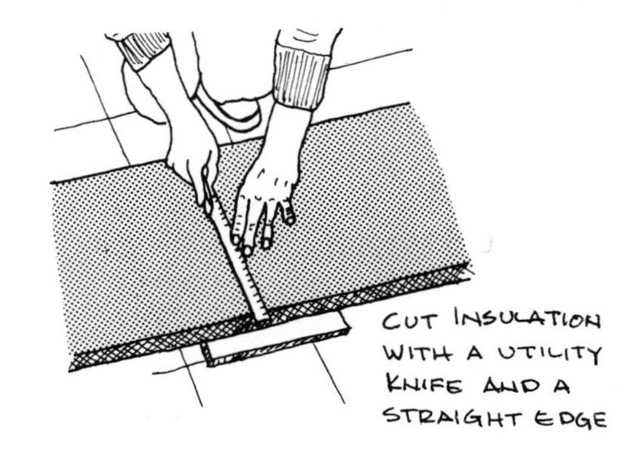

3 Fasten quarter round molding or "edge channels" for wall board to the rafters to hold the gypsum board. Then seal the joint between the perimeter of the gypsum board and rafters. Finish molding pieces (with paint, wood stain, varnish, etc.) before nailing them in place.

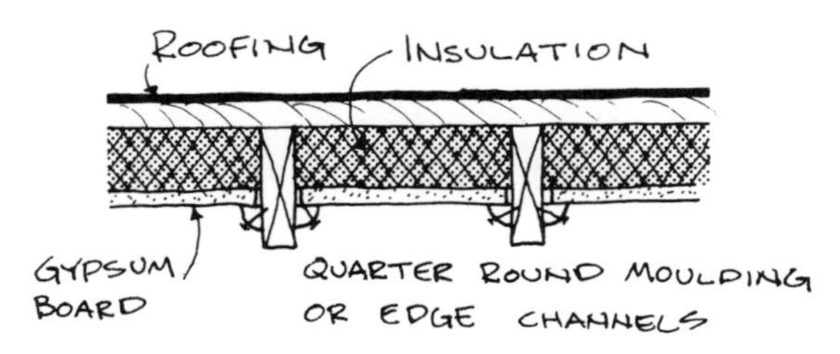

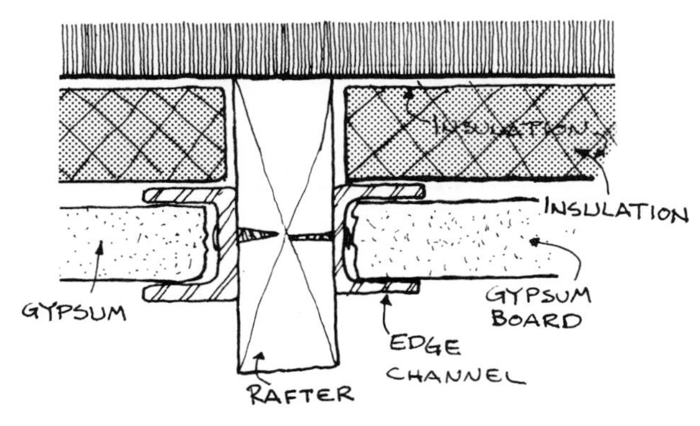

2 Fasten gypsum board with a "15 or 30" minute minimum flame spread rating between beams to cover the exposed insulation. If foil backed gypsum board is used, the foil is to face the insulation. If foil backed gypsum board is not used and vapor protection is required, paint the gypsum board with a vapor retardant paint.

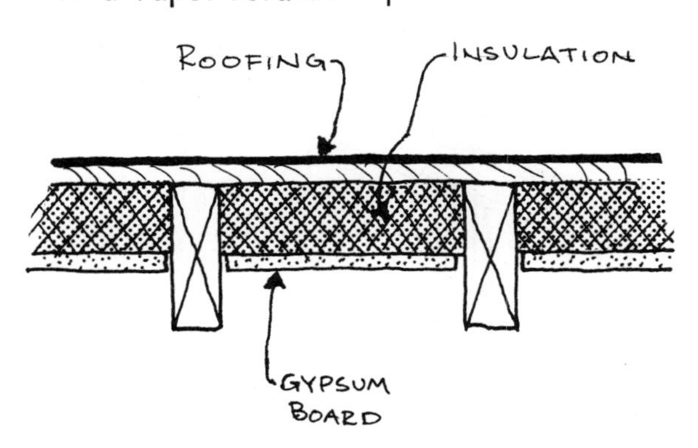

BATT INSULATION

1 Fasten lumber blocking with exposed (downward) edge a distance equal to or greater than the thickness of the batt insulation that you'll be using to insulate the ceiling.

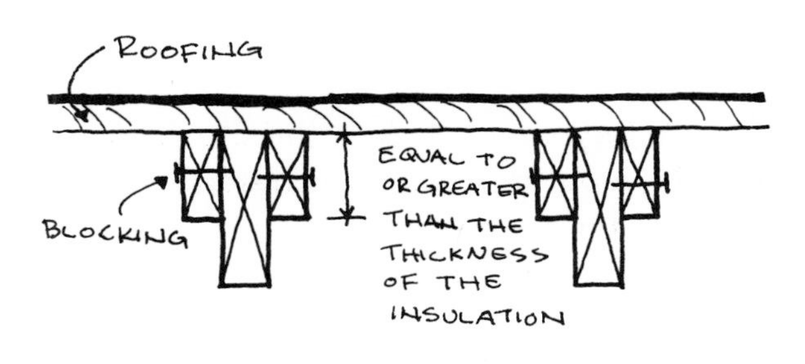

(EXPOSED STRUCTURAL SYSTEM)

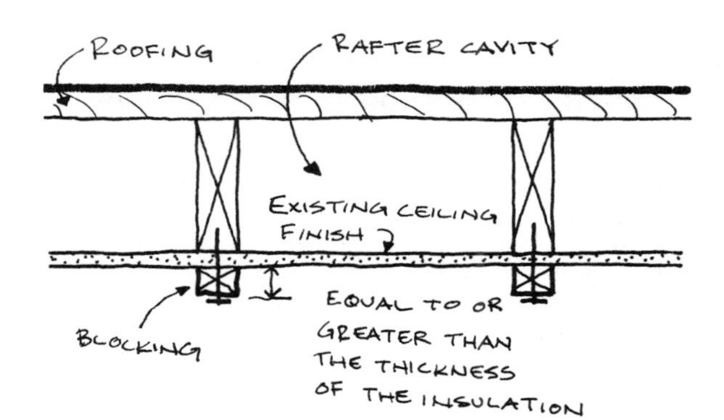

(STRUCTURAL SYSTEM NOT EXPOSED)

2 Install faced insulation by stapling the flanges to the blocking. If the insulation is enclosed on both sides, you can use an adhesive to fasten the insulation to the roof deck.

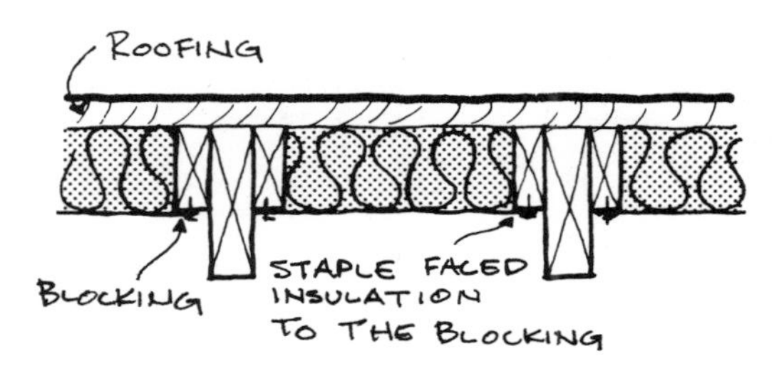

Unfaced insulation can be held in place by stapling polyethylene sheeting to the blocking. Tape over the staples with polyethylene tape. With this method, no additional vapor protection is needed.

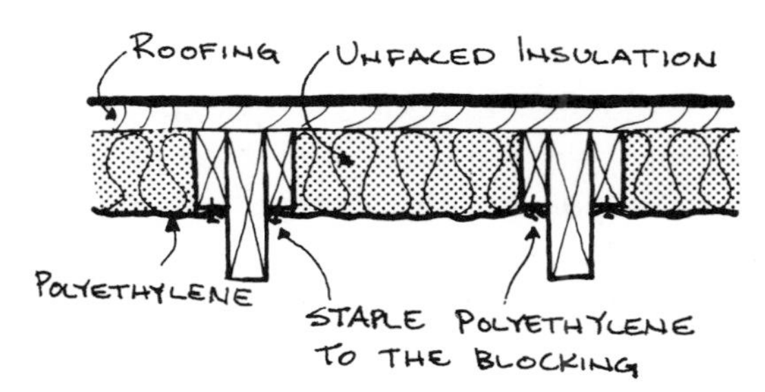

3 Fasten gypsum board to the blocking. Provide vapor protection as described in step 2 for rigid insulation.

Alternatively, use vapor barrier faced insulation as in step 2 and tape over the staples with polyethylene tape.

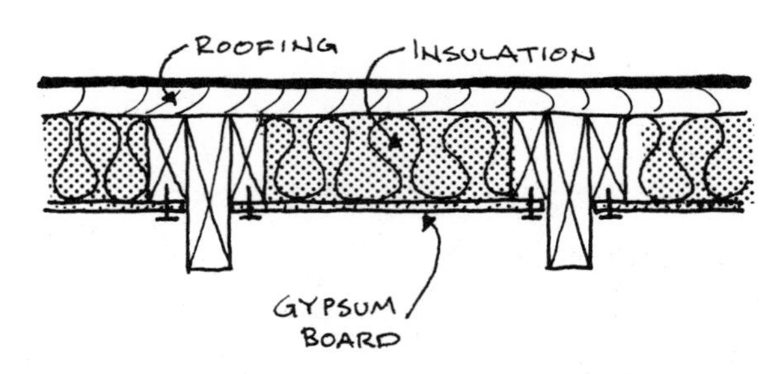

Finish the job with quarter round molding, as described in step 3 for rigid insulation or slip drywall edge channels over the edges for a clean joint.

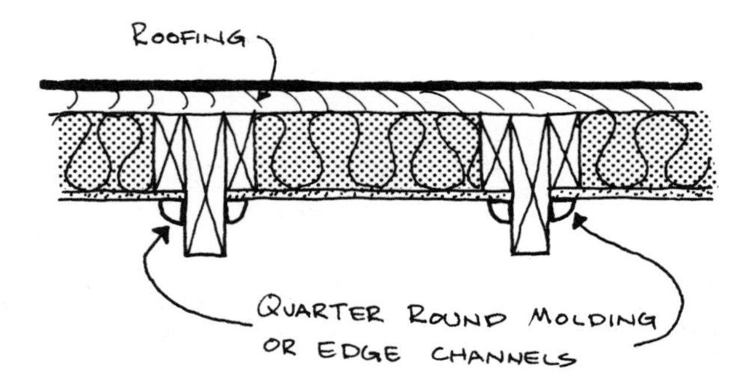

BLOWN IN PLACE INSULATION

1 With blown in place insulation, it is difficult to get a uniform and consistent job while maintaining a 1" air ventilation space between the insulation and the roof deck, especially with pitched roofs. Access holes should be drilled every 5' through the interior ceiling finish along the run of each

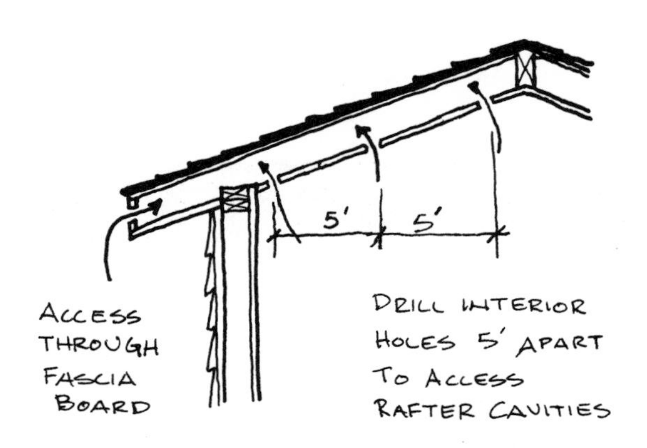

rafter cavity as well as through the facia board between each rafter. Such holes may be necessary even if installing insulation through ridge or strip vents. See Chapter 31, "Insulate Closed Frame Wall", for details on patching these holes.

2 An alternative to step 1 is to remove the facia board and insert the hose to fill the rafter cavity. Start at the top and fill the cavity by withdrawing the hose.

NOTE:

Be very careful when calculating the volume of insulation being installed and the volume of the cavity insulated. Insulation may be installed under reduced pressure or with excess air entrainment so as to create some amount of settling which will provide an air space beneath the roof deck.

3 Or, drill holes through the ceiling finish on both sides of the ridge beam within each rafter cavity. Insert the hose down the cavity and insulate from the bottom. This method in combination with insulating from the facia board (step 2) may work for long spans. See Chapter 33, "Insulate Finished Attic", for patching the ceiling finish that you've removed or drilled near the ridge beam.

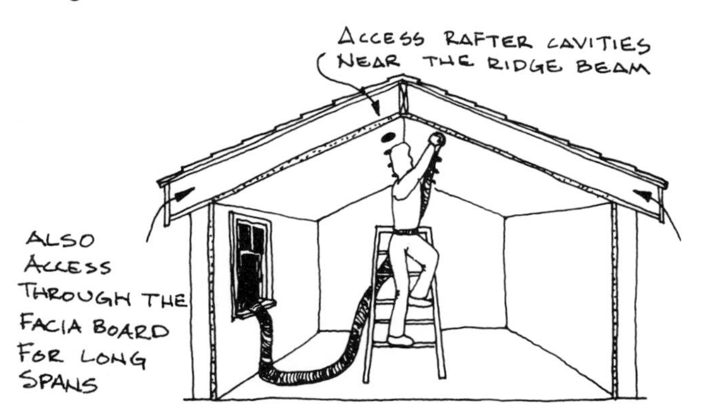

V

hole/crack retrofits

Pack, caulk, and seal structural cracks

Seal air bypasses

Repair structural hole

Seal foundation crawl space vents

Install skirting around building

Insulate exposed ducts and/or pipes

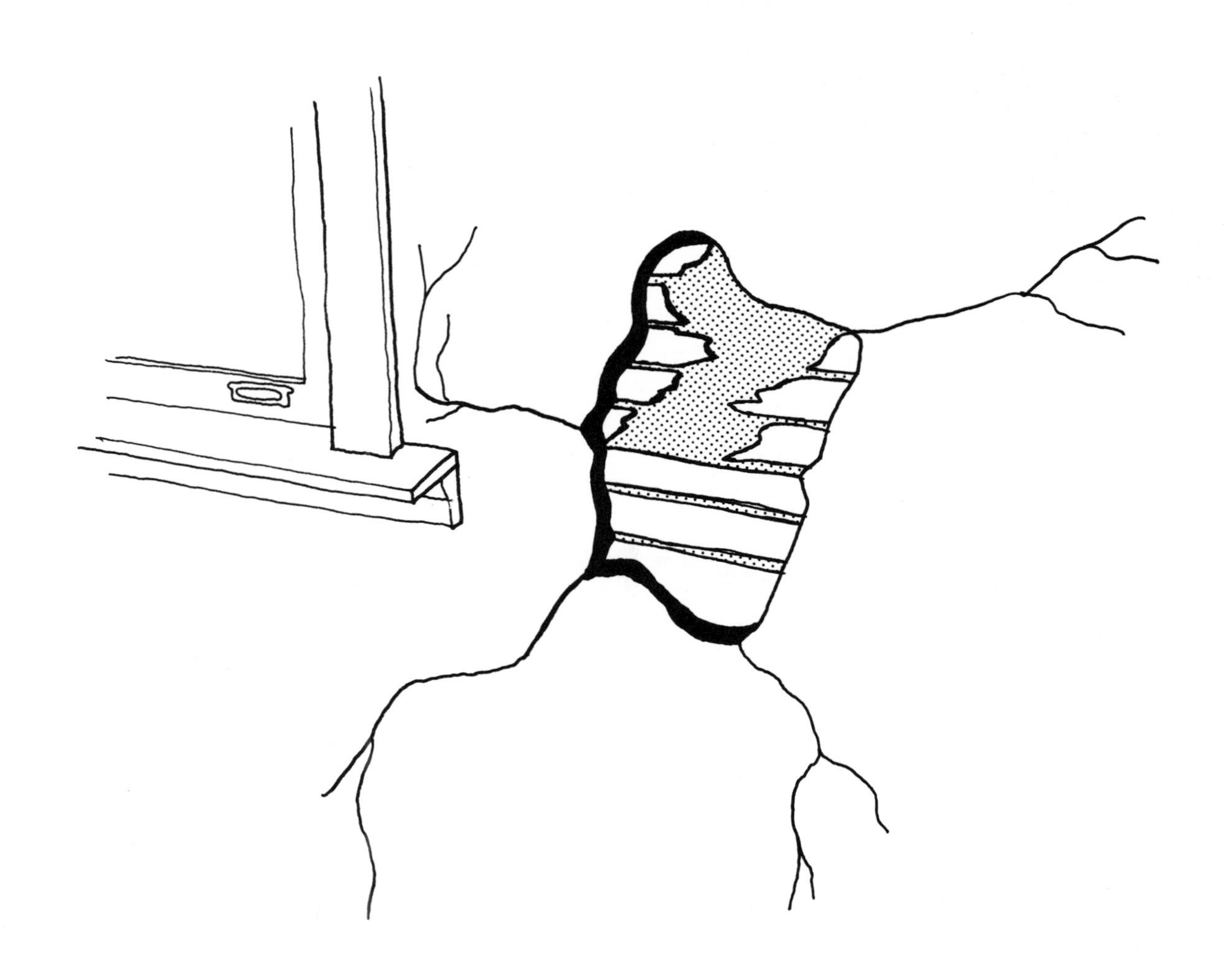

38

pack, caulk, and seal structural cracks

description

Even though windows and doors may have been weatherized (weatherstripped, caulked, storms, etc.), air continues to leak through small cracks in floors, walls, and roofs.

Individually, these small cracks would be insignificant, but collectively, the air leakage through these cracks could make up a significant portion of the air leakage of the home. In fact several studies have shown for existing single family, multifamily, and for mobile homes, that air leakage of windows and doors is seldom more than 20% of the total building air leakage. It is very unlikely that you can make a house "too tight" by sealing and weatherstripping all cracks you find.

Before going further, be sure to read Chapter 5, "A Word about Caulks", to see why caulking is important, where caulking can be used, and the types of caulks available. In addition, see Chapter 10, "Pack and/or Caulk Windows and Doors", for proper caulking installation techniques.

Though caulk is a major material type in this section, you may find cracks in wood siding, stucco, roofs, or masonry work that require sealing with a material other than caulk.

materials

caulk

- oil based caulks
- latex caulks
- solvent based acrylic sealants
- butyl caulks
- polysulfide sealants
- urethane sealants
- silicone sealants
- nitrile rubber
- neoprene
- chlorosulfonated polyethylene (for a detailed description of these caulks, see Chapter 5, "A Word about Caulks".)

packing

- polyethylene (closed cell rod)
- polyurethane (closed cell rod vertical joints only)
- oakum (oil saturated hemp)
- polyethylene bond breaker tape (for joints too shallow for foam rod)

clean solvents

- turpentine
- mineral spirits
- lacquer thinner

other materials

- mortar

- glue

preparation

The importance of preparing the surfaces for caulking must be emphasized. Good adhesion cannot be obtained without proper surface preparation. Preparing the surfaces for some caulks is a tedious process, but is necessary to obtain an effective and long lasting seal. If you feel that stringent preparation requirements will not be met, choose a caulk which does not require much cleaning and surface preparation.

1 Begin by removing all the old caulk. Use a putty knife or screwdriver along with a stiff brush. For dormant cracks, you'll need a gap that is at least 1/4" deep to apply an adequate amount of caulk. Use packing where the width to depth ratio of the crack is greater than 2 to 1. See Chapter 10, "Pack and/or Caulk Windows and Doors", for details.

SCRAPE OUT OLD CAULK AND MORTAR FROM CRACKS AND JOINTS.

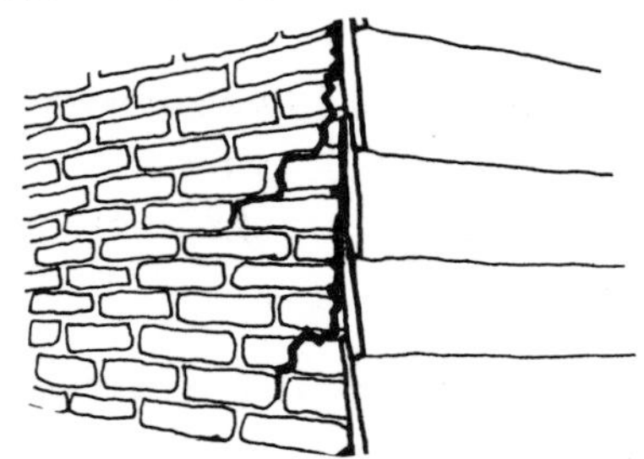

2 Clean wood surfaces of all dirt and loose paint with a stiff brush. Wipe clean with mineral spirits. Aluminum surfaces with protective lacquer should have the lacquer removed with a solvent stronger than mineral spirits, such as xylol, so that the caulk will adhere directly to the metal. Concrete and masonry surfaces which must be cleaned can be done so with a wire brush or by routing.

CLEAN THE CRACKS AND JOINTS WITH A STIFF BRUSH

3 Routed cracks are to be cleaned of all loose material and then rinsed clean with water or masonry cleaner and let dry. Expansion joints should be routed of old sealer to a width and depth of 3/4" to 1" and cleaned as above.

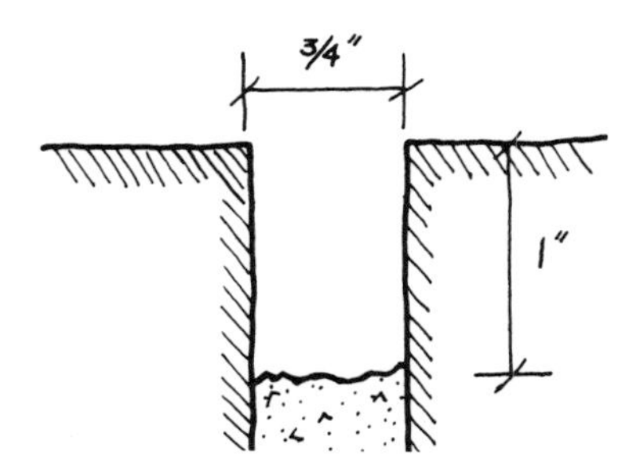

EXPANSION, OR WORKING JOINT

4 Some caulks adhere best to surfaces that have been primed. Paint wood surfaces with primer. Metal surfaces, other than aluminum, should be primed with rust inhibitive primer. Check with the caulk manufacturer's recommendations concerning the necessity of priming.

IF NECESSARY, PRIME THE SURFACES

installation procedures*

1 It would be hard to list all types of cracks and all types of materials you will need to caulk. Generally, the joint formed where two different materials meet is a good candidate for caulk (as pictured on the first page of this chapter).

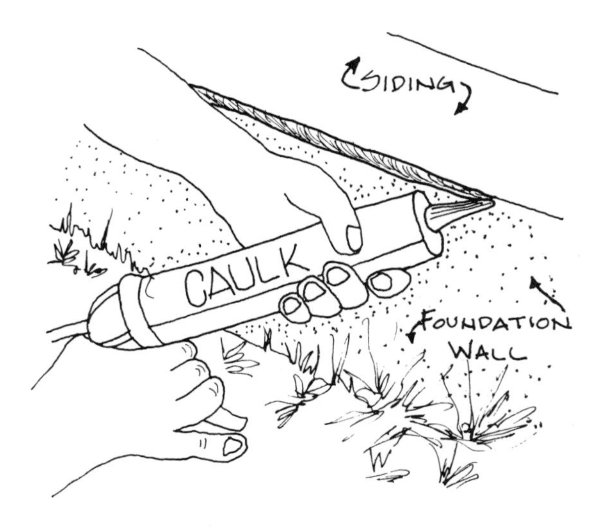

CAULK JOINTS WHERE MATERIALS MEET

*For proper caulking techniques, see Chapter 10, "Pack and/or Caulk Windows and Doors".

2 Other cracks suitable for caulking can be found around openings in walls, floors, and roofs for pipes, wires, and vents. These types of openings may also be found between the living space and unheated spaces, such as basements, enclosed porches, and attics.

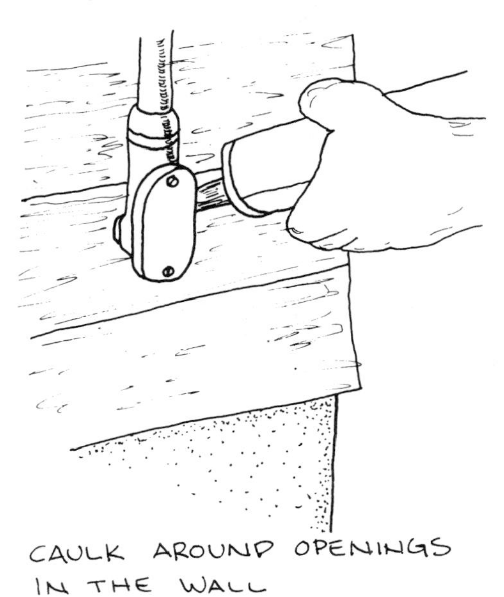

CAULK AROUND OPENINGS IN THE WALL

3 Chapter 10 discusses caulking around exterior window and door frames. This may be ineffective with lap siding as any air flow between siding pieces will bypass this seal. To obtain a "tight" seal around these frames, remove the interior casing and caulk between the frame and rough opening.

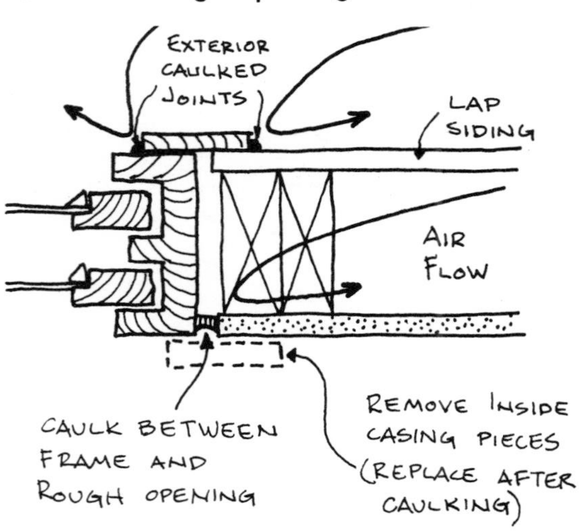

CRACKS IN EXTERIOR FINISHES *

stucco or concrete

4 For cracks less than 1/4" wide, force caulk into the crack until it begins to ooze out. Smooth off the excess while it is still soft to create a neat appearance. For cracks that are wider than 1/4", use patching cement. This cement is available in small bags or boxes and all you have to do is add water. Check the manufacturer's recommendations for the amount of water to add and for what specific purpose the patching cement is intended. For any mortar or cement to set properly it must dry slowly. Wet surfaces as per masonry walls.

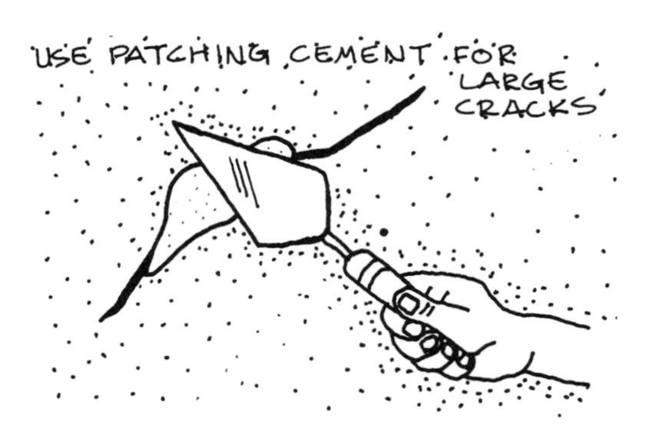

USE PATCHING CEMENT FOR LARGE CRACKS

wood siding

5 Repair cracks in wood siding by prying apart the crack with two or more chisels or screwdrivers spaced along the length of the crack. Using a putty knife, coat the entire length of the crack with glue. Remove the chisels or screwdrivers, and force the crack together

---

*See Chapter 33, "Insulate Finished Attic", for details on patching interior cracks.

by pressing up on the bottom of the board. To keep it closed, nail several galvanized nails at an upward angle along the bottom of the board. Countersink the nails and cover them with putty or wood plastic.

NOTE: 

If there are extensive cracks in the siding, that piece should be removed and replaced. See Chapter 40, "Repair Structural Hole", for details. Also refer to this chapter for replacing wood or asbestos shingles.

roofing

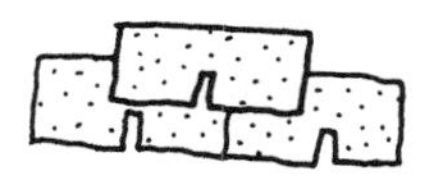

6 Check wherever roofing material meets metal, wood, or masonry for loose or open seams. Also check the flashing around vent pipes, chimneys, and dormers for openings. Seal the joints between roofing and flashing, or where the sealant has dried and cracked, with roofing cement (available in cans or cartridge form).

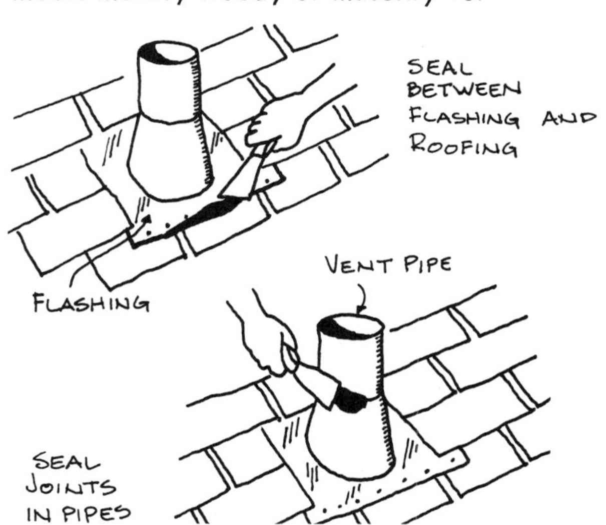

Flashing is usually "seated" in masonry or "tucked under" roofing, siding, parapet coping, etc. Seal between those surfaces and loose flashing, but if flashing is rusted, deteriorated, or is very loose, more extensive roof repair may have to be made.

masonry

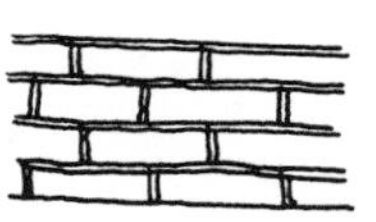

The technique of repairing mortar joints is known as tuck pointing, or repointing. It's relatively easy to do requiring a few basic tools.

7 For cracks less than 1/4" wide, force caulk into the crack until it begins to ooze out. Smooth off the excess while it is still soft to create a neat appearance. For cracks that are wider than 1/4", use patching cement. This cement is available in small bags or boxes and all you have to do is add water. Check the manufacturer's recommendations for the amount of water to add and for what specific purpose the patching cement is intended. For any mortar or cement to set properly it must dry slowly. Wet surfaces as per masonry walls.

SCRAPE OUT THE OLD MORTAR

8 After you clean the joint of mortar, dust, and other loose material, wet the bricks with a hose or splash water on them with a large brush (this keeps the masonry from drawing water out of the fresh mortar). Pack the fresh mortar* into the joints with a small triangular pointing trowel. Place the mortar on a "hawk" and hold the "hawk" beneath the joint to catch the excess mortar. Be sure to pack the mortar in the joints firmly with the tip of your trowel.

*Follow manufacturer's recommendations for mixing mortar.

9 After the mortar begins to stiffen, tool, or finish, it to give it the same appearance as the joints around it.

TYPE OF JOINTS

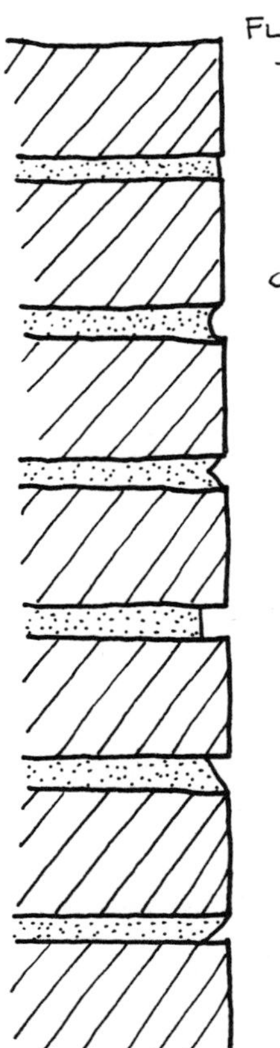

39

seal air bypasses

description

In this chapter, you will be looking for and sealing "bypass" routes. These bypass routes can usually be found in the attic and are simply paths by which warm air escapes from the living space to the attic, bypassing attic insulation. Bypass routes are usually a function of building construction. For example, the sketch on this page shows an attic bypass in a masonry building. The interior finish is fastened to furring strips. Consequently, there is an air space between the finish and the masonry wall that is open to the attic. You can usually find this type of bypass by looking for furring strips that extend up into the attic. As the air behind the interior finish is warmed, it rises into the attic drawing up cooler air from the basement (this is known as a convection current).

Sealing this type of bypass and the other bypasses described in this chapter should be among the first steps that you take to remedy energy use problems in your home and should precede the installation of attic insulation. For example, it was found that despite R-11 insulation in the attic, approximately 35% of winter heat loss was associated with the attic, much of which was attributed to bypasses*.

materials

packing

- closed cell flexible rod (polyethylene, polyurethane)

foam

- one or two part expansion type polyethylene foam

insulation

- roll/batt insulation

caulk

- (see Chapter 5, "A Word about Caulks", for descriptions and properties of caulks)

---

* "Critical Significance of Attics and Basements in the Energy Balance of Twin Rivers Townhouses", Energy and Buildings, Volume 1, Number 3, April 1978; Jan Beyea, Gautam Dutt, and Thomas Woteki; Center for Environmental Studies, Princeton University, Princeton, New Jersey, 08540.

installation procedures

1 Shafts around plumbing and wiring are common bypasses. Frequently, these shafts will extend the full height of the home, from the basement to the attic. Seal this opening with fibrous insulation. Also seal cracks around plumbing and wiring that pass into the attic from inside walls with caulk, foam, or fibrous insulation.

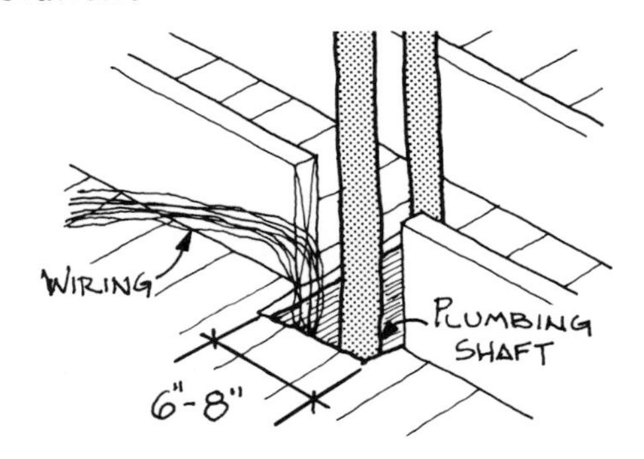

2 Look around chimneys. Any cracks or openings found here should be sealed with non-flammable insulation, such as with pieces of glass fiber insulation.

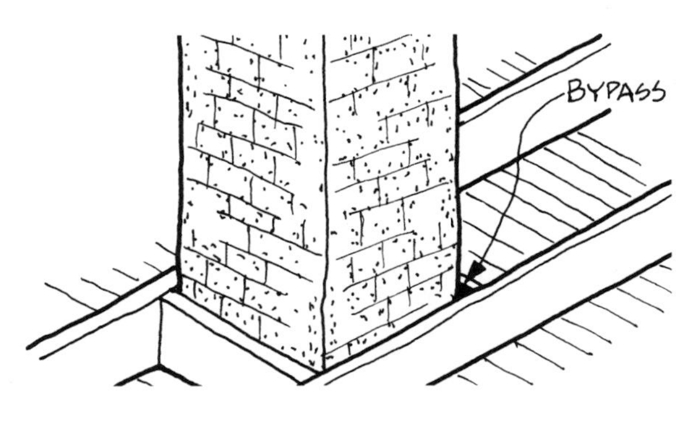

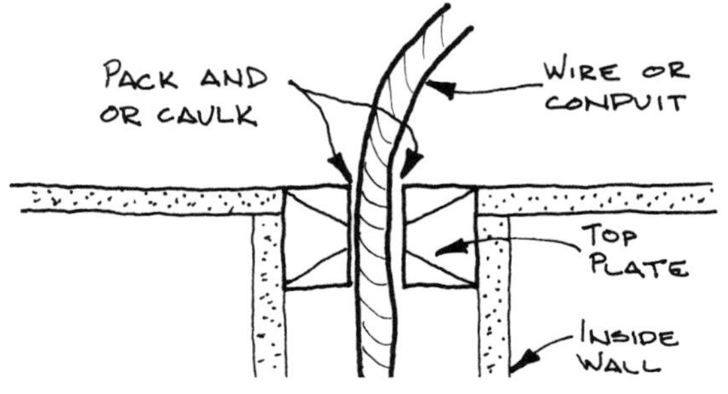

3 Look around the perimeter of balloon framed homes in the attic. Homes that are balloon frame may have stud cavities that open to the attic. If there is no top plate here, these cavities should be sealed with strips of fibrous insulation. This insulation can be packed in tightly from the top of the cavity and should extend down at least 6". The insulation may lose some insulating effectiveness; however, we're more interested in preventing air escape in this situation.

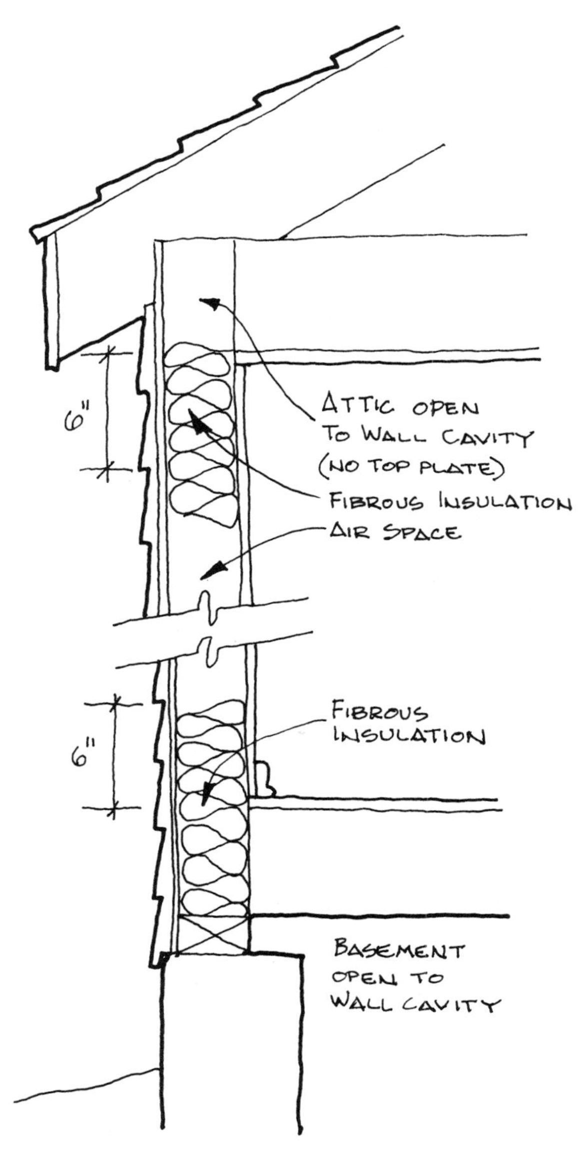

Balloon Framing

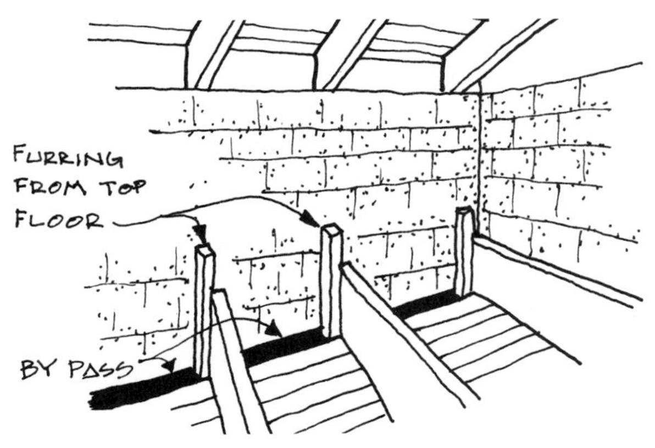

ATTIC

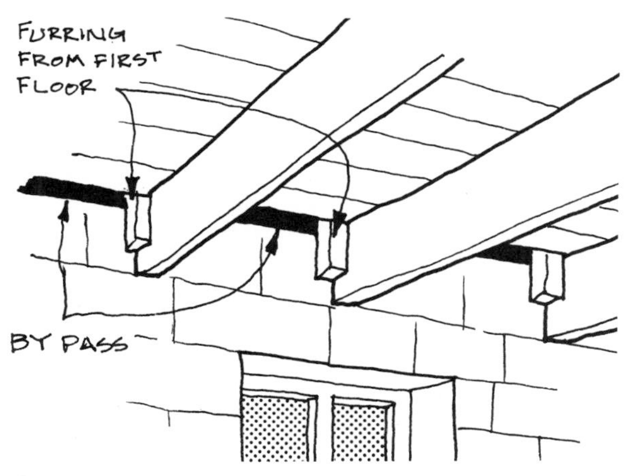

BASEMENT

Masonry buildings, on the other hand, may have furring that extends into the attic with an open fur space. In effect, the building consists of brick walls with "plaster boxes" sitting inside, but not touching the walls. Walls constructed of cinder block that extend into the attic pose another kind of bypass problem. The warm air bypass route is within the cavity of the cinder blocks. Each cavity in these blocks (at the level of the attic floor) must be accessed by knocking a couple of holes in the block and sealed, usually with foam. With infrared equipment, you may "see" this heat coming from the part of the masonry walls that extend into the attic. Unfortunately, there is no easy way to correct this unless the walls are hollow core and can be sealed with foam.

4 Check the tops of interior walls or partitions. These cracks should be sealed with tape or caulk. Some partitions have no top plate; stuff these with fibrous insulation as per perimeter balloon frame walls. Be careful that sealing material does not go into rooms below.

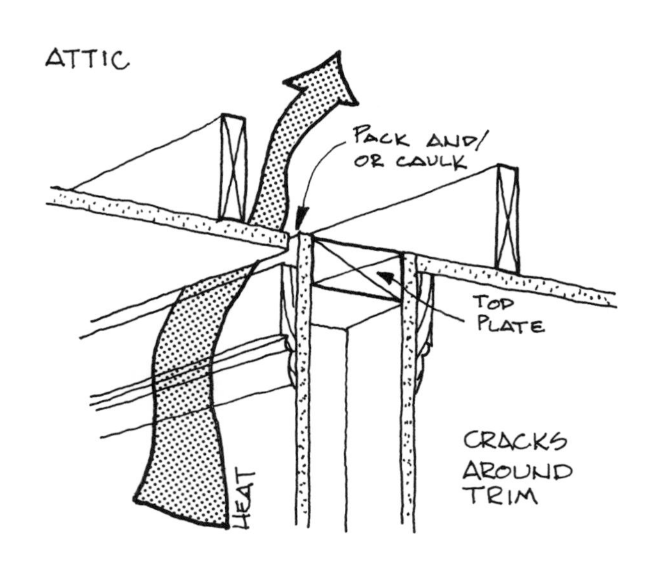

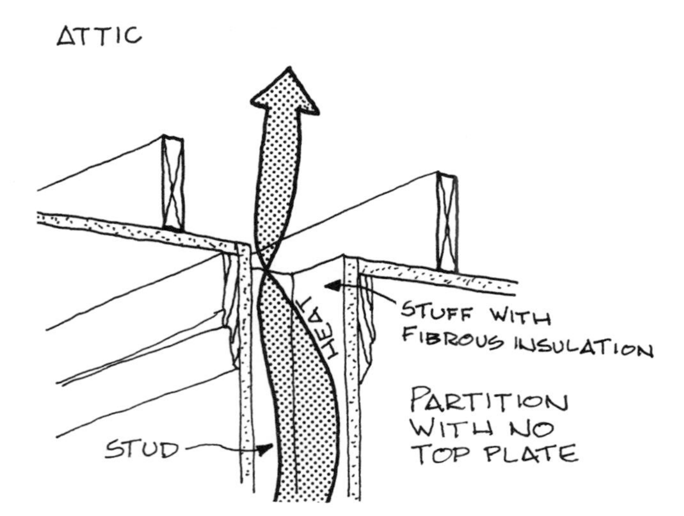

5 Look above staircase ceilings and drop ceilings. These areas can be sealed from the attic by stapling or taping a plastic sheet over the opening (but beneath any insulation on the attic floor).

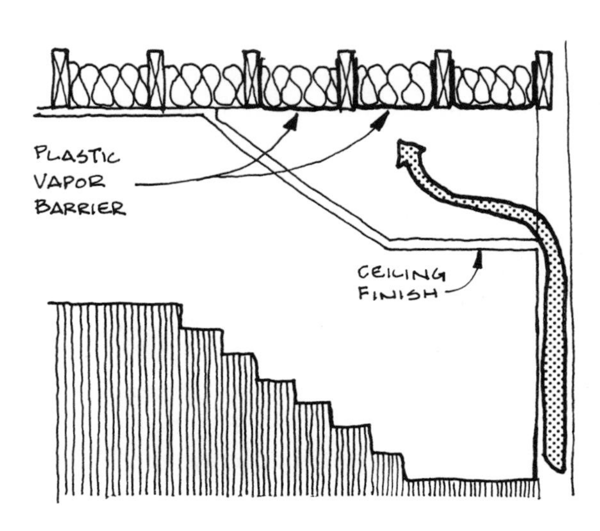

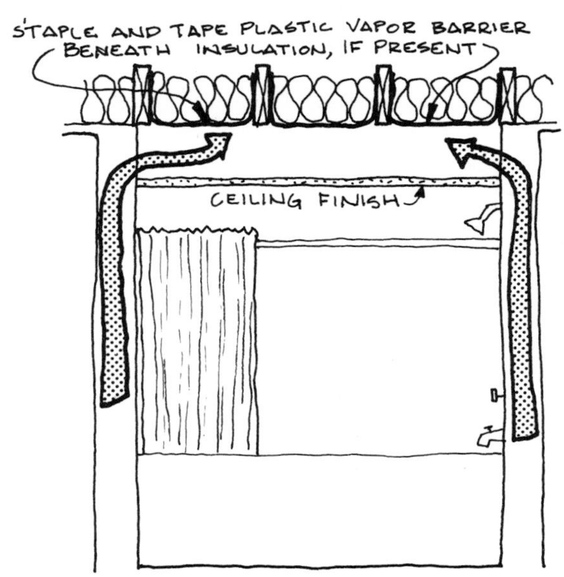

6 Don't forget the attic hatch or entry door. Weatherstripping will seal this leak. If the attic is insulated, the hatch or entry door should also be insulated.

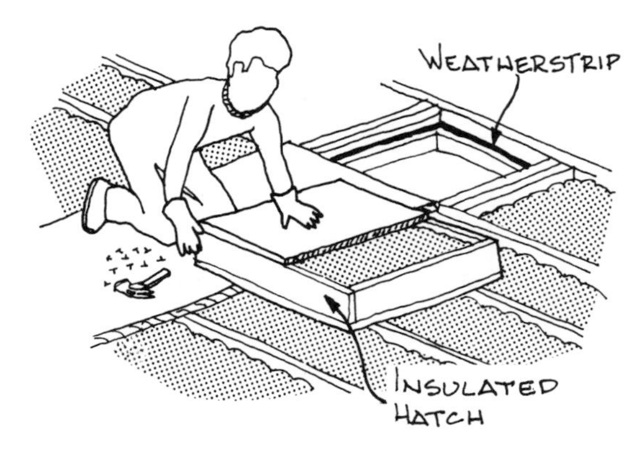

7 Check around light fixtures (except, don't cover the vents of recessed fixture boxes).

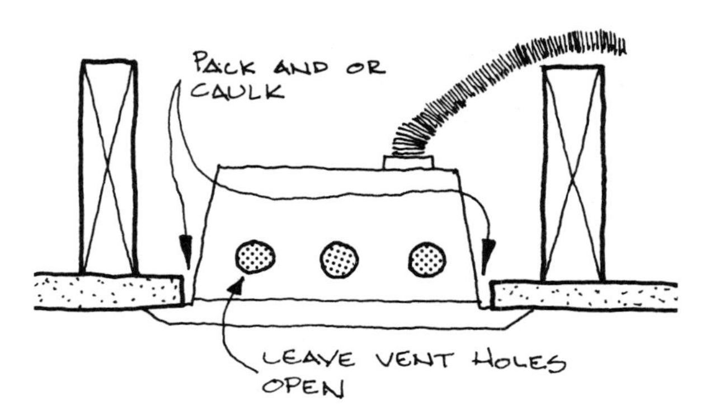

8 If you have a whole-house fan mounted in the ceiling, a tight fitting insulated cover should be installed over the fan during the heating season.

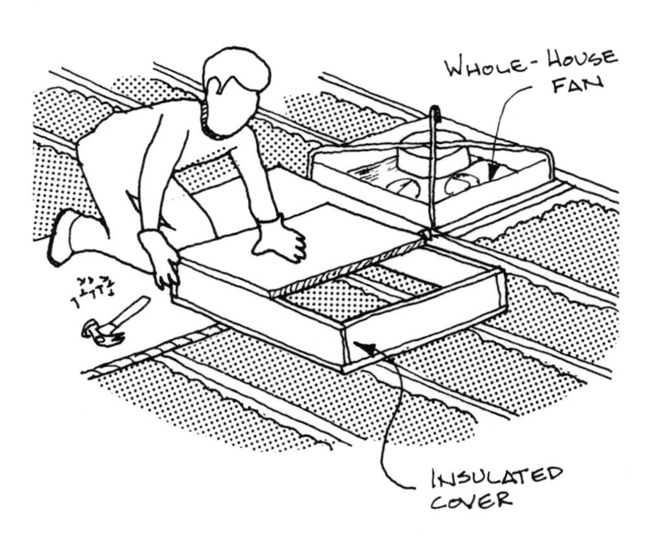

9 Look around ducts that pass into the attic. Seal these cracks with caulk or tape and check the ducts for tightness (tape loose joints with duct tape).

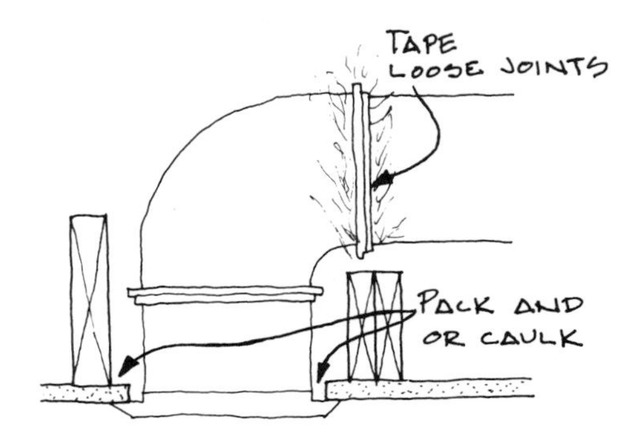

10 Do not seal attic vents! These are required to vent the attic.

40

repair structural hole

description

Exterior structural holes may have started as small structural cracks that were not previously sealed. Cracks allow water to penetrate the finish, hastening the deterioration process of the finish. If water freezes here, it can cause even more damage.

Holes, or extensive cracks, in wood siding, wood shingles, or cement asbestos shingles can be repaired by replacing the damaged piece(s) while missing or severely cracked bricks can be replaced in masonry walls. Patching cement can be used to repair holes in stucco or cement walls (see Chapter 38, "Pack, Caulk, and Seal Structural Cracks", for details).

You also may find structural holes on interior walls. Small holes in gypsum board can easily be patched (see

Chapter 31, "Insulate Closed Frame Wall", for details). For larger holes, however, an entire section of gypsum board may have to be replaced.

Holes that would require structural repairs, such as to wood framing or to steel supporting masonry, are more complex problems which are not covered in this guide.

materials

caulk

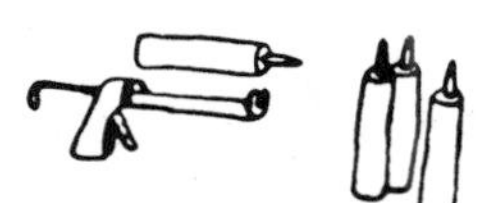

-
- oil based caulks
- latex caulks
- solvent based acrylic sealants
- butyl caulks
- polysulfide sealants
- urethane sealants
- silicone sealants
- nitrile rubber
- neoprene
- chlorosulfonated polyethylene (for a detailed description of these caulks, see Chapter 5, "A Word about Caulks")

packing

- polyethylene (closed cell rod)
- polyurethane (closed cell rod-vertical joints only)
- oakum (oil saturated hemp)
- polyethylene bond breaker tape (for joints too shallow for foam
- rod)

exterior finish materials

- wood siding
- wood shingles
- asbestos shingles
- bricks, mortar

interior finish materials

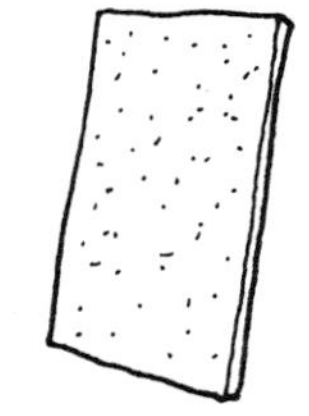

- gypsum board
- gypsum board tape
- gypsum board taping compound
- vapor retardant paint

other materials

- one or two part expansion type
- urethane foam

preparation

When patching structural holes, chances are you'll find cracks that can be sealed with caulk. See the chapters pertaining to caulking* for details.

No special preparation procedures are needed for replacing wood siding and shingles. When replacing brick and gypsum board, however, the bonding surfaces are to be clean, dry, and free of loose material.

---

* "A Word about Caulks" (Chapter 5)
"Pack and/or Caulk Windows and Doors" (Chapter 10)
"Pack, Caulk, and Seal Structural Cracks" (Chapter 38)

installation procedures

WOOD SIDING

1 Pry the damaged board up with a chisel and insert a wedge to the hold the board there.

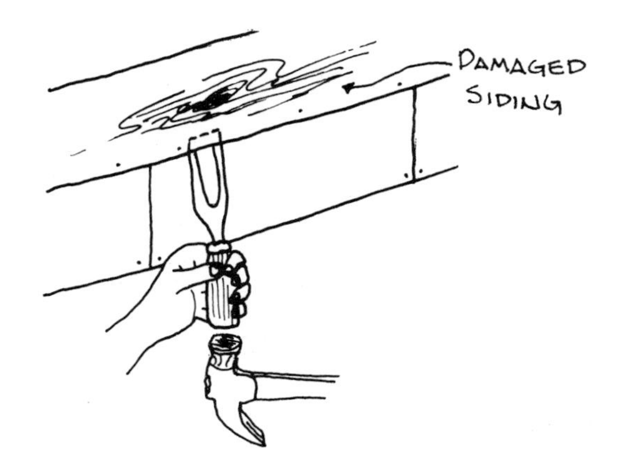

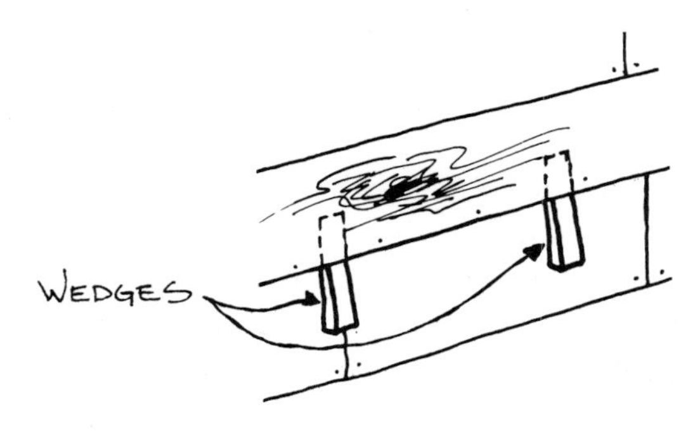

2 If there are nails along the bottom of the damaged board, cut these with a hacksaw blade or push the board back in place and remove the nails with your hammer.

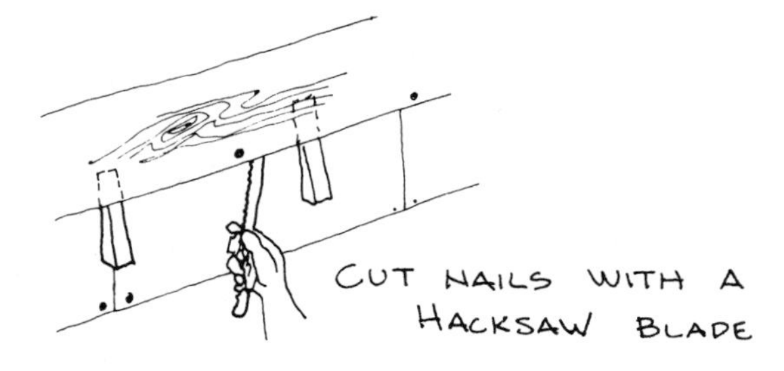

3 Cut out the damaged board by making two vertical cuts with a backsaw.

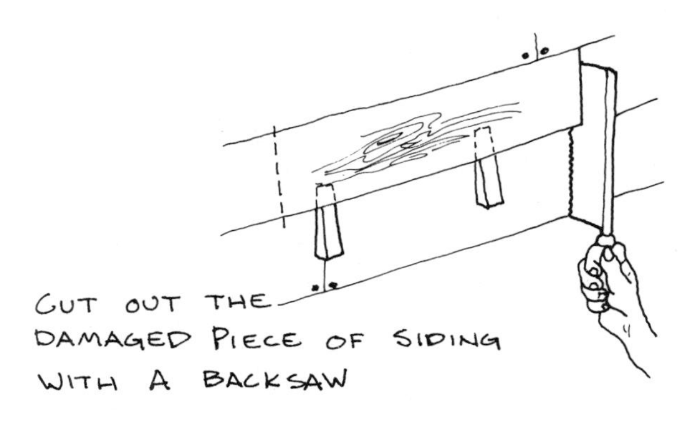

NOTE:

In cases where the finish has been damaged and water may have run into the wall, check for signs of dry rot or other fungus infestation. Infested wood including sheathing and framing should be replaced. This is a more complex problem and is not covered in this guide.

---

4 Insert the replacement piece of siding. The replacement piece should fit snugly between the other pieces of siding. It's a good idea to prime the replacement piece with primer (don't forget to prime the edges).

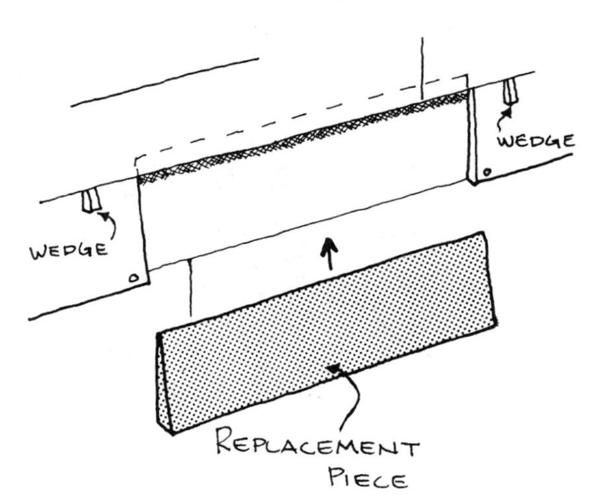

5 After lining up the bottom edge of the board with the boards around it, nail across the bottom edge (be sure to drive these nails through the top edge of the underlying piece and solidly into the framing). Then drive nails along the top (these nails are started through the board above). Finally, paint the replacement piece to match the rest of the siding.

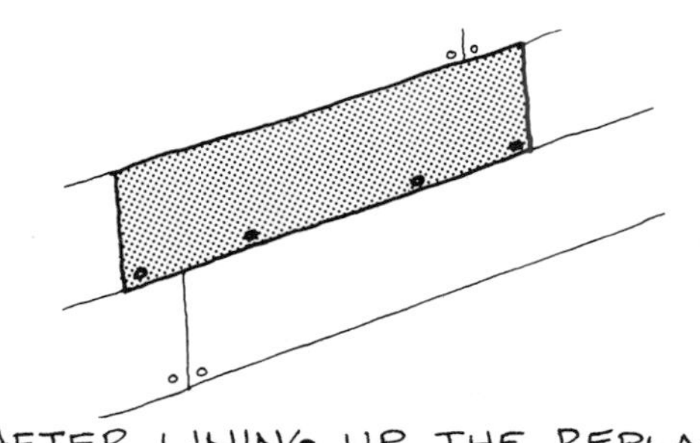

AFTER LINING UP THE REPLACEMENT PIECE, NAIL IT IN PLACE

WOOD OR ASBESTOS SHINGLES

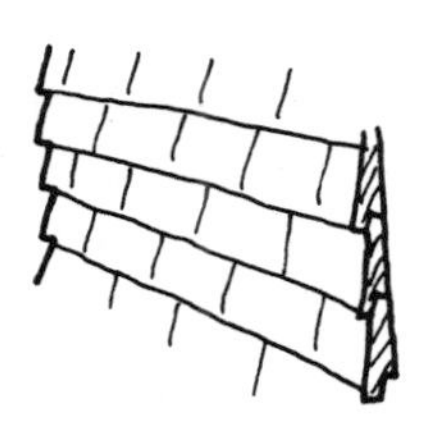

1 Replacing shingles is similar to replacing wood siding, but somewhat easier. Instead of cutting out the damaged piece, pry it out, being careful not to break shingles around it. Pull out any remaining nails.

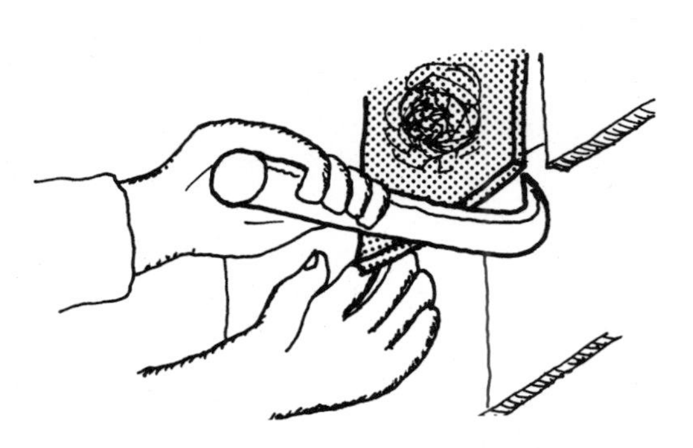

CAREFULLY PRY OUT DAMAGED SHINGLES

2 Slide the new shingle in place and nail the top and bottom as described for wood siding. With asbestos shingles, it may be necessary to pre-drill holes for the nails to avoid shattering the shingle.

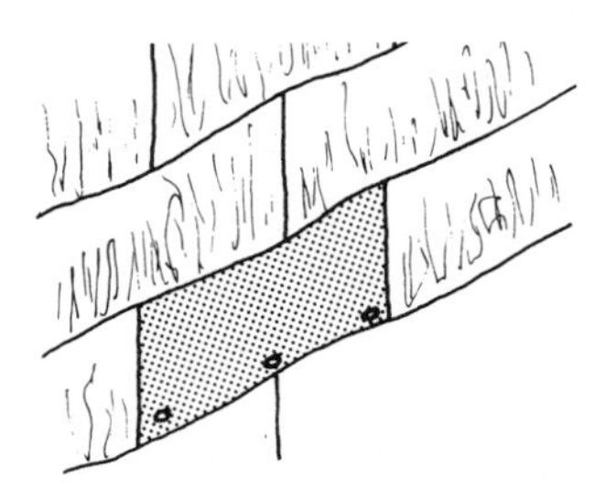

NAIL THE REPLACEMENT PIECE OF SIDING IN PLACE

MASONRY

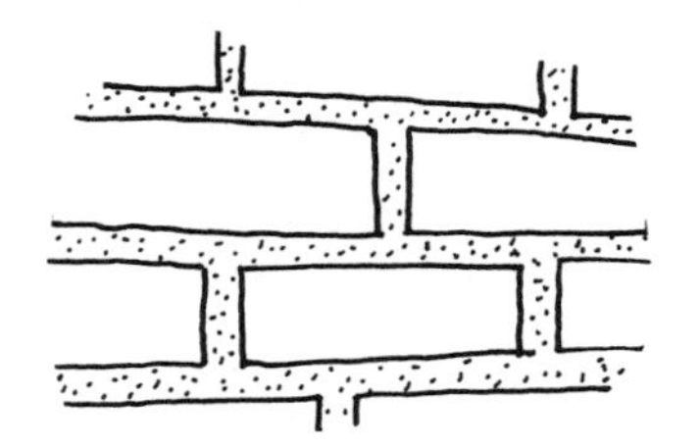

1 To repair or replace loose or missing bricks, you'll need to set it solidly in mortar for a good bond. Chip out all the old mortar, doing it carefully to avoid loosening other bricks. Make sure your mortar joints match existing brick work (see Chapter 38, "Pack, Caulk, and Seal Structural Cracks").

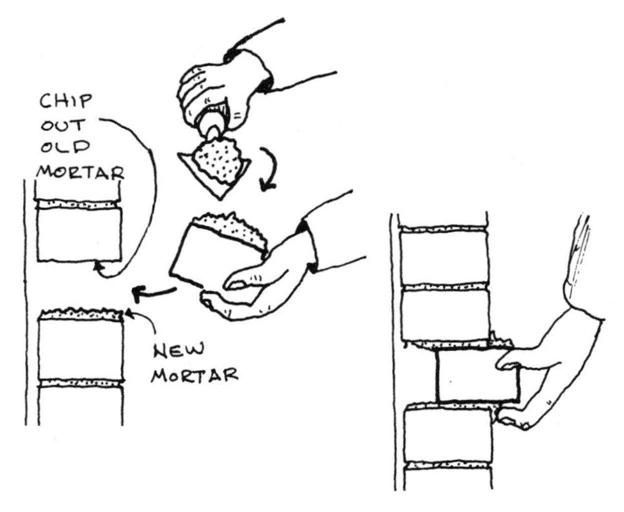

2 If the old mortar appears to be sound (not cracked or crumbling), apply a thin layer of epoxy adhesive to the brick and slide it back in place. Epoxy adhesive bonds well to masonry if the surfaces are clean and dry. Brush all mortar work clean after mortar has been tooled and has set. Use cleaning acid (muriatic acid is commonly used) for stubborn clean-up work.

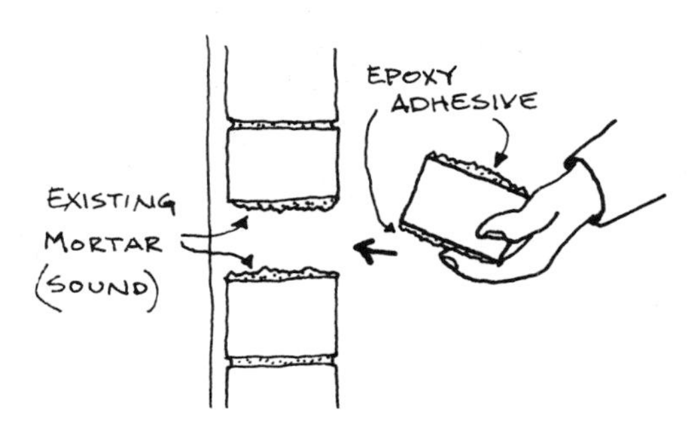

PATCH INTERIOR HOLES *

1 With a keyhole saw, cut out the damaged piece of wallboard. Cut between the studs, exposing 1/2 of each stud face (generally 16" o.c.). The width of the piece should be at least 10" so that the new piece won't buckle when you put it in.

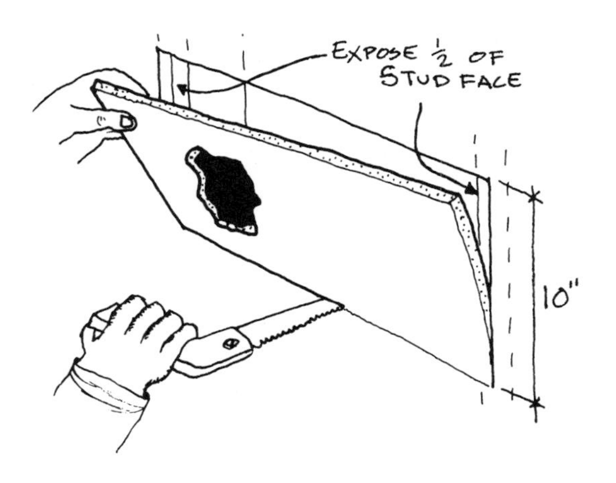

2 Cut the new piece of wallboard to fit in the opening and fasten it by nailing it to the partially exposed studs.

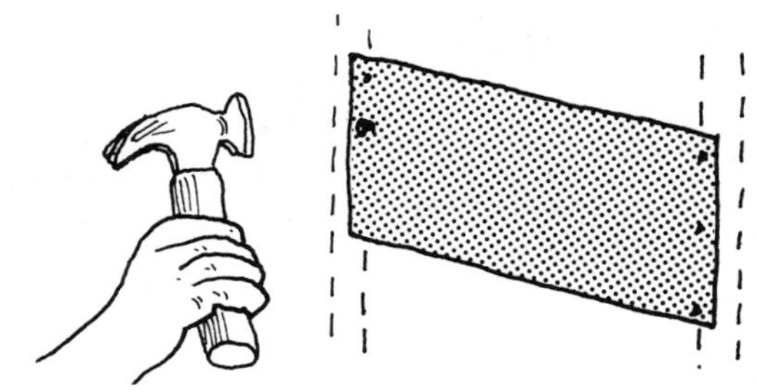

* Applies to walls and ceilings.

3 Seal the crack and nail heads by applying a smooth "bedding" coat of wallboard taping compound over them with a 6" drywall knife. Before the compound dries, apply drywall tape over the compound with an additional thin layer of compound. When this dries, apply a second and third coat of compound (apply this last coat after the second coat has dried). Finally, lightly sand the surface (after it has dried) and then paint to match the interior finish.

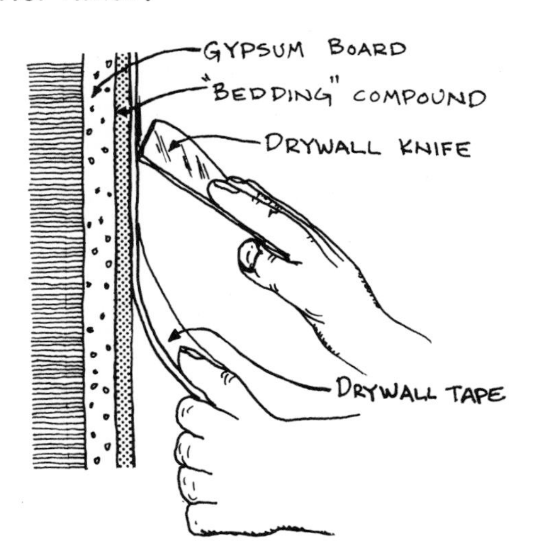

NOTE: 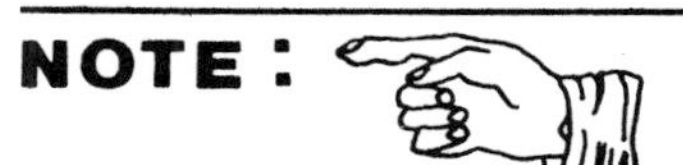

For small interior holes, see Chapter 31, "Insulate Closed Frame Wall".

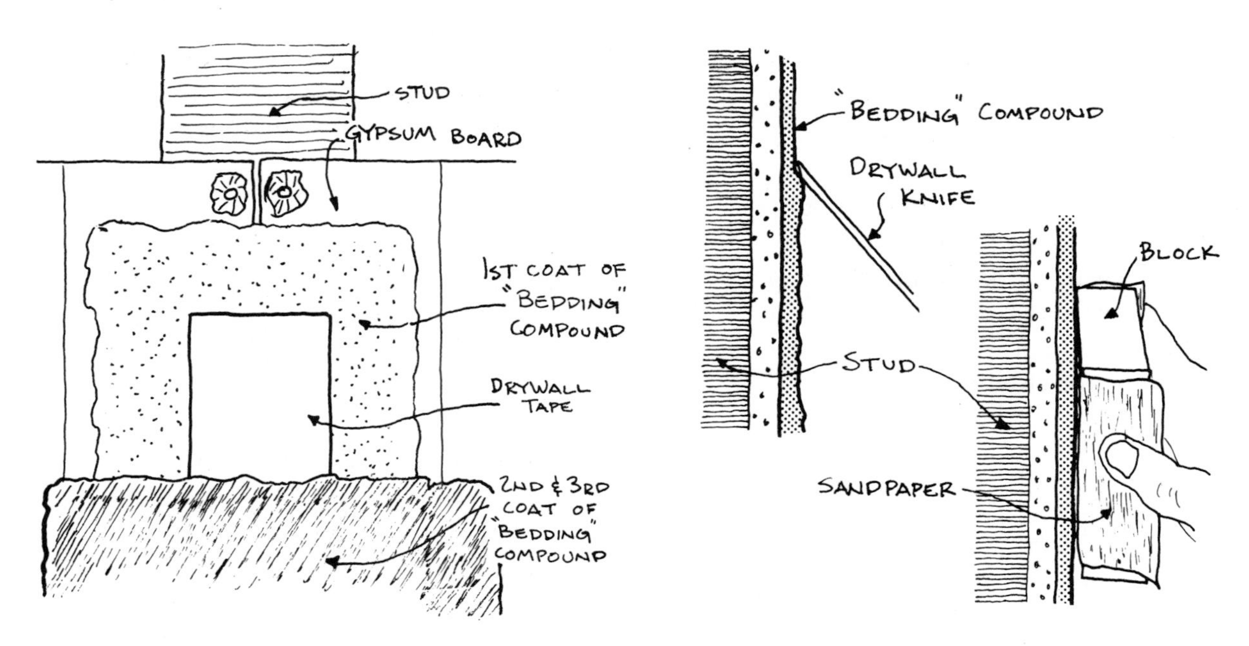
STUD
GYPSUM BOARD
1ST COAT OF "BEDDING" COMPOUND
DRYWALL TAPE
2ND & 3RD COAT OF "BEDDING" COMPOUND
"BEDDING" COMPOUND
DRYWALL KNIFE
STUD
BLOCK
SANDPAPER

41

seal foundation crawl space vents

description

As discussed in Chapter 23, "Insulate Crawl Space Walls on the Interior", crawl spaces need to be vented. Even with a vapor impermeable ground cover, crawl spaces can become damp in summer unless properly vented. During the summer, the crawl space is cooler and may be so cool that humidity from the outside air or from the house could condense in the crawl space (see Chapter 4, "A Word about Condensation"). This moisture can cause structural damage.

During the winter, however, crawl space vents can be sealed. You can even take it a step further by adding rigid insulation over the vents. If you do close the vents, and/or insulate them, check the crawl space once a month for moisture accumulation (frost). If you do find moisture, open the vents and let the space dry out.

NOTE: 

Vents should never be closed off unless there is a vapor impermeable ground cover present.

other material

- caulk 
- sealant tape 

materials

rigid insulation

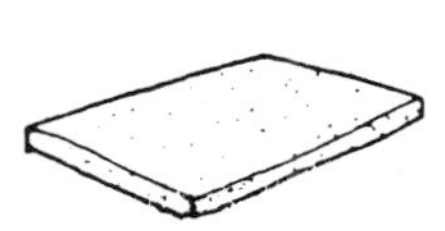

- extruded polystyrene
- polyurethane
- polyisocyanurate

weatherstripping

- foam strips
- rubber or vinyl plastic tubing

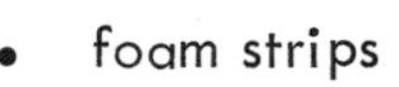

vapor barrier

- vapor impermeable ground cover (6 mil minimum for polyethylene film)

wood

- plywood
- planks
- fur strips

hardware

- bolts
- handles
- hooks
- screws
- fasteners

preparation

If not existing, install a vapor impermeable ground cover. Polyethylene sheeting is commonly used and is to be a minimum 6 mil thick. Lap the edges of the sheets at least 12" and weight down with bricks or wood. Use polyethylene tape to fasten the vapor barrier to the foundation wall.*

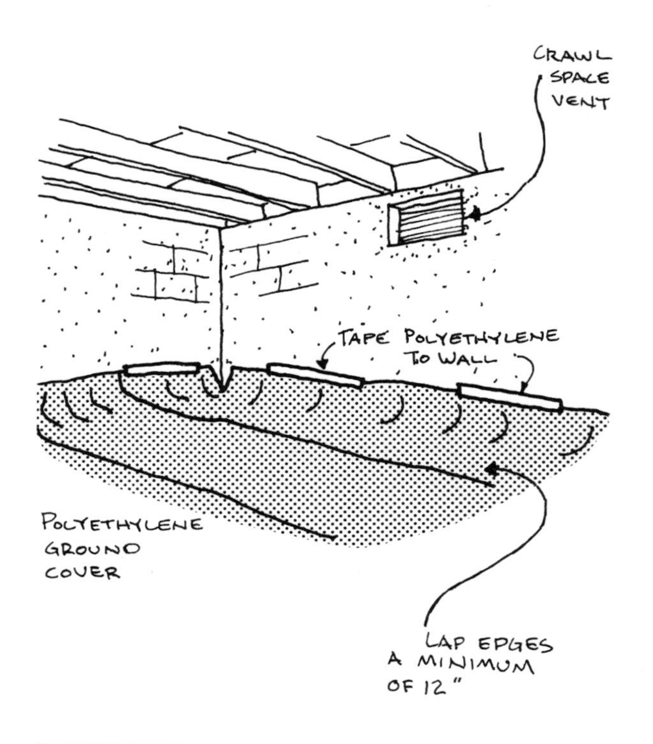

* For additional details, see Chapter 23, "Insulate Crawl Space Walls on the Interior".

installation procedures

SEALING UNSCREENED VENTS-EXTERIOR

1 With furring strips, construct a frame to fit within the foundation wall opening. Fasten wire mesh (screen openings should be no larger than 1/4") to the frame. Secure this frame in the opening (screen to the inside) with masonry nails or lag bolts (tie the frame back to the existing vent with lag bolts). Caulk around the frame.

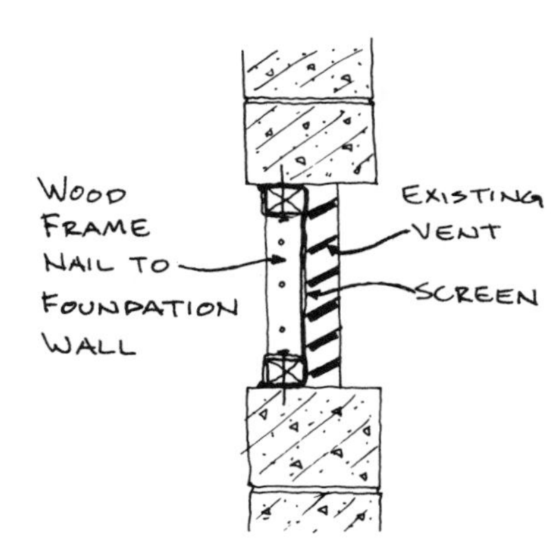

2 Cut a piece of plywood or wood planking to fit across the opening. From the top of the panel, cut off the dimension equal to the thickness of the panel.

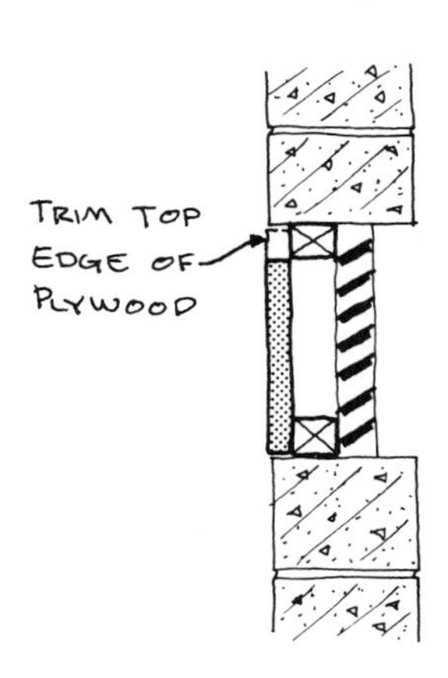

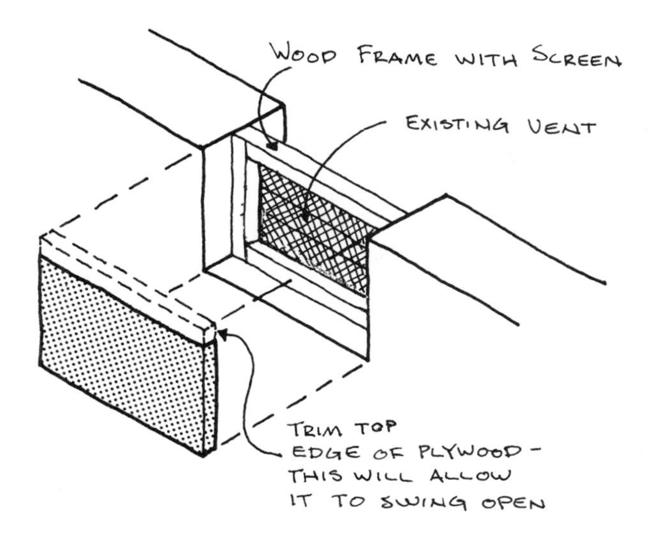

3 Mount two hinges to the top of the plywood and the top member of the frame.

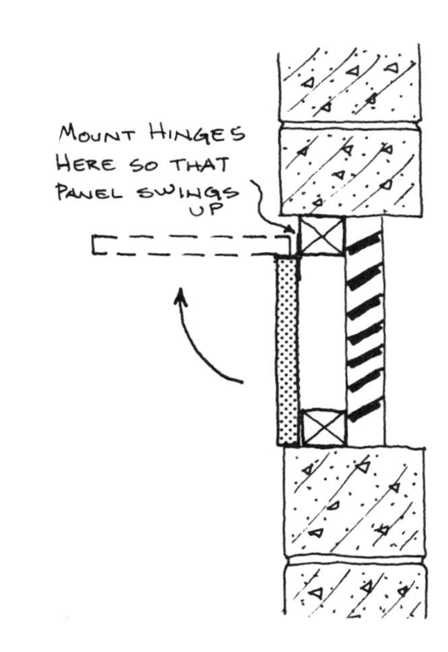

NOTE: 

If vent is flush to exterior foundation wall surface, see step 8 in this chapter.

4 Cut a piece of rigid insulation to fit between the framing members. Glue the insulation to the plywood. The hinge will be "sandwiched" between the plywood and insulation.

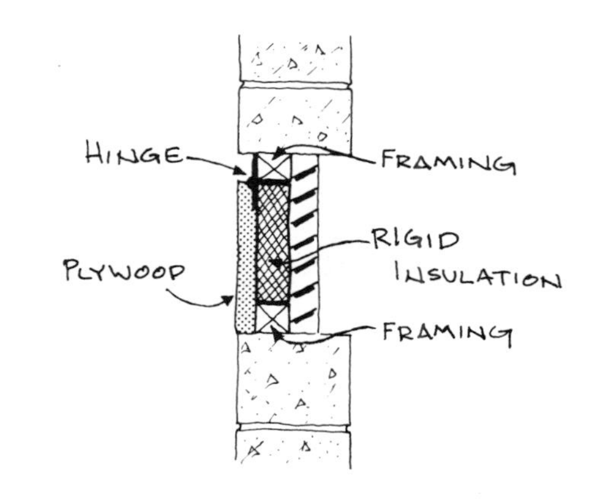

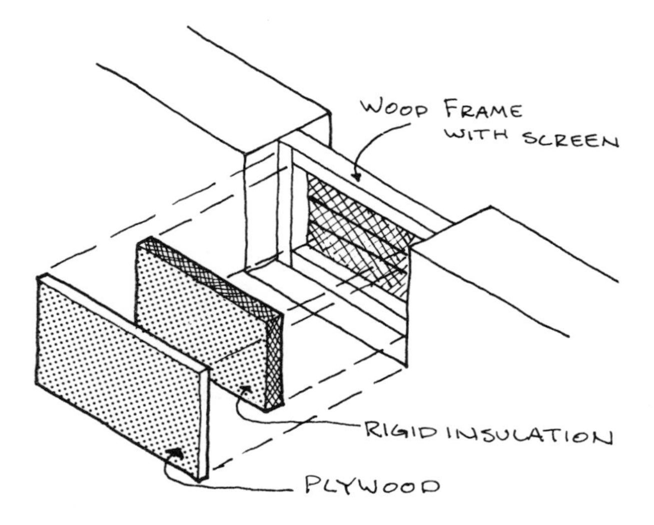

5 Prime and coat the plywood with an exterior paint. Apply foam or weather-strips to the inside face of the framing. Provide a rain drip by attaching sealant tape along the bottom edge of the panel and along the top edge of the foundation opening. Screw a handle or knob on the panel.

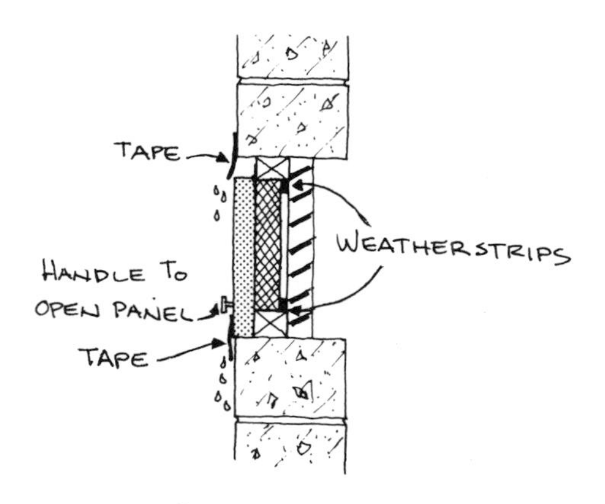

6 The panel can be held open by propping a wood dowel between the bottom framing member and panel (notch a small hole in the insulation to receive the dowel).

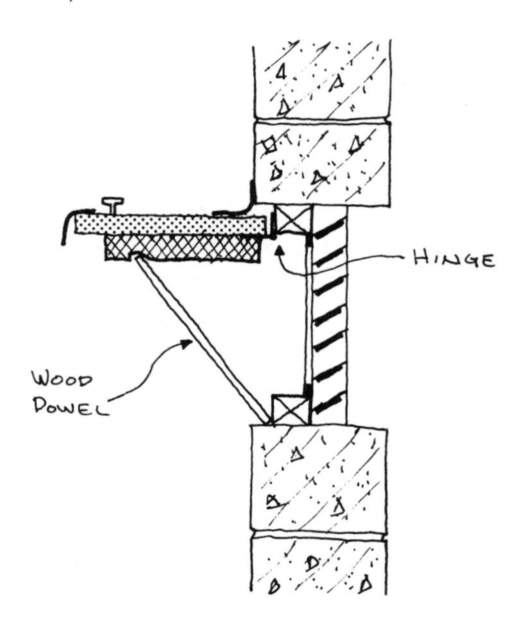

7 The panel can be sealed by a lag bolt protruding through the plywood into the framing. When the panel is closed, it can be secured with a wing nut attached to this bolt.

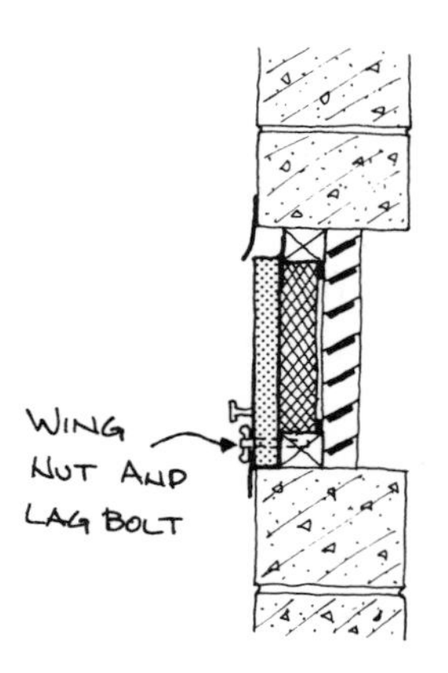

NOTE:

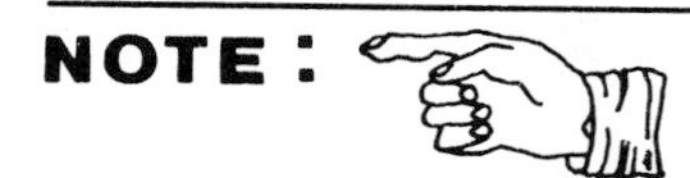

Follow steps 2 through 7 for screened vents.

8 If the vent is flush to the exterior foundation wall surface, follow the same procedure, but mount the framing on the wall instead of within the opening. The panel can be secured in the open position with an eye bolt and hook.

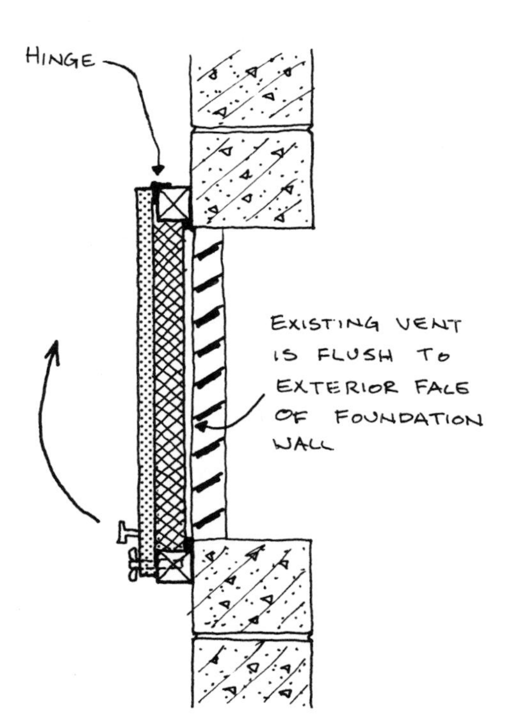

SEALING UNSCREENED VENTS-INTERIOR

9 The procedure for sealing vents on the interior is practically the same as for step 8, except that the insulation is mounted on the outside of the plywood (when the panel is closed, you'll be able to see the insulation from the crawl space). The panel will open fully if the depth of the furring is equal to the depth of the plywood and insulation.

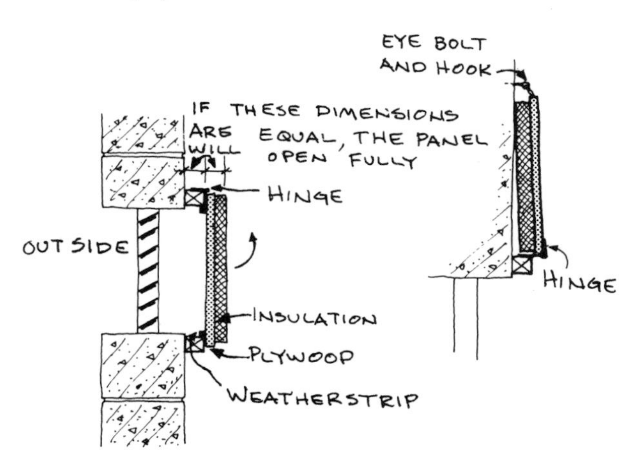

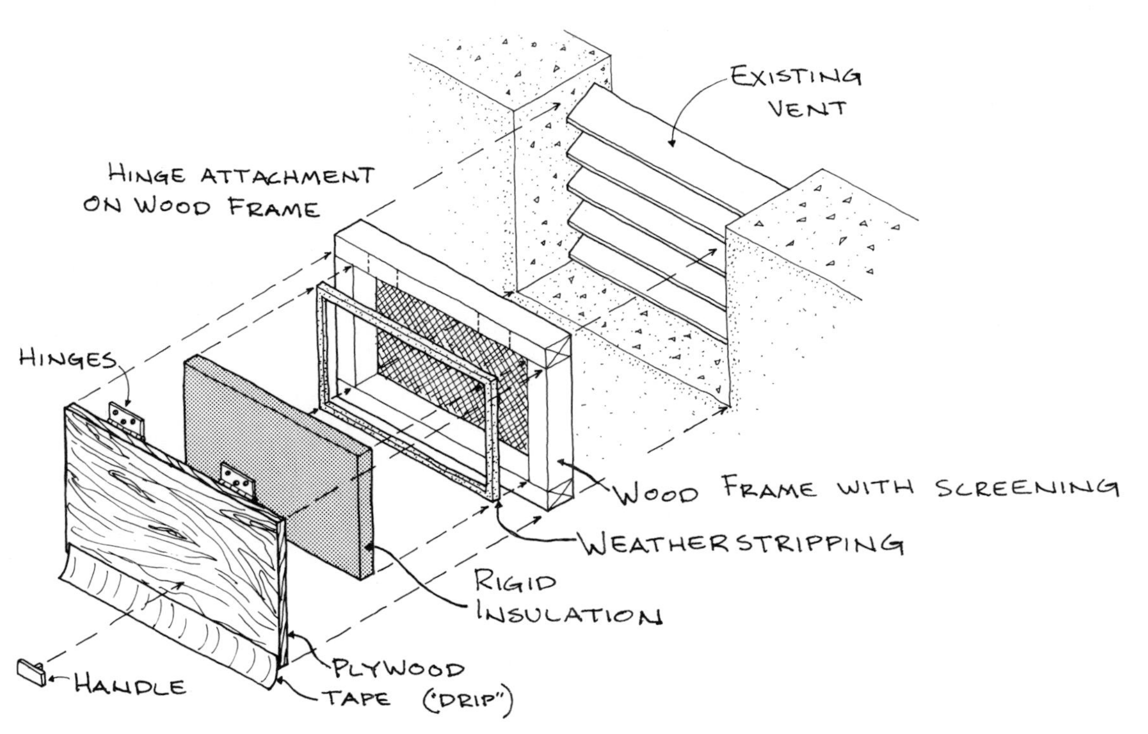

42
install skirting around building

description

This retrofit applies to mobile homes and some home additions that are supported on piers leaving the floors exposed to the cold. Essentially, you're building a crawl space wall, or "skirt", around the open portion of the home.

The skirting is constructed of 2" x 4" framing built on two or more courses of concrete block (or a concrete grade beam) covered with some appropriate finish such as cement asbestos siding, wood, or aluminum siding. An access panel, vents, and ground cover are included in the installation of skirting.

This retrofit involves a high degree of labor skill and should be done by someone skilled in carpentry.

materials

construction lumber

- 2" x 4" pressure treated for water penetration and fungus infestation

rigid insulation

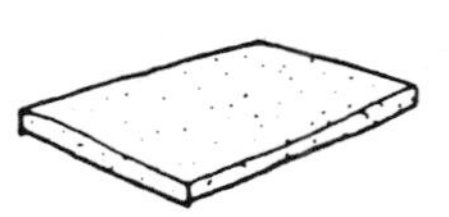

- polystyrene
- polyurethane
- polyisocyanurate

roll/batt insulation

- glass fiber
- rock wool

concrete block

- solid or
- hollow core

exterior finish

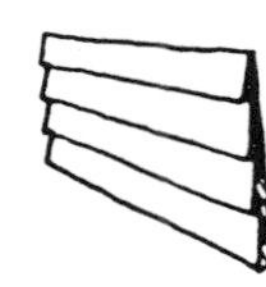

- sheathing
- asbestos cement
- siding (3/16" minimum)
- wood siding
- aluminum siding
- other appropriate and durable siding

other materials

- termite shield

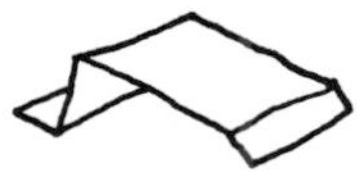

- mortar (see specification in Volume 2 for mixing proportions)
- vapor impermeable ground cover (6 mil minimum polyethylene)
- wall vents (operable type)
- "flexible" flashing

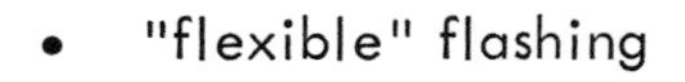

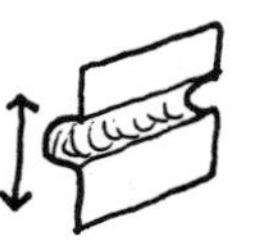

- gravel or crushed stone

- "sill sealer" insulation

A WORD ABOUT SKIRTING

The sole purpose of skirting is to provide a thermal break and a windbreak for the exposed floor of the home. The skirting is not a structural support for the home. This is an important distinction, especially for mobile homes.

Mobile homes are lightweight and are buffeted by high winds. In some cases occupants moving about in a mobile home may cause the home to "teeter" down on one side.

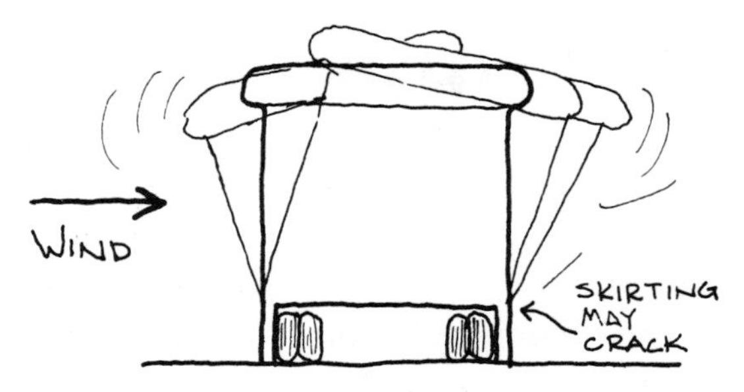

Also, during very cold weather, the ground beneath the skirting may freeze and "heave", damaging the home and/or the skirting.

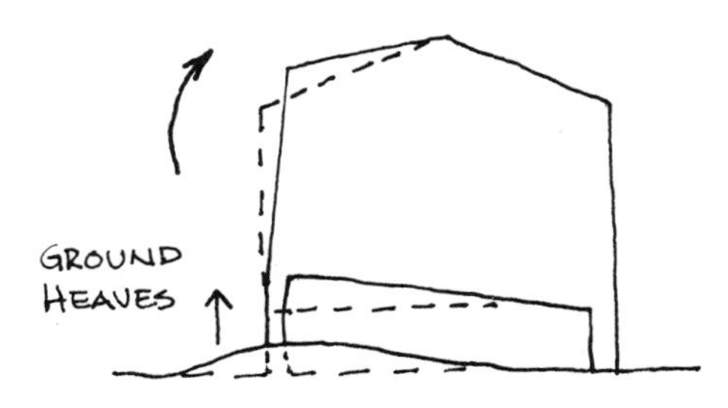

Step 7 in this chapter discusses construction techniques to account for this movement.

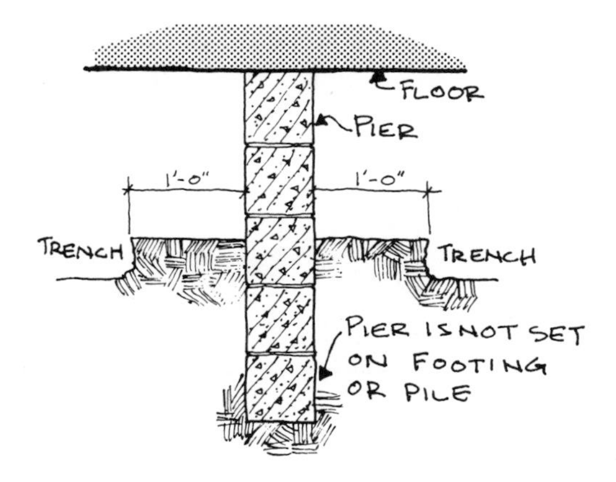

If piers are at perimeter line but are not set on footings or piles, then do not trench closer than 1'-0" to the piers. A tight installation will be much more difficult in this case and you may elect not to install skirting.

installation procedures

1 Dig a trench 8" wide and 16" deep (minimum) around the building perimeter. If the home is supported on piers at the perimeter, trench between these piers. Lay 2 or more courses of concrete block over a 4" gravel bed and backfill the trench with gravel. Set anchor bolts 2'-0" o.c. into grout filled concrete block cavities.

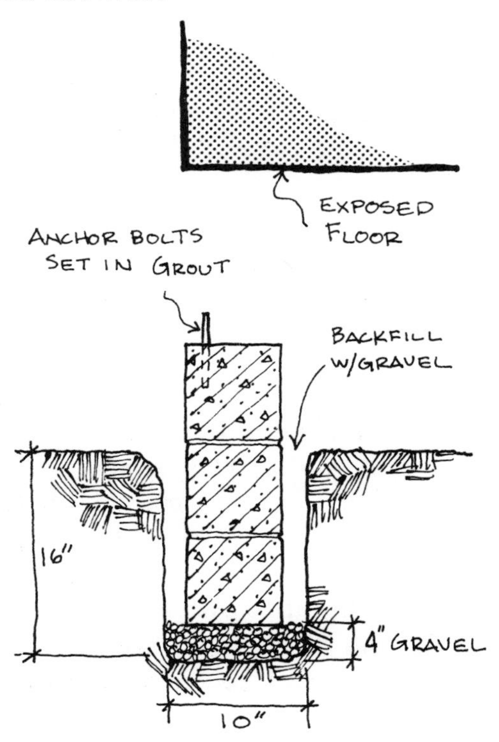

2 An alternative to concrete block is a concrete grade beam. Trench as before and set plywood forms with framing to a height of at least 6" above grade. Set two 1/4" reinforcing bars on intermittent blocks with twisted wire "feet". Use intermittent wood blocks along the top of the form to stabilize it.

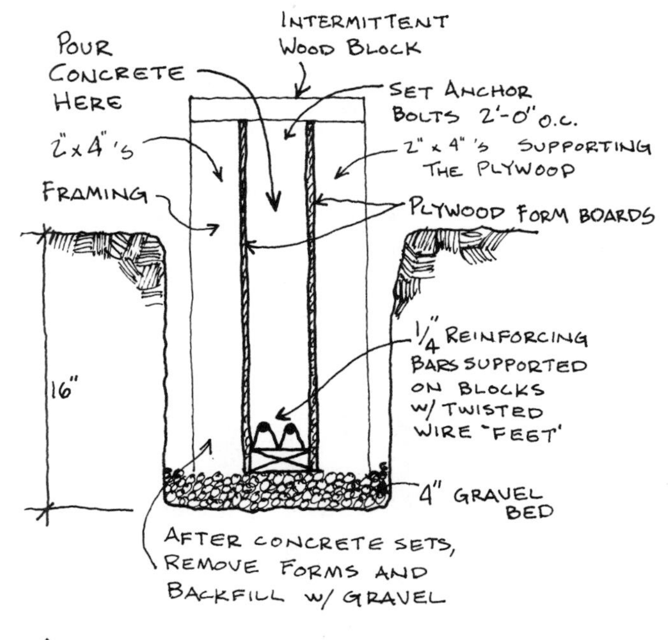

ALTERNATE

3 Some mobile homes are set over concrete pads. Simply lay a course of concrete block on the pad around the perimeter of the home.

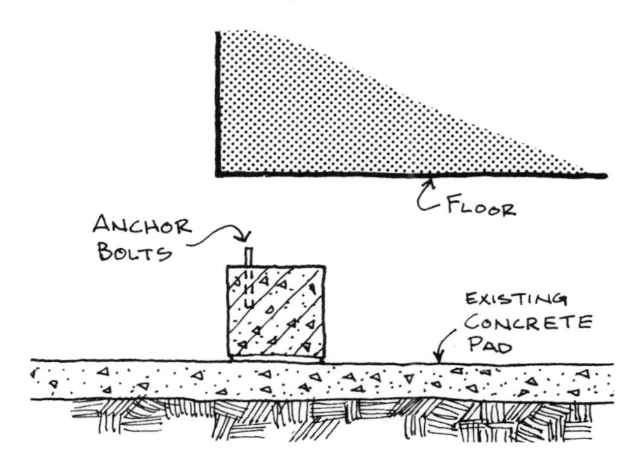

4 If required in your area, set a termite shield over the block or grade beam before fastening a 2" x 4" plate to the anchor bolts. Frame short studs at 16" o.c. between the top and bottom plates. Leave about 1" between the top plate and the bottom of the floor. Be sure to frame an opening for an access panel to the newly enclosed space.

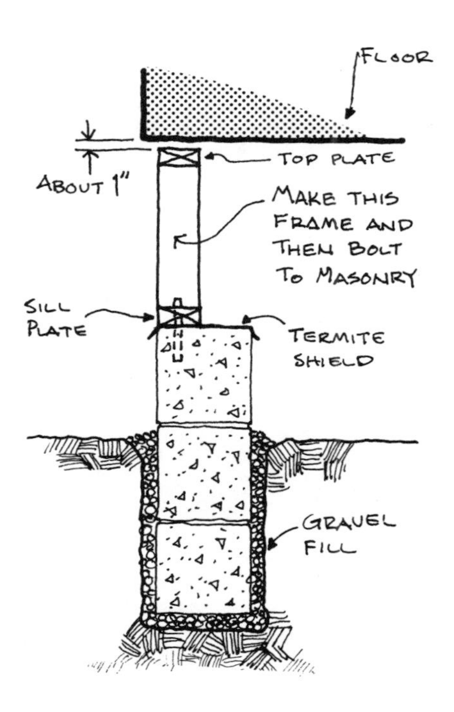

NOTE:

For a more airtight construction, place a "sill sealer" strip over the block or termite shield before fastening the plate in place.

5 Depending on the height between the ground and the underside of the floor, it may be easier to take the concrete block work with a top plate anchored as before to within 2 or 3 inches of the floor and not use wood studs. Vents and an access panel through the masonry must then be provided.

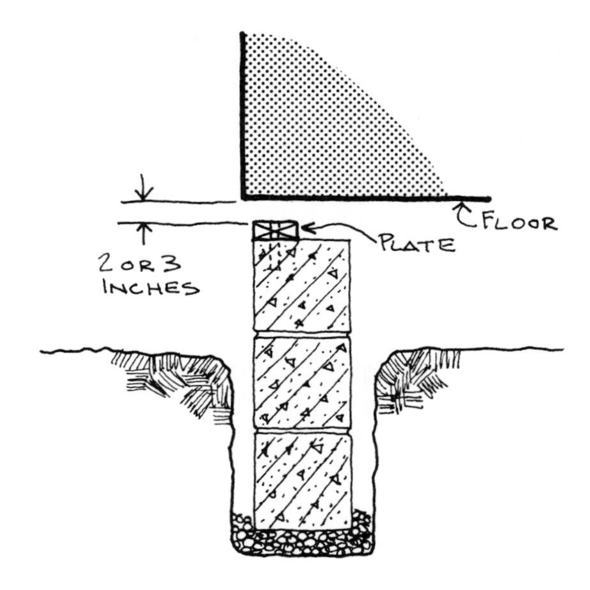

6 Install sheathing and then fasten wood, aluminum, or cement asbestos siding to the studs. If you're using cement asbestos siding, you must pre-drill holes for the nails. In all cases, the siding must be at least 6" above the ground. In addition, install crawl space vents in the skirting (for details, see Chapter 23, "Insulate Crawl Space Walls on the Interior").

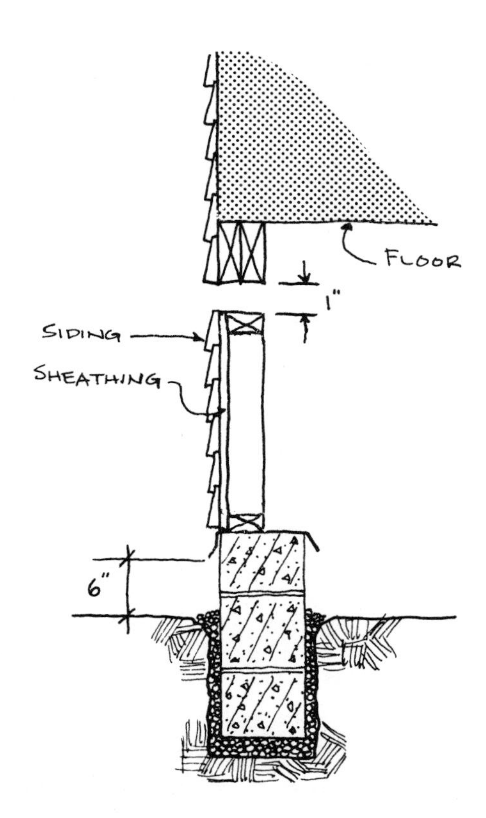

7 Now to seal that crack between the skirting and the floor. Use "flexible" flashing to accommodate any movement by the skirting and/or the home as discussed earlier. Fasten the flashing

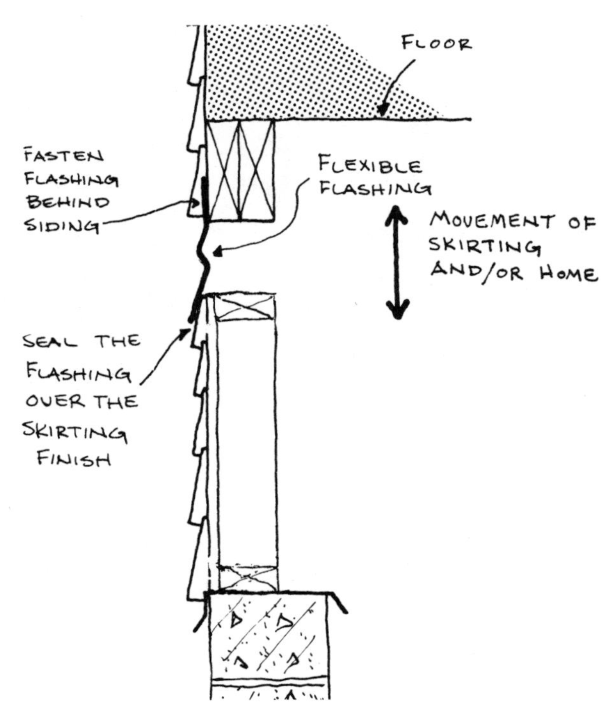

behind the existing sheathing on the home. Fasten the bottom edge of the flashing over the finish applied to the skirting or over the outside edge of the plate where a masonry skirt is provided.

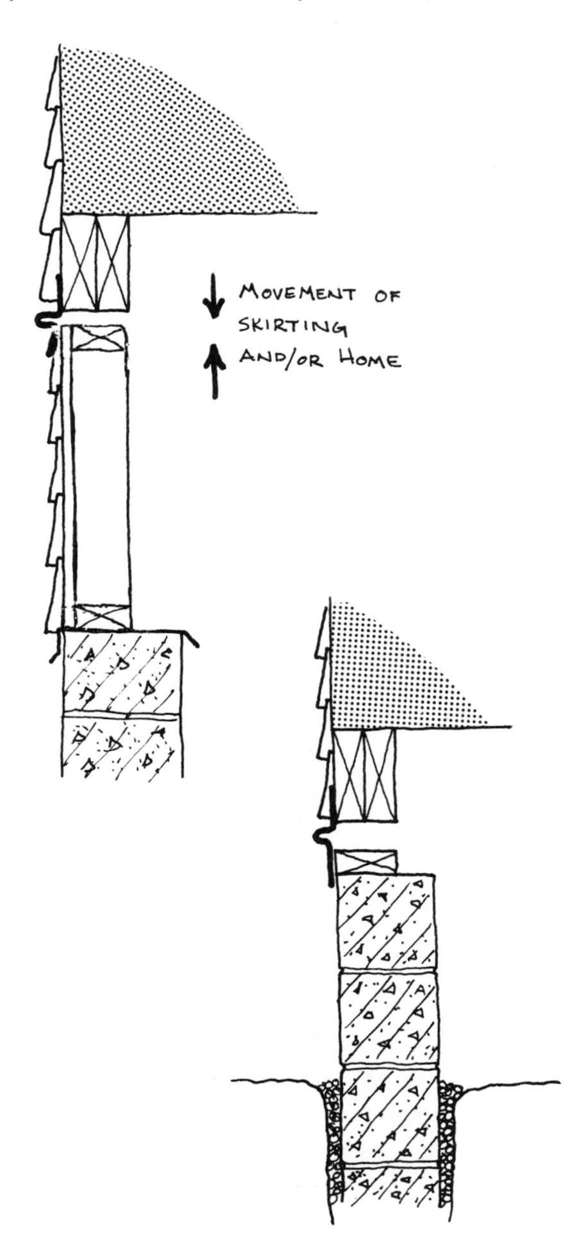

8 Finally, install insulation on the back of the skirting down to the ground. Install batt or rigid insulation as discussed for Chapter 23 with the following differences. If you're using batt insulation, insulate the floor beam (joist) by fastening it to the floor and letting it rest on or along side of the top plate.

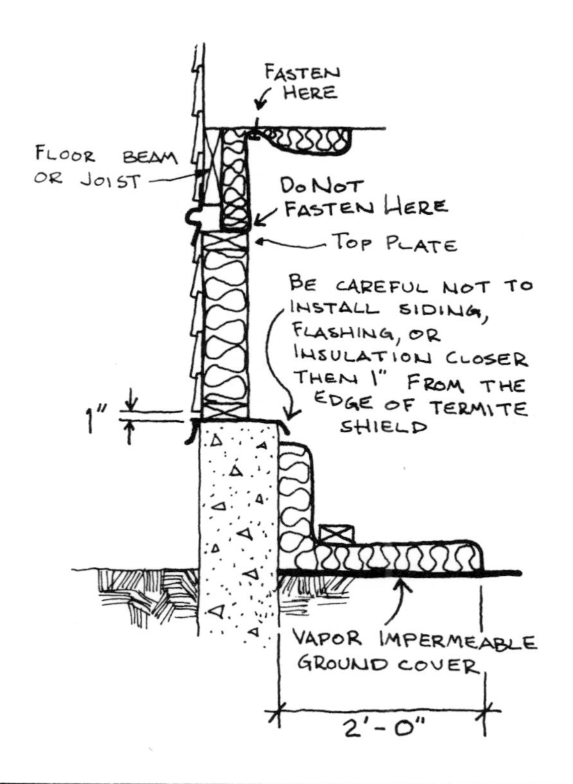

NOTE:

See Chapter 23, "Insulate Crawl Space Walls on the Interior", for details on installing a vapor impermeable ground cover.

9 With rigid insulation, attach a continuous piece alongside the studs until it rests against the floor beam (joist). Do not fasten this insulation to the floor beam (joist). Block out if necessary.

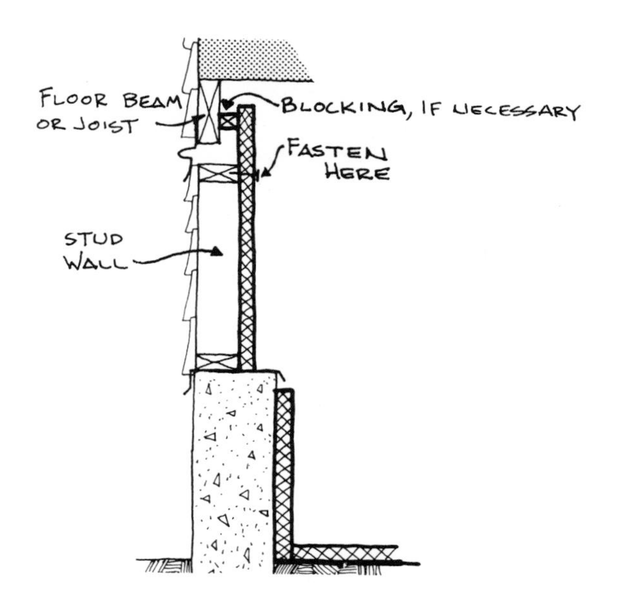

10 Some rigid insulation can be installed below grade and if so, should be let down the side of the masonry and the remaining dimension within the trench. Backfill with gravel as before. Review Chapter 23 ("Insulate Crawl Space Walls on the Interior") and Chapter 24 ("Insulate Crawl Space Walls on the Exterior") for specifics on attachment, finishing, etc. for either interior or exterior application.

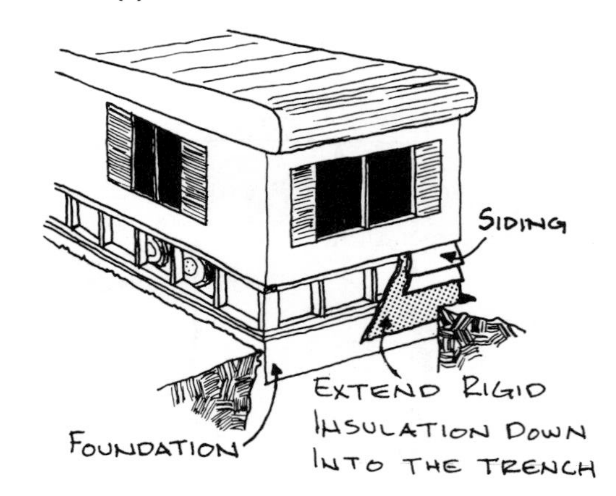

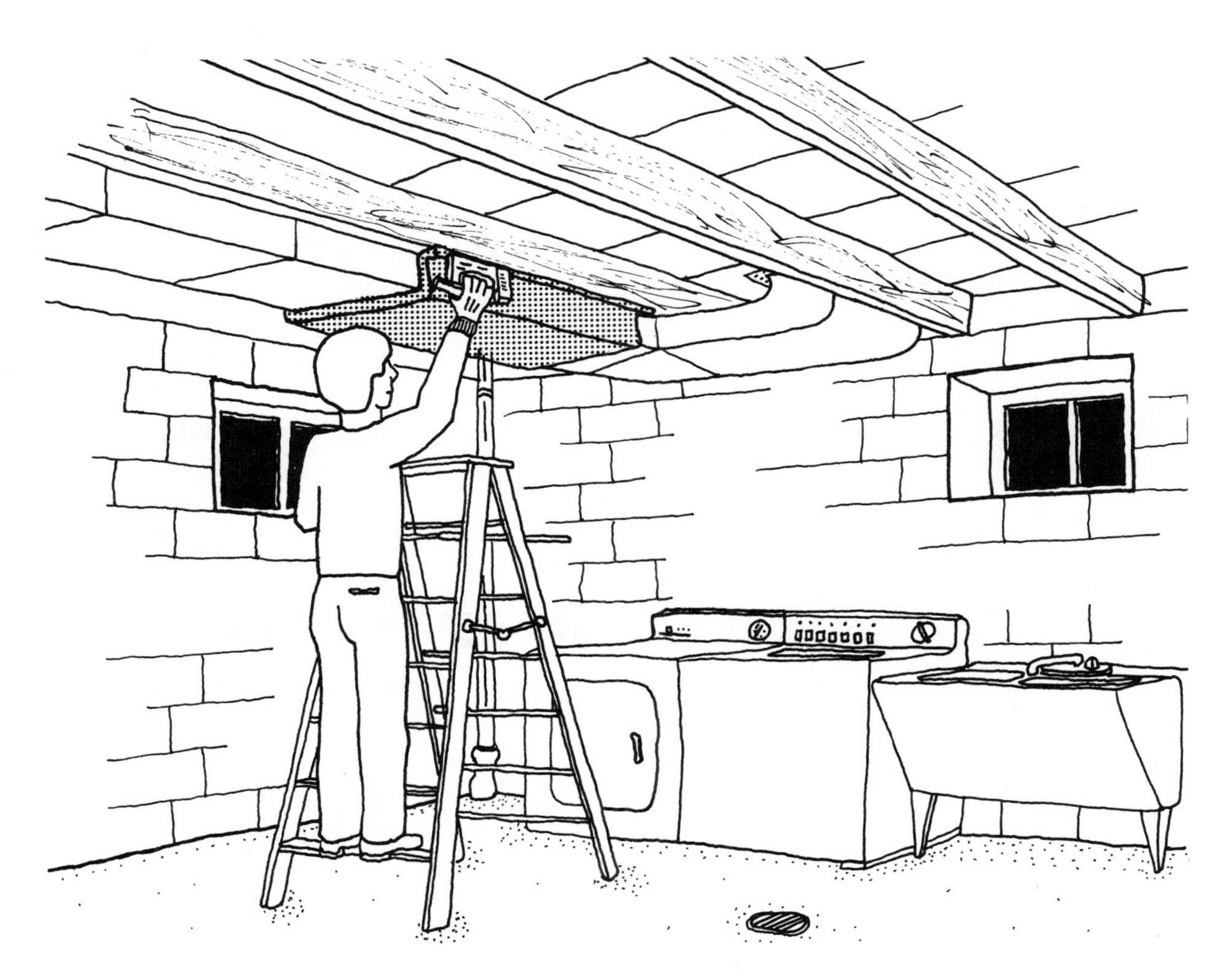

43

insulate exposed ducts and/or pipes

description

Heating ducts in unheated spaces (basements, crawl spaces, attics, garages, etc.) can also be insulated. Special, semi-rigid, duct insulation (usually 1 or 2 inches thick) is available for this. However, the same roll/batt insulation that you used to insulate attics, floors, and exposed frame walls can also be used to insulate ducts. Generally speaking, though, batts greater than 3-1/2" thick may be cumbersome to work with.

Pipes for domestic hot water, hot wate heat, and steam heat can also be insulated to reduce heat loss. Specially shaped insulation is manufactured for standard diameters of pipe. "Wrap around" insulation fits different sizes. Although not related to savings in heating bills cold water pipes can be insulated to prevent "sweating" (condensation) in summer. Depending on the location of other insulation, cold water pipes may be required to be insulated if they're located in unheated spaces.

materials

DUCTS

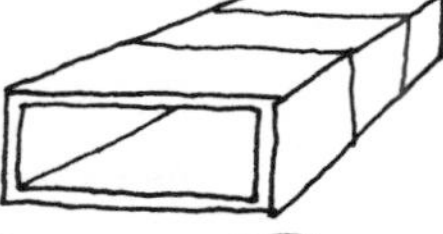

duct tape

polyethylene
vinyl
aluminum

blanket insulation

1"-2" thick
foil or vinyl faced
flexible (for circular ducts and "tight" areas)

adhesive

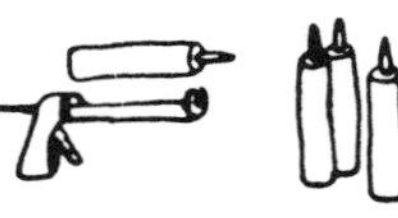

compatible with insulation material (used to join insulation)

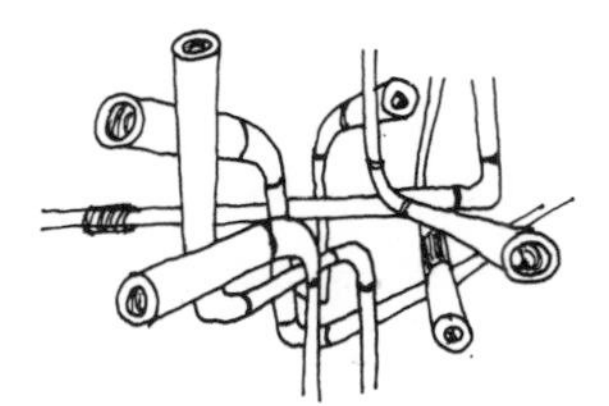

PIPES

sleeves

- polyurethane foam
- expanded rubber

tape

foil backed PVC vinyl foam
glass fiber and plastic

adhesive

compatible with insulation (used to join sleeve insulation)

silica base cement

for steam lines

preparation

Ducts and pipes to be insulated are to be cleaned of dirt, dust, and oil.

installation procedures

DUCT INSULATION

Ducts that are used for both heating and air-conditioning should be insulated with vapor barrier faced insulation with the barrier facing out (the vapor barrier prevents condensation). Heating ducts can be insulated with unfaced insulation, but it's easier to use faced insulation with the facing placed on the outside.

1 Begin by checking the duct joints for leaks. Do this by moving your hands around the joints feeling for drafts. When you feel a draft, wrap the joint with duct tape.

WRAP LEAKY DUCT JOINTS WITH DUCT TAPE

AN IMPORTANT NOTE:

Duct tape is not to interfere with dampers, even though you may feel a draft there.

---

2 Wrap the insulation completely around the duct. If the insulation doesn't fit around the duct because of joists or other framing members, cut the insulation pieces to cover as much of the duct as possible and staple the ends to the joist or other framing member.

WRAP THE DUCTS AS COMPLETELY AS POSSIBLE

STAPLE THE INSULATION TO FRAMING WHERE YOU ARE UNABLE TO WRAP THEM COMPLETELY

3 Butt adjacent pieces of insulation tightly together and tape the joints with duct tape. If necessary, use staples or a compatible adhesive to seal the insulation joints.

TAPE INSULATION JOINTS TOGETHER

4 Cut, fold, and tape the insulation at the end of duct runs.

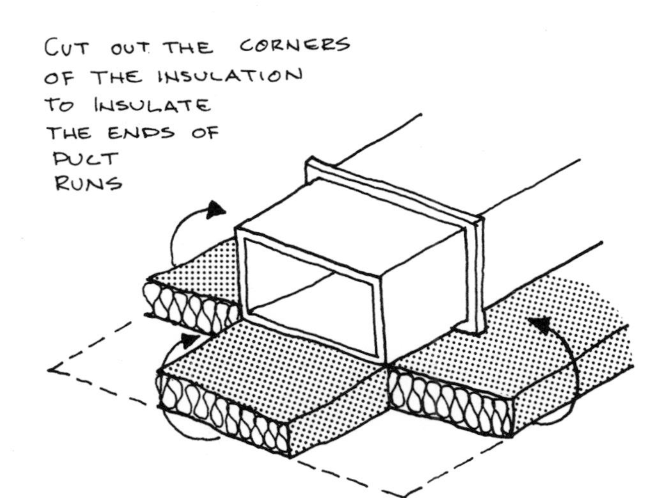

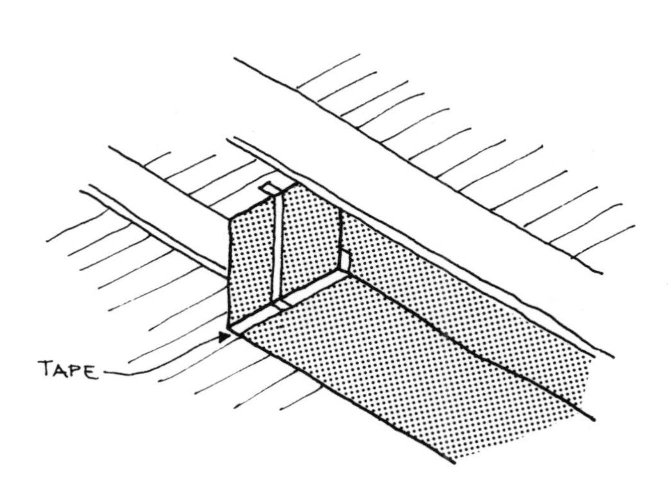

PIPES

Pipes that appear to be leaking are not to be insulated. In addition, pumps, valves, piping in heater rooms, boiler feed lines, pressure relief

NOTE:

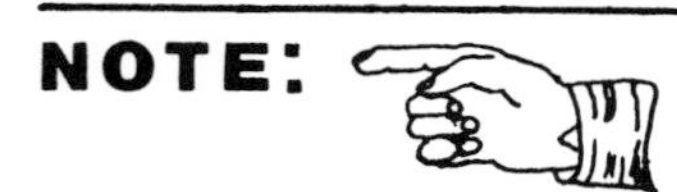

Cylindrical ducts are insulated in the same fashion as rectangular ducts.

5 Return air ducts in unheated spaces car also be insulated in the predescribed fashion. Duct insulation is not to interfere with the operation of register controls, dampers, and blowers. Do not insulate furnace or boiler flue pipes as part of this retrofit.

devices, and vents are not to be insulated. Do not apply insulation within 1" of shut-off or zone valves on pipes.

1 One type of pipe insulation is a roll of glass fiber followed by a roll of plastic tape that holds the glass fiber.

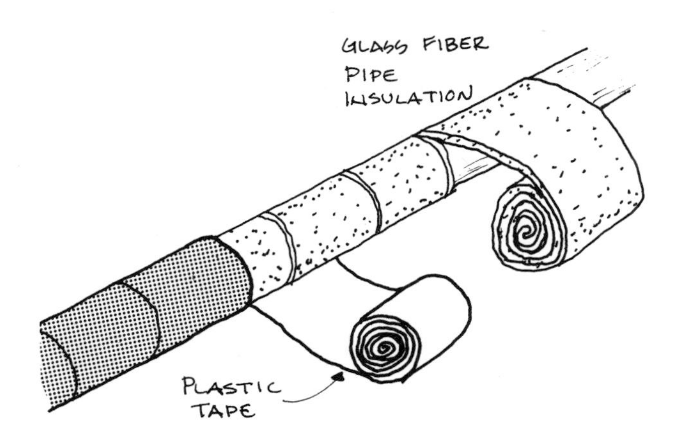

2 Another type of pipe insulation is a self-adhering vinyl foam with aluminum foil.

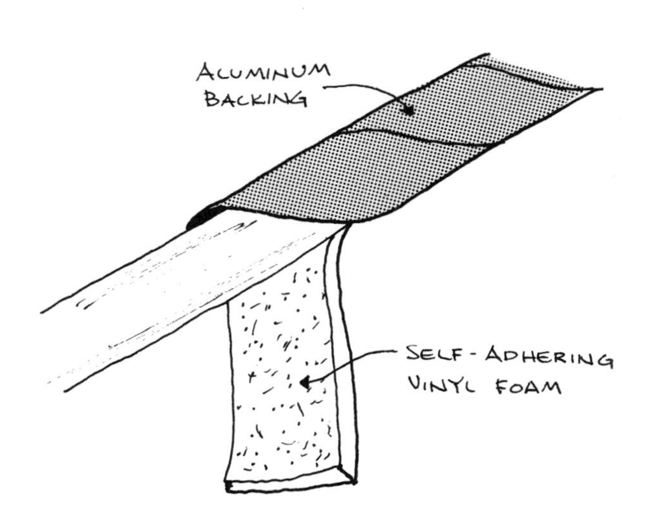

3 With sleeve insulation, open the insulation along the slit and slip it over the pipe. Give the insulation a slight twist. Use polyethylene tape to hold the insulation together if necessary.

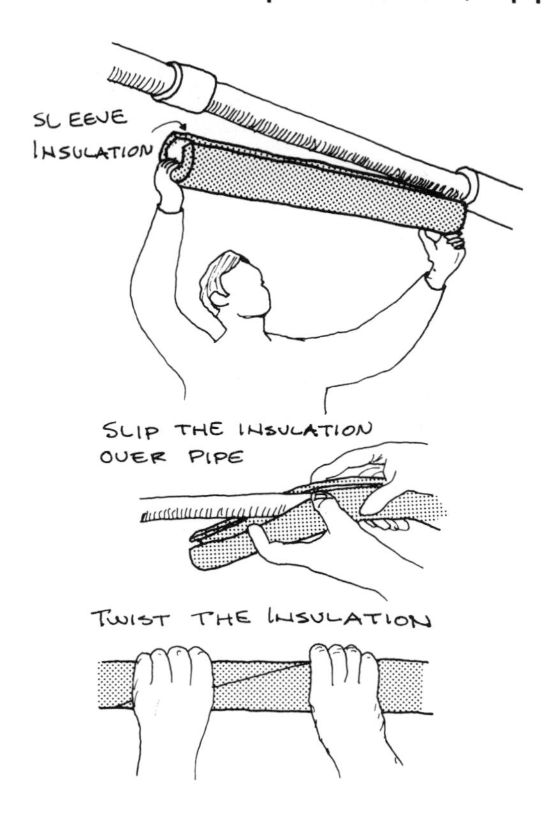

4 At elbows (90° turns), cut the insulation at a 45° angle (called a bias cut).

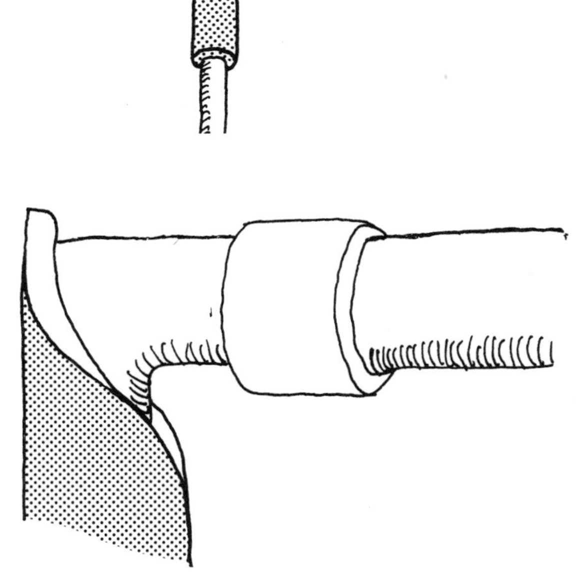

5 At a "T" or "Y" shaped joint, make a V-shaped cut (called a saddle cut).

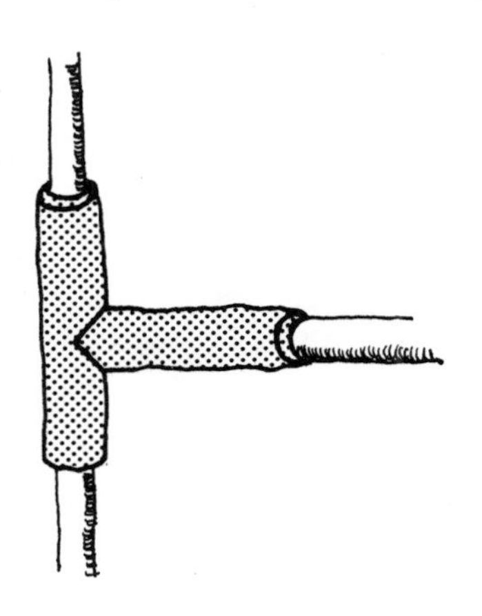

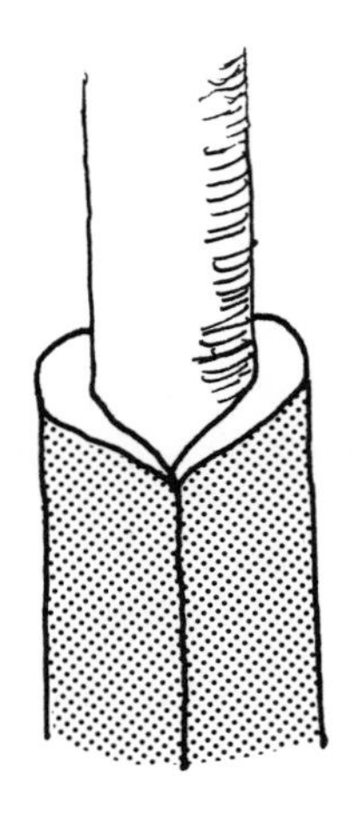

SADDLE CUT

glossary

APRON
That part of the interior window trim below the stool (window shelf).

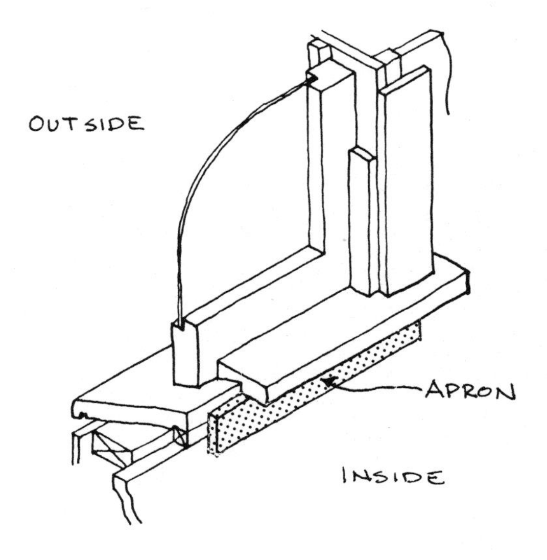

BEAD
A continuous extruded glazing or caulking strip applied around the periphery of a pane of glass or at the interface of two different materials. Used to seal the joint.

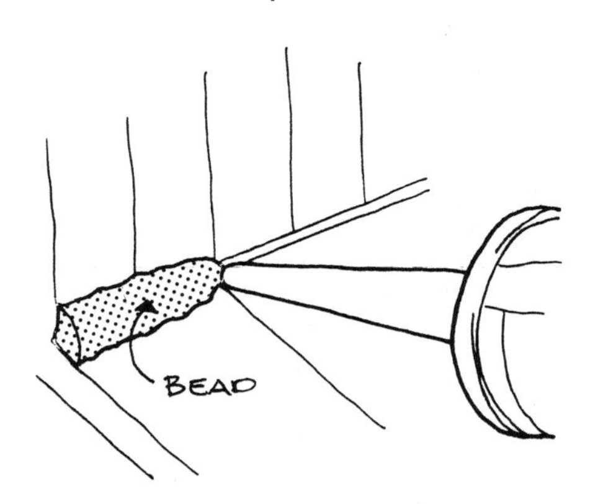

CASING
The trim on a window or door at the sides and at the top of the opening. Is found on the inside and outside.

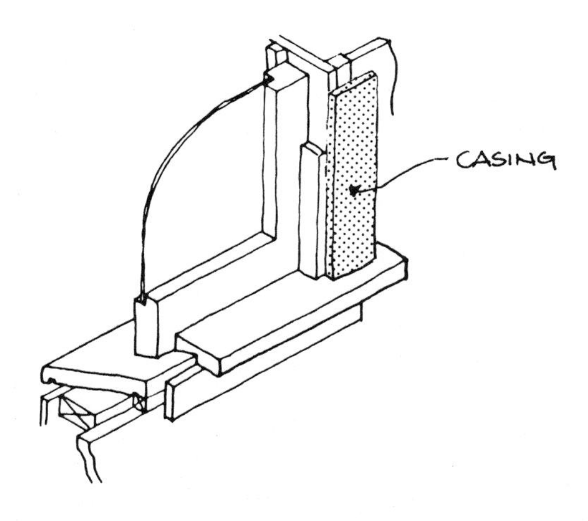

CAULK
A soft pliable material used for sealing joints in buildings and other structures where normal structural movement may occur and where different materials interface. Caulking compound retains its plasticity after application and with wide temperature variances. It is available in forms suitable for application by gun and knife and in extruded preformed shapes.

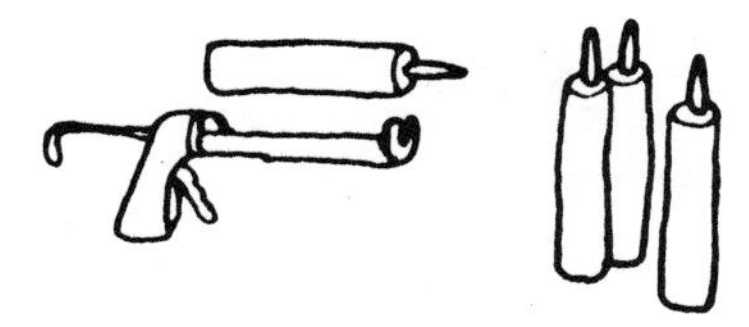

CONDUCTION
A method of heat transfer where heat moves through a solid. Conductivity of a material is the measure of the quantity of heat that flows in a specified unit of time through a given unit area of the surface of the material of a given thickness (Btu/sq.ft./hr.). If a material is relatively non-conductive and is used for insulation, its non-conductive value is expressed as an "R" or "U" value.

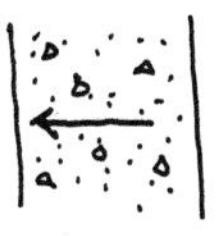

CONVECTION
The transfer of heat by liquids and gases in contact with solids, fluids, or gases.

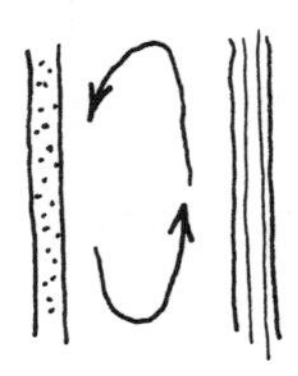

COURSE
A row of masonry units, i.e., bricks, concrete block, etc.

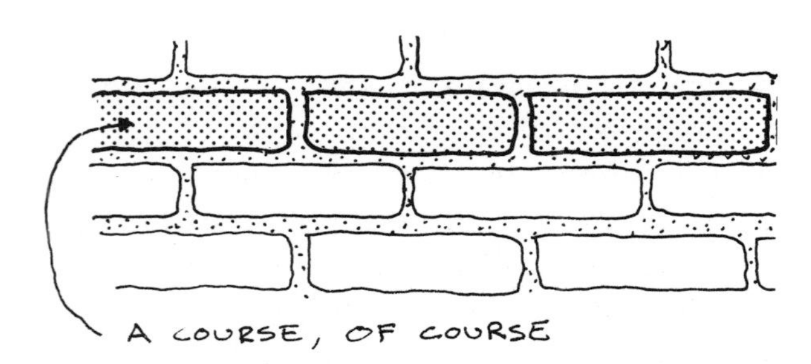

DORMANT CRACK
A crack in a masonry or concrete wall which shows no signs of movement.

FROST LINE
The depth of frost penetration in the soil. This depth varies in different parts of the country.

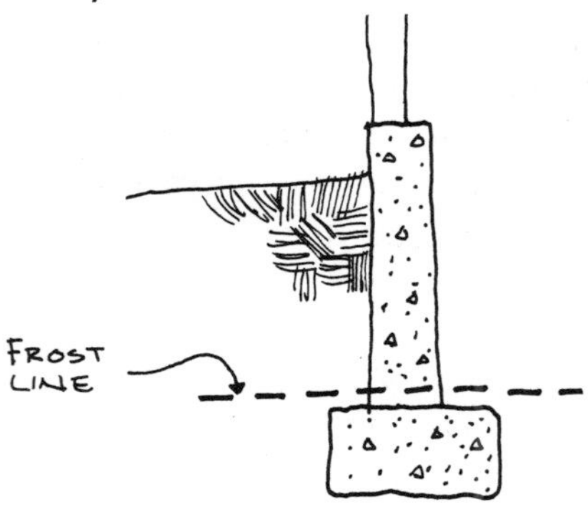

GLAZING COMPOUND
A dough-like material used for sealing window glass in frames. It differs from putty in that it retains its plasticity for an extended period.

GROUT
Cement mortar of pouring consistency used to fill cracks or joints in masonry walls.

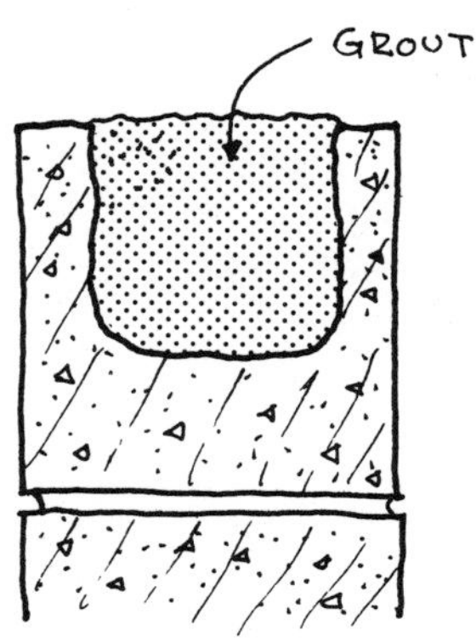

HEADER
The rough wood structural framing over the top (head) of a window, door, or other opening in a wood framed wall. This is also known as a lintel.

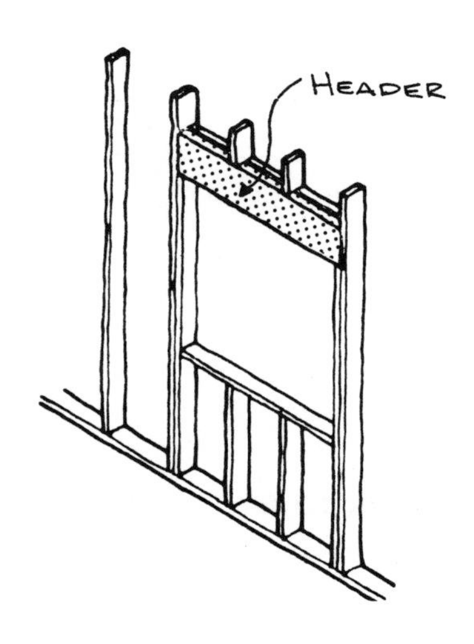

INFILTRATION
The passage of air, dust, or water through cracks, joints, or holes in the exposed surface of a building.

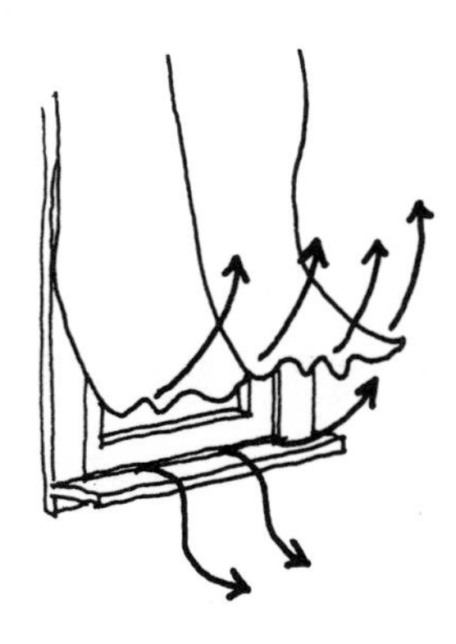

JAMB
The vertical side of a doorway, window, or other opening in a wall.

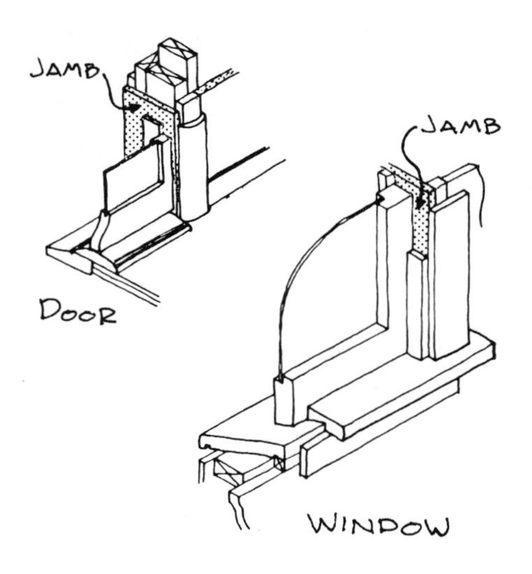

LITE
The divisions of glass within a window frame or door (also called panes).

MITER
The joining of two pieces of beveled material so that, when attached, they adjoin each other at an angle.

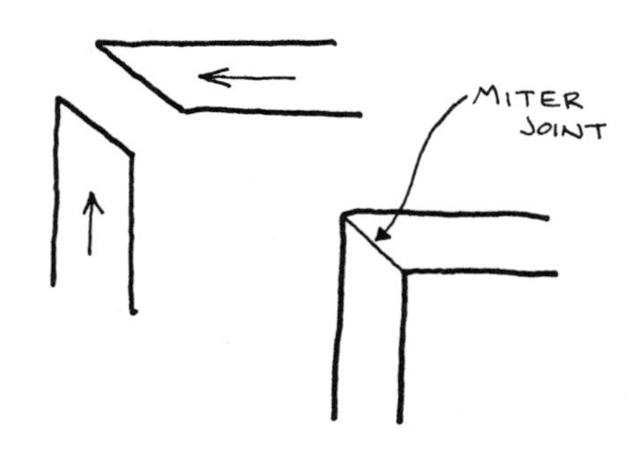

MORTISED
A slot cut into wood, usually edgewise.

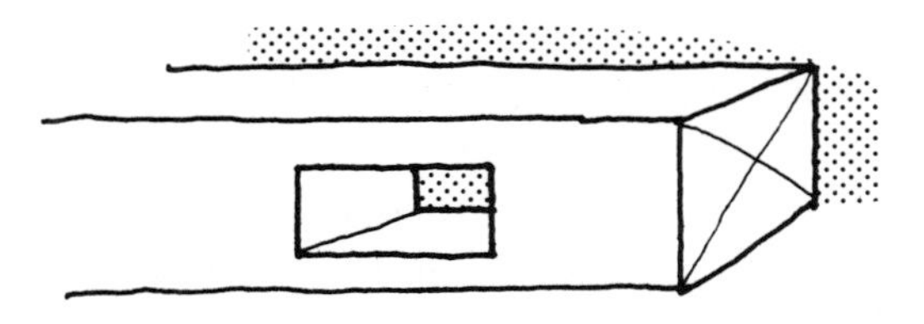

MOVING JOINT
A joint which widens and/or becomes more narrow or whose sides move parallel to the opening.

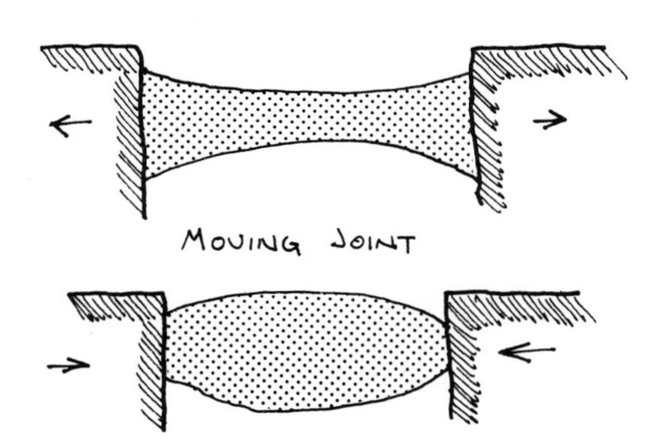

PERMEABILITY (perm)
A measure of water vapor movement through a material per square foot of surface of the material (grains per square

foot, per hour, per inch of mercury difference in vapor pressure).

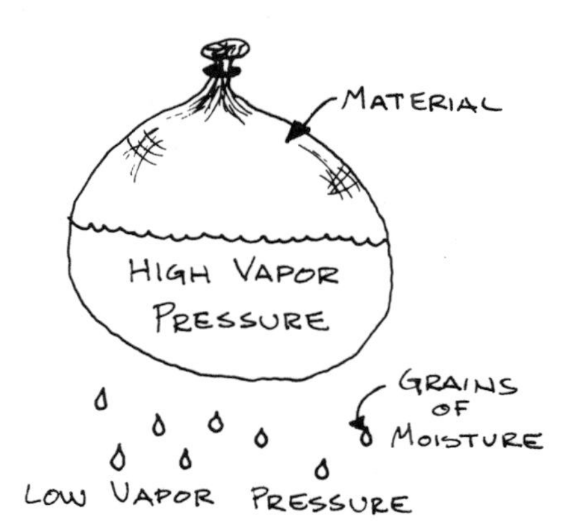

PLATES
Top and bottom nailing strips. A sill plate is a horizontal member anchored to the top of a masonry wall. The sole plate is the bottom horizontal member of a frame wall. The top plate is the top horizontal member of a frame wall that supports joists, rafters, and other members.

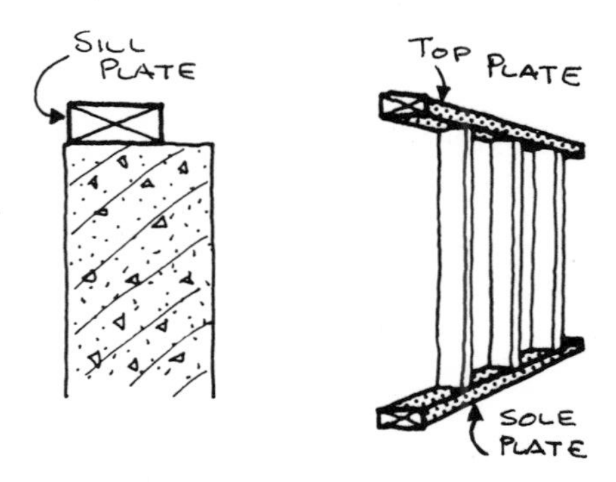

PUTTY
A soft dough-like material with an oil base used to seal window glass in frames. It differs from glazing compound in that it does not remain pliable for an extended period of time.

"R" FACTOR (resistance)
A measure of the insulating value of a material (i.e., the resistance to heat transfer or movement through a material). The higher the "R" value factor, the higher the material's insulation value.

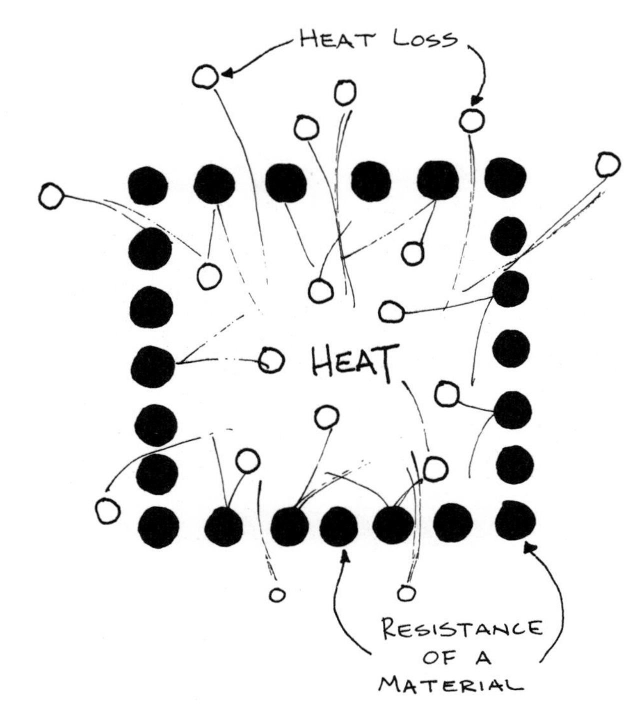

RABBET
A rectangular groove cut into any face or edge of a board. The rabbet of a sash is the groove in the sash frame into which the glass is set.

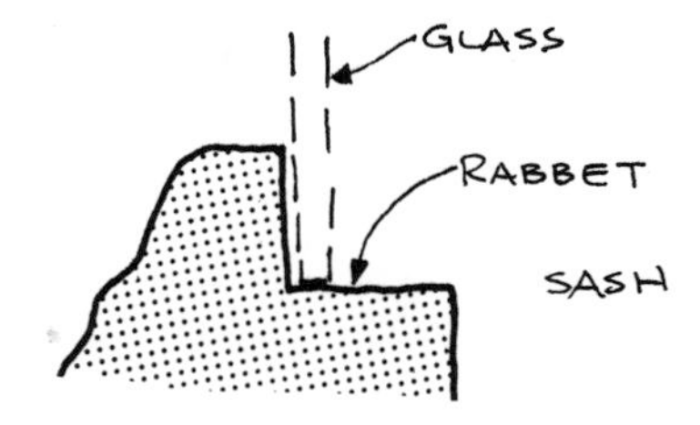

RADIATION
The transfer of heat by electromagnetic waves.

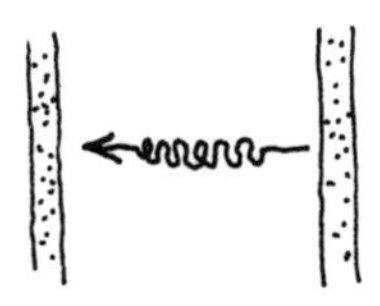

RAIL
Horizontal cross members of panel doors or window sash.

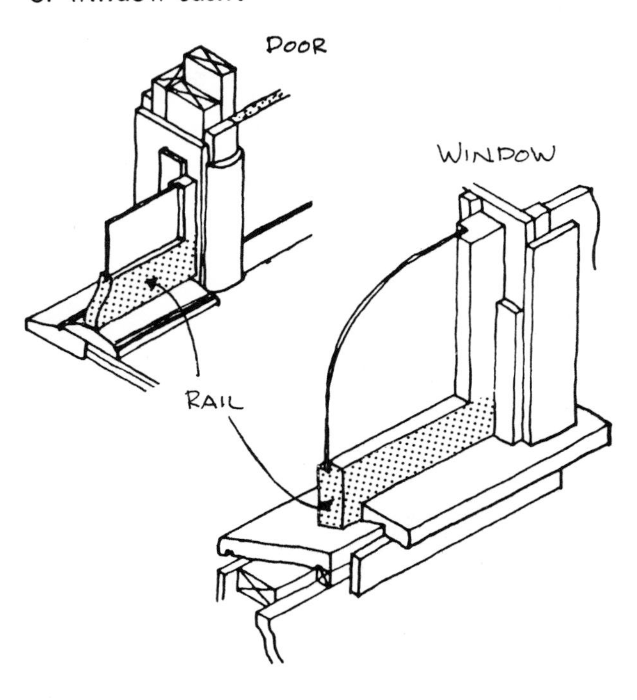

REGLET
A groove cut in masonry or concrete walls for inserting flashing.

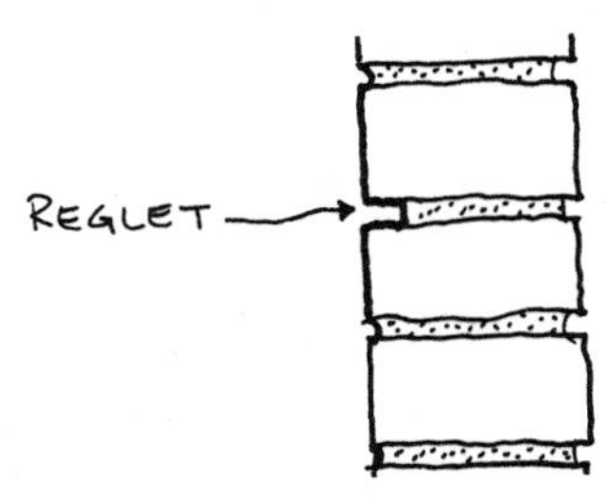

ROUTING
The removal of material by cutting, milling, or gouging to form a groove.

SASH
A framed window pane or panes which may be included within a window opening and may be fixed in position or moveable for ventilating and cleaning. A window may include one or more sash.

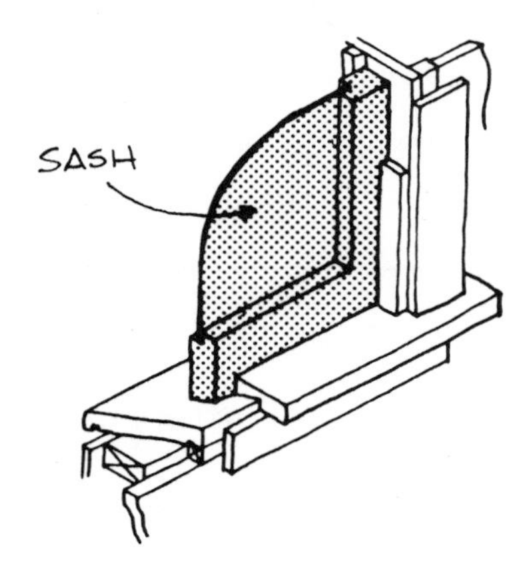

SCORING
To cut a groove in a material, such as plastic glazing or glass, for the purpose of cutting it. After the material has been scored, it is "snapped" over the edge of a hard surface.

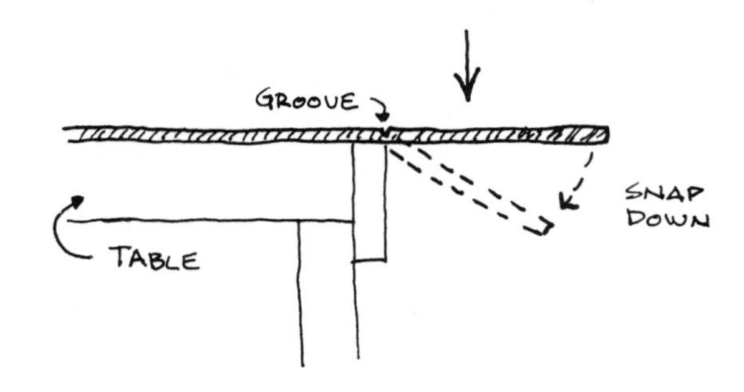

SHIMS
Thin wooden strips or wedges used to fill out or level surfaces.

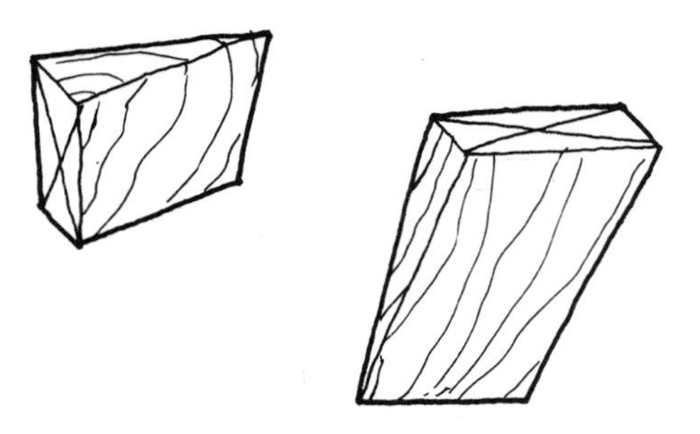

SIDELITE
A fixed window located next to a door.

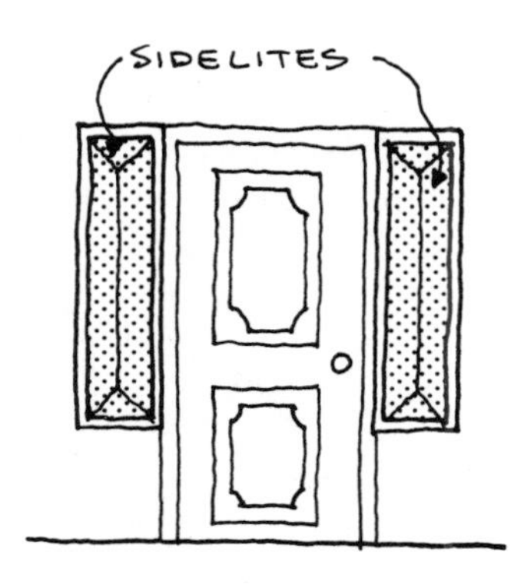

SILL
The member forming the exterior part of the bottom of an opening such as a door sill, window sill, etc.

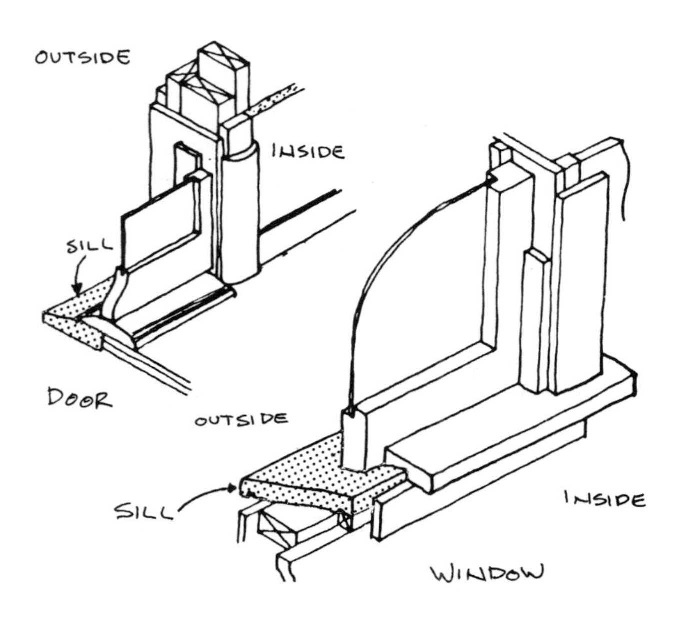

SLUMP TEST
A test of the water content of concrete. A dry mixture will have a small slump and a wet mixture a large slump.

STATIONARY JOINT
The sides of the joint do not move together or apart or move parallel to the opening.

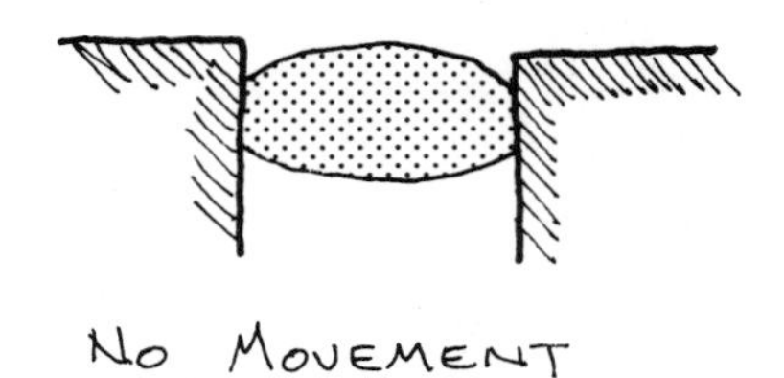

STILE
An upright or side member of a window sash or door.

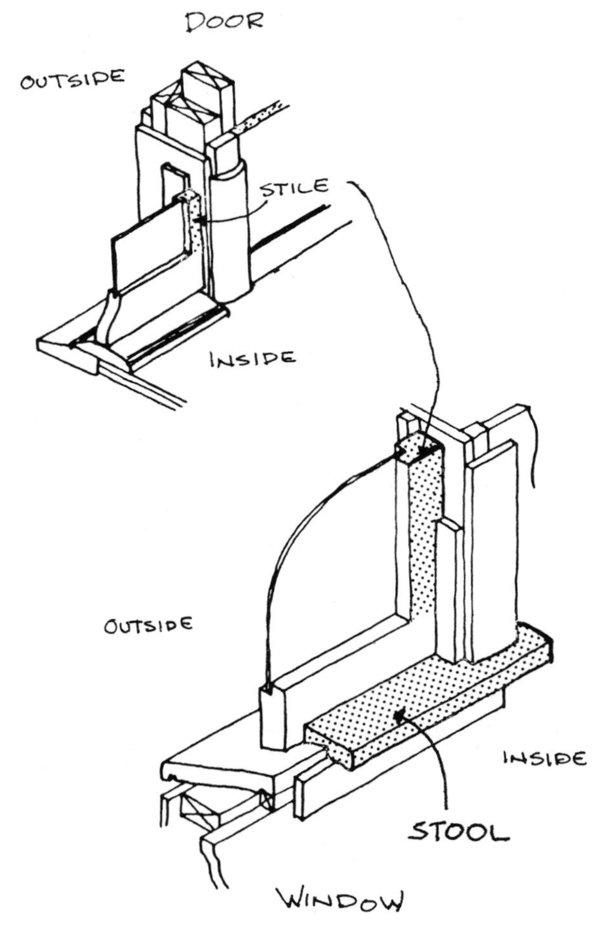

STOOL
A flat molding fitted over the window sill between the jambs which contact the bottom rail of the lower sash.

SWELLING
Wood expansion due to moisture absorption.

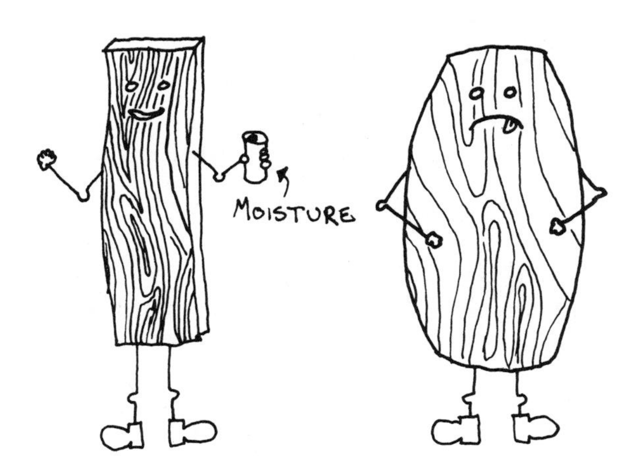

THRESHOLD
A strip of wood or metal with beveled edges used over a finished floor upon which the bottom of a door closes.

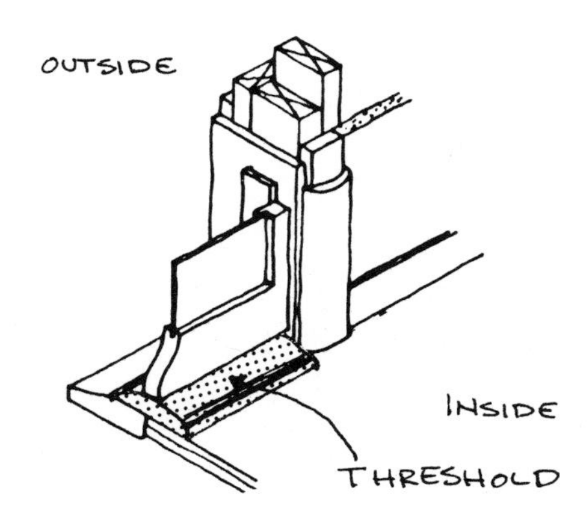

TOENAIL
A nail driven diagonally through the wood instead of perpendicular to its surface.

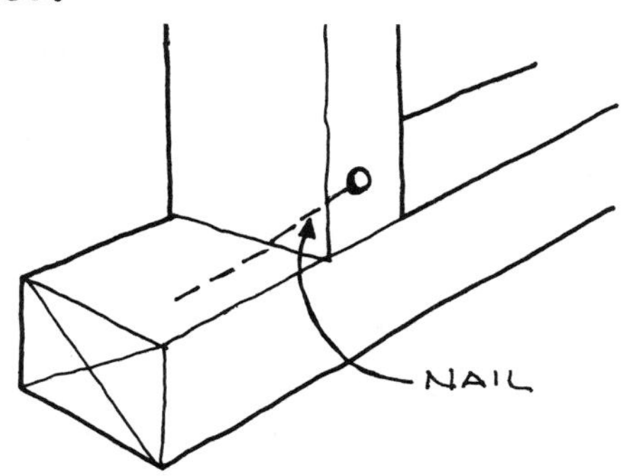

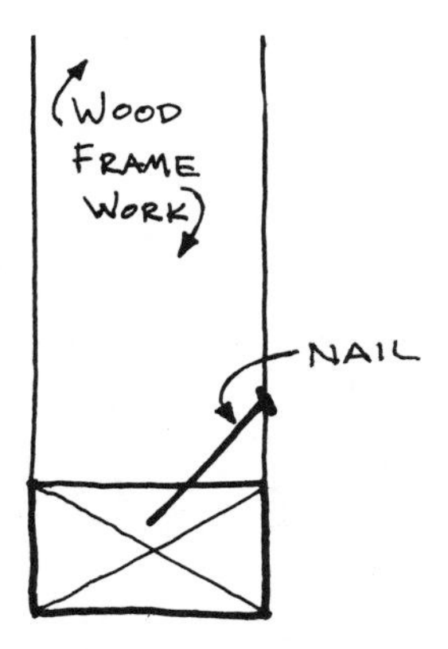

"U" FACTOR
The overall heat transfer coefficient of a material. It indicates the rate of heat flow allowed through a material and is a measure of its insulation value. The lower the value the higher the insulation quality. "U" factor is expressed in BTU's per square foot per hour, per degree fahrenheit.

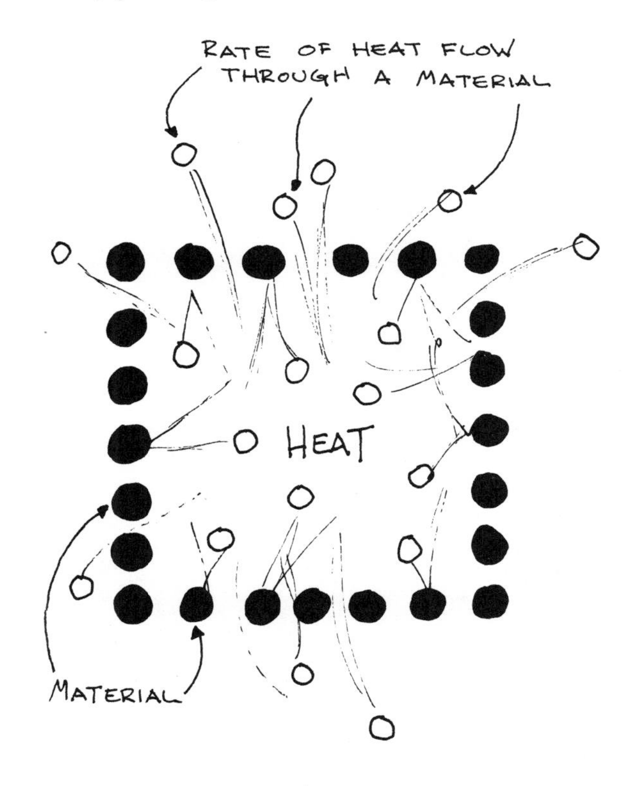

VENT CONE
A cone shaped piece of metal used to keep insulation away from a heat generating source.

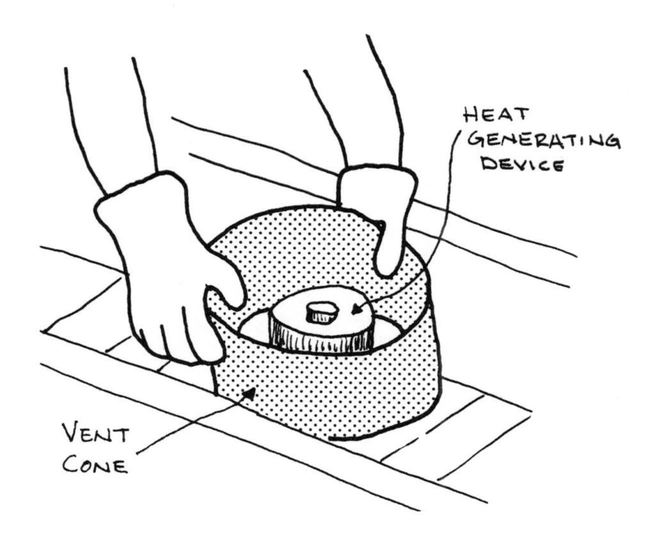

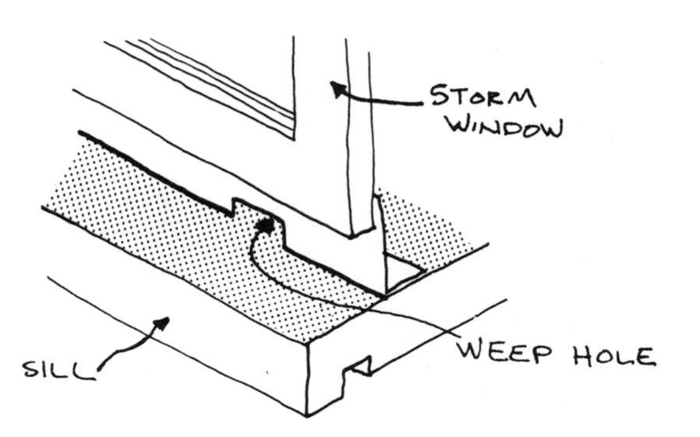

WEEP HOLES
Holes in the bottom rail of a storm window that reduces condensation on the window surfaces by allowing a minimum amount of air infiltration into the air space between the two windows.

WORKING CRACKS
Cracks which show signs of becoming larger or wider.

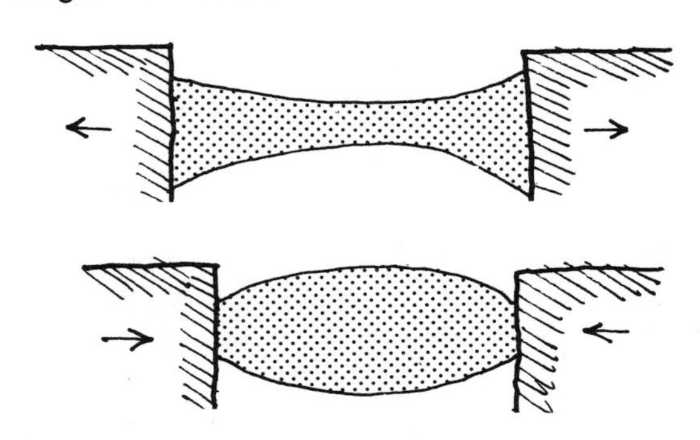

index

Attic access, 274, 284, 287-293

Attic insulation (see Finished Attic and Unfinished Attic)

Attic ventilation (see Ventilation and Vents)

Btu (see British Thermal Unit)

Backfilling, 191, 207, 224-225

Balloon framing, 255, 323

Basement wall insulation, exterior, 189-192

Basement wall insulation, interior:

- blown insulation, 185-186
- rigid insulation, 182-184
- roll/batt insulation, 179-182

Batt/blanket insulation (see Insulation)

Blown fill insulation (see Insulation)

British thermal unit (Btu), 6

Bypasses, 230, 231, 264, 273, 321-325

Cathedral ceiling insulation:

- blown insulation, 310-311
- rigid insulation, 308-309
- roll/batt insulation, 309-310

Caulk, 25-31, 73-79, 317-318

- description of, 25
- preparation, 27, 74-75, 316-317
- types and properties of, 28-29 (table), 30-31
- types of failure, 27
- where to install, 126

Ceiling finished against roof (see Cathedral Ceiling)

Cellulose insulation:

- description of, 10
- R-value of, 12 (table)
- safety, 17
- square foot coverage, 13 (table)

Collar beams (see Finished Attic)

Condensation, 23-24, 296-298

Conduction, 8

- definition of, 4

Convection, 8, 9

- definition of, 4

Crawl space:

- floor hatch installation, 218-220
- wall hatch installation, 216-218

Crawl space floor insulation:

- rigid insulation, 213-214

roll/batt insulation, 210-213

Crawl space wall insulation, exterior, 203-208

Crawl space wall insulation, interior:

rigid insulation, 198-199

roll/batt insulation, 196-198

Crawl space vent insulation, 335-337

Crawl space ventilation (see Ventilation and Vents)

Dew point temperature, 24

Door:

parts of, 140 (illus.)

types of, 166-167

Door sweeps, 143

Door weatherseals, 145

Dormant crack, 26

Duct insulation, 346-348

Excavation, 189, 205, 223

Fiberglass insulation (see Glass Fiber Insulation)

Finished attic insulation

collar beams, 276

end walls, 278

knee walls, 251, 276

outer ceiling joists, 257

outer end walls, 286

outer rafters, 285-286

sloped ceiling, 275, 277

what to insulate, 282 (illus.)

Flashing, 192, 207-208, 225, 244-245, 343-344

Foam sealant, 80

Glass fiber insulation:

description of, 10

R-value of, 12 (table)

safety, 17

square foot coverage, 13-14 (table)

Glazier points, 46, 52

Glazing compound, 46, 49, 52, 55

Glazings:

glass, 45-46

plastic, 88 (table), 89

Ground cover (in crawl spaces), 196-199, 344

Heat gain (see Heat Loss)

Heat loss, 3-9

definition of, 3

example of, 5

measuring, 6

Heat transfer (see Heat Loss)

Hinges, 167-168

Infiltration, 8

definition of, 5

Insulating panel, 123-126

definition of, 123-124

Insulating shutter, 107-122

definition of, 107-108

folding leaf, 109, 110-114

side hinged exterior, 109, 119-122

side hinged interior, 109, 114-119

top hinged, 109, 122

Insulation, 9-17

batts/blankets, 15-16 (table), 17

blowing machines, 269-270

blown fill, 15-16 (table), 17

definition of, 9

loose fill, 15-16 (table), 17

R- values of, 11-12 (table)

recommended R-values, 12 (map)

rigid insulation, 15-16 (table), 17

where to install, 10 (illus.), 176 (illus.)
(see also cellulose, glass fiber, perlite, polystyrene, polyurethane, rock wool, urea-formaldehyde, vermiculite)

Knee wall (see Finished Attic)

Loose fill (see Insulation)

Masonry wall insulation, exterior, 241-245

Masonry wall insulation, interior:

blown insulation, 235-237

rigid insulation, 233-235

roll/batt insulation, 230-233

Moisture problems (see Condensation)

Mortar joints, 320

Packing materials, 74

installation of, 80

Patching:

concrete, 318

gypsum board, 278-279, 331-332

masonry, 319-320, 330-331
plaster, 260
roofing, 319
sheathing, 260
shingles (wood and asbestos), 330
stucco, 318
wood siding, 318-319, 329
Perlite insulation:
description of, 11
R-value of, 12 (table)
Perm
definition of, 19
ratings of various materials, 21
Pipe insulation, 348-350
Planing, 171-172
Polyethylene, 20-21
(see also Vapor Barriers)
Polystyrene insulation:
description of, 11
R-value of, 11 (table)
Polyurethane insulation:
description of, 11
R-value of, 11 (table)
Putty, 46, 49, 52, 55
R-value:
definition of, 6-7
doors, 7-8
floors, 7
insulation types, 11-12 (table)
recommended insulation levels, 12 (map)
roofs, 7
walls, 7
windows, 8
Radiation, 8
definition of, 4
Relative humidity, 23-24
Resistance, 8
definition of, 6-7
(see also R-value)
Rigid board insulation (see Insulation)
Rock wool Insulation:
description of, 10
R-value of, 12 (table)
safety, 17
square foot coverage, 13-14 (table)
Safety, 33-42, 263
gear, 34
guidelines, 33

power tools, 41-42

Sash bead, 46

Sealants (see Caulk)

Sidewall insulation, closed frame:

access, 255-258

blown insulation, 258-260

Sidewall insulation, open frame:

rigid insulation, 252

roll/batt insulation, 249-251

Skirting, 339-344

Slab on grade insulation, 224-225

Storm windows:

aluminum (do-it-yourself type), 101-103

permanent, 99-106

plastic film, 89-92

plastic sheet, 92-97

temporary, 87-97

triple track, 100, 105-106

wood (do-it-yourself type), 103-105

Termite shield, 197, 216

Thermal panel (see Insulating Panel)

Thermal Shutter (see Insulating Shutter)

Unfinished attic insulation:

loose fill insulation, 268-270

roll/batt insulation, 264-267

Urea-formaldehyde insulation:

description of, 11

R-value of, 12

safety, 17-18

Vapor barriers, 19-22, 24

installation: basement walls, 181, 182, 184, 186

cathedral ceiling, 306, 307, 309, 310

closed frame walls, 259

crawl space walls, 196-198, 209-210, 213, 214 (see also Ground Cover)

finished attic, 276, 285

floor over crawl space, 209-210, 213, 214

masonry walls, 230, 232-234, 237

open frame walls, 249-252

unfinished attic, 262, 264, 266-268

with respect to ventilation, 20, 24

Vapor retarders (see Vapor Barriers)

Ventilation, 295-303

in cathedral ceilings, 306-308

in finished attics, 275-277, 285, 286
in unfinished attics, 265
requirements: attics, 262, 298
crawl spaces, 200-201, 335-337
Vents:
types of, 299-302
installation: attic, 302-303
crawl space, 335-337
Vermiculite:
description of, 11
R-value of, 12 (table)

Waterproofing, 179, 189, 195, 206
Weatherization, 8
Weatherstripping:
doors: fixed, 159
garage, 158
hatch, 158
hinged, 154-157
horizontal sliding, 157-158
metal, 157
windows: casement, 68-69
double hung, 60-66
horizontal sliding, 66-67
jalousie, 69-70
metal, 67
tilting, 69
Weatherstrips:
definition of, 57-58
types of, 58-59, 70-71, 152-153
Weep holes, 106
Window:
irrepairable, 129
opening a stuck window, 136-137
parts of, 44 (illus.)
replacement of, 127-135
sash replacement, 134-135
types of, 128
Working crack, 26